W0234421

HEAT PUMP FUNDAMENTALS

NATO ADVANCED STUDY INSTITUTES SERIES

Proceedings of the Advanced Study Institute Programme, which aims at the dissemination of advanced knowledge and the formation of contacts among scientists from different countries.

The series is published by an international board of publishers in conjunction with NATO Scientific Affairs Division

A	Life Sciences	Plenum Publishing Corporation
B	Physics	London and New York
C	Mathematical and Physical Sciences	D. Reidel Publishing Company Dordrecht and Boston
D	Behavioural and Social Sciences	Martinus Nijhoff Publishers The Hague, Boston and London
E	Applied Sciences	
F	Computer and Systems Sciences	Springer-Verlag Berlin, Heidelberg, New York
G	Ecological Sciences	

Series E: Applied Sciences – No. 53

HEAT PUMP FUNDAMENTALS

Proceedings of the NATO Advanced Study Institute on Heat Pump Fundamentals, Espinho, Spain, September 1-12, 1980

edited by

J. Berghmans
Associate Professor
Catholic University of Leuven
Mechanical Engineering Department
Leuven, Belgium

1983

Martinus Nijhoff Publishers
The Hague / Boston / London

Distributors:

for the United States and Canada
Kluwer Boston, Inc.
190 Old Derby Street
Hingham, MA 02043
USA

for all other countries
Kluwer Academic Publishers Group
Distribution Center
P.O.Box 322
3300 AH Dordrecht
The Netherlands

Library of Congress Cataloging in Publication Data

NATO Advanced Study Institute on Heat Pump Fundamentals (1980 : Espinho, Portugal)
Heat pump fundamentals.

(NATO advanced study institutes series. Series E, Applied sciences ; no. 53)
1. Heat pumps--Congresses. I. Berghmans, J. II. Title. III. Series: NATO advanced study institutes series. Series E, Applied sciences ; v. 53.
TJ262.N37 1980 621.402'5 82-24517
ISBN-13: 978-94-009-6820-2 e-ISBN-13: 978-94-009-6818-9
DOI: 10.1007/978-94-009-6818-9

TABLE OF CONTENTS

PREFACE

This book contains the texts of the lectures which were given at the Nato Advanced Study Institute on Advanced Heat Pumps which was held at Espinho, Portugal in September 1980.

A previous NATO Advanced Study Institute on the topic of heat pumps had been held in 1975. The significance of heat pumps with respect to energy conservation was the main topic of this Institute. In 1980 it was felt that considerable research had to be done in order to be able to produce more energy efficient, less costly and more widely applicable heat pumps. This requires a good understanding of the functioning of the types of heat pumps available. The simultaneous coverage of the basic fundamentals of heat pumps of different drive in one lecture series therefore was the goal of the 1980 Advanced Study Institute. Only a few lectures were devoted to heat pump applications.

The lectures on heat pump applications were intended to give only a short overview. They were supplemented by lectures on the latest developments on vapour compression as well as sorption systems.

Because of time limitations the number of topics which could be treated in depth had to be restricted. It is hoped however that the present book will provide the reader with a clear insight into the phenomena upon which the operation of mechanically and thermally driven heat pumps is based and thus may assist him in his research.

In the first place the editor wishes to thank all the authors for their collaboration. The fine efforts of mrs. F. Decoster who typed the manuscript and mr. J. Beutels who assisted in the editing are gratefully acknowledged. Particular thanks are also due to dr. T. Kester whose assistance was invaluable when organizing the Institute.

J. BERGHMANS
Heverlee, Belgium
September 1982.

INTRODUCTION

J. Berghmans

Katholieke Universiteit Leuven, Heverlee, Belgium

It has become common practice now to call a heat pump any device which extracts heat from a source at low temperature and gives off this heat at a higher temperature level where it can be used beneficially. It should be pointed out that the purpose of the device is to create heat at a high temperature level. If it is the purpose to extract heat from a low temperature source, the device is called a refrigeration system.

In most processes the production of heat at a high temperature level can be traced back directly to the consumption of primary energy (fossil or nuclear fuels, wind or solar energy...). A heat pump only upgrades heat and does not necessarily directly require a source of primary energy. Thus the heat pump shows to be an important instrument in the fight for energy conservation. This also explains the sudden breakthrough of heat pumps in several areas since the advent of the energy crisis.

However the second law of thermodynamics requires the presence of a third energy source in order for the heat pump to be able to achieve the "unnatural" transfer of heat from low to high temperature. If this auxiliary source delivers mechanical energy then the heat pump is of the mechanically driven type. If the auxiliary reservoir delivers heat then the heat pump is of the thermally driven type. Almost all heat pumps belong to one of these two classes.

First of all it should be pointed out that there is a fundamental difference between mechanically and thermally driven heat pumps. The former rely on high grade energy. The efficiency with which this driving energy can be produced from primary energy is low.

Primary energy conservation with mechanically driven heat pumps therefore very often is difficult to achieve unless special heat recuperation techniques can be applied.

Thermally driven heat pumps on the other hand can be operated directly by means of heat, without the inefficient conversion to mechanical energy. In addition there is some added flexibility from the side of the temperature level of the auxiliary heat source. It is possible in fact to drive a heat pump with waste heat : i.e. with a source at a temperature lower than the temperature of the useful heat to be produced (the heat transformer).

To arrive at a good understanding of the nature of the different heat pumps, the thermodynamic aspects of the upgrading of heat are discussed in the first two lectures of this course. In particular it will be shown how the concept of exergy can be used to compare heating systems.

The following lectures are devoted to mechanically driven heat pumps. First the thermodynamic cycles of vapour compression heat pumps are presented and analysed. It is shown how the performance of the heat pump depends upon the parameters defining the thermodynamic cycle (source temperatures, compressor performance...).

This is followed by a discussion of the requirements which the working fluid of a vapour compression heat pump has to satisfy. Suitable choices of working fluids are presented.

Once the thermodynamic cycle is defined, the different components of a vapour compression heat pump are dealt with. In particular evaporator, condenser and compressor are discussed in detail. Special attention is given to the sizing and optimisation of the first two components. The compressors used in heat pumps are discussed subsequently. The elements leading to the choice of a particular compressor are presented.

Vapour compression heat pumps presently are encountered in a multitude of applications. Many of these are characterized by widely varying temperatures of the heat sources (e.g. domestic heating). A good design has to take such changes into account. For this reason considerable attention is given in this course to heat pump optimiation and design. An important step towards optimization of a heat pump system is the modeling of the heat pump itself. It is by means of accurate models of the heat pump system that off-design performance can be predicted.

Presently most vapour compression heat pumps, as they are used in the domestic sector for instance, are driven by electric motors.

Due to the low efficiency with which electricity is produced it is difficult to achieve primary energy conservation with these heat pumps compared to alternative heating systems (gas, oil boilers).

Obviously the heat pump compressor can be driven by other prime movers than electric motors. In this course possible prime movers for vapour compression heat pumps are discussed. Special attention is also given to heat pumps driven by internal combustion engines (gas, oil). Such heat pumps make it possible to achieve high primary energy efficiency by recuperation of engine waste heat.

Improved performance of vapour compression heat pumps can be achieved also by decreasing the temperature differences occurring in condenser and evaporator. This can be achieved by the use of non-azeotropic mixtures as working fluids. The advantages of this technique are discussed.

Only the most interesting applications of heat pumps in the domestic as well as in the industrial sector are described. The problems involved are briefly mentioned. The temperature at which most heat pumps are able to operate is too low for most industrial processes (e.g. process steam production). For domestic applications the problems are located at the low temperature source (e.g. outside air) and the use of costly electricity.

In recent years thermally driven heat pumps have received considerable interest due to a number of features :
- lack of compressor;
- possibility for direct firing;
- high primary energy efficiency;
- long life expectancy due to the lack of components with moving parts;
- no vibration or noise problems.

For these reasons intense research efforts are being undertaken presently to develop such heat pumps.

Most attention presently is going to the absorption heat pump. The phenomenon of vapour absorption in a liquid has been applied successfully in refrigeration equipment. However only very few absorption heat pumps presently are available on the market. The transition from cooling device to heating device, in contrast with vapour compression systems, has not been achieved thus far.

It is the purpose of this Institute to provide researchers with a better insight in the processes which determine the performance of absorption heat pumps in order to promote the large scale application of these devices.

A profound understanding of the operation of an absorption heat pump requires acquaintance with the thermodynamics of binary mixtures. Based upon the properties of these mixtures, thermodynamic absorption cycles are developed and discussed. Distinction here is made with heat transformer cycles, the latter being characterized by the utilization of low temperature heat as driving heat source. Special attention is devoted to the requirements to be filled by the working pairs suited for absorption heat pumps and transformers.

Absorption heat pumps, in addition to condensers and evaporators, require components such as : generators, rectifyers and absorbers. The principles underlying the operation of each of these components are discussed in detail. It is shown that design and optimization of these items are very complex and require extensive knowledge of the thermodynamic and thermal properties of the working pairs. Such knowledge often is lacking.

Furthermore it will become apparent how a number of limitations originating from the working pair influence the performance of the heat pump. It is felt that more suitable working pairs have to be developed.

Absorption systems basically consist of a series of components in which heat and mass transfer occur. The size of these components necessary to achieve the required power level often is very large and leads to costly devices. Therefore it can be stated that optimization of heat and mass transfer together with the development of more suitable working pairs are the main areas in which research to promote the development of absorption systems should be performed.

Resorption systems offer interesting prospects to be used as heating systems. The principles of operation of these systems are explained. Recent developments in this field are discussed.

Throughout these lectures the need for further research and development work will be pointed out. This need is particularly urgent in the field of sorption heat pumps, where even basic problems as the choice of working pairs have not received conclusive answers. The enormous energy conservation potential which these heat pumps offer should be sufficient stimulation to undertake this task.

ENERGY ANALYSES - THERMODYNAMIC PERFORMANCE CRITERIA

H.F. Sullivan

University of Waterloo, Waterloo, Canada

1. INTRODUCTION

All engineering undergraduates, irregardless of their particular discipline, are usually subjected to at least one course in thermodynamics in which attention is focused on devices in which energy is converted or transformed or transported. Obviously, living now in a period in which energy costs are subject to such rapid escalation (as has been experienced since the oil embargo of 1973), it is not surprising that the engineering community has put renewed emphasis on the "energy aspects" of devices and components along with the usual considerations of materials, fabrication costs etc. which have always been part of engineering design.

It is appropriate to begin our deliberations with a brief consideration of the concepts governing the movements, transformations and conversions of energy.

1.1. The First and Second Laws of Thermodynamics

The First and Second Laws of Thermodynamics which will form the basis for much of our discussion can be expressed in a number of ways. For a system closed to mass transfer we can express the first law as follows (1) :

$$dU = dQ_{in} + dW_{in}$$

where we have adopted the sign convention that work done on the

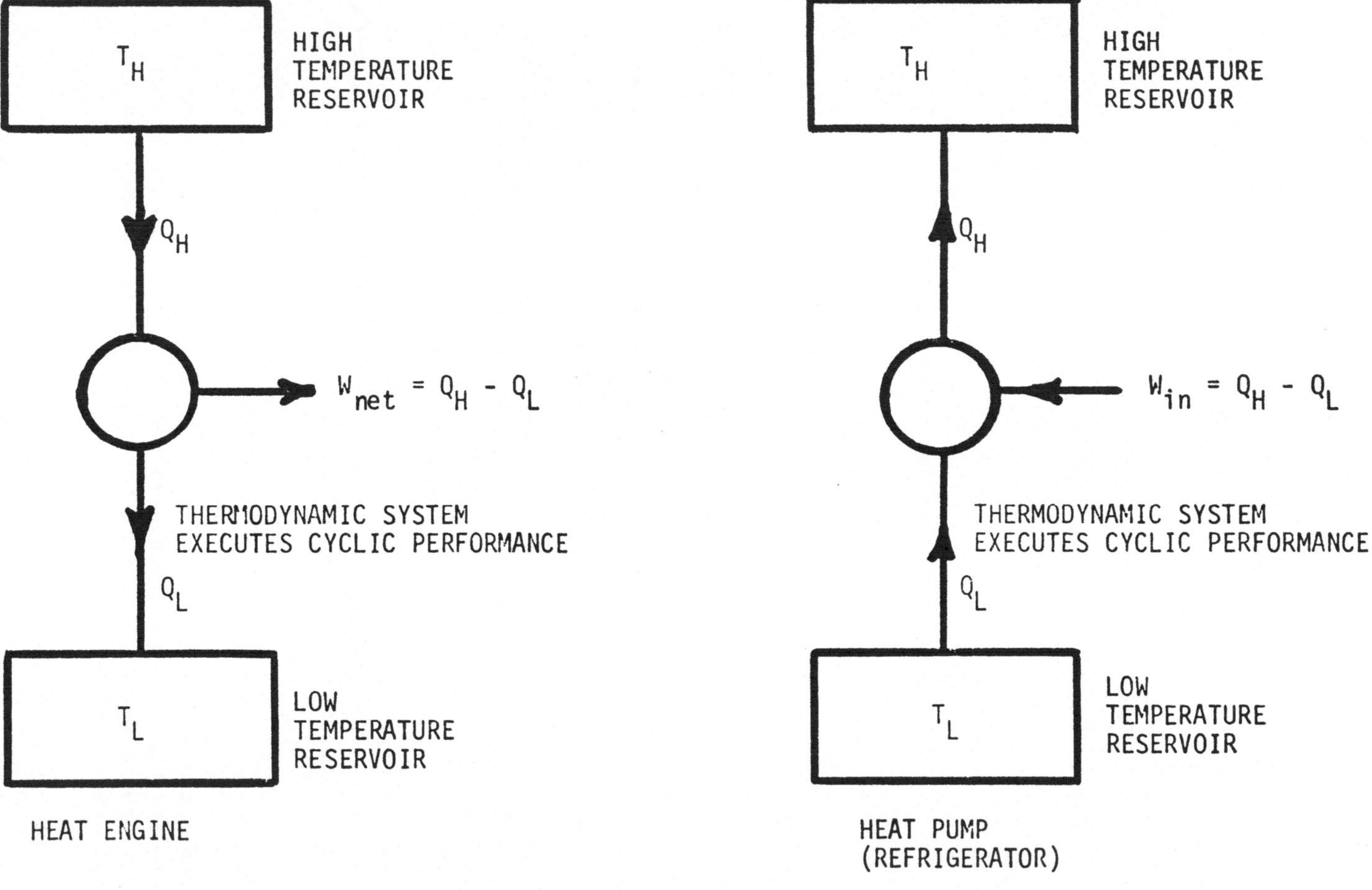

Figure 1. Schematic Representations of a Heat Engine and a Heat Pump (or Refrigerator)

system and heat transferred to the system are positive and dU represents the increase in the internal energy of the system.

The steady state form of the First Law for a system open to mass transfer can be expressed as follows (1) :

$$\dot{m}_{in} (u + pv)_{in} + \dot{Q}_{in} + \dot{W}_{in} = \dot{m}_{out} (u + pv)_{out}$$

where the dots above the symbols designate quantities expressed as rates (e.g. $\dot{W}_{in}$ represents power or the rate at which work is transferred to the system). In this last expression we have made the assumption that the changes in kinetic and potential energies can be neglected.

The Second Law of Thermodynamics can be expressed as follows :

$$dS - \frac{dQ_{in}}{T} \geq 0$$

for systems closed to mass transfer, and

$$\sum_{out} (s\dot{m}) + \sum_{out} \frac{\dot{Q}_{out}}{T} - \sum_{in} (s\dot{m}) - \sum_{in} \frac{\dot{Q}_{in}}{T} \geq 0$$

for systems open to mass transfer.

The latter expression is expressed on a rate basis and assumes steady-state operation of the system.

1.2. Heat Engines, Heat Pumps and Refrigerators

Among the many classes of devices to which we apply thermodynamic analyses we have heat engines, heat pumps and refrigerators. The schematic representations of Figure 1 illustrate the flows of heat and work for these devices as well as the thermal energy reservoirs (sources and sinks).

1.3. Definitions of Thermal Efficiencies, Coefficients of Performance

For a heat engine, the useful function performed is the delivery of a quantity of work. In order to supply this output, the heat engine requires an input of heat at the high temperature Q_H. A measure of performance for this heat engine is the ratio of these two quantities which we call a thermal efficiency (sometimes also called a cycle efficiency or energy conversion efficiency). Using the First Law we can express this efficiency as follows :

$$\eta \equiv \frac{W_{net}}{Q_H} = \frac{Q_H - Q_L}{Q_H} = 1 - \frac{Q_L}{Q_H}$$

Application of the Second Law leads to

$$\eta \leq 1 - \frac{T_L}{T_H}$$

The maximum value of this efficiency is obtained for an ideal reversible heat engine operating between the two constant temperature reservoirs and is commonly referred to as the Carnot efficiency but it applies equally well to any other reversible heat engine operating between the same two reservoirs.

Similar analyses can be applied to the heat pump and refrigerator. As in the case of the heat engine, our measure of performance is the ratio of the useful function provided by the device (Q_H for the heat pump and Q_L for the refrigerator) to the required input to the device (W_{in} for both the heat pump and the refrigerator). Since this ratio exceeds unity, it is termed a coefficient of performance (COP) rather than an efficiency. For these two devices, application of the First Law leads to the following expressions :

$$COP_{hp} = \frac{Q_H}{W_{in}} = \frac{Q_H}{Q_H - Q_L} = \frac{1}{1 - \frac{Q_L}{Q_H}}$$

$$COP_{ref} = \frac{Q_L}{W_{in}} = \frac{Q_L}{Q_H - Q_L} = \frac{1}{\frac{Q_H}{Q_L} - 1}$$

Application of the Second Law leads to the following :

$$COP_{hp} \leq \frac{1}{1 - \frac{T_L}{T_H}}$$

$$COP_{ref} \leq \frac{1}{\frac{T_H}{T_L} - 1}$$

Once again, the equal sign associated with maximum values of the COP's corresponds to operation of ideal reversible heat pumps and refrigerators. Devices operating on the Carnot cycle satisfy this requirement and the maximum values of the COP's are thus usually designated as "the Carnot COP's".

1.4. Other Efficiencies

In addition to the energy conversion efficiencies (or coefficients of performance) discussed above, we also frequently define other efficiencies for individual components such as compressors, turbines etc. which are parts of the heat engine or heat pump system. Usually, these efficiencies compare the performance of the device to the performance of a hypothetical ideal device operating between similar specified conditions at inlet and outlet. For example in the case of a compressor operating between inlet and outlet pressures P_1 and P_2, an efficiency for the compressor can be defined as the ratio of the work required by an isentropic compressor (operating between the same pressures P_1 and P_2) to the work required by the actual compressor, i.e.,

$$\eta_{comp} \equiv \frac{W_{isentropic}}{W_{actual}}$$

This efficiency is usually termed an isentropic compressor efficiency (also called the adiabatic efficiency).

In the case of heat pumps and refrigerators it is possible to compare the COP's of actual devices with the Carnot COP (for the same temperature limits of operation) and thus define an efficiency as follows :

$$\eta_R = \frac{COP_{actual}}{COP_{Carnot}}$$

2. MAXIMIZING ENERGY PERFORMANCE OF COMPONENTS AND SYSTEMS

In the introduction, we indicated the need to pay increasing attention to the performance of devices. The thermodynamic efficiencies outlined above provide us with expressions for the performance of a system or a component but do not present the complete picture. In particular, the "quality" of the energy supplied or rejected by the devices undergoing various processes is not specifically addressed by these performance measures.

2.1. Exergy

An important principle in the energy efficient design of systems and components is to try to match the quality of the energy supplied with the quality of the energy required by a particular application. A thermodynamic function which provides a measure of the energy quality as well as quantity, is termed exergy. Brzustowski and Golem (2), (3) point out that "all the predictions and insights gained by using the exergy are corollaries of the First and Second Laws of Thermodynamics. All results of exergy calculations can also be obtained by applying the appropriate forms of the First Law and by calculating the detailed balances of entropy increase, decrease and generation". The exergy formulation "incorporates the subtle concept of irreversibility and its consequences into a routine engineering calculation". Other thermodynamic functions such as the availability are closely related to the exergy.

The reader interested in more detail on the topic of exergy is referred to the papers by Brzustowski and Golem (2), (3) or a number of other literature sources noted in their references.

Exergy is defined (2) as "the maximum work which can be extracted from a given system in a given state in any process which brings the system into equilibrium with its environment". Whereas the Second Law is usually thought of in terms of entropy and the <u>increase</u> in entropy associated with any irreversible process, an alternate viewpoint would consider the exergy <u>decrease</u> associated with all irreversible processes.

Both Brzustowski and Golem (2) and Borel (4) use the theory of exergy to present some interesting comparisons of different systems for supplying electricity and heat to a dwelling. In order to follow these comparisons we require the following.

2.2. Measures of Performance : Effectiveness (Exergetic Efficiency)

Brzustowski and Golem define an "effectiveness, ε" for a system as the ratio of the exergy gained to the exergy expended by the system during a process. Borel terms this same quantity an "exergetic efficiency, η" (to be denoted by ε in this text).
In order to use these performance parameters we must be able to determine the exergy associated with an amount of mechanical or electrical energy (or power), with a quantity of heat (or heat flow rate) supplied at a particular temperature and with the chemical energy in a fuel. Following Borel, we have the following :

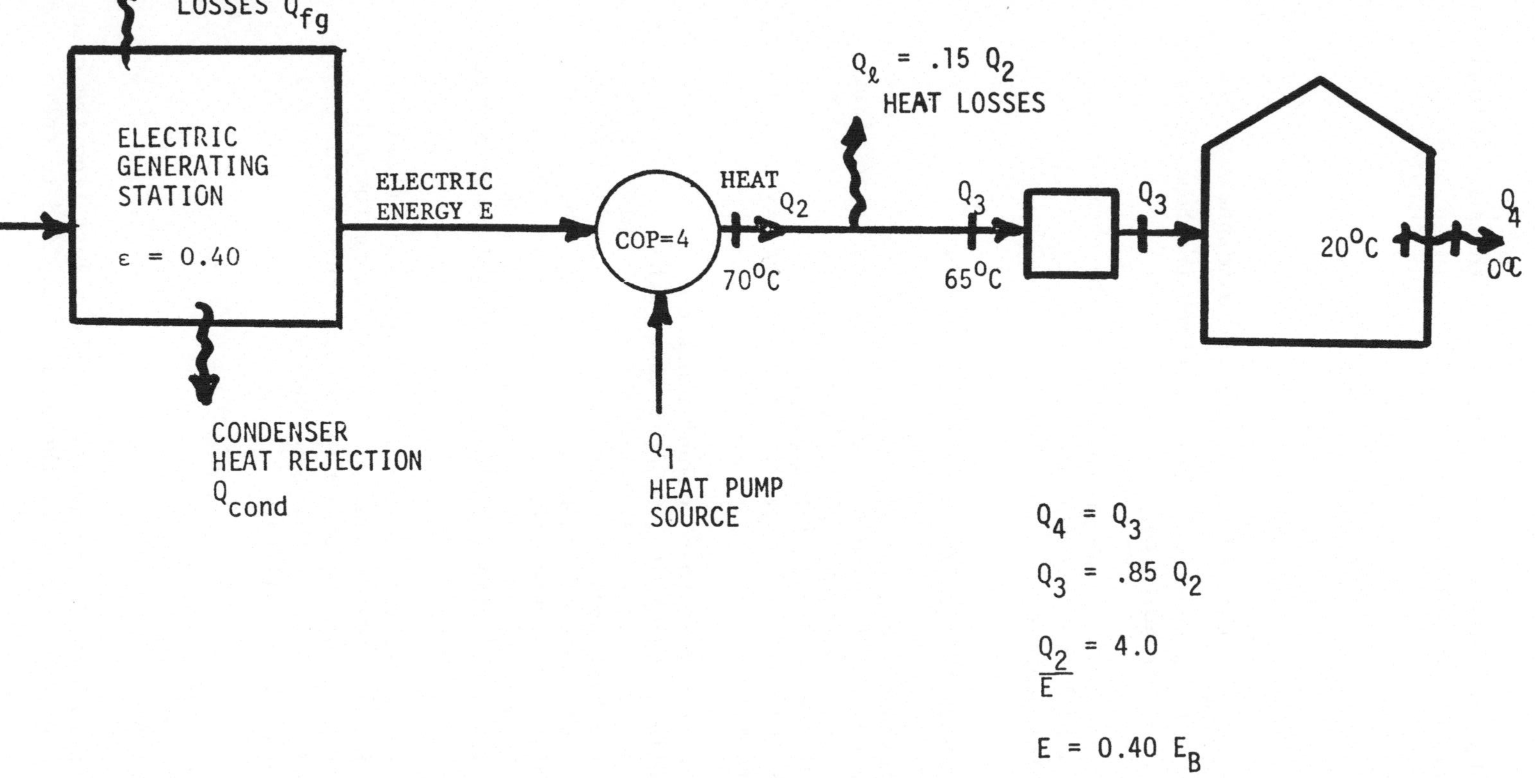

Figure 2. Schematic of Energy Flows for an Electric Heat Pump - District Heating System

(1) Mechanical or electrical exergy E is the same as mechanical or electrical energy;

(2) Heat exergy E_q is defined by

$$E_q = Q\left(1 - \frac{T_a}{T}\right)$$

where T_a is the temperature of the atmosphere

T is the temperature at which the heat is being transferred

(3) For energy supplied in the form of the chemical energy of a fuel, we assume that the exergy E_W is equivalent to the heating value of the fuel.

3. COMPARISONS OF ALTERNATIVE SYSTEMS FOR HEATING

Borel (4) presents exergetic efficiency calculations for the individual components of systems to supply heat and electricity (an electric power station, a boiler, an electric heat pump, an electric radiator, a district heating pipeline etc.) as well as a summary of the overall exergetic efficiency of five different heating systems.

As an example of one of these calculations consider the case of an electrically driven heat pump operating with a COP of 4 delivering heat at 70 °C to a district heating line which is 85 % efficient. The heat from the line is in turn supplied to the house at a temperature of 65 °C to maintain the house at 20 °C when the ambient temperature is 0 °C. The electricity to run the heat pump is received from an electric generating station operating with an efficiency of 40 % and assumed to be a fossil-fired station.

The overall exergetic efficiency of this system depicted schematically in Figure 2 can be derived by calculating the ratio of the heat exergy out of the system (from the house at 20 °C to the atmosphere at 0 °C) to the fuel exergy into the electric generating station as follows :

$$\varepsilon = \frac{Q\left(1 - \frac{Ta}{T}\right)}{E_B}$$

The relationship between the quantity of heat lost from the house to the surroundings in terms of the energy supplied to the electric generating station can be calculated from an energy balance using the various component efficiencies. Thus

$$Q = E_B \times 0.40 \times 4.0 \times .85 = 1.36\ E_B$$

$$\varepsilon = \frac{1.36\ E_B\ (1 - \frac{273 + 0}{273 + 20})}{E_B}$$

$$= .092$$

An alternate formulation of the system exergetic efficiency can be obtained by forming the product of the individual component exergetic efficiencies determined as shown in the Table I.

Thus the exergetic efficiency of the overall system is as follows :

$$\varepsilon = .40\ (.80)(.81)(.37)$$

$$= .096$$

Table II presents the summary of the calculations by Borel (4) for the supply of electricity and heat for five different heating systems and for three values of heating rate defined as the amount of energy for space heating as a percentage of the total amount of energy for space heating and electric power for purposes other than heating.

Brzustowski and Golem (2) presented a similar example of the effectiveness of domestic energy systems using electricity and natural gas (which they treated as being pure methane for the calculations).

The example (2) considered a house which required 3 kW of electric power and heat at the rate of 7 kW. The house was maintained at 20 °C and the ambient conditions were 0°C and 20 % relative humidity. It was assumed that the electric energy supplied to the house was generated and transmitted to the house at an effectiveness of 25 %.

The output exergy flow from this domestic energy system was calculated as follows :

$$\dot{E}_{out} = 3 + 7\ (1 - \frac{273}{293})$$

$$= 3.478 \text{ kW}$$

TABLE I

Component Exergetic Efficiencies (4)

Component	Exergetic Efficiency
electric generating station	$\frac{E}{E_B} = 0.40\,\frac{E_B}{E_B} = 0.40$
electric heat pump	$\frac{Q_2\,(1 - \frac{273}{343})}{E} = \frac{4E\,(.20)}{E} = 0.80$
district heating line	$\frac{Q_3\,(1 - \frac{273}{338})}{Q_2\,(1 - \frac{273}{343})} = .85\,\frac{(.19)}{(.20)} = 0.81$
internal heating system	$\frac{Q_4\,(1 - \frac{273}{293})}{Q_3\,(1 - \frac{273}{338})} = \frac{.07}{.19} = 0.37$

TABEL II

Overall Exergetic Efficiencies (4)

Method of Distribution of Energy	Heating Rate		
	0 %	65 %	100 %
Electric Energy Individual Heating	40 %	23 %	5 %
Electric Energy Collective Heating	40 %	23 %	5 %
Electric Energy Electric Heating	40 %	16 %	3 %
Electric Energy Heating by Heat-Pump	40 %	30 %	10 %
Electric Energy Forced-Draft Heating	40 %	32 %	-

The input exergy flow $\dot{E}_{in}$ appears in two forms : the exergy flow in the natural gas supply and the electrical exergy flow. The latter is the electrical exergy flow delivered to the house divided by the effectiveness with which it is supplied.

The effectiveness ε (equivalent to the exergetic efficiency used by Borel (4)) of the home energy system is computed as the ratio of the exergy flows out of and into the system, respectively.

$$\varepsilon \equiv \frac{\dot{E}_{out}}{\dot{E}_{in}}$$

Table III summarizes the results of the calculations of Brzustowski and Golem (2) for seven different cases including all electric energy provided from the grid as well as various combinations including furnaces, heat pumps and fuel cells. The seven cases are arranged in order of increasing effectiveness. The low value of 0.087 is for an electric heating system whereas, when a fuel cell is used to provide the complete requirements (both electricity and heat) the effectiveness is 0.348.

3.1. Discussion of Results

In these two different studies (2), (4) the temperatures assumed for the dwelling and the atmosphere were identical although the efficiencies assumed for various components in the systems were quite different (e.g. the electric power plant). However, both studies draw similar conclusions about the overall effectiveness (or efficiency) of systems. The electric heating process in which high quality chemical energy in the fuel is converted first into electrical energy in a generating station (with a rather low energy conversion efficiency) and then back into low quality heat energy at the dwelling is the least effective of the various systems studied. Furnace or boiler systems and electrically-driven heat pumps have intermediate values of effectiveness, the values of which depend on the assumed heat pump COP and boiler and furnace efficiencies. The highest ranking system utilizes the fuel cell, a device in which the high quality chemical energy in the fuel undergoes a direct conversion to electrical energy.

Although these examples are illustrative of the Second Law analysis applied to alternate systems, similar calculations can be performed on the separate components within a system to attempt to determine where the largest irreversibilities or departures from ideal performance are occurring.

Example calculations for refrigeration systems performed using the

TABLE III

Effectiveness of Domestic Energy Systems Using Electricity and Natural Gas (2)

System Configuration	Effectiveness
electric heating, electrical power from grid	.087
gas heating, 60 % furnace thermal efficiency, electrical power from grid	.143
electric powered heat pump of $\varepsilon = 0.5$ provides 7 kW of heat, electrical power also from grid	.220
space heating with hot water at 70 °C from district heating system, electrical power from grid	.259
gas supplied to fuel cell of $\varepsilon = 0.5$ to produce electricity and 3 kW of waste heat, 4 kW of heat supplied by furnace	.268
gas supplied to fuel cell of $\varepsilon = 0.5$ to produce 5 kW electricity, and 5 kW waste heat, 2 kW of electricity used for electric heating	.348
gas supplied to fuel cell of $\varepsilon = 0.5$ to produce 7 kW electricity and 7 kW waste heat, 4 kW electricity fed to grid	.534

concepts of availability and irreversibility (rather than exergy) are presented in Chapter 1 of the ASHRAE Handbook of Fundamentals (5). The analyses presented illustrate further the application of the Second Law in attempts to maximize performance and obtain optimum designs for heat pump or refrigeration cycles.

REFERENCES

1. Reynolds, W.C. and Perkins, H.C., *Engineering Thermodynamics* (New York, McGraw-Hill, 1977).

2. Brzustowski, T.A. and Golem, P.J., *Second-Law Analysis of Energy Processes Part I : Exergy - An Introduction*", Transactions of the CSME 4 (1976-77) 209-218.

3. Golem, P.J. and Brzustowski, T.A., *Second-Law Analysis of Energy Processes Part II : The Performance of Simple Heat Exchangers*", Transactions of the CSME 4 (1976-77) 219-226.

4. Borel, L., "*Energy Economics and Exergy - Comparison of Different Heating Systems Based on the Theory of Exergy*", in Heat Pumps and Their Contribution to Energy Conservation, edited by Camatini, E. and Kester, T., (Alphen aan den Rijn : Sijthoff & Noordhoff, 1980) 51-96.

5. *ASHRAE Handbook and Product Directory, 1977 Fundamentals*, American Society of Heating, Refrigerating and Air-Conditioning Engineers, Inc. 1977.

PRINCIPLES OF VAPOUR COMPRESSION HEAT PUMPS

H.F. Sullivan

University of Waterloo, Waterloo, Canada

1. INTRODUCTION

The basic principles of operation of heat pumps and refrigeration devices are available in a great many references, several of which [1-4] were utilized in the preparation of this material. It is obvious that the basic principle of operation for a heat pump is the same as that of a refrigeration device except that for a heat pump attention is focused on the heat delivered by the device rather than the heat picked up at the evaporator. Definitions of the coefficients of performance for these two devices reflect this difference (see chapter : Energy Analysis - Thermodynamic Performance Criteria).

The heat delivered by the heat pump is equal to the heat absorbed from the low temperature source plus the work input to the compressor. The COP as a heat pump is greater by one than the COP as a refrigerator.

$$COP_{ref} = \frac{Q_L}{W_{in}}$$

$$COP_{hp} = \frac{Q_H}{W_{in}} = \frac{Q_L + W_{in}}{W_{in}} = COP_{ref} + 1$$

Figure 1 shows a schematic representation of a heat pump including only the major components of the device, namely the evaporator, compressor, condenser and throttling device (expansion valve or capillary tube). In addition to these components, a practical heat pump has other components such as accumulators, receivers, various controls etc.

There are a number of possible sources for the low temperature supply of heat including ambient air, water from a lake, river or well, the soil, a fluid rejected from an industrial process, a solar heated source (air, water etc.), internal building heat etc. In the application of heat pumps to large buildings with different demands in various zones, frequently provision is made for simultaneous supply of heating and cooling [5].

Although when discussing heat pumps, heat engines etc. we normally speak of the temperatures of the source and sink reservoirs, we must remember that in order to provide sufficient heat transfer with moderately sized heat exchanger areas for the evaporator and condenser, we require a reasonable temperature difference between the refrigerant and the medium to/from which heat is to be transferred. Figure 2 demonstrates the effect of this temperature difference in the case of a Carnot refrigerator [4].

2. OPERATING CHARACTERISTICS

2.1. Heating Capacity

The heating capacity of the heat pump is equal to the rate at which heat is rejected by the condenser which, by the First Law, is equal to the rate of heat absorbed in the evaporator plus the power required by the compressor.

2.2. Coefficient of Performance

The coefficient of performance for the heat pump, which is analogous to an efficiency for a heat engine, is defined as the ratio of the useful output (the heat delivered at the high temperature) to the required input (the work supplied to the compressor). In terms of the nomenclature of Figure 1

$$\mathrm{COP} \equiv \frac{Q_{out}}{W_{in}} = \frac{Q_H}{Q_H - Q_L} = \frac{1}{1 - \dfrac{Q_L}{Q_H}}$$

Application of the Second Law of thermodynamics leads to the following result :

$$COP \leqslant \frac{1}{1 - \frac{T_L}{T_H}} = \frac{T_H}{T_H - T_L}$$

It is obvious from this expression that the highest COP values are attainable when the source and sink temperatures are close to each other. As the spread between these temperatures increases, the COP of the heat pump decreases rapidly.

2.3. Heat Pump Efficiency

An efficiency can be defined for a heat pump as the ratio of the COP for an actual heat pump to the COP for an ideal reversible (Carnot) heat pump operating between the same two temperatures.

$$\eta_{id} \equiv \frac{COP_{act}}{COP_{Carnot}}$$

3. SINGLE-STAGE VAPOUR COMPRESSION HEAT PUMPS

Figure 3 is a schematic representation of an ideal vapour compression device and its corresponding process representation on temperature and entropy coordinates. Figure 4 presents the same process representation except in terms of pressure and enthalpy coordinates.

Although it has been common to use pressure-enthalpy coordinates for heat pump and refrigeration cycles, the temperature-entropy coordinates are particularly useful in illustrating departures from ideal reversible performance.

3.1. Ideal Cycle

The description of the operation of the ideal or theoretical cycle begins at state point "a" which is downstream of the throttling valve at the inlet to the evaporator. State point "a" is in a mixed-phase region (wet-vapour region). In the evaporator the process is idealized as constant pressure (and constant temperature) heat addition to the refrigerant which leaves the evaporator at state point "b". In these process schematics, point "b" is shown as corresponding to a saturated vapour state. In real devices, it is common to have some degree of superheat for the vapour at point "b" although increasing the amount of superheat will cause decreased efficiency because of increased compressor work.

The compression of the refrigerant vapour occurs from state point "b" to "c". In this ideal process representation, the compression is assumed to be adiabatic and reversible, i.e. isentropic. On the T-s diagram, therefore, point "c" is on a line vertically above point "b" and at a pressure corresponding to the compressor delivery pressure. State point "c" thus generally lies in the superheat region of the process diagram.

The rejection of the heat in the cycle from "c" to "d" is assumed to occur at constant pressure in the condenser. The first part of the heat rejection corresponds to a desuperheating of the compressed vapour (from state point "c" tot "g" on the T-s diagram) followed by a condensation of the vapour. State point "d" representing the exit state of the refrigerant from the condenser is shown as a saturated liquid state on the process representations for this ideal cycle although in an actual device it is normal to have some degree of subcooling of the liquid leaving the condenser.

After the high pressure liquid refrigerant leaves the condenser it passes through a throttling valve or capillary tubing and exits at state point "a". If the valve or capillary tubing is assumed to be adiabatic, then the process occurring from "d" to "a" is an isenthalpic (constant enthalpy) process. The location of state point "a" is thus visualized more easily on the p-h coordinate representation.

Figures 5a and 5b are process representations which illustrate the superheating of the vapour at state point "b" entering the compressor and the subcooling of the liquid leaving the condenser at state point "d". These modifications are made for practical reasons. If the fluid leaving the condenser corresponded to a saturated liquid, then any friction losses occurring in the piping would result in some vapour formation thus increasing the specific volume and reducing the capacity of the piping and of the throttling device [4].

The superheating of the vapour at point "b" entering the compressor is performed to prevent the possibility of liquid refrigerant entering the compressor. However, the amount of superheat must be kept to a minimum because the increased specific volume reduces the capacity of the compressor, the peak temperature following compression is higher, and more work is required by the compressor.

The schematic representation of Figure 6 shows that the two operations of subcooling the liquid at "d" and superheating the vapour at "a" can be accomplished by the addition of a heat exchanger to the basic components of the device.

3.2. Equations for the Single-Stage Compression Heat Pump

Application of the First Law to the various components of the basic heat pump results in very simple expressions for the amounts of heat transfer in the evaporator and condenser and the amount of work required by the compressor. The equations are presented for the assumed situation of steady-state and steady flow; negligible changes in kinetic and potential energies across the various components; no heat losses or pressure drops in the piping between components etc.

The equations are as follows :

Evaporator $\dot{Q}_{ab} = \dot{m}\,(h_b - h_a)$

Isentropic compressor $\dot{W}_{bc} = \dot{m}\,(h_c - h_b)$

Actual (non-isentropic) compressor $\dot{W}_{bc}\,(\text{act}) = \dfrac{\dot{m}\,(h_c - h_b)}{\eta_s\,\eta_m}$

where η_s is the isentropic (adiabatic) efficiency and η_m is the mechanical efficiency

Condenser $\dot{Q}_{cd} = \dot{m}\,(h_d - h_c)$

Throttling valve $h_d = h_a$

The mass flow rate of refrigerant, for a positive-displacement type of compressor, depends upon the volumetric efficiency η_v of the compressor (which will change with changing state points at the compressor inlet) as well as the piston displacement and speed of rotation.

$$\dot{m} = \frac{\eta_v\,(PD)N}{v_b}$$

where η_v = "mass of vapour actually pumped by the compressor divided by the mass of vapour which the compressor could pump if it handled a volume of vapour equal to its piston displacement and if no thermodynamic state changes occurred during the intake stroke" [2]

The heat pump COP is obtained as follows :

$$COP = \frac{Q_H}{W_{in}} = \frac{h_c - h_d}{h_c - h_b} \cdot \eta_s \cdot \eta_m$$

The efficiency of the heat pump defined relative to the performance of the Carnot heat pump is therefore :

$$\eta_{id} = \frac{COP_{actual}}{COP_{Carnot}} = \frac{h_c - h_d}{h_c - h_b} \cdot \eta_s \cdot \eta_m \cdot (1 - \frac{T_d}{T_a})$$

The COP and efficiency of a heat pump decrease significantly as the temperature spread between the refrigerant in the evaporator and condenser widens. Figure 7 demonstrates the changes in COP and efficiency both for a Carnot refrigeration device and for the performance of ideal cycles using various refrigerants [2]. The data are presented as a function of the evaporator temperature for a fixed condenser temperature.

3.3. Volume Flow Rate of Refrigerant

The volume flow rate of refrigerant to be handled by the compressor for a specified heating rate depends both on the nature of the refrigerant and the operating conditions of the heat pump. This volume flow rate will be proportional to the specific volume of the vapour at state point "b". The higher the vapour pressure of the refrigerant and consequently the lower the normal boiling point, the smaller will be the specific volume and the volume flow rate of the compressor. As an illustration Duminil [6] calculates the heating effect per unit of swept volume for R11, R12 and R22 refrigerants for the following conditions : evaporator temperature 20 °C, condenser temperature 80 °C, 10 °C of superheat at state point "b" and 5 °C of subcooling at state point "d". The results are as follows :

R11 : 859.2 KJ/m^3
R12 : 3581.8 KJ/m^3
R22 : 5633.4 KJ/m^3

Ratios of these numbers will be inversely proportional to the relative vapour volume flow rates for these refrigerants.

4. MULTISTAGE VAPOUR COMPRESSION

The single-type compression cycle discussed above is the simplest and the most commonly used cycle. However, under some circumstances it becomes advantageous to consider devices in which the compression of the vapour occurs in two or more stages, termed "multi-stage systems". Distinctions are made between "compound systems" in which the two separate refrigeration systems are interconnected in a way that the heat rejected from the condenser of one cycle is the heat absorbed by the evaporator of the second cycle [5]. In the case of cascade systems it is possible to use different refrigerants in the two cycles, each refrigerant selected as the optimum for the particular temperature range of operation. Figures 8-11 illustrate schematically several of the many possible configurations of multistage compression devices. As indicated in the schematic diagrams, some of these devices employ intercoolers to cool the vapour between stages of compression resulting in decreased compressor work requirements.

Figure 12 graphically demonstrates on T-s coordinates how the use of a two stage compression cycle with intercooling eliminates both the increased throttling losses and high peak temperatures following compression associated with large differences in pressure and temperature between the evaporator and condenser [4].

Generally the thermodynamic analysis of these cycles proceeds in a relatively straightforward manner following the format used for the single stage compression cycle. The major difference involves the determination of the different mass flow rates of refrigerant in the various components.

Duminil [6] presented calculations of the power required for several variations of two-stage and a one-stage compression heat pump all using R-12. For the calculations he assumed for a heat source water cooled from 50 to 25 °C. The data are presented as a function of the condenser temperature for a range from 60 to 90 °C. For the conditions assumed, he concludes that considerable energy savings are possible with two-stage heat pumps in comparison with the single stage device. These energy savings were about 11 % and 25 % at condenser temperatures of 60 °C and 90 °C respectively.

The ASHRAE Handbook of Fundamentals [1] also performs a series of example calculations for four different cases : a Carnot cycle, an ideal single-stage compression cycle, a non-ideal single-stage compression cycle, and an ideal two-stage compression cycle. All calculations are performed for R-12 for similar operating conditions. In addition to the First Law results, which indicate the amount of work required for a given refrigeration effect or heating effect, these calculations also include an availability analysis. It is clear from the calculations that the use of the two-stage compression cycle reduces the irreversibilities and brings about energy savings. The determination of an optimum system design for given operating conditions must balance the improvements in energy performance associated with multi-staging etc. with the increased complexity and increased component costs.

REFERENCES

1. *"Thermodynamics and Refrigeration Cycles"*, American Society of Heating, Refrigerating and Air-Conditioning Engineers, Inc. New York, *ASHRAE Handbook and Product Directory, 1977 Fundamentals*, 1977.

2. Threlkeld, James L., *Thermal Environmental Engineering* Ed. 2, New Jersey, Prentice Hall, 1970.

3. McQuiston, F.C. and Parker, J.D., *Heating Ventilating, and Air Conditioning, Analysis and Design*, New York, John Wiley and Sons, U.S.A. 1977.

4. Wood, B.D., Applications of Thermodynamics, Reading, Addison-Wesley, 1969.

5. "System Practices for Multistage Applications", American Society of Heating, Refrigerating and Air-Conditioning Engineers, Inc. New York, *ASHRAE Handbook and Product Directory, 1980 Systems*, 1980.

6. Duminil, M., "Basic Principles of Thermodynamics as Applied to Heat Pumps : Thermodynamic Cycles in Heat Pumps, in *Heat Pumps and Their Contribution to Energy Conservation*, Camatini, E. and Kester, J. (Alphen aan de Rijn : Sythoff & Noordhoff, 1976), 97 - 154.

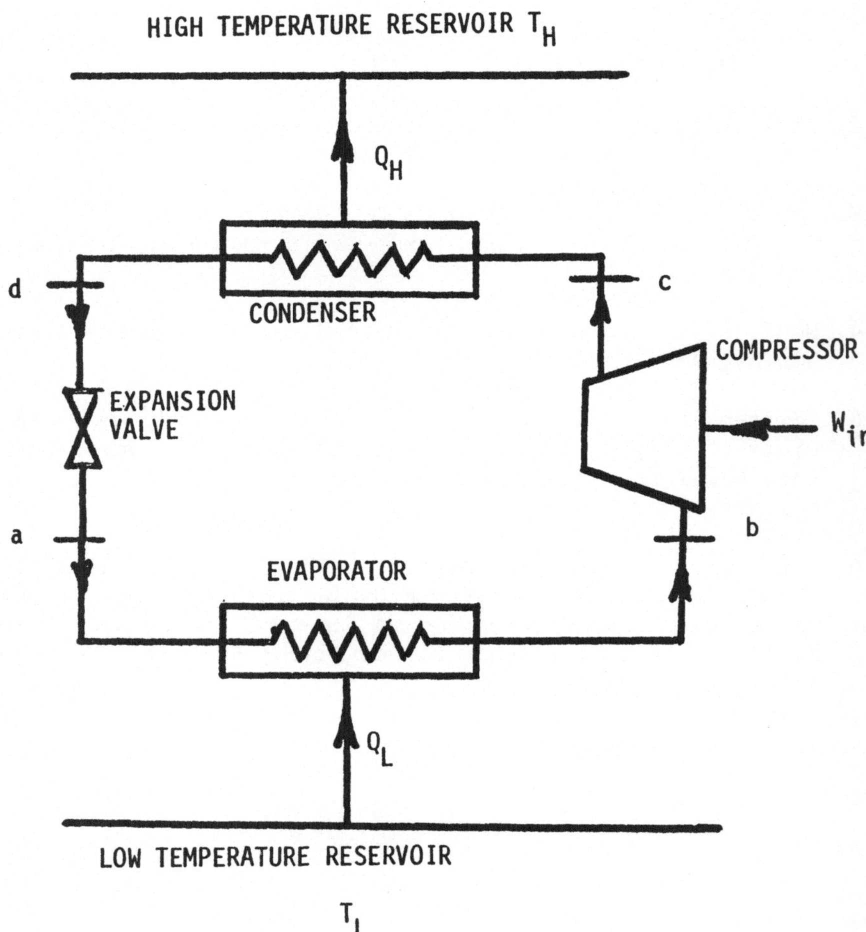

Figure 1. Schematic Representation of Heat Pump

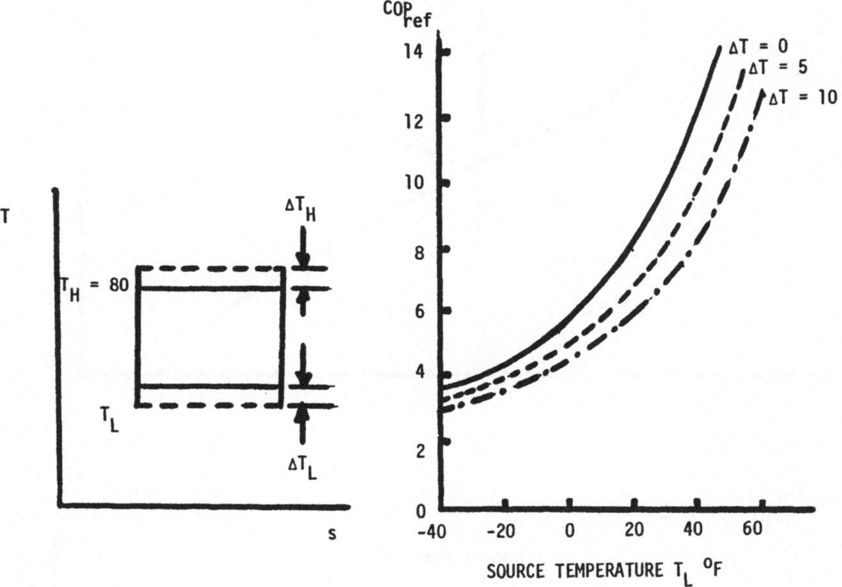

Figure 2. Effect of Source Temperature and Heat Exchanger Temperature Difference on COP

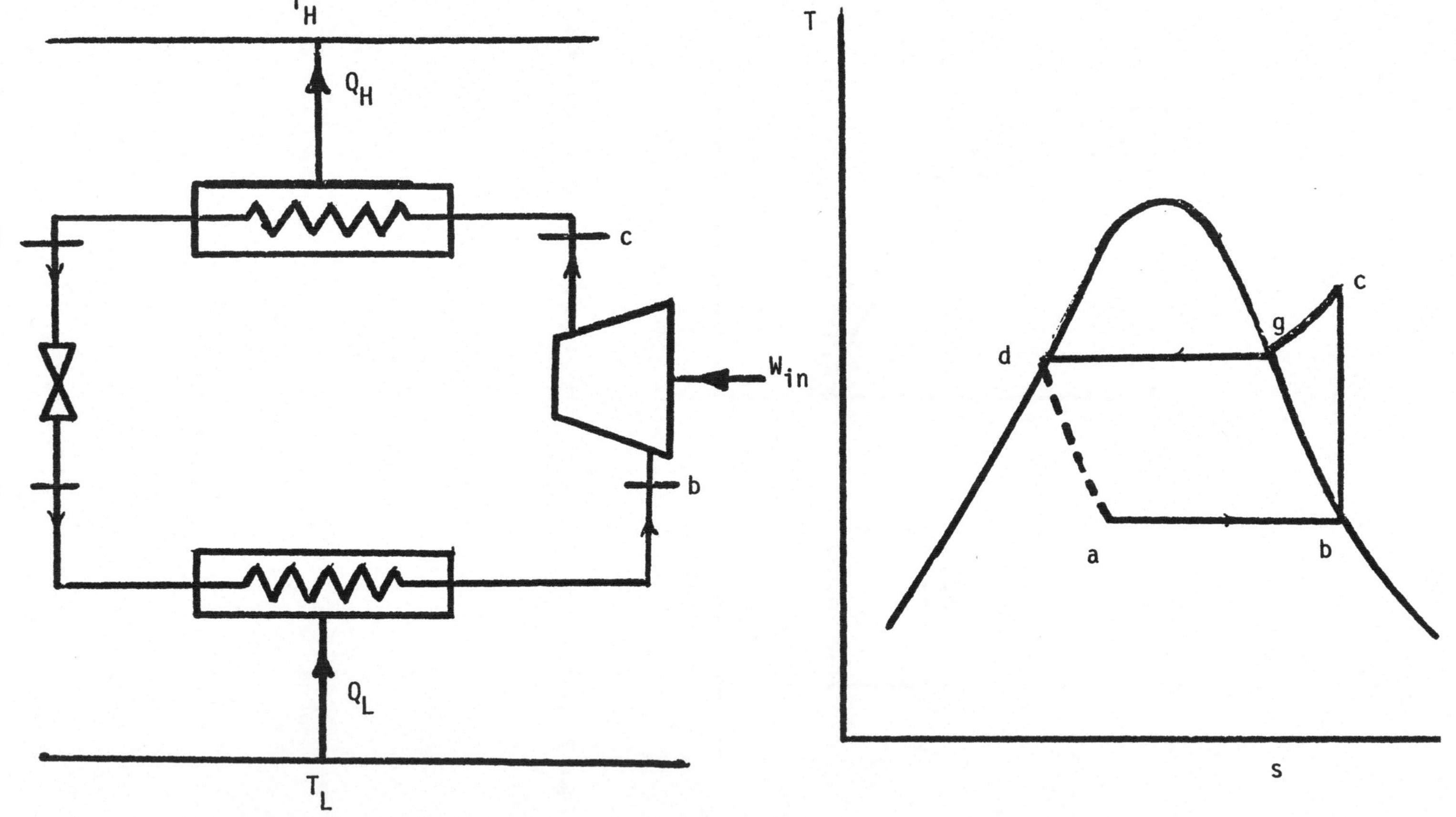

Figure 3. Heat Pump Schematic and T-s Process Representation

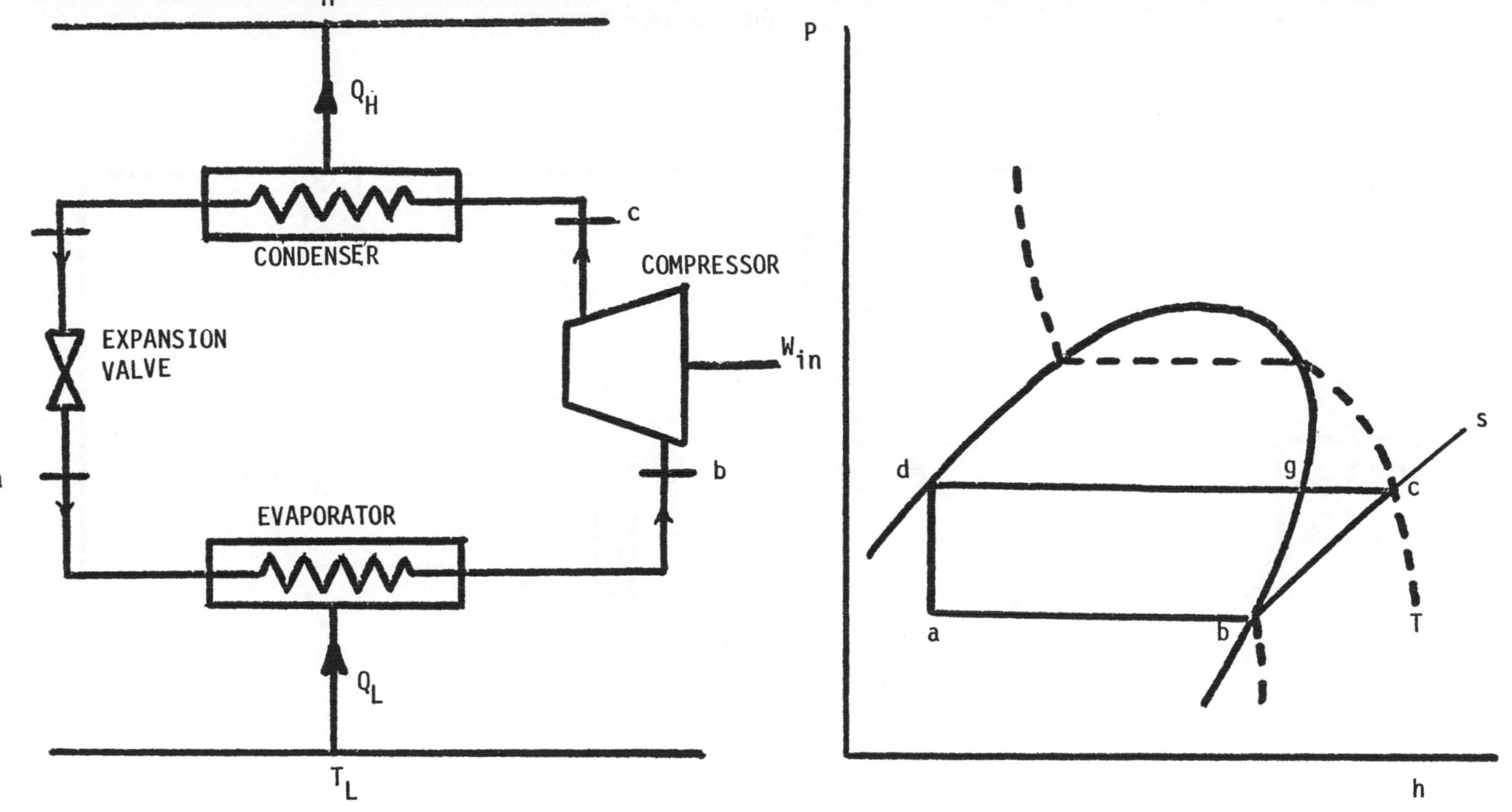

Figure 4. Heat Pump Schematic and P-h Process Representation.

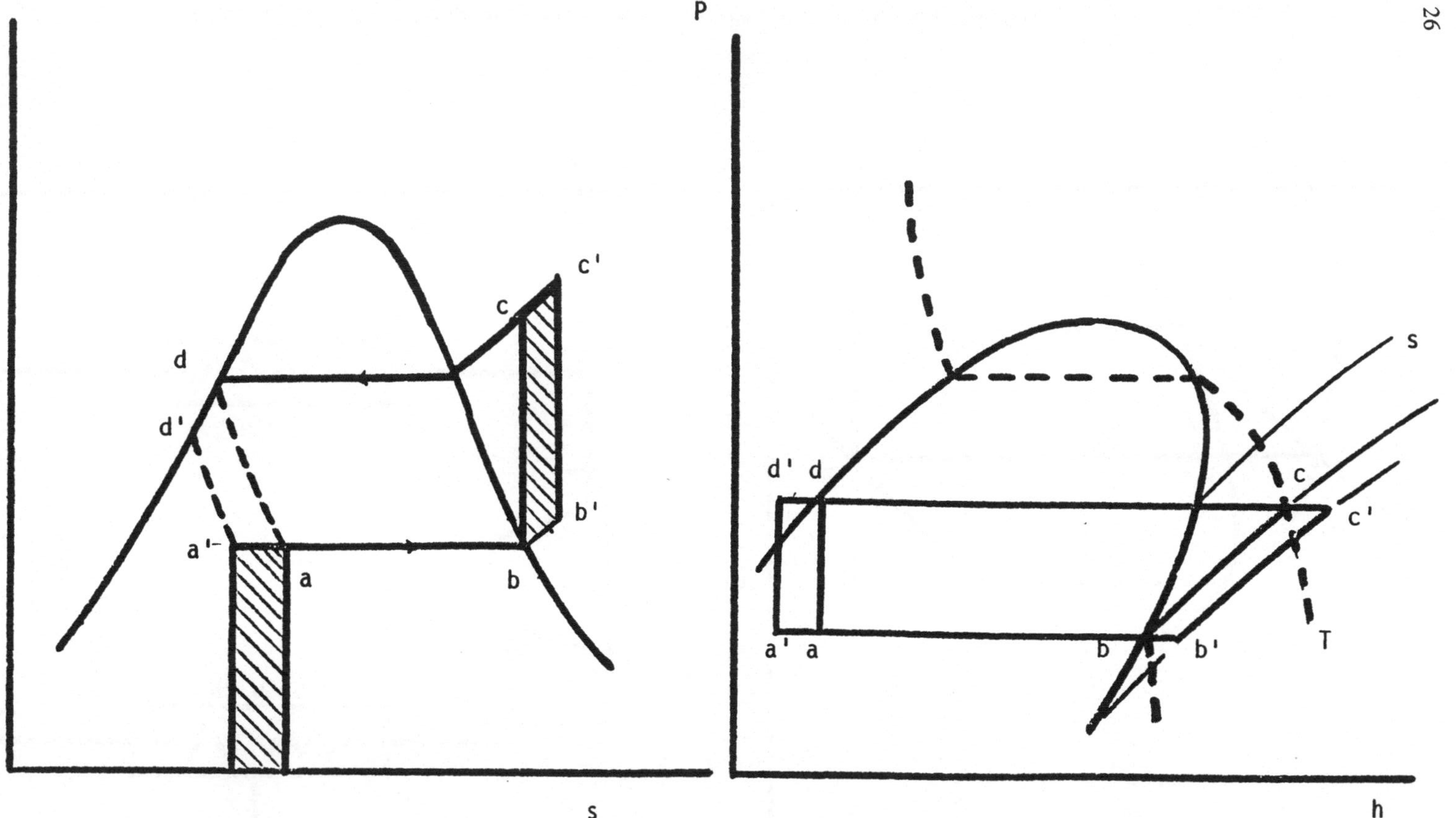

Figure 5. Subcooling and Superheating

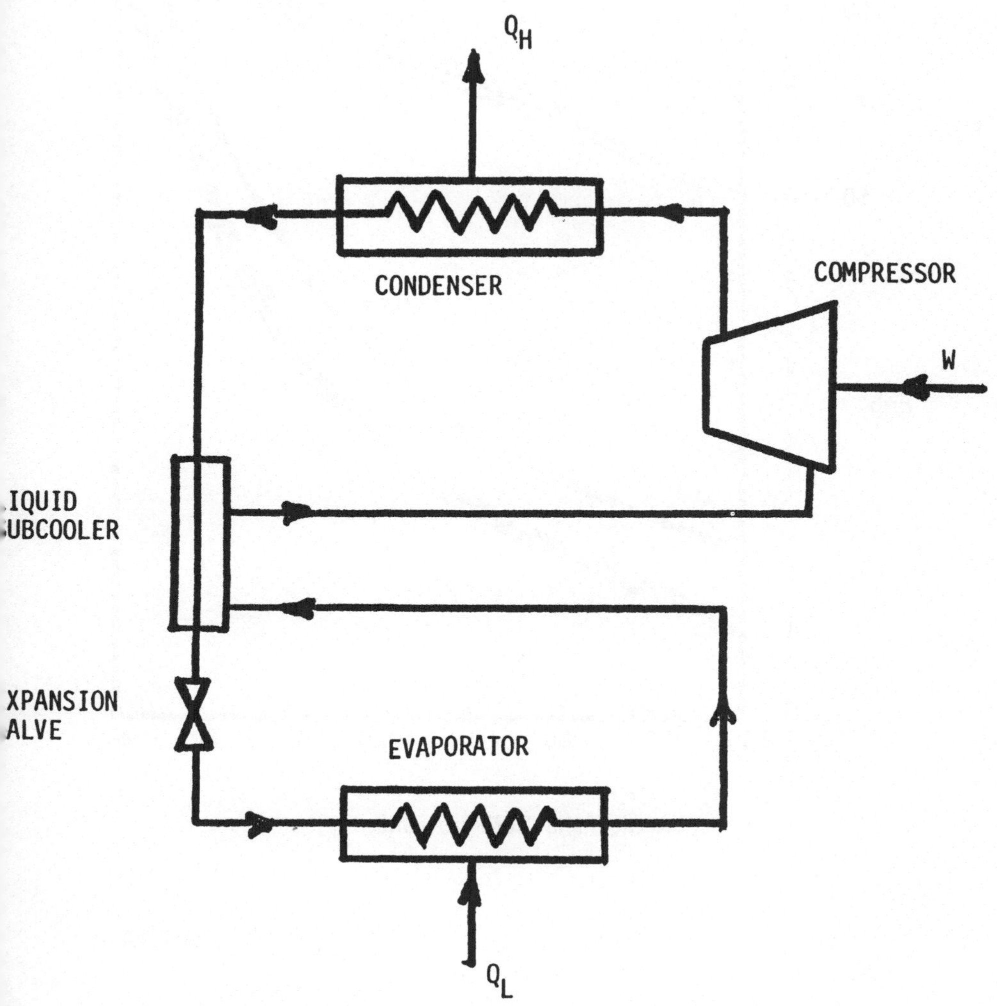

Figure 6. Liquid Subcooling Using Suction Vapour

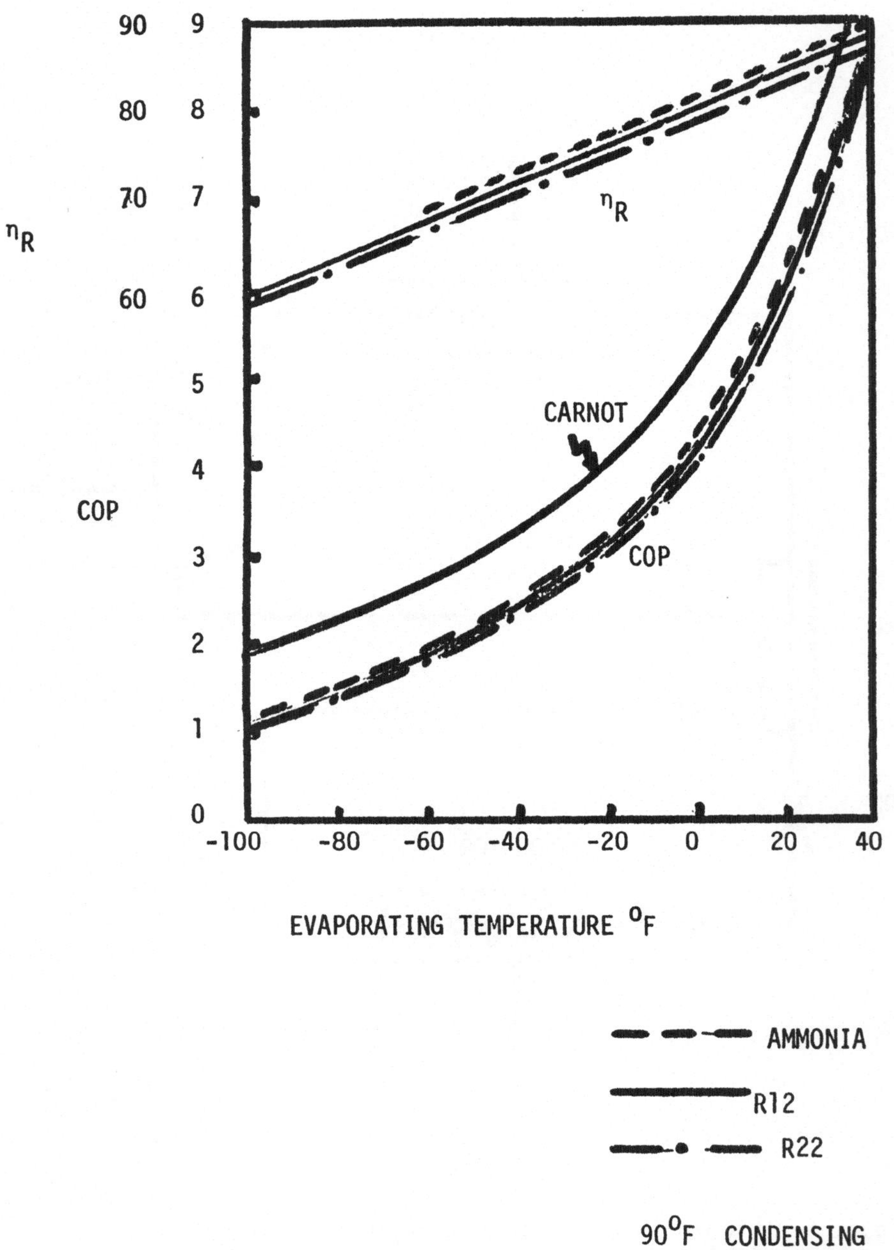

Figure 7. COP and Refrigeration Efficiency for Ideal Single-Stage Cycl

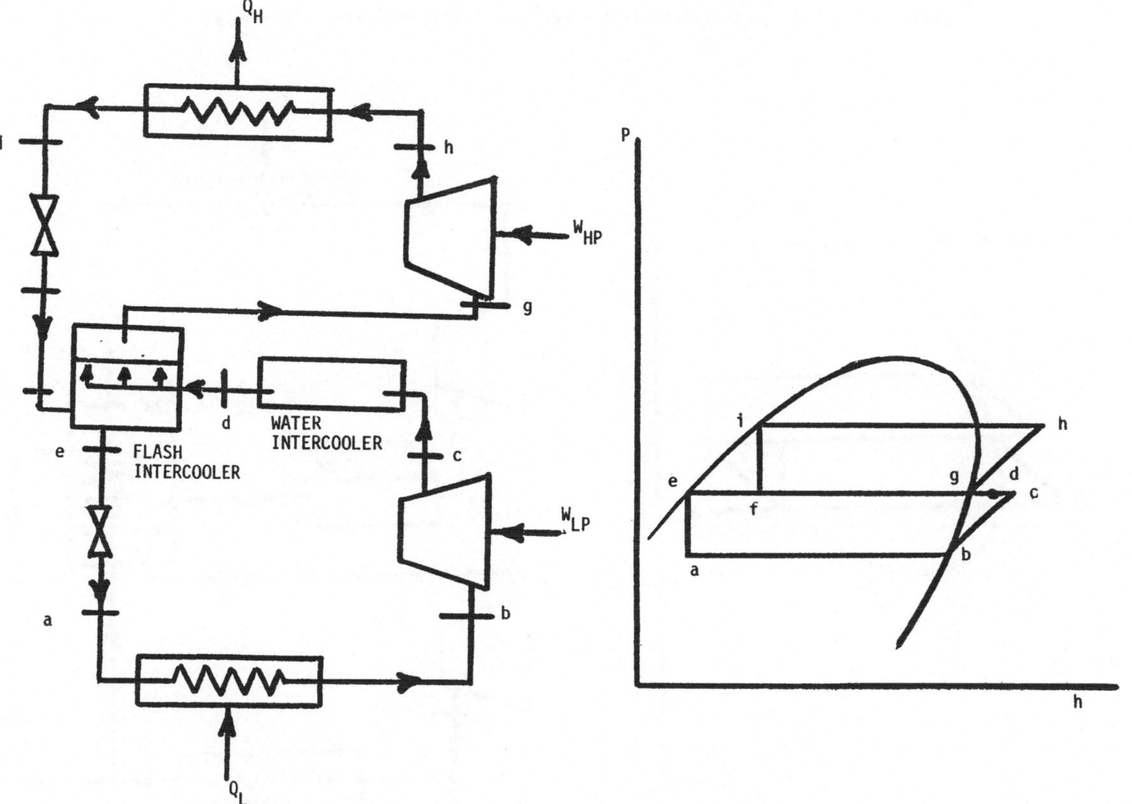

Figure 8. Two-stage Compressor, Water Intercooler and Flash Intercooler

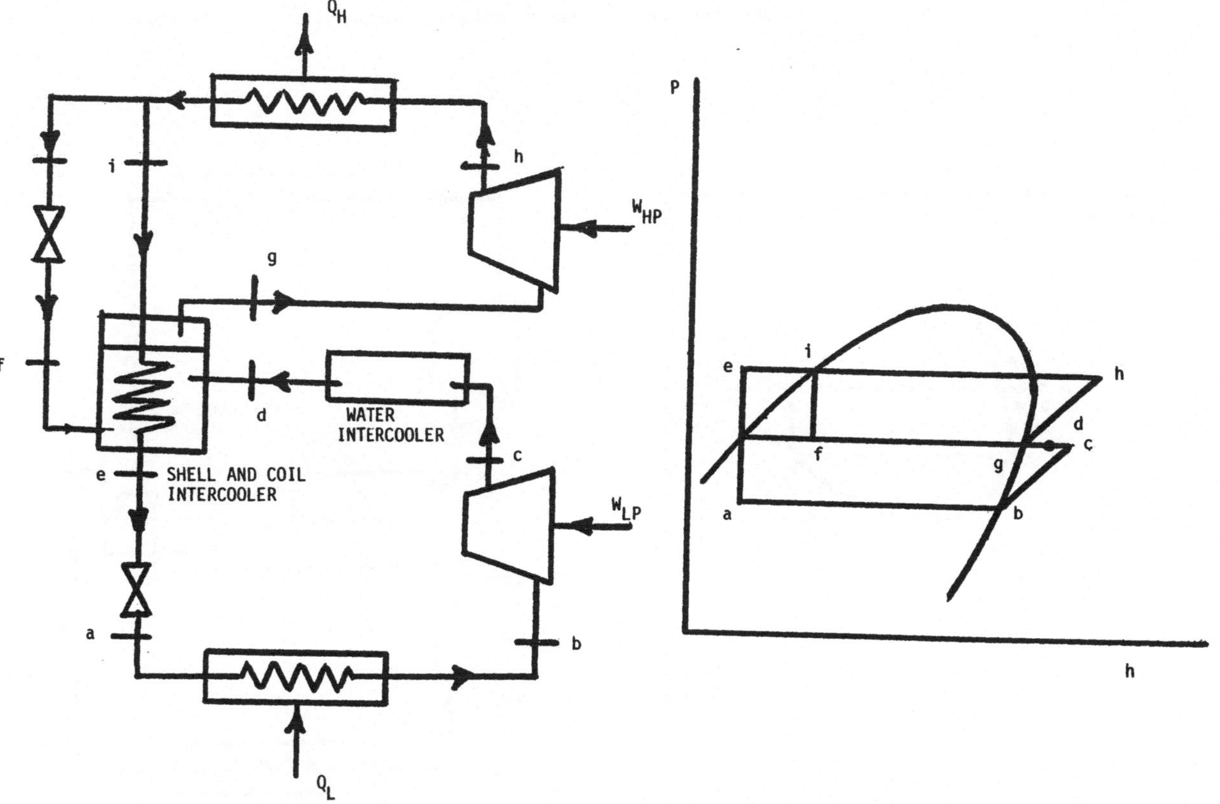

Figure 9. Two-Stage Compressor, Water Intercooler, Shell and Coil Intercooler

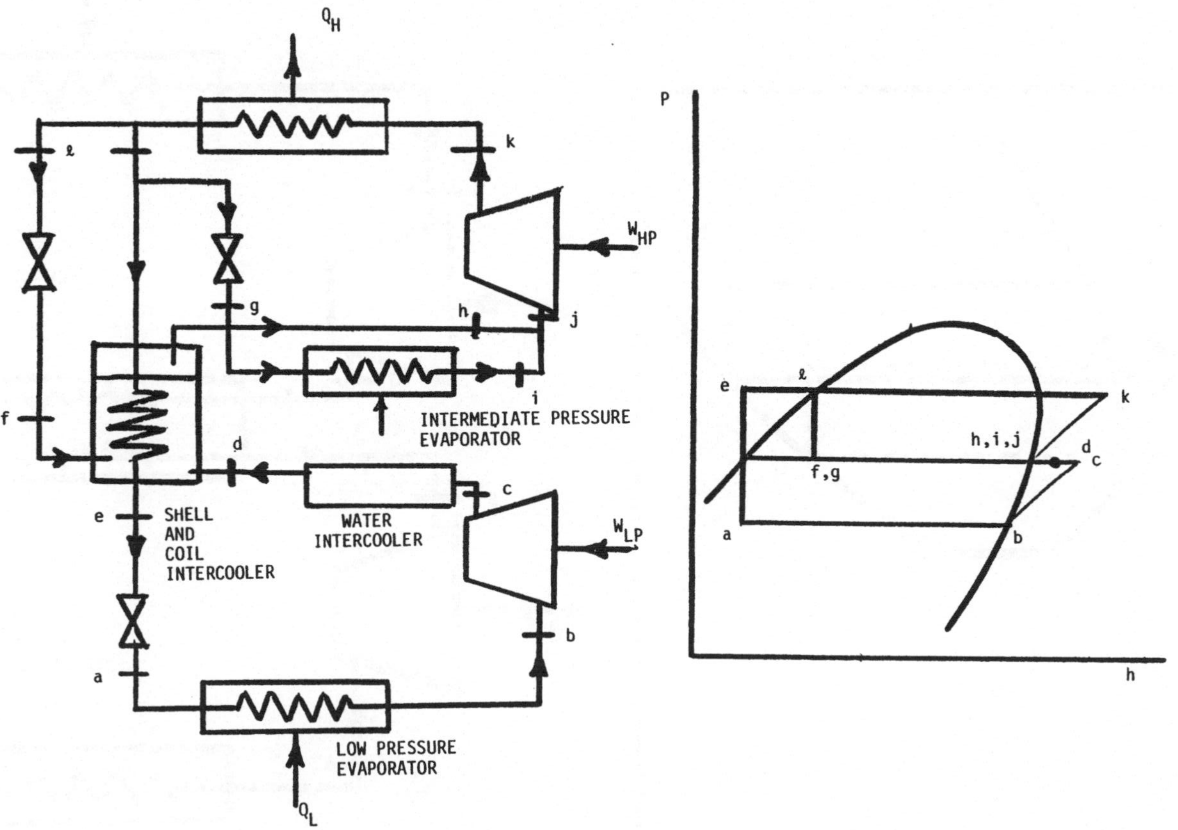

Figure 10. Two-stage Compressor, Low Pressure and Intermediate Pressure Evaporators

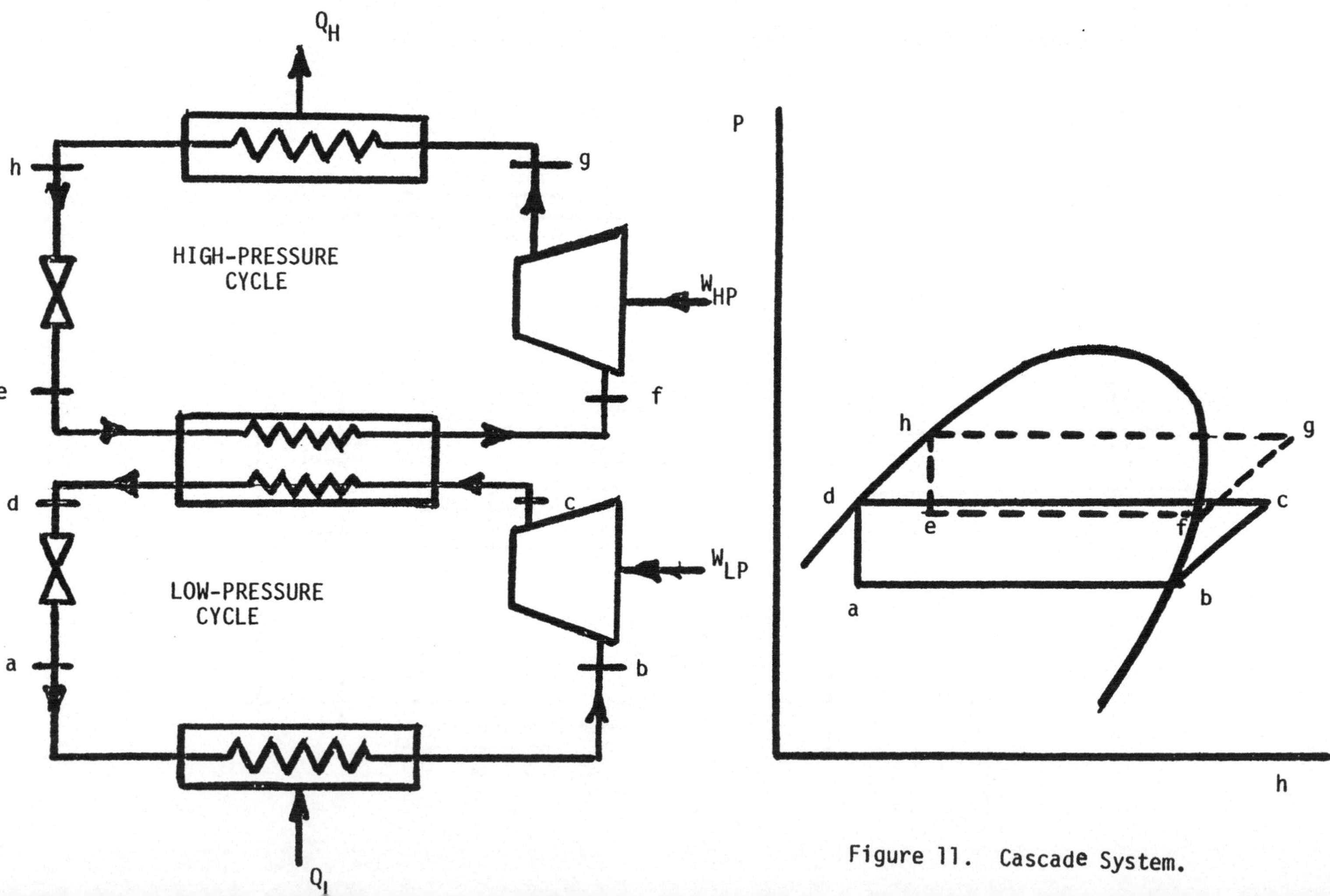

Figure 11. Cascade System.

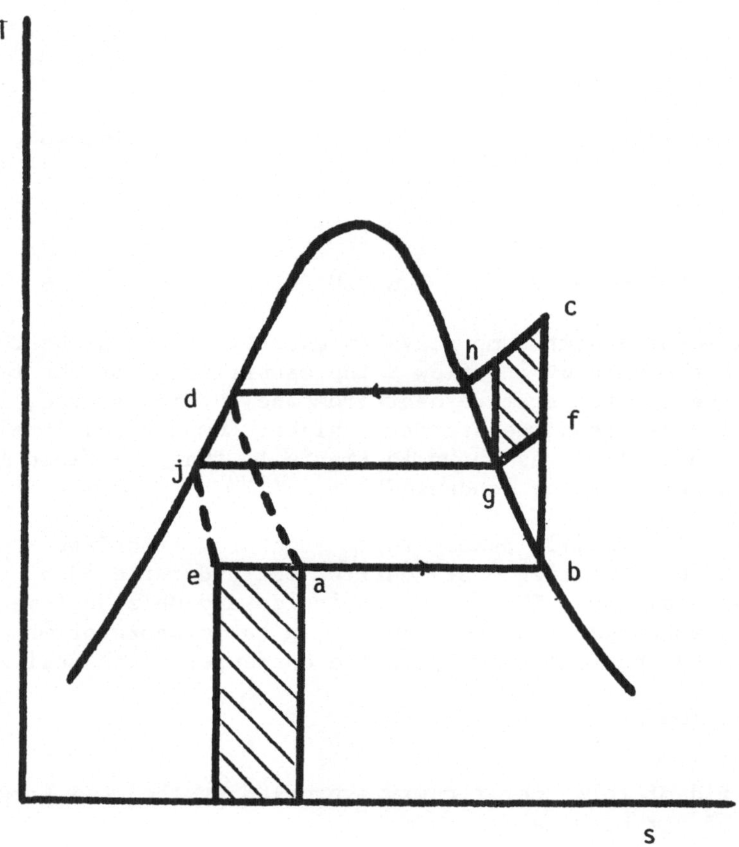

Figure 12. Two-stage Compression Cycle with Intercooling.

VAPOUR COMPRESSION HEAT PUMP FLUIDS PROPERTIES AND SELECTION

J. Durandet

Institut Français du Pétrole, Rueil-Malmaison, France

1. HEAT PUMP - FUNCTION OF THE FLUID

The fluid of a heat pump plays an essential role as heat is moved through the pump by its flow : the vaporization of the fluid collects the low temperature heat from the outside source, its condensation releases the heat at a higher temperature level. This operation implies, as shown in Figure 1, that the fluid is compressed before being condensed.

To simplify, we will assume that both heat transfers - vaporization and condensation - are carried out under the ideal conditions of reversibility. Therefore the inside and outside temperatures in each exchanger are the same (T_o in the evaporator which transfers Q_0 to the fluid and T_1 in the condenser which delivers Q_1).

An enthalpy balance gives :

$$Q_1 = Q_0 + W \tag{1}$$

The yield of this operation is expressed by the well known coefficient of performance :

$$COP = \frac{Q_1}{W} \tag{2}$$

The simplest and most efficient thermodynamic cycle that a fluid could theoretically follow in a heat pump is the CARNOT's cycle (see Figure 2) whose heat exchanges are isothermic and whose compression as well as expansion steps are isentropic.

In Figure 2, Q_0, Q_1 and W are respectively expressed by the areas FADE, FBCE and ABCD; the COP may then be evaluated by the area ratio

$$COP = \frac{FBCE}{ABCD} = \frac{T_1}{T_1 - T_0} \tag{3}$$

It will be therefore noted that the COP of a CARNOT cycle is not depending on the fluid nature : all fluids give the same COP provided they operate in the same temperature range.

2. PHASE TRANSITION OF A PURE SUBSTANCE (THERMODYNAMIC REFRESHER)

2.1. PVT-diagram

The phase rule requires that a system of 2 coexistent phases is defined by one variable (say p or T) whereas a system of one phase is defined by 2 variables (p and T, for instance). The graphical representation of a pure substance phase transition may then be a pVT surface. Figure 3 presents such a surface the projection of which on a p-T plane gives the variation of the fluid vapour pressure with temperature (see Figure 4).

The maximum temperature and pressure at which liquid and vapour can coexist are those of the critical point (CP). At temperatures lower than the triple point (TP) a solid phase appears. A similar projection on the p-v plane (Figure 5) shows the evolution of a mole of a substance at a given temperature : it goes usually from undercooled liquid (1') to a saturated one (1) in equilibrium with a saturated vapour (v) and then to dry superheated vapour (v').

2.2. Enthalpy (H) and entropy (S) charts

T-H diagram

At a given pressure, the heat content (enthalpy) of a mole of fluid changes with temperature, as shown in Figure 6, from undercooled to saturated liquid (1), saturated vapour (v) and finally to dry vapour (v'). We note that the molar enthalpy of vaporization ($H_v - h_1 = L$) decreases when T and p increase, and becomes nil at the critical point.

P-H diagram

A typical pressure-enthalpy diagram is given in Figure 7. It is valid for dichlorodifluoromethane.

T-S diagram

The assumption of isentropic steps in many thermodynamic cycles favors the use of diagrams relating temperature to entropy. Two examples of such charts are given in Figures 8 and 9 in order to illustrate the different shapes that the saturated phase curve may have. We will pay a special attention to :

$(\frac{\delta T}{\delta S})_{liq}$, slope of the liquid branch of this curve

and

$(\frac{\delta T}{\delta S})_{vap.}$, slope of the vapour branch .

The latter may either be positive (as it is frequently encountered with large molecules such as benzene - see Figure 9 -, toluene and certain halogenated hydrocarbons) or negative (see Figure 8 relative to water).

Let us calculate each of these slopes :

For the vapour phase $(\frac{\delta T}{\delta S})_{vap.}$

$$dS_{vap.} = C_p \frac{dT}{T} - R \frac{dp}{p} \tag{4}$$

p here is the vapour pressure of the fluid, which may be evaluated from Clapeyron's relation :

$$(\frac{dp}{p})_{sat.} = \frac{L}{R} \frac{dT}{T^2} \tag{5}$$

if we assume that the vapour phase is an ideal gas. Therefore

$$dS_{vap.} = \frac{dT}{T} (C_p - \frac{L}{T}) \tag{6}$$

and

$$(\frac{\delta T}{\delta S})_{vap.} = \frac{T}{C_p - \frac{L}{T}} \tag{7}$$

For the liquid phase $(\frac{\delta T}{\delta S})_{liq}$

$$S_1 = S_{vap.} - \frac{L}{T} \tag{8}$$

$$dS_1 = dS_{vap.} + \frac{L}{T} \cdot \frac{dT}{T} \tag{9}$$

By introducing the value of $(dS)_{vap.}$ from (6) one finds :

$$dS_1 = C_p \frac{dT}{T}$$

Finally

$$\left(\frac{\delta T}{\delta S}\right)_1 = \frac{T}{C_p} \qquad (10)$$

2.3. Equations of state - data predictions

All the previously considered thermodynamic properties of pure components may be established from pVT data together with data on the heat capacity at constant pressure (C_p).

An equation of state of the fluid has then to be available when these calculations must be carried out, that is to say when the data cannot be found in the specialized litterature [1] [2].

The well known perfect gas model is too approximate to be used for these calculations particularly in the high pressure range and, amongst the different equations of state, we will select the modified REDLICH and KWONG relation :

$$p = \frac{RT}{v-b} - \frac{a(T)}{V(V+b)} \qquad (11)$$

The thermodynamic data (p, V, T, H, S...) needed for selecting a heat pump fluid may then be established, for instance, by using the method proposed by ASSELINEAU [3]; other methods are also available [4].

3. HEAT PUMP RANKINE CYCLE

For practical reasons, CARNOT's cycle cannot be applied in vapour compression heat pumps. Instead, the RANKINE cycle is used.

In this cycle :

- the compression step is still adiabatic and isentropic but only vapour flows through the compressor as the formation of liquid droplets has to be avoided;
- the expansion step is no more isentropic since it is not economically justified to recover the energy released by this expansion. The expansion is performed adiabatically.

The consequences of these changes are emphasized in Figure 10 (a and b) :

- the vapour leaving the compressor (B') is dry, superheated and therefore, at a temperature (T_1') above the temperature (T_1) of the saturated vapour (B") at the same pressure P_1;

- the expansion step produces a mixture of saturated liquid (D") and vapour (A). The quality of this mixture may be denoted by the mole fraction (x) of vapour in the mixture. Then

$$1 - x = \frac{n}{N} = \frac{\overline{D'A}}{\overline{D''A}} \tag{12}$$

N and n being the molar flow rates of the total mixture and of the liquid phase respectively.

- Referring to the enthalpy balance we may express Q_0 and Q_1 respectively by the areas FAD'E' and FB'B"CE, therefore W corresponds to the area EE'D'AB'B"CE.

Figure 11 shows that the yield of the RANKINE cycle (COP_R) is lower than the yield of the CARNOT (COP_C). It also shows that this yield depends, in contradiction with what is the case for the Carnot Cycle, on the nature of the fluid. The example of figure 11 corresponds to a heat pump operated with 4 different fluids (3 fluorocarbons, one hydrocarbon) and with a heat source at 70 °C, the condensing temperature being variable.

4. HEAT PUMP FLUID PROPERTIES (selection of the fluid)

Heat pump fluids have first to be compatible with the required thermodynamic cycles, particularly with their temperature levels, and have then to offer such properties that the heat pump operation is economically optimal and safe altogether.

4.1. Thermodynamic Properties

The most important fluid property is its vapour pressure as the fluid must boil and condense in the heat pump.

4.1.1. Vapour pressure

With our simplifying assumptions of perfect heat exchanges, the fluid boils in the evaporator at T_0, which is also the heat source temperature, and condenses at T_1, which also corresponds to the temperature of the delivered heat. Both temperatures T_0 and

T_1 are dictated by the level at which heat is available and at which it is required. The fluid to be selected for this operation must have a critical temperature above T_1 and a triple point temperature lower than T_0

$$T_t < T_0 < T_1 < T_C \tag{13}$$

Furthermore, the following requirements are to be met :

- Evaporators are operated at pressure above atmospheric. Thus, air entry in the pump is avoided. This condition implies that the fluid must have a boiling temperature under atmospheric pressure (T_B) lower than T_0

$$T_B < T_0 \tag{14}$$

- Better heat pump yield is obtained with fluids offering higher critical temperatures. Figure 12 shows for instance that the pump efficiency decreases when T_1 comes closer to T_C. On this figure, we express the pump efficiency by the ratio of the COP's (Rankine to Carnot) in order to eliminate the effect of temperature variation on the yield. We also use a reduced operating temperature T_r (ratio of T_1 to T_C, both values being absolute temperatures). From this figure it appears that it is preferable to work at reduced temperatures lower than .9 for obtaining high COP.

$$T_C > \frac{T_1}{.9} \tag{15}$$

Figure 13 points out the effect of these 2 specifications on the preliminary selection of fluids to be used in a pump working between T_0 = 70 °C and T_1 = 90 °C (or 363 °K). For instance, we see that 2 hydrocarbons may be used (n butane and n pentane) as well as several fluorocarbons (R11, R113, R114, R133, R216) and sulfur dioxyde.

4.1.2. Fluid properties and heat pump yield

In order to point out the main fluid properties upon which the COP of the pump depends, we can calculate it in an oversimplified way. It is assumed that the vapour is a perfect gas and the compression is both adiabatic and isentropic.

As

$$COP = \frac{Q_1}{W} = 1 + \frac{Q_0}{W} \tag{16}$$

in which $Q_0 = n\ L_0$

and as :

$$W = N\ R\ \ln \frac{p_1}{p_0} \cdot \frac{(T_1' - T_0)}{\ln \frac{T_1'}{T_0}} \tag{17}$$

one obtains :

$$COP = 1 + \frac{n}{N} \cdot \frac{L_0}{R} \cdot \frac{\ln \frac{T'_1}{T_0}}{(T_1' - T_0)} \cdot \frac{1}{\ln \frac{p_1}{p_0}} \tag{18}$$

If we furter assume that the liquid specific volume is negligible compared to the gaseous one and that the molar enthalpy of vaporization is constant, so that

$$\ln \frac{p_1}{p_0} = \frac{L_0}{R} \left(\frac{T_1 - T_0}{T_0\ T_1}\right) \tag{19}$$

we get the final simplified relation :

$$COP = 1 + \frac{n}{N} \cdot \frac{T_0\ T_1}{T_1 - T_0} \cdot \frac{\ln \frac{T_1'}{T_0}}{T_1' - T_0} \tag{20}$$

We will only retain from this very approximate relation that high COP's are favored by high values of n/N and $\ln T_1'/T_0\ /\ (T_1' - T_0)$ respectively. Let us examine each of these points.

- Effect of the liquid mole fraction n/N

  High COP's correspond to high liquid mole fractions as shown in Figure 14. As previously pointed out, this value is bound to the energy losses occuring during the expansion step.

  From Figure 10, an isentropic expansion gives the maximum n/N ratio corresponding to $\overline{DA}/\overline{D''A}$; an isenthalpic step gives a lower liquid mole fraction which may be evaluated by $\overline{D'A}/\overline{D''A}$.

  Figure 10-b will be used to evaluate the n/N ratio. In this figure :

$$\overline{D''A} = h_A - h_{D''} = L_0$$

$$\overline{D'A} = \overline{D''A} - \overline{D'D''}$$

and $\overline{D'D''} = h_c - h_{D''}$,

In the considered operating conditions, far from critical :

- L is constant,
- the 2 branches of the saturated curve are therefore parallel, which implies that $C_{p,liquid} = C_{p,vapour}$
- the vapour phase is assumed to be in the ideal state, so its C_p is independent of pressure.

Consequently $\overline{D'D''} = C_p\ (T_1 - T_0)$

and $$\frac{n}{N} = 1 - \frac{C_p}{L}\ (T_1 - T_0) \qquad (21)$$

This relation shows that the best fluids should have low C_p values and high heats of vaporization.

- Effect of the molar latent heat of vaporization ($\Delta H_v = L_0$)

Fluids of higher heat of vaporization give best COP's. In the same way, for a given fluid, better results are obtained at lower operating temperatures to which correspond higher heats of vaporization. Therefore the fluids to be selected must offer high critical temperatures and large L_θ. The latter may be roughly evaluated with the help of TROUTON's rule which states that the entropy of vaporization of non-associated fluids at their normal boiling point is approximately given by

$$\Delta S_v = \frac{\Delta H_v}{T_B} \sim 20 \text{ cal/g mole} \qquad (22)$$

This value is higher for polar associated molecules which consequently may, from this point of view, offer some advantage. Table 1 gives some values of enthalpies and entropies of vaporization for some hydrocarbons and for water.

- Effect of heat capacity (C_p)

Fluids with small C_p are the best. That is to say the lightest members of given chemical families must be selected as the C_p increases with the fluid molecular weight (see Table 1).

To low C_p correspond high $(\delta T/\delta S)_l$ as indicated by relation (10).

Therefore, the higher $(\delta T/\delta S)_l$, the larger will be n/N. Consquently fluids offering small $(\delta T/\delta S)_l$ values should not be selected as heat pump fluids. If they are still to be used, a cycle modification will be required in order to keep the n/N ratio at a convenient value. Figure 15 shows such a modified cycle : the liquid leaving the condenser is subcooled (from C to C') previously to its expansion (from C' to D'''). In this way, the liquid mole fraction may be sufficiently increased (D'' close to D) to justify the use of the considered substance in a heat pump. This liquid subcooling may be carried out by means of an auxiliary cooling or, better, by heat exchange with the vapour leaving the evaporator as will be studied later.

- Effect of vapour superheating

From relation (20) high COP's require high values of

$$\frac{\ln \frac{T_1'}{T_0}}{T_1' - T_0} .$$

To the isentropic compression corresponds (see relation (4))

$$\ln \frac{p_1}{p_0} = \frac{C_p}{R} \ln \frac{T_1'}{T_0} \tag{23}$$

Therefore, by combining (19) and (23) one obtains :

$$\ln \frac{T_1'}{T_0} = \frac{L_0}{C_p} \left(\frac{1}{T_0} - \frac{1}{T_1}\right) \tag{24}$$

Which finally gives :

$$\frac{\ln \frac{T_1'}{T_0}}{T_1' - T_0} = \frac{L_0}{C_p} \cdot \frac{T_1 - T_0}{T_1 \ T_0} \cdot \frac{1}{(T_1' - T_0)} \tag{25}$$

This relation shows that efficient heat pump fluids offer :

- a high (L_0/C_p) ratio as previously stated
- a low superheating effect expressed by a minimum value of $(T_1' - T_1)$.

This second effect is shown in Figure 16 relative for R133 used in a heat pump working between 70 °C and either 90, 110, 120 or 130 °C.

The temperature difference $(T_1' - T_1)$ may be evaluated with the help of Figure 10-a and with the simplifying assumptions of linear relations, one finds since :

$$(T_1' - T_1) = \overline{BB'}$$

and as $\overline{BB'} = \overline{B''B} \left(\frac{\delta T}{\delta S}\right)_{P_1}$

and

$$\left(\frac{\delta T}{\delta S}\right)_{P_1} = \frac{T_1}{C_p} ,$$

$$\overline{B''B} = (T_0 - T_1) \left(\frac{\delta S}{\delta T}\right)_{vap.}$$

Finally

$$(T'_1 - T_1) = \frac{(T_0 - T_1)\, T_1}{C_p} \left(\frac{\delta S}{\delta T}\right)_{vap.} \tag{26}$$

With the help of relation (7), we may change (26) into :

$$(T'_1 - T_1) = \frac{(T_o - T_1) T_1}{T_0} \cdot \left(1 - \frac{L_0}{C_p\, T_0}\right) \tag{27}$$

This equation shows the importance of the ratio $L_0/C_p\ T_0$ which should be as close as possible to unity in order to reduce the superheating effect. We note that the value and the sign of the slope $(\delta T/\delta S)_{vap.}$ depend on this ratio which we may write as $L_0/T_0\ /\ C_p$ in order to point out that this is the ratio of Trouton's constant to the fluid C_p. Table 1 shows that C_p increases with molecular weight for substances of a given chemical family and that (L_0/T) is almost constant for non-associated substances.

For associated ones, such as water, the Trouton constants are higher. Therefore, polar molecules generally give T-S diagrams of the Figure 8 type, because of their high Troutons. Non-associated substances of high molecular weight present a T-S diagram of the Figure 9 type due to their high C_p.

An optimal heat pump fluid therefore has an infinite $(\delta T/\delta S)_{vap.}$ slope and good fluids should give Trouton constants close to their heat capacity C_p. Figure 17 shows the interest offered, from this point of view, by normal butane.

Should the slope $(\delta T/\delta S)_{vap.}$ be too low or positive, it might be necessary to transform the thermodynamic cycle as indicated by Figures 18 and 19.

In Figure 18-a, valid for benzene, the slope is positive, therefore an isentropic compression of the vapour (A) would cause unwanted liquid formation. In order to avoid this perturbating effect, the vapour is slightly superheated (from A to A') under p_0 before being compressed (from A' to B', the latter being as close as possible to B"). As previously pointed out this vapour superheating may be coupled with the subcooling of the liquid leaving the condenser (see Figure 18-b).

In Figure 19-a, valid for water, the saturated vapour curve is rather flat (small $(\delta T/\delta S)$). In this case a direct compression of the vapour (A) leads to a high degree of superheating (expressed by $T_1' - T_1 = \overline{BB'}$) responsible of a low COP. In order to reduce this superheating effect, a multistage compression cycle with intercooling may be used. The principle of this cycle is shown in Figures 19-a and b. The liquid (C) leaving the condenser expands to I (mixture of liquid I' and vapour I") in an economiser kept at an intermediate pressure P_I. The liquid (I') then expands to E while the vapour (I") is mixed with the vapour (B_I) leaving the first stage of the compressor to give the vapour (F) which goes into the second stage of the compressor. Such an arrangement corresponds obviously to rather complex equipment of high investment which should only be justified by high capacity heat pumps.

4.2. Fluid Properties and Kinetic of Exchange

Heat and momentum transfers will be the object of this section. We will, however, restrict it to the compressor study as an exhaustive study including the heat exchanges in the evaporator and the condenser is obviously beyond the scope of this text. We will therefore only consider here the fluid properties which determine the choice of the compressor and its sizing. This choice is based, as we know, on the inlet volumetric flow rate ($\dot{V}$) of the fluid and on its outlet pressure (p_1) as recalled by Figure 20.

4.2.1. Suction volumetric flowrate ($\dot{V}$)

Fluids which minimize the volumetric flowrate at the suction of the compressor are economically interesting as it will be shown later.

With our notations, and for a heat pump of given power (Q_1) and COP, we have :

$$\dot{V} = N\, v_o \tag{28}$$

where v_o is the fluid molar volume in the compressor suction conditions.

$$Q_O = n\, L_O = Q_1 \left(\frac{COP - 1}{COP}\right) \tag{29}$$

so
$$\dot{V} = \left[Q_1 \,.\, \frac{COP - 1}{COP}\right] . \frac{N}{n} . \frac{v_O}{L_O} \tag{30}$$

which with (21) becomes :

$$\dot{V} = [Q_1 \cdot \frac{COP - 1}{COP}] \cdot \frac{v_0}{L_0 - C_p\ (T_1 - T_0)} \tag{31}$$

This relation shows that <u>fluids of high volumetric heat content</u> (L_0/v_0) <u>give minimum compressor inlet fluid flowrate</u>.

For instance, butane and freons 133 or 114 are, from this point of view, to be preferred to R 216 or R 113 as fluids to be used in a heat pump working between 70 and 90 °C (see Figure 21).

4.2.2. Suction and discharge pressures and compression ratio $(r = p_1/p_0)$

To avoid the need of costly equipment the compression ratio must be kept at reasonable values. For instance, ratios over 5 should not be selected in order to avoid the need of several compression stages.

The cost of a compressor depends on the adiabatic head (or theoretical adiabatic power) defined by (see [5]) :

$$HA = \frac{P_0\ \dot{V}}{26,5} \cdot \frac{\gamma}{\gamma - 1} \cdot [r^{\frac{\gamma-1}{\gamma}} - 1] \tag{32}$$

where HA is expressed in (HP)

P_0 " " " bar

$\dot{V}$ " " " m^3/h

and $\gamma = C_p/C_v$ is the specific heat ratio.

γ may be roughly evaluated by giving the following value to the constant volume specific heat, which is correct for ideal gases:

$$C_v = C_p - 2 \tag{33}$$

With this approximation, (32) becomes :

$$HA = a\ P_0 \cdot \dot{V} \cdot C_p \cdot [r^{\frac{2}{C_p}} - 1] \tag{34}$$

Therefore, fluids which do not require high compressor investment are to be looked for among substances offering <u>low values of P_0, v_0/L_0, C_p and r</u>.

However, and for practical reasons, evaporators usually work at pressures over 2 atmospheres. We must also note that small compression ratios correspond to low molar vaporization latent enthalpies (see relation (19)). This specification is opposite to previously stated optimal fluid properties.

As an example, we show in Figure 22 the operating pressures and their corresponding ratios, established for different fluids pumping heat from 70 °C to 90 °C.

4.3. Miscellaneous Fluid Properties

Besides the already stated basic properties, heat pump fluids have to fullfill some others requirements. Such fluids must be :

- stable and compatible with the various materials at operating conditions. The latter requirement may become severe when the vapour superheating is high. Here one should also mention fluid compatibility with the lubricating oil. The miscibility of these products should be such that both compressor lubrication and good heat transfer are achieved.
- non corrosive either by itself or by the products resulting from its possible degradation. For instance, fluorhydric acid resulting from fluorocarbon degradation must be avoided.
- non toxic. Though fluorocarbons are widely used in refrigeration and air conditioning, they are presently suspected both of atmospheric ozone reduction and of carcinogenic properties. Their use may be restricted in the future.

With respect to safety, highly flammable products must be proscribed, at least for domestic uses.

- available and , of course, not too costly.

5. CASE STUDY

Instead of a conclusion, we will give an example of selection of fluids to be used in a heat pump for which operating conditions are :

- fluid evaporating temperature : 70 °C
- fluid condensing temperature : 90 °C

This preliminary choice has, of course, to be confirmed by an economical evaluation.

The selection is practically carried out by means of a computer simulation which takes simultaneously into account the different fluid properties - with their possibly opposite effect on the COP - as well as the complexity and selection of the thermodynamic cycle.

The computer programme calculates first the thermodynamic data relative to each selected pure substance, the method being similar to Asselineau's [3] and then carries out the cycle evaluation. A printout of this programme is given in appendix.

Under " Fluide " , are given the thermodynamic characteristics of the products. Here :

- CA is defined by $\log_{10} \frac{P}{P_C} = (CA) \left(\frac{T_C}{T} - 1\right)$ (35)
- m is the SOAVE's slope (see [3])
- CIDAL (1) - (2) - (3) are the temperature coefficients of the constant pressure heat capacity of the fluid in the ideal gas state :

$$C_p = C_1 + C_2T + C_3T^2 + C_4T^3 \quad (36)$$

- CHIDAL and CSIDAL are respectively the enthalpy and the entropy constants defined by the choice of the pure saturated liquid at its normal boiling temperature as the reference state.

The computer programme selects the thermodynamic cycle and introduces, whenever needed, a vapour superheating (from 1 to 2, see Figure 23) which results from a subcooling of the liquid (from 5 to 6). This heat exchange has to be carried out with an imposed minimum Δt between the 2 fluids at each level of the exchanger (in particular between 6 and 1).

The results of these calculations are given for each fluid on tables entitled "caractéristiques de la pompe à chaleur". These tables give the COP_s (CARNOT and RANKINE with the corresponding compressor polytropic yield) as well as W and Q_1 expressed in calories/mole of circulating fluid. They also indicate the values characterizing the different steps of the cycle (i.e. values relative to points 1, 2, 3, 4, 5, 6 and 7 of Figure 23).

The different steps of the preliminary fluid selection are then as follows :

- from T_B and T_C values (see Figure 13), 8 substances may be used,
- SO_2 is discarded because of its corrosive properties,
- n pentane is not selected as n butane will give better

results due to the Trouton to C_p relation (see Figure 17),

- R11 is rejected by the computer as the superheating-subcooling heat exchange cannot be carried out under the imposed condition of temperature difference,
- the 5 fluids which have not been rejected by this preselection give all the same COP (roughly 10.5 - see Figure 11). Among them, freon R 133 has the advantages first of offering a high volumetric heat content [(L/v) see Figure 21] and then of not requiring vapour superheating as its T-S diagramme is of the Figure 19 type (with an interesting high $(\delta T/\delta S)_{vap}$ slope). Therefore, R 113 appears as the most promising fluid to use though it gives high operating pressures (see Figure 2).

REFERENCES

1. Canjar, L.N., and Manning, F.S. *Thermodynamic Properties and Reduced Correlations for Gases* (1967) Ed. Gulf Publishing Corp., Houston, Texas.

2. A.S.H.R.A.E. *Handbook of Fundamentals* (1972) Ed. American Society of Heating, Refrigerating and Air Conditioning Engineers, New York.

3. Asselineau, L., Bogdanic, G., and Vidal, J. *Calculation of Thermodynamic Properties and Vapor Liquid Equilibria of Refrigerants* (1978) Chem. Eng. Sci. 33 p. 1269-1276.

4. Reid, R.C., Prausnitz, J.M., and Sherwood, Th. K. *The properties of Gases and Liquids* (1977) McGraw-Hill New York.

5. Chauvel, A., Leprince, P., Barthel, Y., Raimbault, Cl., and Arlie, J.P. *Manuel d'Evaluation économique des Procédés* (1976) Technip Paris.

TABLE I

Properties of Fluids (atmospheric pressure)

	Boiling point T_o(K)	C_p° Cal/mole.K	L_o Cal/mole	L_o/T_o Cal/mole.K
C_2 normal paraffine	185	10.2	3500	19.06
C_3 normal paraffine	231	14	4500	19.4
C_4 normal paraffine	273	22	5350	19.6
C_5 normal paraffine	306	28.8	6160	19.9
C_6 normal paraffine	342	38	6900	20.2
C_7 normal paraffine	372	47	7600	20.4
Benzene	353	23	7350	20.8
Water	373	8.2	10800	29

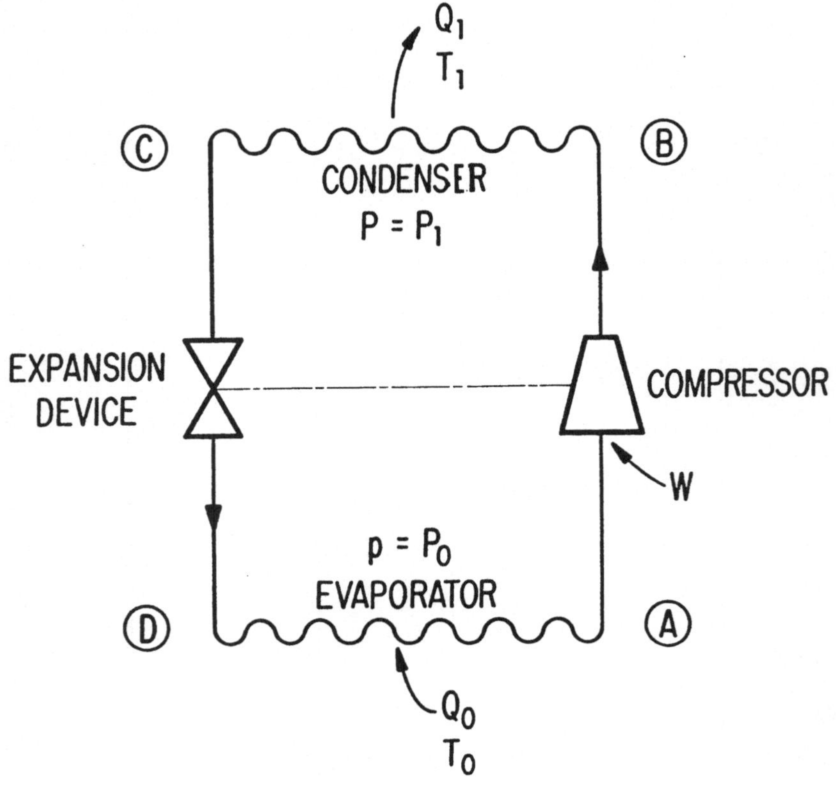
Q1
T1
C
B
CONDENSER
P = P1
EXPANSION
DEVICE
COMPRESSOR
W
p = P0
EVAPORATOR
D
A
Q0
T0
BASIC VAPOR COMPRESSION
HEAT PUMP

Figure 1

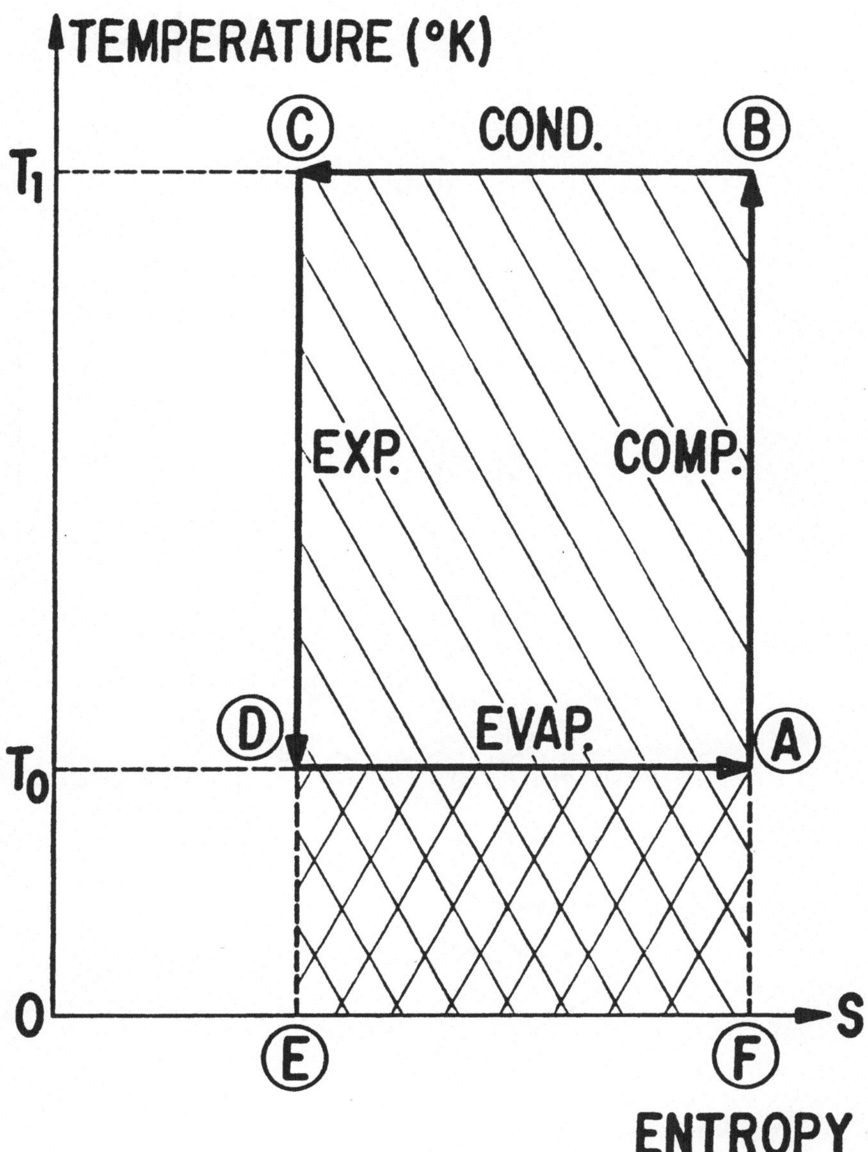

CARNOT CYCLE

Figure 2

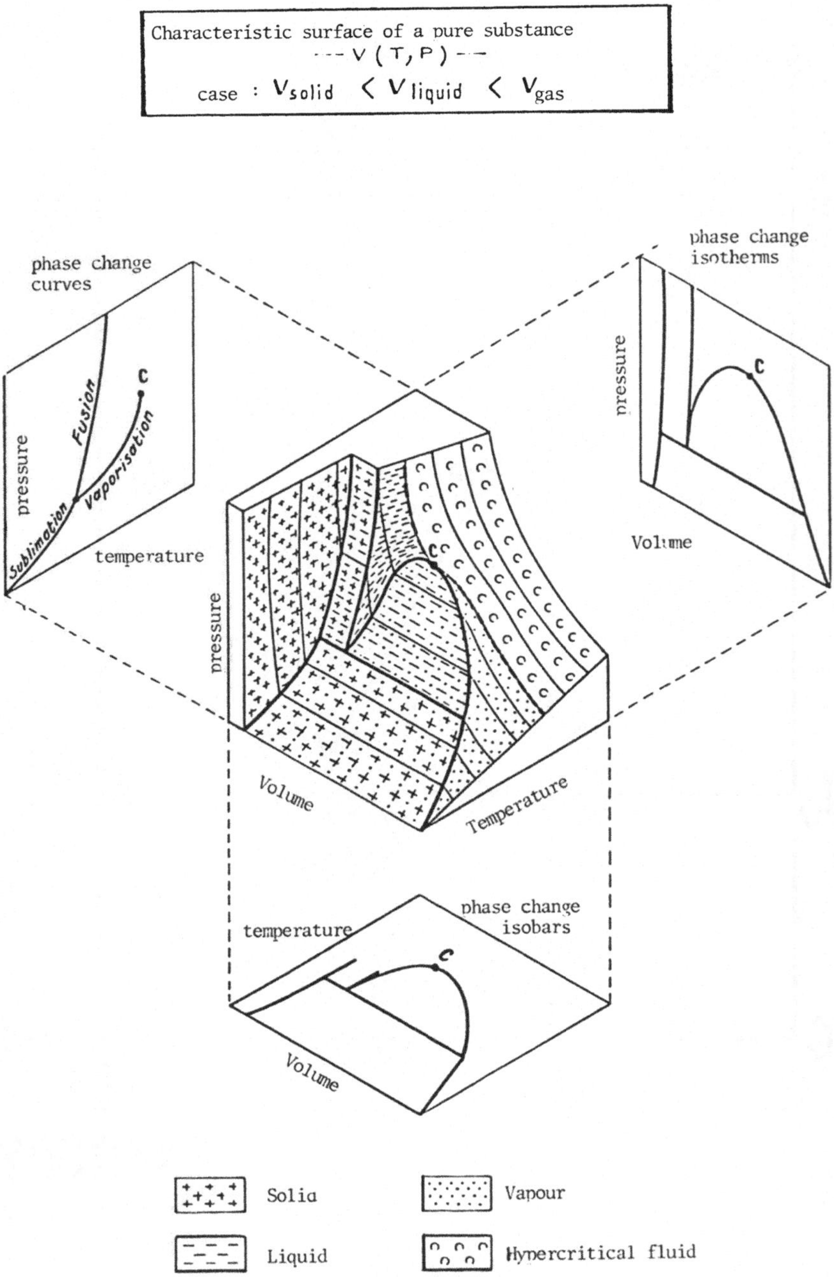

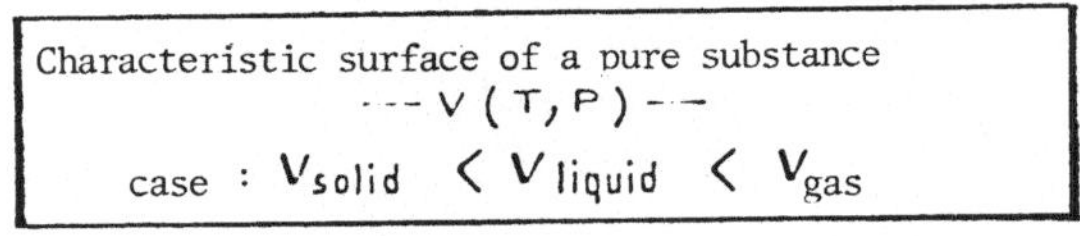
Characteristic surface of a pure substance
--- V (T, P) ---
case : $V_{solid} < V_{liquid} < V_{gas}$
phase change curves
pressure
temperature
Fusion
Vaporisation
Sublimation
C
phase change isotherms
pressure
Volume
C
pressure
Volume
Temperature
C
temperature
phase change isobars
Volume
C
Solid
Liquid
Vapour
Hypercritical fluid

Figure 3

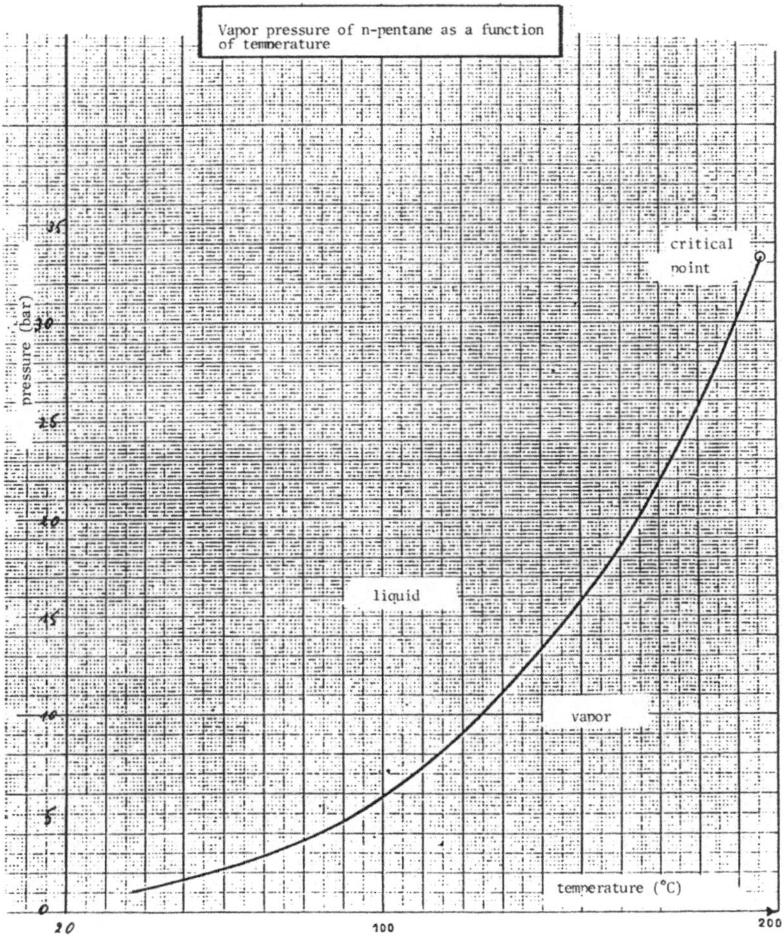

Vapor pressure of n-pentane as a function of temperature

pressure (bar)

35

30

25

20

15

10

5

0

critical point

liquid

vapor

temperature (°C)

20

100

200

Figure 4

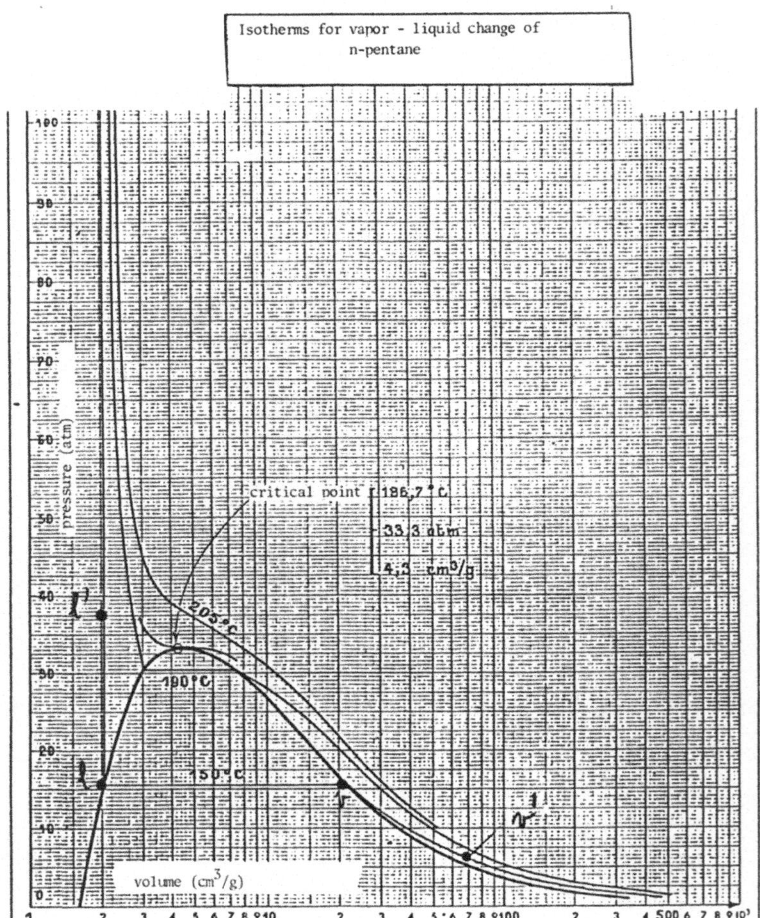
Isotherms for vapor - liquid change of n-pentane
pressure (atm)
critical point
186,7 °C
33,3 atm
4,3 cm³/g
205 °C
190 °C
150 °C
volume (cm³/g)

Figure 5

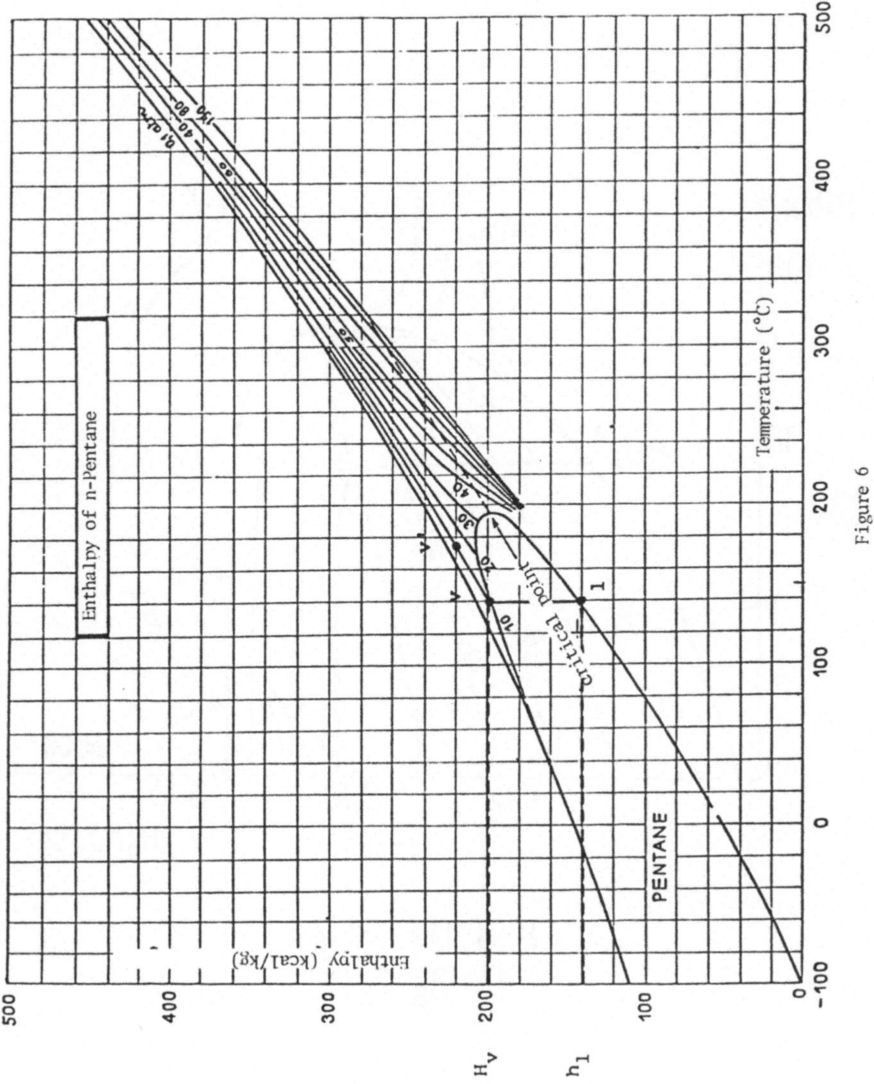

Enthalpy of n-Pentane
Enthalpy (kcal/kg)
Temperature (°C)
PENTANE
critical point
-100
0
100
200
300
400
500
H_v
h_1

Figure 6

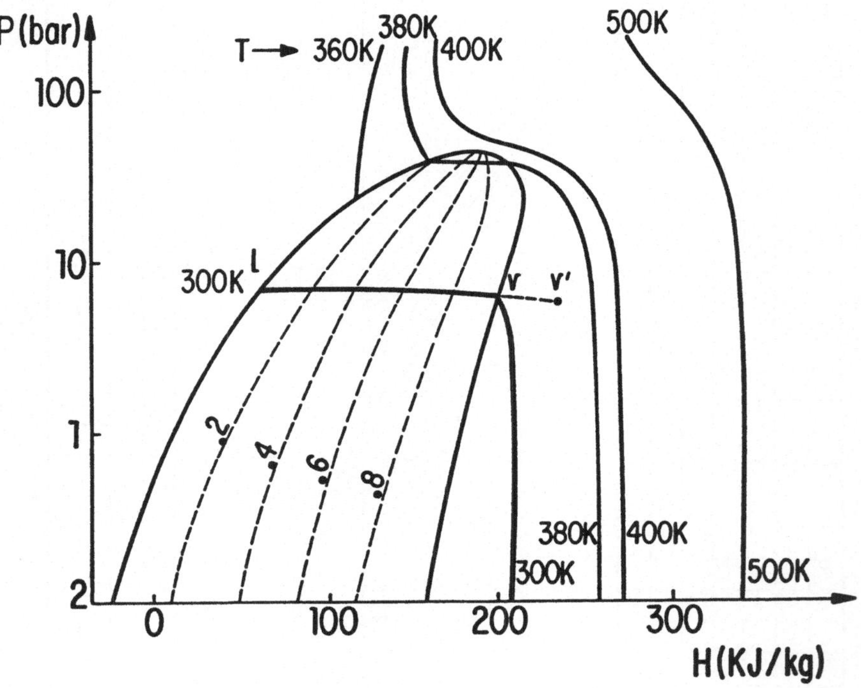

P-H DIAGRAM
($C\,Cl_2\,F_2$)

Figure 7

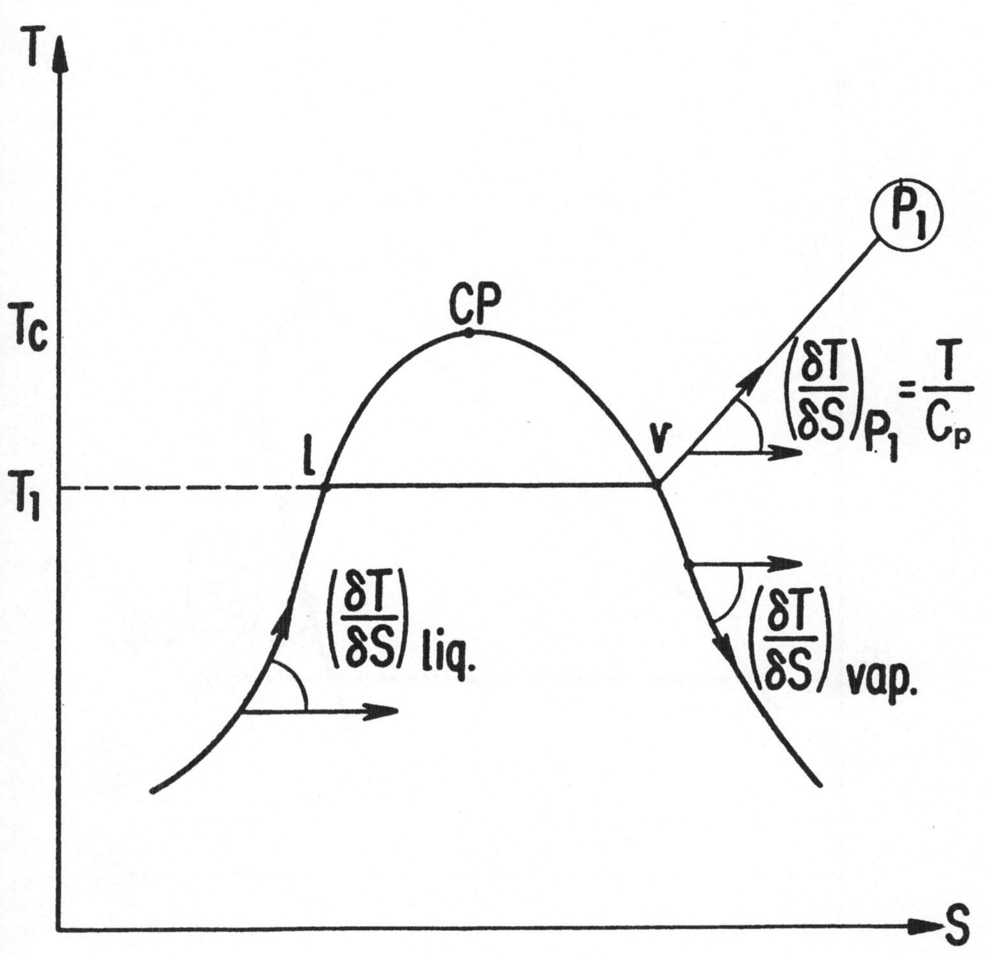

T-S Chart (water)

MOLLIER DIAGRAM

Figure 8

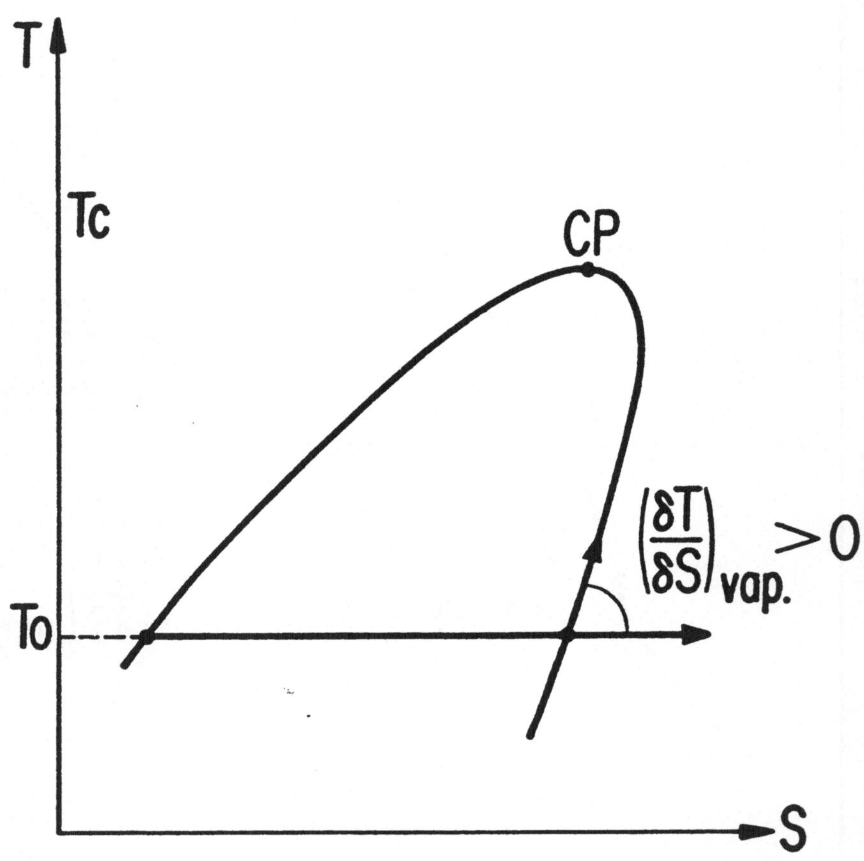

T-S Chart (benzene)

MOLLIER DIAGRAM

Figure 9

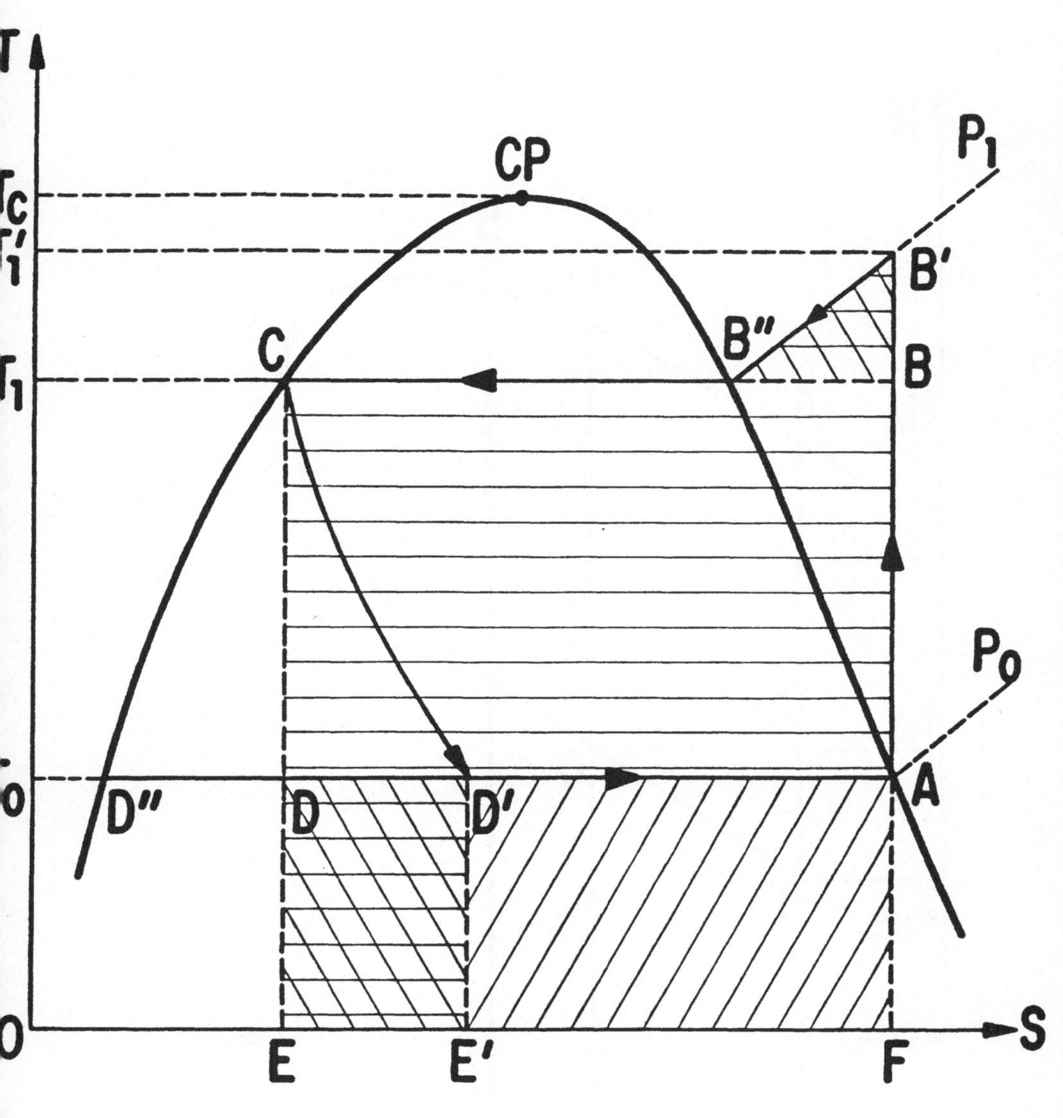

FIGURE 10(a)

RANKINE CYCLE

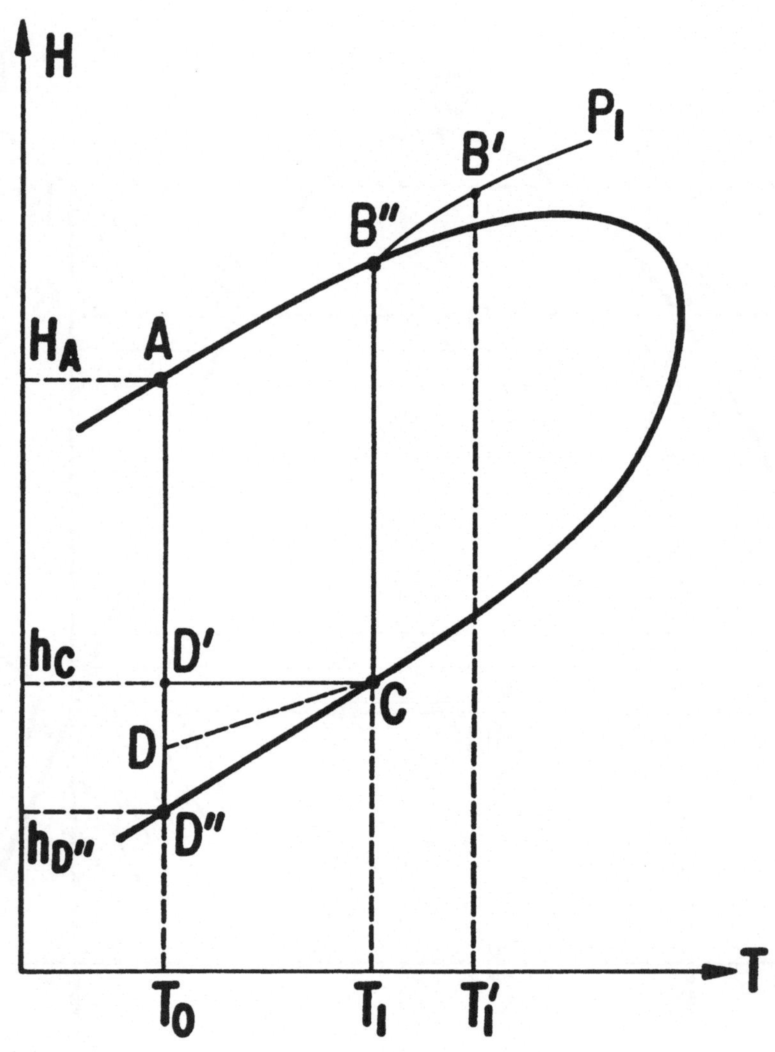

FIGURE 10 (b)

RANKINE CYCLE

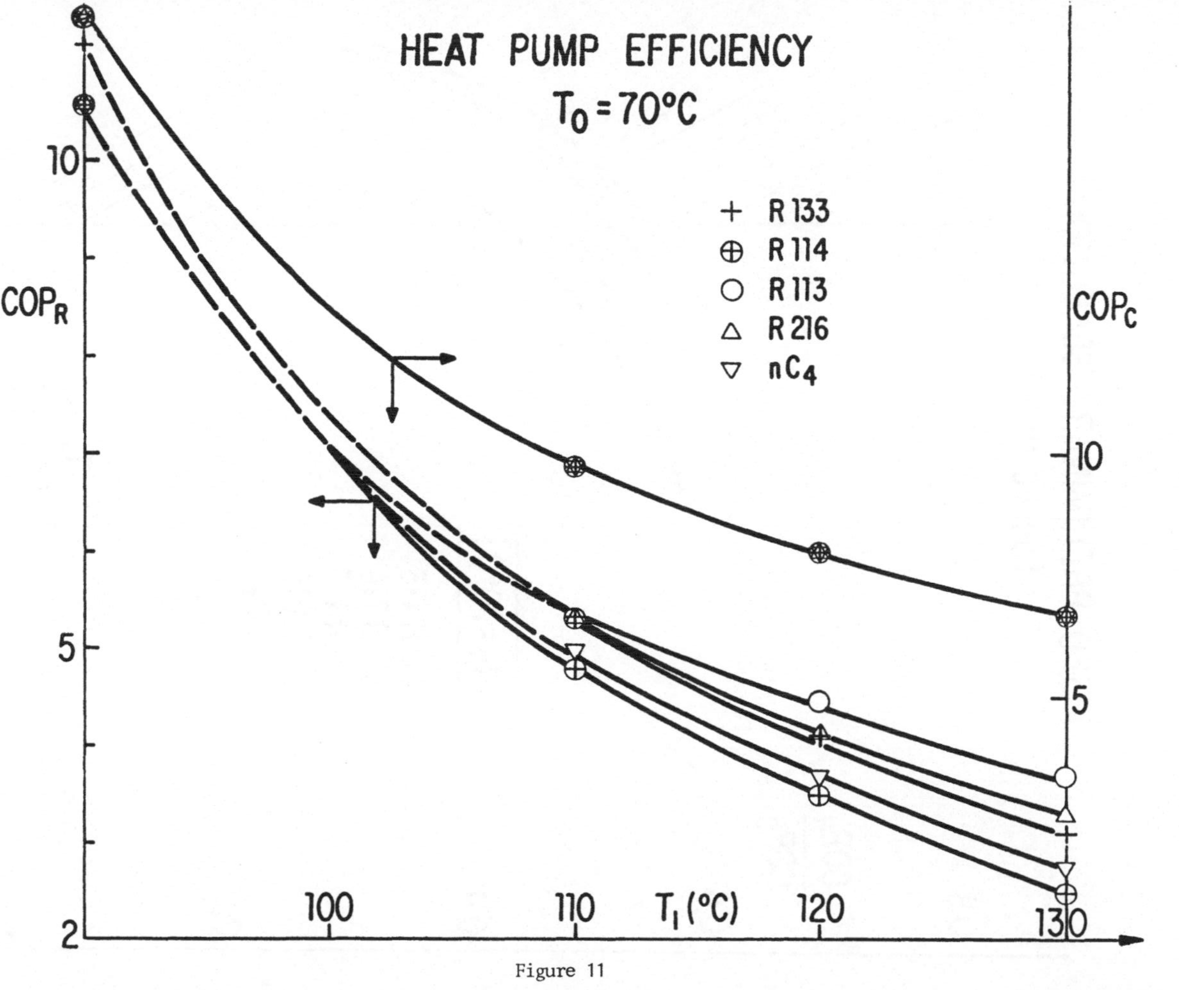
HEAT PUMP EFFICIENCY
$T_0 = 70°C$
+ R 133
⊕ R 114
○ R 113
△ R 216
▽ nC_4
COP_R
10
5
2
COP_C
10
5
100
110
T_1 (°C)
120
130

Figure 11

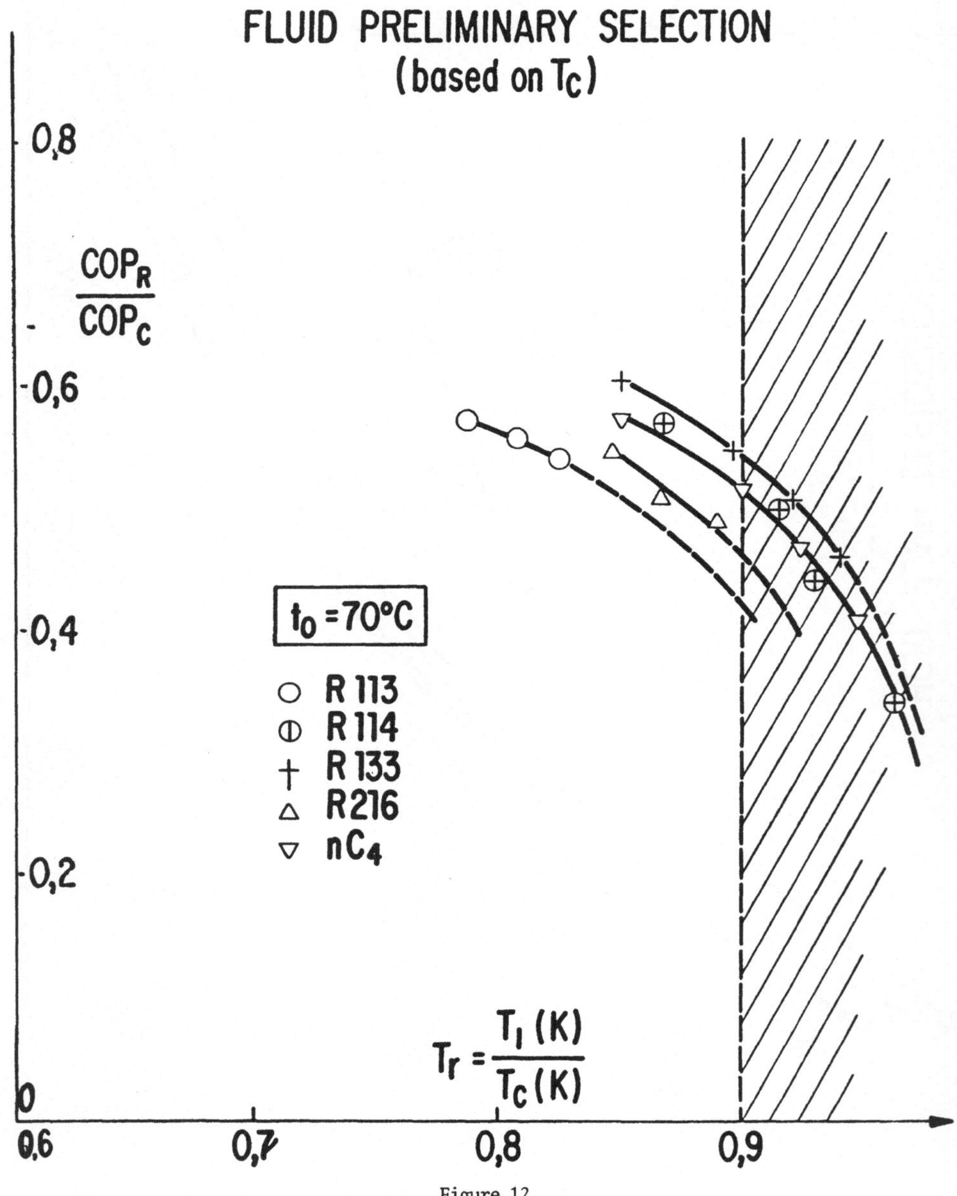

FLUID PRELIMINARY SELECTION
(based on T_C)
0,8
$\frac{COP_R}{COP_C}$
0,6
0,4
0,2
0
t_0 = 70°C
R 113
R 114
R 133
R 216
nC_4
$T_r = \frac{T_l (K)}{T_C (K)}$
0,6
0,7
0,8
0,9

Figure 12

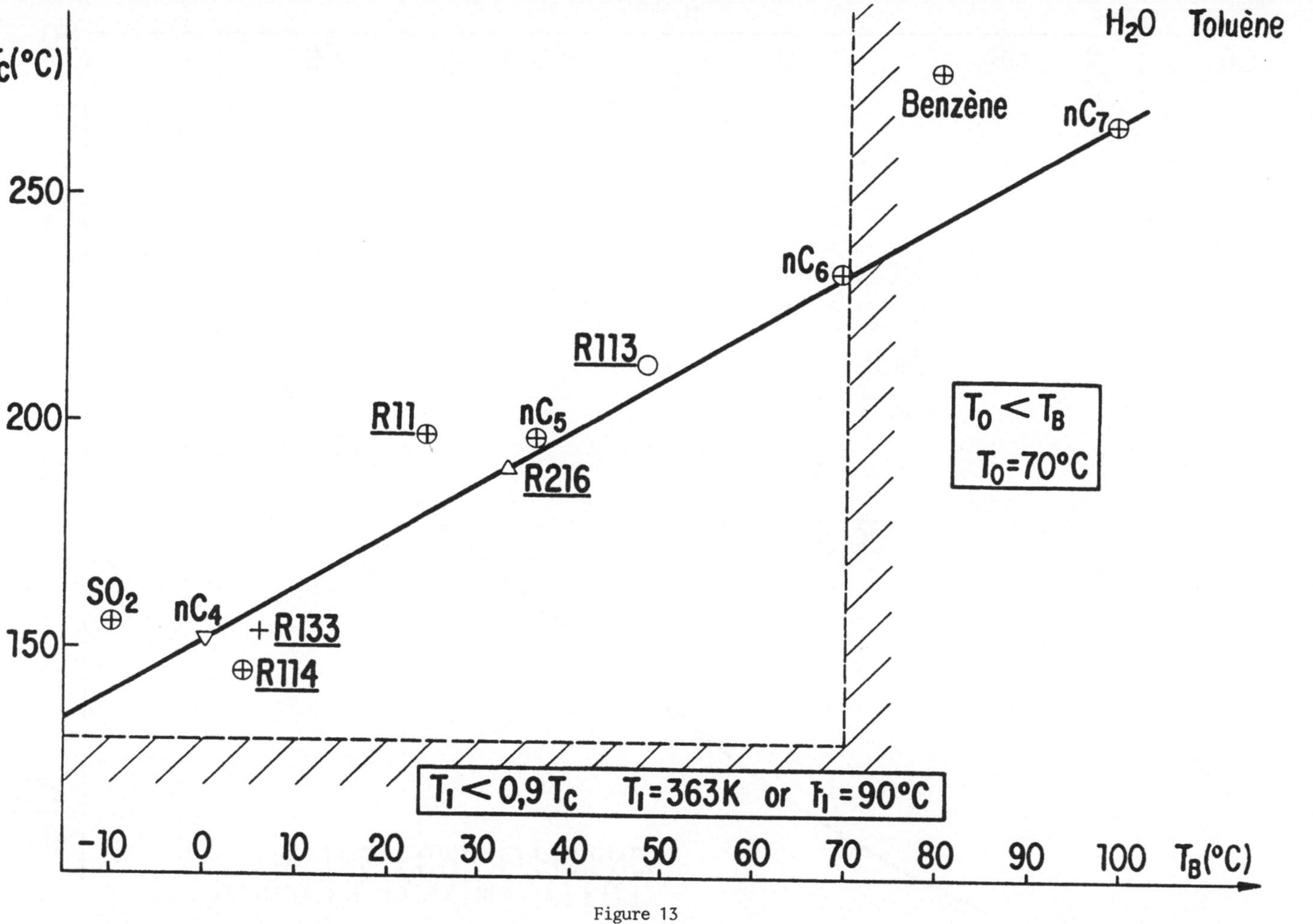

T_C(°C)
250
200
150
T_B(°C)
-10
0
10
20
30
40
50
60
70
80
90
100
H_2O
Toluène
Benzène
nC_7
nC_6
nC_5
nC_4
R113
R216
R11
R133
R114
SO_2
$T_0 < T_B$
$T_0 = 70°C$
$T_1 < 0,9\,T_C$ $T_1 = 363K$ or $T_1 = 90°C$

Figure 13

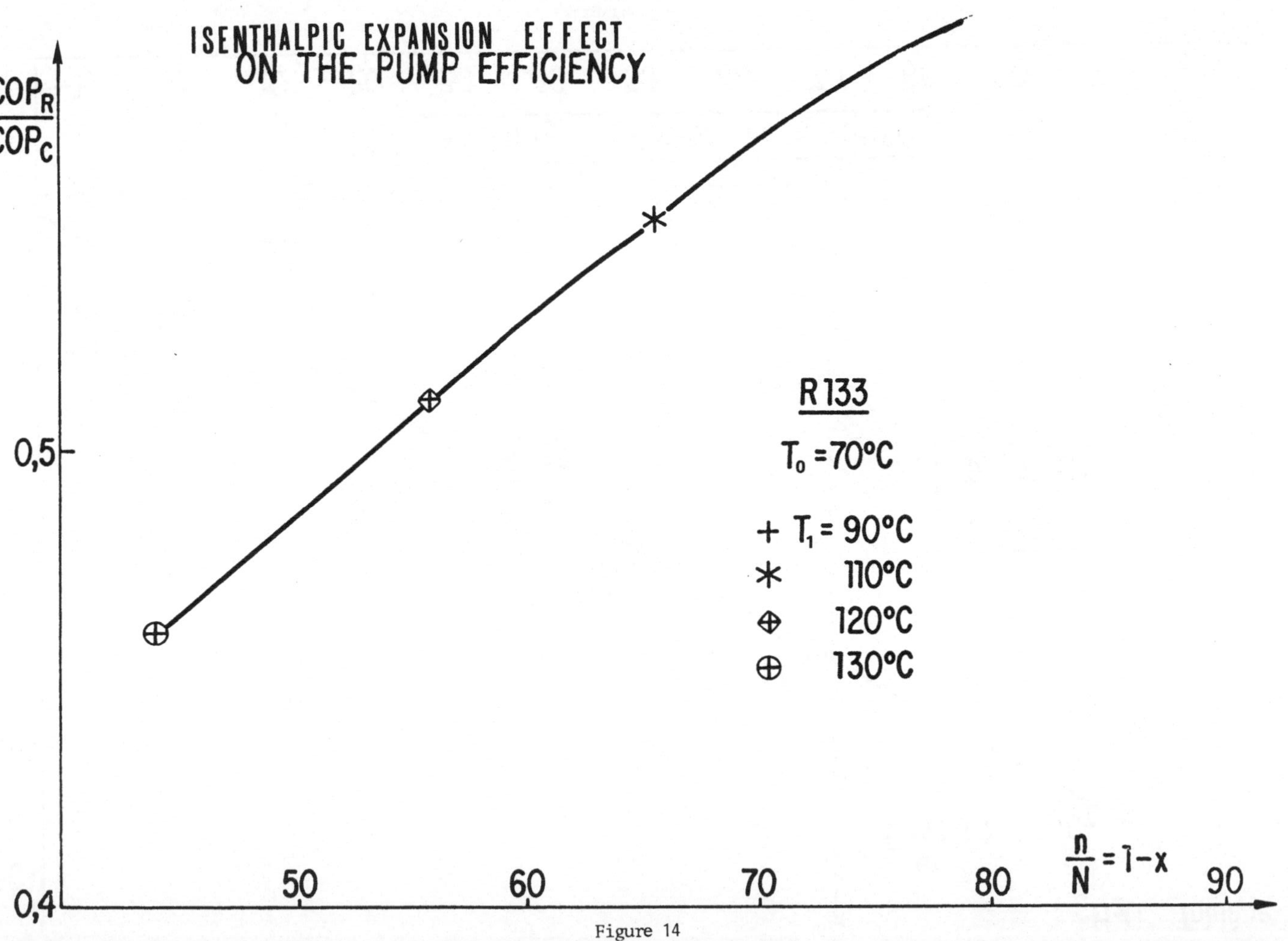

ISENTHALPIC EXPANSION EFFECT
ON THE PUMP EFFICIENCY
$\frac{COP_R}{COP_C}$
0,5
0,4
50
60
70
80
90
$\frac{n}{N} = 1 - x$
R 133
$T_0 = 70°C$
$+ \; T_1 = 90°C$
110°C
120°C
130°C

Figure 14

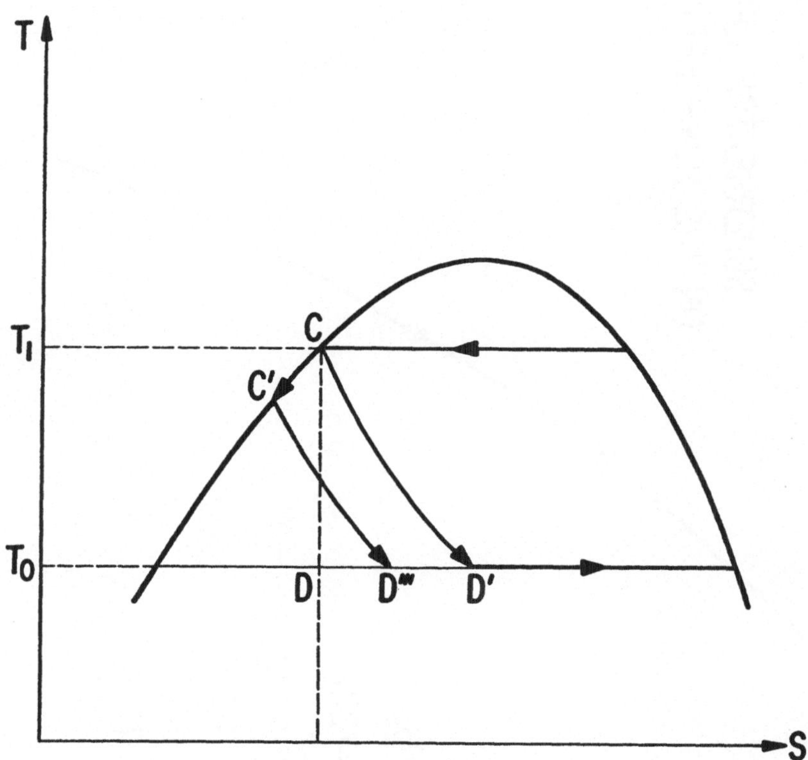

LIQUID SUBCOOLING BEFORE EXPANSION
T
T1
T0
C
C'
D
D'''
D'
S

Figure 15

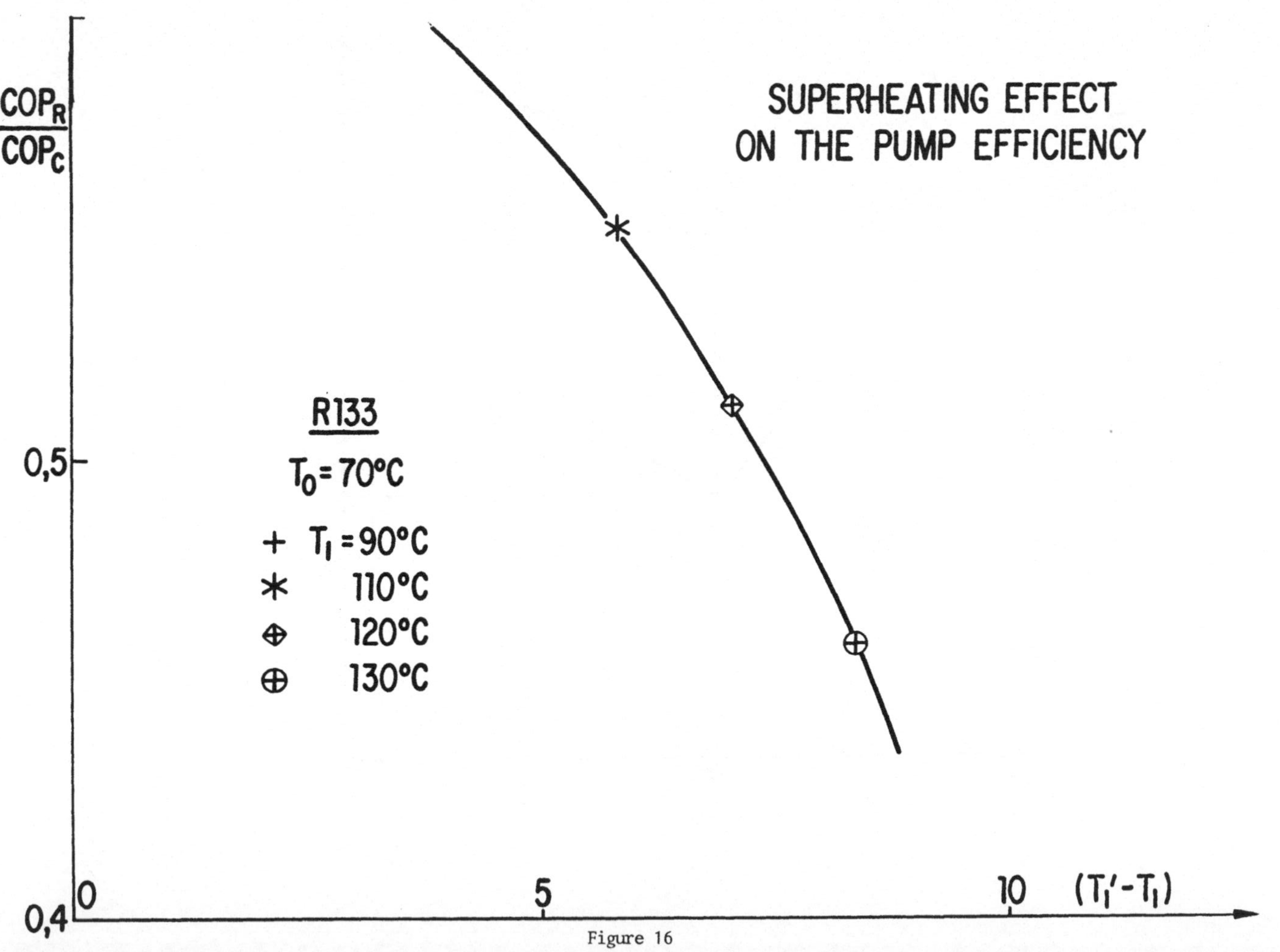
SUPERHEATING EFFECT
ON THE PUMP EFFICIENCY
COPR / COPC
0,5
0,4
R133
T0 = 70°C
+ T1 = 90°C
110°C
120°C
130°C
0
5
10
(T1' - T1)

Figure 16

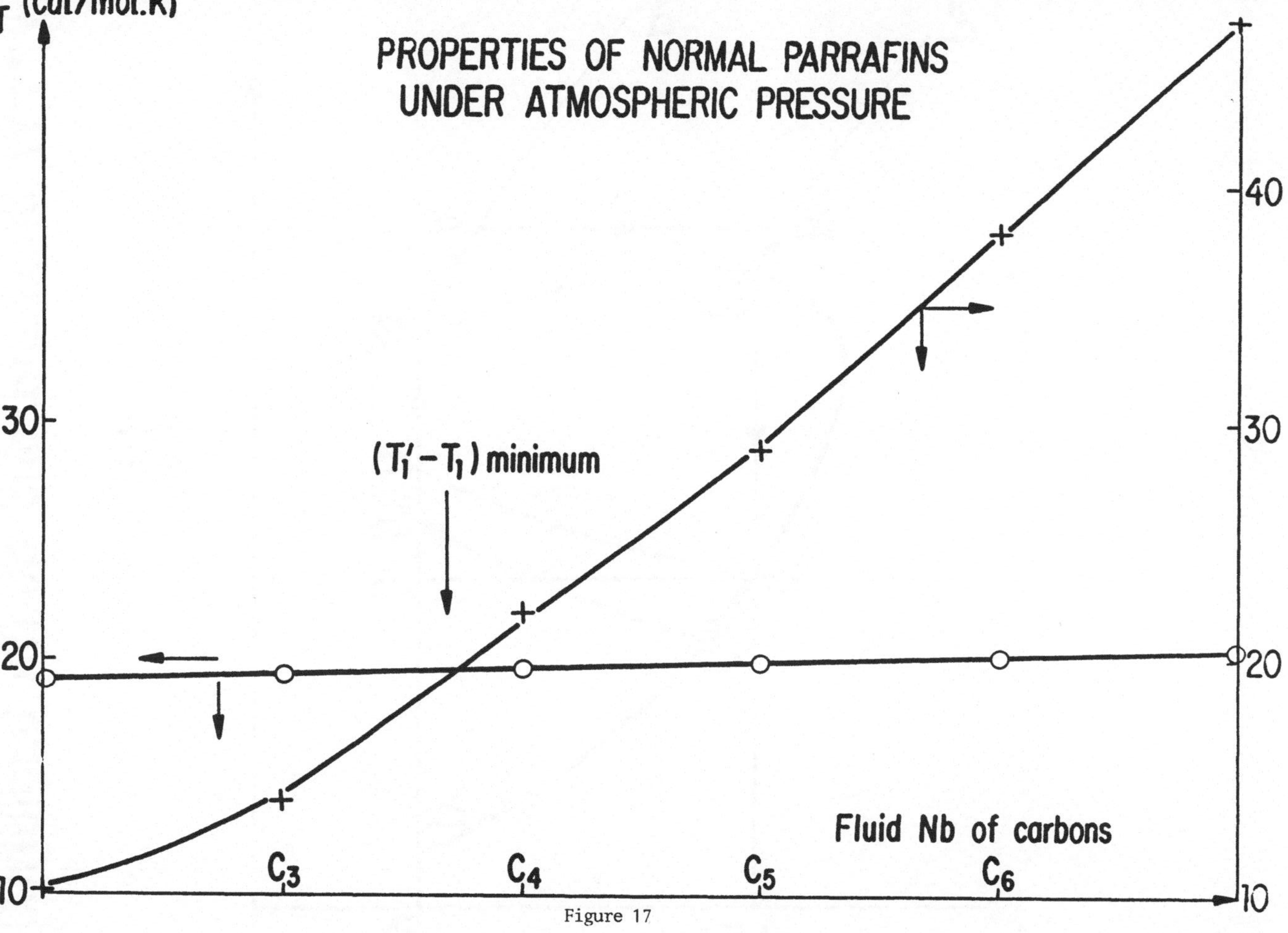
PROPERTIES OF NORMAL PARRAFINS
UNDER ATMOSPHERIC PRESSURE
T (cal/mol.K)
30
20
10
40
30
20
10
(T1' − T1) minimum
Fluid Nb of carbons
C3
C4
C5
C6

Figure 17

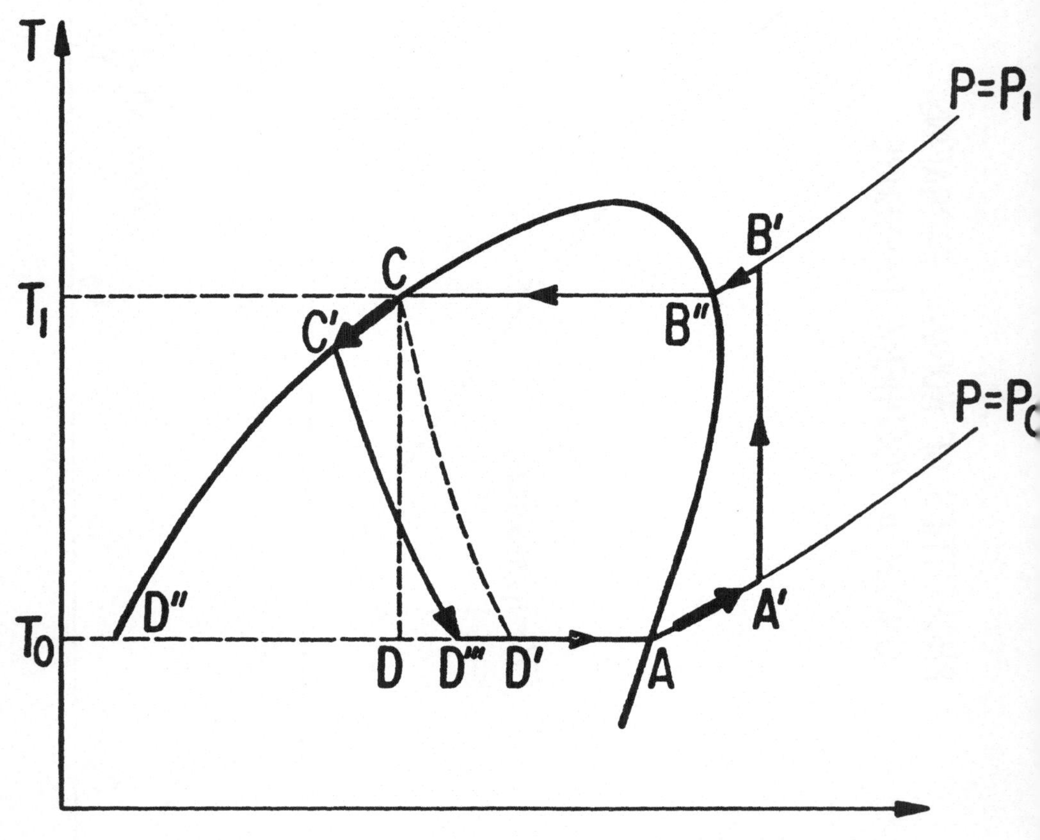

FIGURE 18(a)

COMBINED VAPOR SUPERHEATING (A→A')
AND LIQUID SUBCOOLING (C→C')

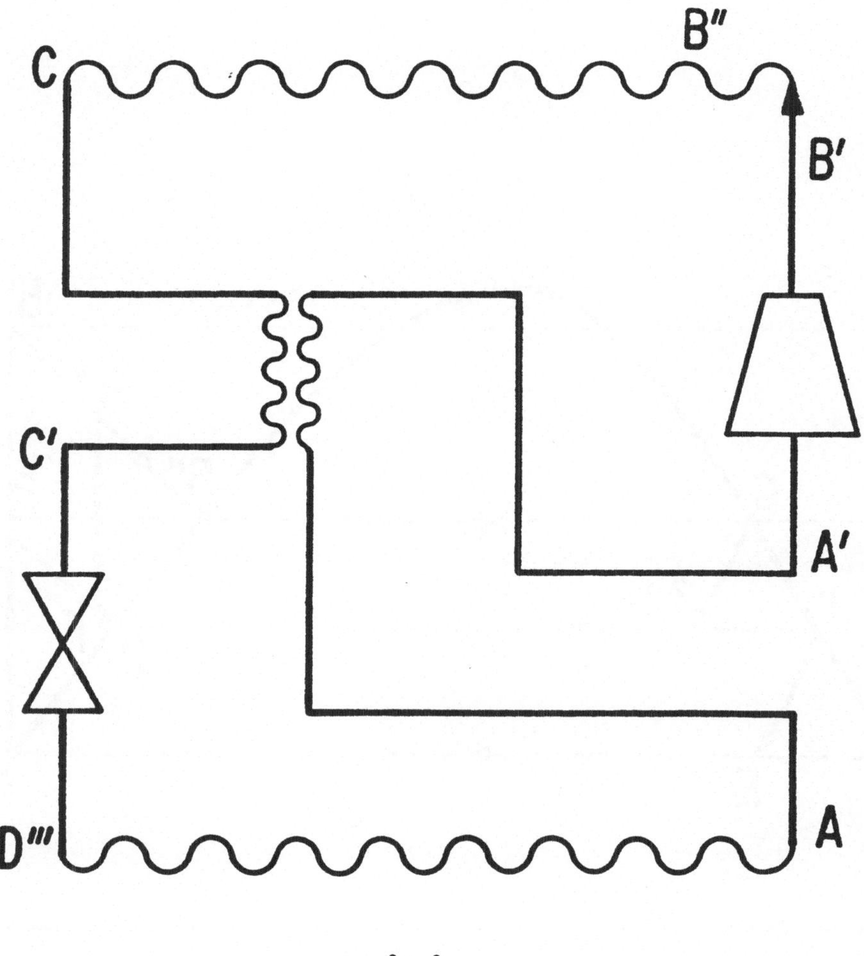

FIGURE 18(b)

COMBINED VAPOR SUPERHEATING (A→A′)
AND LIQUID SUBCOOLING (C→C′)

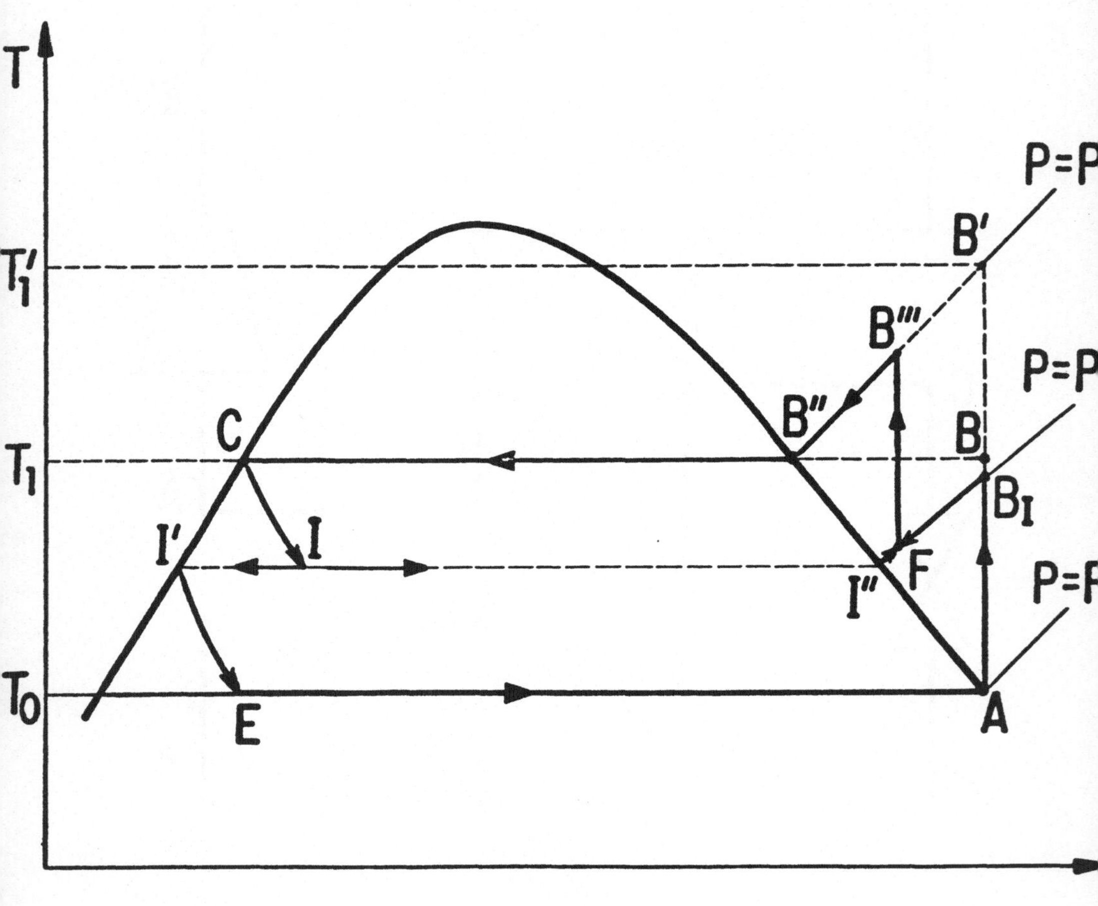

FIGURE 19 (a)

2 STAGES VAPOR COMPRESSION CYCLE WITH INTERCOOLING

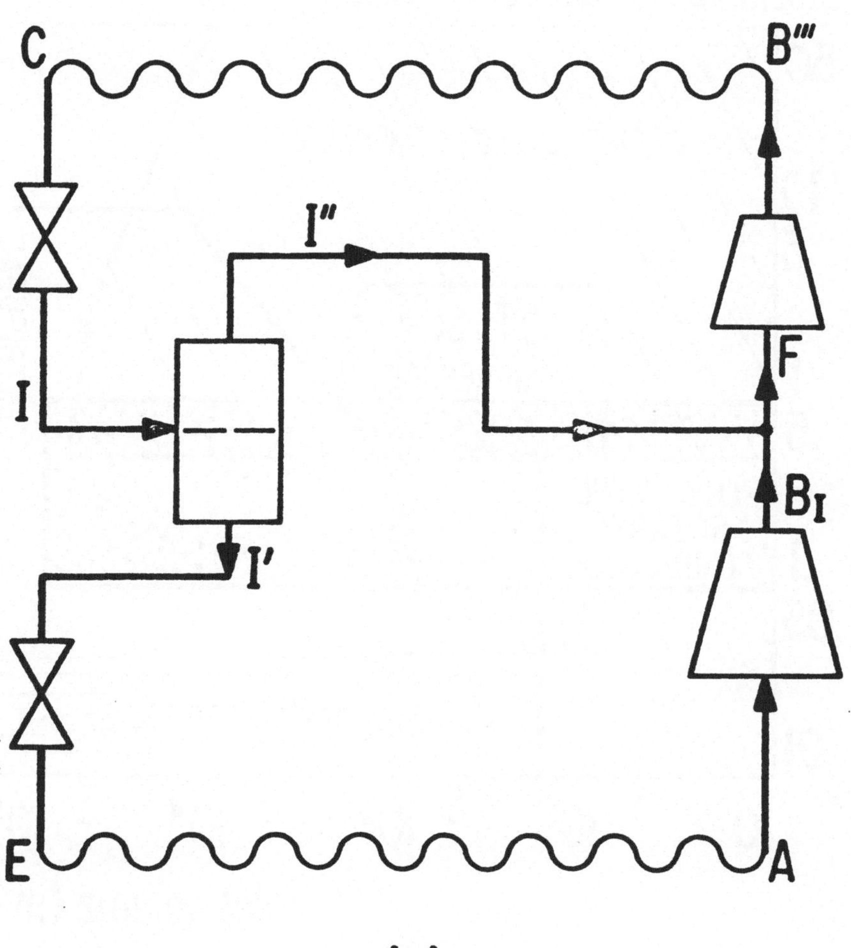

FIGURE 19 (b)

2 STAGES VAPOR COMPRESSION CYCLE WITH INTERCOOLING

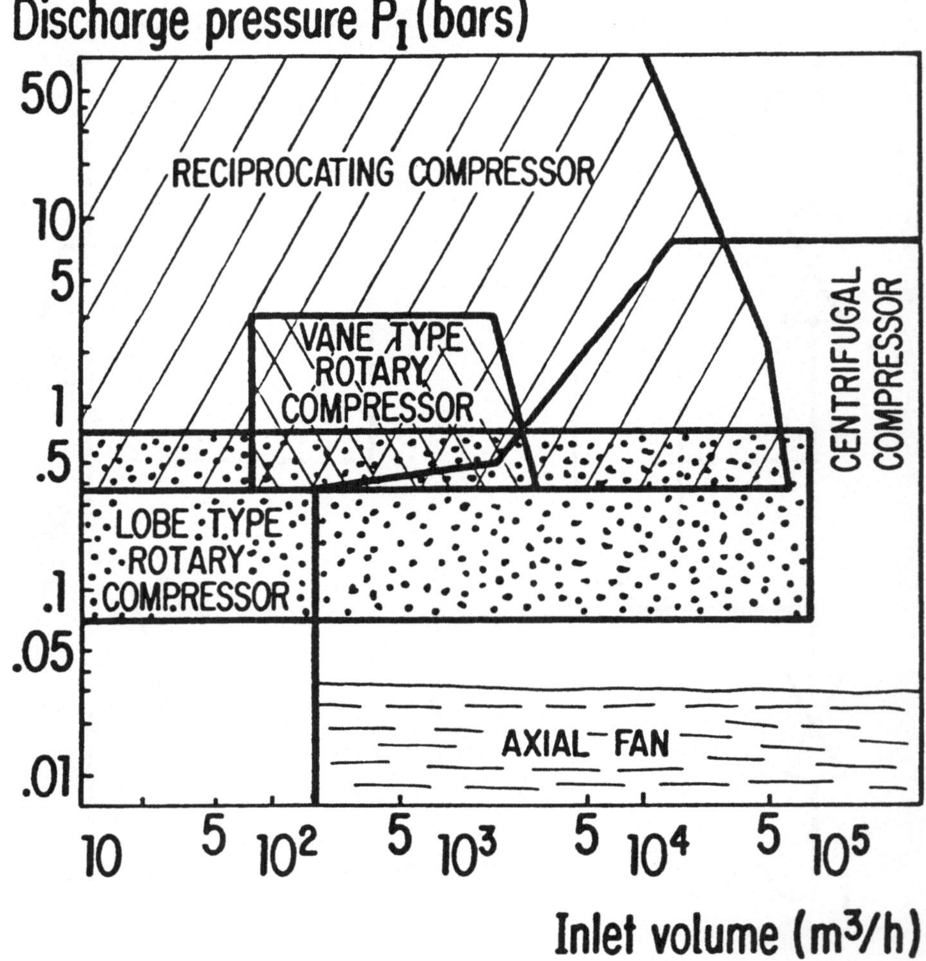

DUTIES OF VARIOUS COMPRESSORS (5)

Figure 20

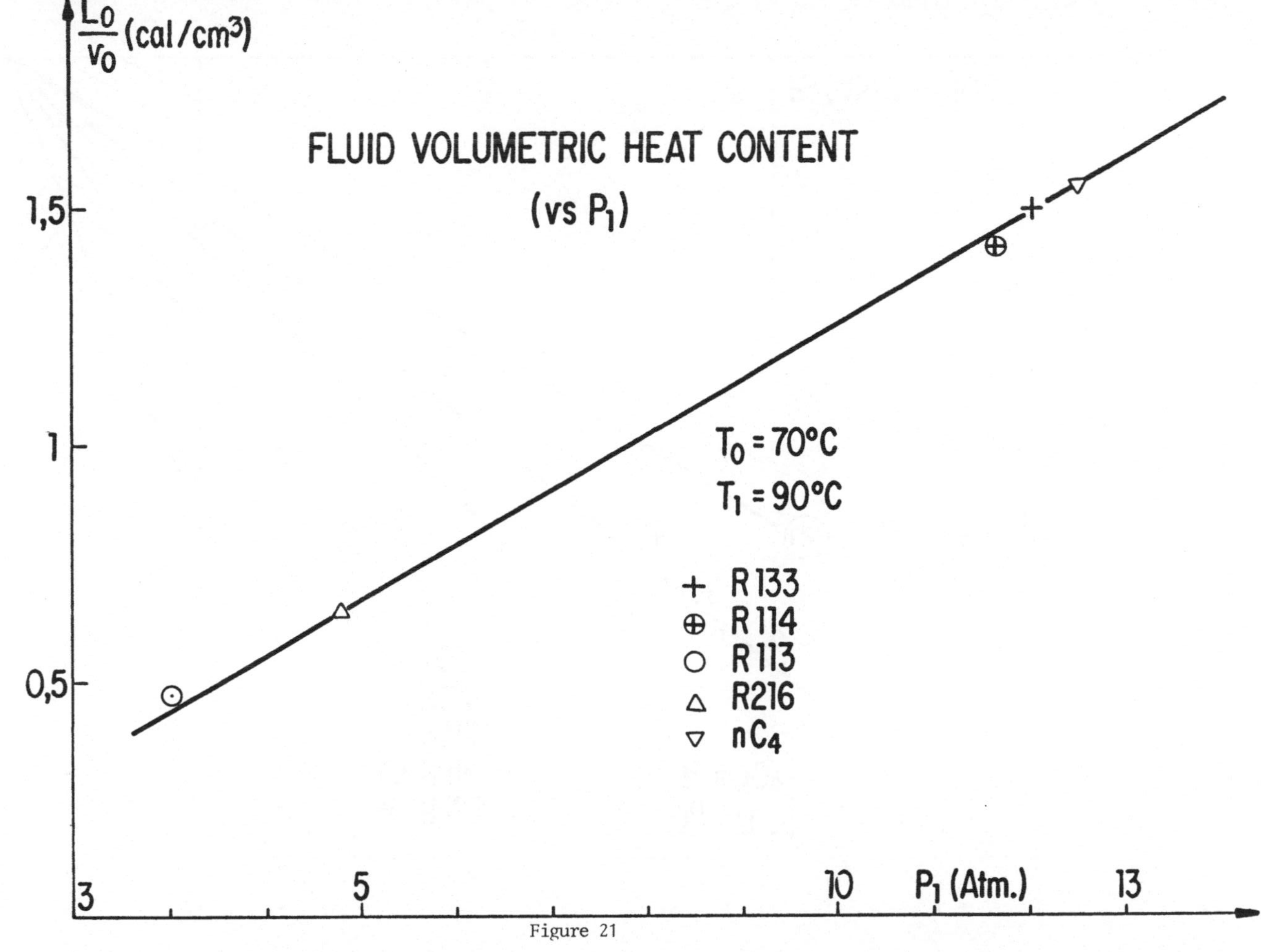

FLUID VOLUMETRIC HEAT CONTENT
(vs P_1)
$\frac{L_0}{v_0}$ (cal/cm^3)
1,5
1
0,5
3
5
10
13
P_1 (Atm.)
$T_0 = 70°C$
$T_1 = 90°C$
+ R 133
⊕ R 114
○ R 113
△ R216
▽ nC_4

Figure 21

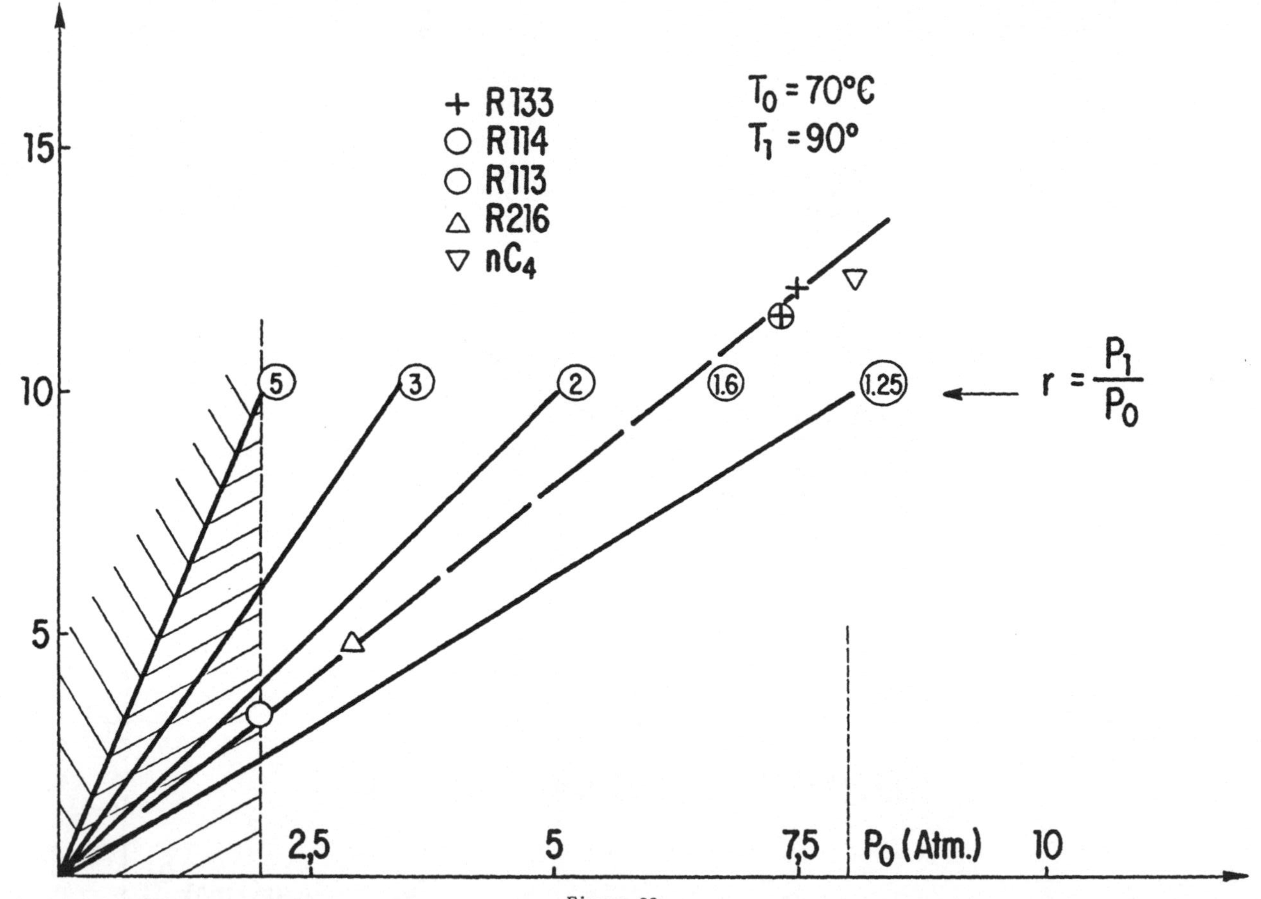

+ R133
○ R114
○ R113
△ R216
▽ nC_4
$T_0 = 70°C$
$T_1 = 90°$
5
3
2
1.6
1.25
$r = \frac{P_1}{P_0}$
15
10
5
2,5
5
7,5
P_0 (Atm.)
10

Figure 22

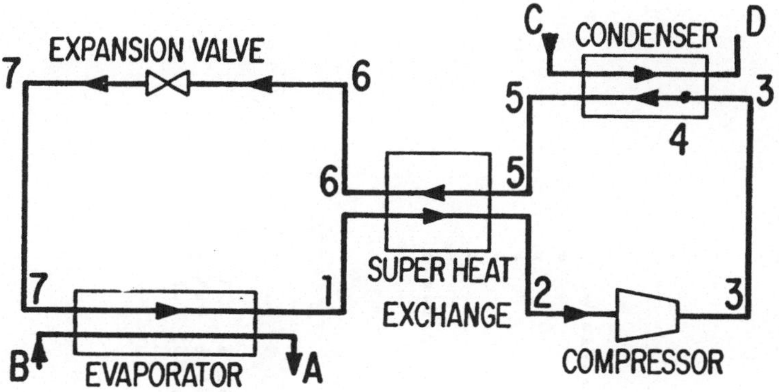

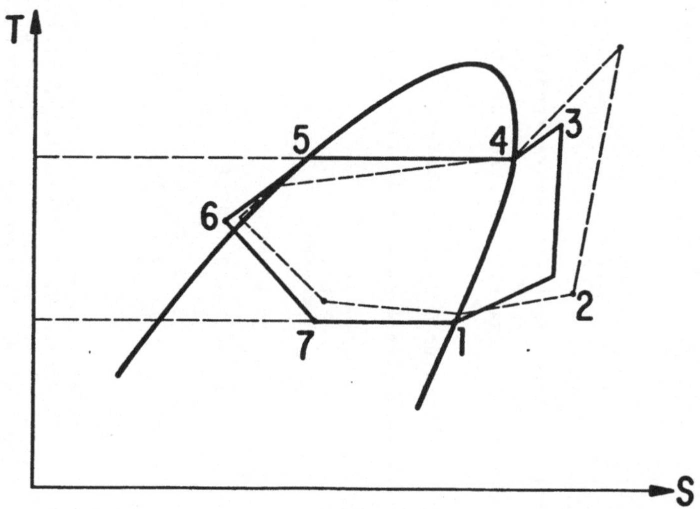

Full line = Compressor yield = 1, $\Delta P_i = 0$

Dotted line = Compressor yield<1, $\Delta P_i \neq 0$

COMPUTER PROGRAM BASIS

Figure 23

```
*****************************************************************
*                                                               *
*   CALCUL D UNE POMPE A CHALEUR A 1 ETAGE DE COMPRESSION       *
*                   POUR UN CORPS PUR                           *
*                                                               *
*              (EQUATION D ETAT DE SOAVE)                       *
*                                                               *
*****************************************************************

                          FLUIDE
                          ******

                         N-BUTANE

POIDS MOLECULAIRE            PM =  58.12 G/MOLE
PRESSION CRITIQUE         PCRIT =  37.47 ATM    =  37.96 BARS  = 550.66 PSI
TEMPERATURE CRITIQUE      TCRIT = 152.01 DEG.C = 425.16 DEG.K = 305.62 DEG.F
VOLUME CRITIQUE           VCRIT = 310.35 CM3/M =  5.340 CM3/G =  .0855 C.F/LB
PENTE TENSION VAPEUR         CA = -2.815
PENTE DE SOAVE            PENTE =  .7845

COEFFICIENTS DE TEM-   CIDAL(1) = -.585433E+00 CAL/MOLE/DEG.K
PERATURE DE LA CAPA-   CIDAL(2) =  .935864E-01 CAL/MOLE/DEG.K**2
CITE CALORIFIQUE DU    CIDAL(3) = -.484829E-04 CAL/MOLE/DEG.K**3
GAZ PARFAIT            CIDAL(4) =  .974318E-08 CAL/MOLE/DEG.K**4

CONSTANTE ENTHALPIQUE   CHIDAL = 2483.564 CAL/MOLE
CONSTANTE ENTROPIQUE    CSIDAL =   -.4384 CAL/MOLE/DEG
```

CARACTERISTIQUES DE LA POMPE A CHALEUR

TEMPERATURE INFERIEURE	(POINT 1)	70.00 DEG.C
TEMPERATURE SUPERIEURE	(POINT 4)	90.00 DEG.C
INCREMENT DE SURCHAUFFE		1.00 DEG.C
SURCHAUFFE	(ENTRE POINTS 1 ET 2)	1.00 DEG.C
ECART MINIMAL DE TEMPERATURE		3.00 DEG.C
PRESSION MINIMALE DU POINT 2		1.00 ATM
PRESSION MAXIMALE DU POINT 3		50.00 ATM
PERTE DE CHARGE (ENTRE POINTS 1 ET 2)		.10 ATM
PERTE DE CHARGE (ENTRE POINTS 3 ET 4)		.10 ATM
PERTE DE CHARGE (ENTRE POINTS 4 ET 5)		.10 ATM
PERTE DE CHARGE (ENTRE POINTS 5 ET 6)		.10 ATM
PERTE DE CHARGE (ENTRE POINTS 7 ET 1)		.10 ATM
RENDEMENT DU COMPRESSEUR		.70

RESULTATS DES CALCULS

UNITES USUELLES

POINT	ETAT DU FLUIDE	TEMPERATURE DEG.C	PRESSION ATM	VOLUME MOL CM3/MOLE	ENTHALPIE CAL/MOLE	ENTROPIE CAL/MOLE/DEG
1	VAPEUR SATUREE	70.00	8.11	2901.44	6841.20	20.71
2	VAPEUR SURCHAUFFEE	71.00	8.01	2960.55	6874.45	20.83
3	VAPEUR SURCHAUFFEE	91.28	12.65	1828.65	7241.89	21.13
4	VAPEUR SATUREE	90.00	12.55	1834.55	7207.30	21.05
5	LIQUIDE SATURE	89.61	12.45	134.92	3324.19	10.35
6	LIQUIDE SOUSREFROIDI	88.85	12.35	134.43	3290.94	10.26
7	LIQUIDE ET VAPEUR	70.53	8.21	* 17.66	3290.94	10.36

DANS LE DOMAINE BIPHASIQUE (LIQUIDE ET VAPEUR), LE NOMBRE FIGURANT DANS LA COLONNE DES VOLUMES (ET PRECEDE D UN ASTERISQUE) INDIQUE LE POURCENTAGE VAPORISE

UNITES S.I

POINT	ETAT DU FLUIDE	TEMPERATURE DEG.C	PRESSION PASCALS	VOLUME SPEC M3/KG	ENTHALPIE JOULES/KG	ENTROPIE J/KG/DEG
1	VAPEUR SATUREE	70.00	821681	.049922	492667	1491
2	VAPEUR SURCHAUFFEE	71.00	811551	.050939	495062	1499
3	VAPEUR SURCHAUFFEE	91.28	1281669	.031463	521523	1521
4	VAPEUR SATUREE	90.00	1271739	.031565	519031	1515
5	LIQUIDE SATURE	89.61	1261609	.002321	239390	745
6	LIQUIDE SOUSREFROIDI	88.85	1251479	.002313	236996	739
7	LIQUIDE ET VAPEUR	70.53	831811	* 17.66	236996	745

DANS LE DOMAINE BIPHASIQUE (LIQUIDE ET VAPEUR), LE NOMBRE FIGURANT DANS LA COLONNE DES VOLUMES (ET PRECEDE D UN ASTERISQUE) INDIQUE LE POURCENTAGE VAPORISE

```
TRAVAIL DE COMPRESSION ISENTROPIQUE  RDT.(H3-H2)=  257.21 CAL/MOLE   SOIT   4.4255 CAL/G   OU  18522.9 J/KG
TRAVAIL DE COMPRESSION REELLE        W  = H3-H2 =  367.44 CAL/MOLE   SOIT   6.3221 CAL/G   OU  26461.3 J/KG
QUANTITE DE CHALEUR FOURNIE          Q  = H3-H5 = 3917.71 CAL/MOLE   SOIT  67.4072 CAL/G   OU 282132.8 J/KG

COEFFICIENT DE PERFORMANCE  COP = Q/W =10.662            TAUX DE COMPRESSION  P3/P2   =  1.58
COEFFICIENT DE CARNOT  COC =T5/(T5-T1)=18.497            TEMPERATURE LOGARIT. MOYENNE = 18.73 DEG
```

```
*************************************************************************
*                                                                       *
*        CALCUL D UNE POMPE A CHALEUR A 1 ETAGE DE COMPRESSION          *
*                         POUR UN CORPS PUR                             *
*                                                                       *
*                     (EQUATION D ETAT DE SOAVE)                        *
*                                                                       *
*************************************************************************


                            FLUIDE
                            ******

 111TRIFLUORO 2CHLORO ETHANE    R133 RHP

POIDS MOLECULAIRE           PM = 118.48 G/MOLE
PRESSION CRITIQUE        PCRIT =  39.96 ATM    =  40.48 BARS   = 587.25 PSI
TEMPERATURE CRITIQUE     TCRIT = 154.00 DEG.C = 427.15 DEG.K = 309.20 DEG.F
VOLUME CRITIQUE          VCRIT = 292.38 CM3/M =   2.468 CM3/G =   .0395 C.F/LB
PENTE TENSION VAPEUR        CA = -3.022
PENTE DE SOAVE           PENTE =   .9133

COEFFICIENTS DE TEM-  CIDAL(1) =  .502700E+01 CAL/MOLE/DEG.K
PERATURE DE LA CAPA-  CIDAL(2) =  .655000E-01 CAL/MOLE/DEG.K**2
CITE CALORIFIQUE DU   CIDAL(3) = -.358000E-04 CAL/MOLE/DEG.K**3
GAZ PARFAIT           CIDAL(4) = 0.           CAL/MOLE/DEG.K**4

CONSTANTE ENTHALPIQUE   CHIDAL =    0.000 CAL/MOLE
CONSTANTE ENTROPIQUE    CSIDAL =   0.0000 CAL/MOLE/DEG
```

CARACTERISTIQUES DE LA POMPE A CHALEUR

TEMPERATURE INFERIEURE	(POINT 1)	70.00 DEG.C
TEMPERATURE SUPERIEURE	(POINT 4)	90.00 DEG.C
INCREMENT DE SURCHAUFFE		1.00 DEG.C
SURCHAUFFE	(ENTRE POINTS 1 ET 2)	0.00 DEG.C
ECART MINIMAL DE TEMPERATURE		3.00 DEG.C
PRESSION MINIMALE DU POINT 2		1.00 ATM
PRESSION MAXIMALE DU POINT 3		50.00 ATM
PERTE DE CHARGE	(ENTRE POINTS 1 ET 2)	.10 ATM
PERTE DE CHARGE	(ENTRE POINTS 3 ET 4)	.10 ATM
PERTE DE CHARGE	(ENTRE POINTS 4 ET 5)	.10 ATM
PERTE DE CHARGE	(ENTRE POINTS 5 ET 6)	.10 ATM
PERTE DE CHARGE	(ENTRE POINTS 7 ET 1)	.10 ATM
RENDEMENT DU COMPRESSEUR		.70

RESULTATS DES CALCULS

UNITES USUELLES

POINT	ETAT DU FLUIDE	TEMPERATURE DEG.C	PRESSION ATM	VOLUME MOL CM3/MOLE	ENTHALPIE CAL/MOLE	ENTROPIE CAL/MOLE/DEG
1	VAPEUR SATUREE	70.00	7.50	3204.31	4793.88	45.09 —
2	CONFONDU AVEC LE POINT 1					
3	VAPEUR SURCHAUFFEE	93.46	12.11	1982.77	5188.04	45.42
4	VAPEUR SATUREE	90.00	12.01	1965.99	5097.67	45.18 —
5	LIQUIDE SATURE	89.62	11.91	123.92	771.97	33.27
6	CONFONDU AVEC LE POINT 5					
7	LIQUIDE ET VAPEUR	70.53	7.60	* 16.09	771.97	33.37

DANS LE DOMAINE BIPHASIQUE (LIQUIDE ET VAPEUR), LE NOMBRE FIGURANT DANS LA COLONNE DES VOLUMES (ET PRECEDE D UN ASTERISQUE) INDIQUE LE POURCENTAGE VAPORISE

UNITES S.I

POINT	ETAT DU FLUIDE	TEMPERATURE DEG.C	PRESSION PASCALS	VOLUME SPEC M3/KG	ENTHALPIE JOULES/KG	ENTROPIE J/KG/DEG
1	VAPEUR SATUREE	70.00	760001	.027045	169351	1593
2	CONFONDU AVEC LE POINT 1					
3	VAPEUR SURCHAUFFEE	93.46	1226752	.016735	183276	1604
4	VAPEUR SATUREE	90.00	1216622	.016593	180083	1596
5	LIQUIDE SATURE	89.62	1206492	.001046	27271	1175
6	CONFONDU AVEC LE POINT 5					
7	LIQUIDE ET VAPEUR	70.53	770131	* 16.09	27271	1178

DANS LE DOMAINE BIPHASIQUE (LIQUIDE ET VAPEUR), LE NOMBRE FIGURANT DANS LA COLONNE DES VOLUMES (ET PRECEDE D UN ASTERISQUE) INDIQUE LE POURCENTAGE VAPORISE

TRAVAIL DE COMPRESSION ISENTROPIQUE	RDT.(H3-H2)= 275.91 CAL/MOLE	SOIT 2.3288 CAL/G	OU 9747.1 J/KG
TRAVAIL DE COMPRESSION REELLE	W = H3-H2 = 394.16 CAL/MOLE	SOIT 3.3268 CAL/G	OU 13924.4 J/KG
QUANTITE DE CHALEUR FOURNIE	Q = H3-H5 = 4416.07 CAL/MOLE	SOIT 37.2727 CAL/G	OU 156005.0 J/KG

COEFFICIENT DE PERFORMANCE COP = Q/W =11.204 — TAUX DE COMPRESSION P3/P2 = 1.61

COEFFICIENT DE CARNOT COC =T5/(T5-T1)=18.487 — PAS D ECHANGEUR DE SURCHAUFFE

```
****************************************************************************
*                                                                          *
*         CALCUL D UNE POMPE A CHALEUR A 1 ETAGE DE COMPRESSION            *
*                          POUR UN CORPS PUR                               *
*                                                                          *
*                     (EQUATION D ETAT DE SOAVE)                           *
*                                                                          *
****************************************************************************


                                FLUIDE
                                ******

        1-3-DICHLORO 1-1-2-2-3-3-HEXAFLUORO PROPANE (FREON 216)

POIDS MOLECULAIRE           PM = 220.93 G/MOLE
PRESSION CRITIQUE        PCRIT =  27.18 ATM    =  27.53 BARS  = 399.43 PSI
TEMPERATURE CRITIQUE     TCRIT = 180.00 DEG.C = 453.15 DEG.K = 356.00 DEG.F
VOLUME CRITIQUE          VCRIT = 456.02 CM3/M =  2.064 CM3/G =  .0331 C.F/LB
PENTE TENSION VAPEUR        CA = -3.070
PENTE DE SOAVE.          PENTE =  .9526

COEFFICIENTS DE TEM-   CIDAL(1) = -.116163E+03 CAL/MOLE/DEG.K
PERATURE DE LA CAPA-   CIDAL(2) =  .111724E+01 CAL/MOLE/DEG.K**2
CITE CALORIFIQUE DU    CIDAL(3) = -.268994E-02 CAL/MOLE/DEG.K**3
GAZ PARFAIT            CIDAL(4) =  .225002E-05 CAL/MOLE/DEG.K**4

CONSTANTE ENTHALPIQUE   CHIDAL =10254.087 CAL/MOLE
CONSTANTE ENTROPIQUE    CSIDAL = 447.6116 CAL/MOLE/DEG
```

CARACTERISTIQUES DE LA POMPE A CHALEUR

TEMPERATURE INFERIEURE	(POINT 1)	70.00 DEG.C
TEMPERATURE SUPERIEURE	(POINT 4)	90.00 DEG.C
INCREMENT DE SURCHAUFFE		1.00 DEG.C
SURCHAUFFE	(ENTRE POINTS 1 ET 2)	6.00 DEG.C
ECART MINIMAL DE TEMPERATURE		3.00 DEG.C
PRESSION MINIMALE DU POINT 2		1.00 ATM
PRESSION MAXIMALE DU POINT 3		50.00 ATM
PERTE DE CHARGE	(ENTRE POINTS 1 ET 2)	.10 ATM
PERTE DE CHARGE	(ENTRE POINTS 3 ET 4)	.10 ATM
PERTE DE CHARGE	(ENTRE POINTS 4 ET 5)	.10 ATM
PERTE DE CHARGE	(ENTRE POINTS 5 ET 6)	.10 ATM
PERTE DE CHARGE	(ENTRE POINTS 7 ET 1)	.10 ATM
RENDEMENT DU COMPRESSEUR		.70

RESULTATS DES CALCULS

UNITES USUELLES

POINT	ETAT DU FLUIDE	TEMPERATURE DEG.C	PRESSION ATM	VOLUME MOL CM3/MOLE	ENTHALPIE CAL/MOLE	ENTROPIE CAL/MOLE/DEG
1	VAPEUR SATUREE	70.00	2.89	8815.16	7545.00	22.25
2	VAPEUR SURCHAUFFEE	76.00	2.79	9380.43	7807.98	23.07
3	VAPEUR SURCHAUFFEE	91.26	4.94	5200.19	8316.75	23.49
4	VAPEUR SATUREE	90.00	4.84	5299.01	8268.22	23.39
5	LIQUIDE SATURE	89.14	4.74	176.97	2947.73	8.74
6	LIQUIDE SOUSREFROIDI	84.64	4.64	174.40	2684.75	8.01
7	LIQUIDE ET VAPEUR	71.25	2.99	13.49	2684.75	8.08

DANS LE DOMAINE BIPHASIQUE (LIQUIDE ET VAPEUR), LE NOMBRE FIGURANT DANS LA COLONNE DES VOLUMES (ET PRECEDE D UN ASTERISQUE) INDIQUE LE POURCENTAGE VAPORISE

UNITES S.I

POINT	ETAT DU FLUIDE	TEMPERATURE DEG.C	PRESSION PASCALS	VOLUME SPEC M3/KG	ENTHALPIE JOULES/KG	ENTROPIE J/KG/DEG
1	VAPEUR SATUREE	70.00	292667	.039900	142939	421
2	VAPEUR SURCHAUFFEE	76.00	282537	.042459	147921	437
3	VAPEUR SURCHAUFFEE	91.26	499999	.023538	157560	445
4	VAPEUR SATUREE	90.00	489869	.023985	156640	443
5	LIQUIDE SATURE	89.14	479739	.000801	55844	165
6	LIQUIDE SOUSREFROIDI	84.64	469609	.000789	50862	151
7	LIQUIDE ET VAPEUR	71.25	302797	13.49	50862	153

DANS LE DOMAINE BIPHASIQUE (LIQUIDE ET VAPEUR), LE NOMBRE FIGURANT DANS LA COLONNE DES VOLUMES (ET PRECEDE D UN ASTERISQUE) INDIQUE LE POURCENTAGE VAPORISF

TRAVAIL DE COMPRESSION ISENTROPIQUE	RDT.(H3-H2)= 356.14 CAL/MOLE	SOIT	1.6120 CAL/G	OU 6747.0 J/KG
TRAVAIL DE COMPRESSION REELLE	W = H3-H2 = 508.77 CAL/MOLE	SOIT	2.3029 CAL/G	OU 9638.6 J/KG
QUANTITE DE CHALEUR FOURNIE	Q = H3-H5 = 5369.02 CAL/MOLE	SOIT	24.3019 CAL/G	OU 101715.6 J/KG

COEFFICIENT DE PERFORMANCE COP = Q/W =10.553
COEFFICIENT DE CARNOT COC =T5/(T5-T1)=18.930

TAUX DE COMPRESSION P3/P2 = 1.77
TEMPERATURE LOGARIT. MOYENNE = 13.87 DEG

```
****************************************************************************
*                                                                          *
*        CALCUL D UNE POMPE A CHALEUR A 1 ETAGE DE COMPRESSION             *
*                          POUR UN CORPS PUR                               *
*                                                                          *
*                      (EQUATION D ETAT DE SOAVE)                          *
*                                                                          *
****************************************************************************


                                  FLUIDE
                                  ******

          1-1-2-TRICHLORO 1-2-2-TRIFLUORO ETHANE (FREON 113)

POIDS MOLECULAIRE            PM = 187.38 G/MOLE
PRESSION CRITIQUE         PCRIT =  33.70 ATM    =  34.14 BARS   = 495.25 PSI
TEMPERATURE CRITIQUE      TCRIT = 214.10 DEG.C = 487.25 DEG.K = 417.38 DEG.F
VOLUME CRITIQUE           VCRIT = 395.47 CM3/M =  2.111 CM3/G =   .0338 C.F/LB
PENTE TENSION VAPEUR         CA = -2.942
PENTE DE SOAVE            PENTE =  .8669

COEFFICIENTS DE TEM-   CIDAL(1) = -.151043E+02 CAL/MOLE/DEG.K
PERATURE DE LA CAPA-   CIDAL(2) =  .247878E+00 CAL/MOLE/DEG.K**2
CITE CALORIFIQUE DU    CIDAL(3) = -.434752E-03 CAL/MOLE/DEG.K**3
GAZ PARFAIT            CIDAL(4) =  .283343E-06 CAL/MOLE/DEG.K**4

CONSTANTE ENTHALPIQUE    CHIDAL = 2713.344 CAL/MOLE
CONSTANTE ENTROPIQUE     CSIDAL =  47.3686 CAL/MOLE/DEG
```

CARACTERISTIQUES DE LA POMPE A CHALEUR

TEMPERATURE INFERIEURE	(POINT 1)	70.00 DEG.C
TEMPERATURE SUPERIEURE	(POINT 4)	90.00 DEG.C
INCREMENT DE SURCHAUFFE		1.00 DEG.C
SURCHAUFFE	(ENTRE POINTS 1 ET 2)	1.00 DEG.C
ECART MINIMAL DE TEMPERATURE		3.00 DEG.C
PRESSION MINIMALE DU POINT 2		1.00 ATM
PRESSION MAXIMALE DU POINT 3		50.00 ATM
PERTE DE CHARGE	(ENTRE POINTS 1 ET 2)	.10 ATM
PERTE DE CHARGE	(ENTRE POINTS 3 ET 4)	.10 ATM
PERTE DE CHARGE	(ENTRE POINTS 4 ET 5)	.10 ATM
PERTE DE CHARGE	(ENTRE POINTS 5 ET 6)	.10 ATM
PERTE DE CHARGE	(ENTRE POINTS 7 ET 1)	.10 ATM
RENDEMENT DU COMPRESSEUR		.70

RESULTATS DES CALCULS

.UNITES USUELLES

POINT	ETAT DU FLUIDE	TEMPERATURE DEG.C	PRESSION ATM	VOLUME MOL CM3/MOLE	ENTHALPIE CAL/MOLE	ENTROPIE CAL/MOLE/DEG
1	VAPEUR SATUREE	70.00	2.00	13222.68	7129.72	20.86
2	VAPEUR SURCHAUFFEE	71.00	1.90	14017.16	7166.96	21.06
3	VAPEUR SURCHAUFFEE	92.22	3.51	7732.81	7734.23	21.53
4	VAPEUR SATUREE	90.00	3.41	7920.80	7668.98	21.40
5	LIQUIDE SATURE	88.82	3.31	144.22	1746.33	5.10
6	LIQUIDE ET VAPEUR	87.62	3.21	* .27	1709.10	4.99
7	LIQUIDE ET VAPEUR	71.72	2.10	* 11.42	1709.10	5.05

DANS LE DOMAINE BIPHASIQUE (LIQUIDE ET VAPEUR), LE NOMBRE FIGURANT DANS LA COLONNE DES VOLUMES (ET PRECEDE D UN ASTERISQUE) INDIQUE LE POURCENTAGE VAPORISE

UNITES S.I

POINT	ETAT DU FLUIDE	TEMPERATURE DEG.C	PRESSION PASCALS	VOLUME SPEC M3/KG	ENTHALPIE JOULES/KG	ENTROPIE J/KG/DEG
1	VAPEUR SATUREE	70.00	202272	.070566	159256	465
2	VAPEUR SURCHAUFFEE	71.00	192142	.074806	160088	470
3	VAPEUR SURCHAUFFEE	92.22	355804	.041268	172759	480
4	VAPEUR SATUREE	90.00	345674	.042271	171301	478
5	LIQUIDE SATURE	88.82	335544	.000770	39007	113
6	LIQUIDE ET VAPEUR	87.62	325414	* .27	38175	111
7	LIQUIDE ET VAPEUR	71.72	212402	* 11.42	38175	112

DANS LE DOMAINE BIPHASIQUE (LIQUIDE ET VAPEUR), LE NOMBRE FIGURANT DANS LA COLONNE DES VOLUMES (ET PRECEDE D UN ASTERISQUE) INDIQUE LE POURCENTAGE VAPORISE

TRAVAIL DE COMPRESSION ISENTROPIQUE RDT.(H3-H2)= 397.09 CAL/MOLE SOIT 2.1192 CAL/G OU 8869.8 J/KG
TRAVAIL DE COMPRESSION REELLE W = H3-H2 = 567.27 CAL/MOLE SOIT 3.0274 CAL/G OU 12671.1 J/KG
QUANTITE DE CHALEUR FOURNIE Q = H3-H5 = 5987.90 CAL/MOLE SOIT 31.9559 CAL/G OU 133751.5 J/KG

COEFFICIENT DE PERFORMANCE COP = Q/W =10.556
COEFFICIENT DE CARNOT COC =T5/(T5-T1)=19.231

TAUX DE COMPRESSION P3/P2 = 1.85
TEMPERATURE LOGARIT. MOYENNE = 17.72 DEG

CALCULATION AND OPTIMIZATION OF CONDENSERS AND EVAPORATORS FOR HEAT PUMPS

P. Paikert

GEA-HAPPEL, Bochum, BRD

SUMMARY

Starting from the universally valid physical laws of heat and mass transfer with phase change, the main computation steps and the basic ideas for condenser and evaporator optimization are explained.

Common unit types are compared with one another and their favourite application is pointed out. Decision influencing criteria for choosing one or the other special type, such as flooded or dry evaporator, tube bundle or coil system, thermostatic expansion valve or capillary tube type etc. are presented.

Apart from the effort to clearly present the internal phenomena in condenser and evaporator and link them to the overall process, calculation procedures for a condenser and for an evaporator are given and as an example a step-by-step calculation is performed and commented.

1. SURVEY OF TYPICAL CONDENSER AND EVAPORATOR TYPES

Condensers and evaporators are in most cases manufactured according to standard design data. Shell-and-tube type, double pipe, fintube-coils or plate type heat exchangers are the commonly used types of such units. The evaporating or condensing fluid flow inside or around the tubes, as is most convenient. "Most convenient" means efficient from the point of view of practical use, regarding ease of cleaning on the side of heat carrying or heat discharging fluids and regarding the choice of materials required for reasons of corrosion or material thickness required for reasons of strength.

From the thermodynamic viewpoint the working fluid can equally well be made to flow around the tubes as through the tubes.

The basic thermodynamic principles for calculating the phenomena of condensation and evaporation have been established by now and are grouped under mathematical relationship with dimensionless characteristics, so that design and optimization of what kind of unit so ever can be realized according to but one formula and making use of the same dimensionless characteristics.

Figure 1 shows a shell and tube type unit designed as condenser. The working medium is condensed on the tube outside in the shell area of the unit, whereas the cooling water flows through the tubes. This type of design is best suited if cleaning facilities on the cooling water side are required, as the water headers at both sides are easily detached. The unit can also be operated in inverse direction of phase change i.e. as evaporator, with only minor modifications being required.

Equipped with short fin tubes, performance density can be augmented by a factor of 2 or 3, noting that the performance increase is often more than correspondes to the surface increase by finning. This is due to the augmentation of the heat transfer coefficient at phase change depending upon the geometrical shape of the surface. As dimensioning of the shell wall thickness is dictated by the highest possible vapor pressure of the working medium to be reached under ambient- or process-temperature, rather thick walles may be required. The wall thickness in return can be the reason for transporting the working medium through the tubes, as is shown in Figure 2.

Figure 2 is a schematic diagram of a continuous flow (shear force) evaporator with evaporating fluids inside the tubes. The number of tubes is increased after each pass and thus with increasing steam generation. This ensures almost uniform and sufficiently high speeds in all passes. The return headers in particular are to be dimensioned in such a way that complete separation of liquid and vapor is avoided. Generally, separation can be prevented if the mass velocity even in the return header is held higher than 3 m/s. The return headers are then reduced to thick plates with only narrow return channels. In this design the vapor pressure of the working fluid no longer influences the wall thickness for any shell size.

Figure 3 shows typical types of double-pipe coils. The working fluid generally flows within the outer ring area of the double pipe while water flows through the bare tube. This unit can function not only as evaporator but also as condenser. The bare tube is often designed as low fin tube or grooved because of the considerable increase of performance involved.

The plate-type evaporator ensures good adaptation of the volume flow of the 2-phase mixture to the individual flow cross section (see Figure 4). Units of this type are especially suited for heat extraction from open pools or flowing waters, noting that the plate pitch may be choosen to allow for cleaning from coarse deposits during operation and without disturbance.

Figure 5 shows a typical unit of the coil-type condenser or -evaporator of finned tube design. This type is mainly choosen in case of heat exchange with an air flow, the working medium flowing through the coil. Due to the much lower heat transfer coefficient or flowing air, the outside is provided with fins. The surface increasing factor of condensers is in most cases between 10 and 20, with fin pitches from 2 to 4 mm. Designed as evaporator, a larger fin pitch is necessary taking into account the risk frosting, unless frosting during operation can be ruled out.

2. CHARACTERISTICS FOR DIMENSIONING AND OPTIMIZATION OF EVAPORATORS AND CONDENSERS

2.1. Dimensionless Characteristics

Notwithstanding the considerable geometric differences of the various types, dimensioning can be achieved according to the same dimensionless characteristics. From these characteristics the required heat exchange surfaces can be determined for known or given mass flow of heat exchanging liquids. By repeating the design- and calculation procedure while varying certain unfixed values, optimization of unit size and unit cost can be derived.

Figure 6 gives the temperature distributions during evaporation and condensation as well as the dimensionless characteristics required for dimensioning. The latter being independent of the unit design.

2.1.1. Heating or cooling number

The temperature distribution is determined by the task to be performed and the possible limitations of the realization of this task. It is characterized by the heating or cooling number Φ :

$$\Phi = \frac{\Delta T}{\Delta \theta_o} \qquad (1)$$

which includes the maximum temperature difference between the phase changing process (condensation or evaporation) and the inlet temperature of the fluid to be heated or cooled ($\Delta\theta_o$) as well as the possible or requested temperature change of same fluid (ΔT).

2.1.2. Heat exchanger characteristics

From the dimensionless characteristic Φ heat exchange characteristic κ can be derived, which is a measure for the efforts to be performed to reach the requested or desired heating or cooling number.

$$\kappa = \frac{U \cdot A}{\dot{M} \cdot c_p} = - \ln (1 - \Phi) \qquad (2)$$

2.2. Dimensioning and Optimization Characteristics

2.2.1. Specific exchanger performance

From the dimensionless heat exchanger characteristic κ the specific exchanger performance U.A follows :

$$U \cdot A = \frac{\dot{Q}}{\Delta \theta_o} \cdot \frac{\kappa}{\Phi} = \frac{\dot{Q}}{\Delta \theta M}$$

which in addition to the already known characteristics includes the requested heat load $\dot{Q}$.

2.2.2. Heat exchange surface

The required exchange surface can be calculated, independent of the choosen unit type by :

$$A = \frac{U \cdot A}{U} \qquad (4)$$

noting that U must be determined first, it being highly dependent on the choosen unit type and several other parameters, described in detail below.

2.2.3. Heat transmission coefficient

The heat transmission coefficient U is mainly composed of the heat transfer coefficients h_o (outside) and h_i (inside). Of influence is also the geometry and material of the heat transfer surface choosen as both aspects have an effect on the heat flux through the exchanger surface.

U always refers to the exchange surface A and has, according to Equation (3), a definable value only together with A. For fin tubes, and referring to the total finned surface, its value is given by :

$$\frac{1}{U} = \left(\frac{1}{h_o} + r_o\right) + \frac{A_o}{A_m} \cdot r_w + \frac{A_o}{A_i}\left(\frac{1}{h_i} + r_i\right) \qquad (5)$$

The heat transfer coefficients h_o and h_i are determined according to section 4, whereas the heat resistance of the fin tube r_w as well as the geometry parameters A_o/A_m and A_o/A_i must be obtained from the fin tube manufacturer. The values r_o and r_i stand for external and tube side fouling factors as possible additional heat transfer resistances.

3. THE GENERAL MATHEMATICAL INTERRELATIONSHIP OF FLUIDS IN SINGLE AND TWO-PHASE FLOW WITH AND WITHOUT PHASE CHANGE

The heat transfer of flowing single- and two-phase fluids always takes place in the viscinity of the walls of the heat emitting or absorbing surfaces. The same is true for the phase change during condensation or evaporation of a fluid. For this reason a good understanding of the phenomena occurring requires the knowledge of the flow regimes from process beginning to end. The flow near the walls is of primary interest. The transport of particles from the wall area to the center of the flow is only of secondary interest.

Figure 7 shows the 2-phase flow patterns for flow with phase change in a horizontal tube. The patterns are observed to succeed each other from left to right during condensation and from right to left during evaporation of a fluid. The upper figure shows the behaviour of a large mass flow influenced mainly by shear forces. The lower figure shows the case of a small mass flow rate influenced mainly by gravity forces.

This demonstrates that, in the wall viscinity, heat transfer always takes place between a boundary layer of a single-phase liquid and the wall, e.g. in the left part between a gas boundary layer (vapor) and the wall and in the center and right part between a liquid boundary layer and the wall. The center flow characterizes only the handling of the fluid elements separated from the wall area and may thus influence handling of the heat transferred in the wall area.

From this general understanding regarding the flow conditions, it has been proved possible to establish generally valid, non-dimensional laws for heat and mass transfer in tube and duct flow. They are all similar or identical in their principal make-up and allow for a uniform treatment of phenomena as different from one another as single-phase flow, evaporation and condensation.

The flow of two-phase fluids on the tube outside, with and without phase change, is governed by these same laws, as far as axial flow in ducts of arbitrary cross-section is concerned (e.g. double pipe-evaporators or -condensers). In an approximate way these laws also are valid for evaporation or condensation on the outside of horizontal tubes in shell-and-tube units, if the flow pattern near the wall area corresponds to that of axial flow, i.e. if a continuous film with boundary layer is formed.

Figure 8 shows the flow pattern during film condensation around horizontal tubes in a double pipe condenser and in a horizontal shell-and-tube condenser. A laminar or turbulent condensate film covers the whole tube and flows downward under the effect of gravity.

Phenomena of the above type can be described by dimensionless mathematical relationships and represented as exponential laws, in which construction or geometry effects are taken into account by constants or exponents.

From the general mathematical relationship the heat transfer coefficients of the individual one- or two-phase fluids (with or without phase change) can then be determined.

4. CALCULATION OF THE HEAT TRANSFER COEFFICIENTS FOR CONDENSERS AND EVAPORATORS

Condensers of heat pumps transfer heat to water- or air flows mainly while their evaporators extract heat from water- or air flows.

The transfer of heat between a liquid and a wall always requires a driving temperature difference to overcome the heat transfer resistance. The reciprocal value of the heat transfer resistance is called heat transfer coefficient.

4.1. Heat Transfer Coefficients Single-phase Liquids

The calculation of heat transfer coefficients on the water or air side of condensers or evaporators is generally known. Use is made of the laws of heat transfer for single-phase liquids for which literature is available in comprehensive form for many fundamental geometries of the exchange surface (see e.g. [1]). For special types, e.g. complicated fin tube systems, information is to be requested from the manufacturer.

For axial flow of water or air in smooth tubes or ducts sufficient accuracy can be obtained by using expressions of the type :

$$Nu = \frac{h \cdot D}{k} = C \cdot Re^m \cdot Pr^n \tag{6}$$

with

$$Re = \frac{4 \cdot \dot{M}}{D \cdot \pi \cdot \mu} = \frac{w \cdot D \cdot \rho}{\mu} \tag{7}$$

$C = 0{,}024$

$m = 0{,}8$

$n = m/2$

(axial flow in tubes and ducts without shedding of vortices).

4.2. Heat Transfer Coefficients Condensing Vapors

Even for condensing pure vapors the general Nu-correlation according to Equations (6) and (7) can be employed if, dependent on Re range and flow pattern, appropriate values for the constant C and exponents m and n are used. The characteristic diameter D in particular is differently defined in the Nu-number for two-phase than for flowing single-phase fluids. Its designation is consequently D_{Nu}.

$$Nu_C = \frac{h_c \cdot D_{Nu}}{k} = C \cdot Re^m \cdot Pr^n \tag{6a}$$

For laminar and turbulent film condensation, as long as the

condensate film flow is governed by gravity, the expression for D_{Nu} is :

$$D_{Nu} = \left(\frac{\nu^2}{g}\right)^{1/3} = \left(\frac{\mu^2}{\rho^2 \cdot g}\right)^{1/3} \qquad (8)$$

The characteristic diameter in the Nu_C-number is consequently a pure fluid property parameter with the dimensions of a length. At high Re-numbers (Re > 2000) the film is moved mainly by shear forces due to the vapor flow. In such a case a tube diameter corrected by the relative density of liquid and vapor has to be used as characteristic geometric parameter in the Nu_C-number. It is given by [2] :

$$D_{Nu} = \frac{D}{F_{TP} \cdot 0{,}305} \qquad (9)$$

with

$$F_{TP} = \frac{1}{2} \left[\left(\frac{\rho_L}{\rho_V}\right)^{1/2} + 1\right] \qquad (10)$$

as two-phase parameter for complete condensation.

Table I lists the values of characteristic lengths constants and exponents in Equation (6a) for different Re-ranges and film flow patterns.

The justification for employing the same mathematical correlation for procedures as differing from one another as the flow of single-phase liquids and vapor condensation is that all important heat resistances occur in the laminar film near the wall which is formed by the viscous forces from liquid or vapor flow.

The above becomes even more apparent in shear controlled turbulent film condensation, where the flow types and the coefficients of the Nu-correlation according to Table I do not differ from that of single-phase turbulent flow.

In Figure 9 the Nu-numbers are represented graphically according to the above mathematical correleations. The Nu_C-numbers of laminar condensation decreasing with the Re-number, are followed by increasing values when reaching turbulent flow, gravity controlled, condensation which are highly dependent on the Pr-number.

The Nu-number of turbulent, shear-force controlled, condensation increase even more with the Re-number. The very high values in this diagram are, however, partly due to a different definition of the characteristic diameter D_{Nu} used in the Nu-number.

Knowing that the transition from one type of condensation to the other is not well defined, some caution is advised in the region 1000 < Re < 3000. The higher of the Nu-numbers determined according to the various procedures is always the one to be used.

The heat transfer coefficient for condensation results from the transformation of Equation (6a) :

$$h_C = Nu_C \cdot \frac{k}{D_{Nu}} \tag{11}$$

For axial flow Nu_C can also be determined from Figure 9 directly, in which the Re-number is determined from Equation (7).

Out of these generally applicable mathematical correlations the heat transfer coefficients can then be determined separately for the individual single or two-phase fluids, with or without phase change.

4.3. Heat Transfer Coefficients of Evaporating Liquids

4.3.1. Mean heat transfer coefficients

For the heat transfer coefficients of evaporating liquids flowing in tubes, similar general laws are applicable which can be represented in form of a Nu-equation.

Pierre [3] found the following mathematical correlation for the evaporation of refrigerants, particularly R 12, R 22 and R 502 in tubes of common diameter :

$$Nu_E = \frac{h_E \cdot D}{k} = C \cdot Re^m \cdot K_f^n \tag{12}$$

with Re according to Equation (7) and :

$$K_f = \frac{\Delta h}{l \cdot g} \quad \text{(boiling number)} \tag{13}$$

Δh = enthalpy difference
l = characteristic length

This equation demonstrates the same mathematical interrelationship to be valid for evaporation as for convection and condensation. Only the boiling number is replaced by the Pr-number.

Generally, however, the evaporation process is described by 2 different mathematical expressions. Each being valid for one of the physically differing types of evaporation of liquids flowing in tubes : namely nucleate boiling and convective boiling.

During nucleate boiling, where vapor bubbles penetrate from the tube wall into the flow core and to the liquid surface, without condensing, the heat transfer coefficient is highly dependent on the heat flux density :

$$\dot{q} = h_E \cdot \Delta\theta , \tag{14}$$

and only little dependent on the mass flow density $\dot{m}$:

$$\dot{m} = \frac{\dot{M}}{f} \tag{15}$$

During convective boiling, where the bubbles formed near the wall condense in the flow center, dependence is the opposite. The heat flux density $\dot{q}$ has only a minor, the mass flow density $\dot{m}$ a major influence.

Figure 10 shows the typical curve of the mathematical interrelationship of the two different types of boiling as well as the general interrelationship according to Pierre which gives, at least for the transition region, useful values. Figure 10 is applicable for R 22 at a heat flux density $\dot{q}$ = 5000 W/m^2, the following constants and exponents being valid in the Nu-equation according to Pierre.

$$C = 0{,}012$$

$$m = 0{,}8$$

$$n = 0{,}4 .$$

The up to date knowledge of heat transfer coefficients of boiling liquids flowing in tubes or ducts has been summarized by B. Slipccevic [4] for both nucleate boiling and convective boiling. On the basis of these data, the mean heat transfer coefficient for the field of nucleate boiling can be calculated by :

$$h_E = \frac{0{,}9 \cdot B \cdot (\Delta x)^{0{,}1}}{D_i^{0{,}5}} \cdot \dot{m}^{0{,}1} \cdot \dot{q}_i^{0{,}7} \tag{16}$$

with B = property value according to Table II

Δx = vapor content change in the evaporator

$\dot{m}$ = mass flow density

$\dot{q}_i$ = heat flux density

D_i = tube inner diameter

As to convective boiling the mean heat transfer coefficient is

$$h_E = K \cdot \frac{\dot{m}^{1,4}}{D_i^{0,5}} \tag{17}$$

with K = property value according to Table 3

all other values as in Equation (16).

Schlünder, Chawla and Thomé [5] have elaborated useful working diagrams as shown in Figure 11, enabling to directly determine the mean heat transfer coefficients for various refrigerants, for tube flow and complete evaporation, as a function of heat flux- and mass flow-density.

Equations (12), (16) and (17) also give such mean values of heat transfer coefficients for the whole range of evaporation for various mass flow- and heat flux-densities.

4.3.2. Path of local heat transfer coefficients

The comprehensive work of Chawla and others [6] showed that during evaporation a significant maximum of the heat transfer coefficient at $x \simeq 0,8$ is found and that, independent of mass flow and heat flux-density, the last tube section operates at very low heat transfer coefficients, provided complete evaporation occurs.

The local heat transfer coefficient decreases slightly at constant heat flux density with increasing vapor content and then, upon reaching a critical value for x, increases considerably to a maximum value between x = 0,75 and 0,85, before decreasing abruptly. The critical vapor content x, however, is not constant but a function of mass flow and heat flux density.

The mean value of the heat transfer coefficient in a dry evaporator can be found from Figure 12 by integration of the local h_E-values via the vapor content between x = 0 and x = 1 for the individual heat flux density $\dot{q}$ and the mass flow density $\dot{m}$. Figure 11 shows the result of such an integration for various mass flow densities at constant heat flux density lines.

Depending on construction features and operation, the heat flux density $\dot{q}$ varies along the path of evaporation and, consequently, over the vapor content as is shown in Figure 12 for counter flow and parallel flow of refrigerant and water in a double pipe evaporator or in a tube injection evaporator with only one pass for the refrigerant. The parallel flow evaporator starts with high heat flux density on the water and refrigerant inlet side, causing high heat transfer coefficients. With continuing evaporation $\dot{q}$- and h_E-values decrease, first during nucleate boiling due to falling local temperature differences. As soon as convective boiling is reached, the heat flux density looses its influence and the h_E-values increase again, dropping abruptly upon exceeding their optimum value at $x \simeq 0{,}8$.

In case of counter current evaporation, the process starts with the smallest temperature difference and, consequently, with small $\dot{q}$ and h_E values due to the low vapor content. The $\dot{q}$ and h_E values rise continuously along the tube length and drop abruptly upon reaching the value of $x \simeq 0{,}8$.

Figure 12 is a schematic representation of the variation of the heat transfer coefficient during counter flow and parallel flow evaporation and Figure 13 shows the pertinent temperature diagrams.

Small starting temperature differences between water and refrigerant flow provided, the parallel flow process is more favourable and ensures better performance than does the counter flow process for the same exchange surface.

As shown in Figure 13, this apparent contradiction to well known heat exchanger theories - starting that counter current operation always results in the best exchange of heat - is due to the temperature drop of the evaporating refrigerant when flowing through the exchanger caused by the inevitable pressure loss.

For an evaporation temperature of + 3°C at the evaporator inlet, a temperature drop of the refrigerant of approximately 2°C as well as a heating water temperature change from + 10°C to + 5°C result in an effective mean temperature difference at parallel flow operation of $\Delta\theta_M = 5{,}36$ K and at countercurrent flow of $\Delta\theta_M = 4{,}65$ K. The difference is about 15 %.

At the same time, the local heat transfer coefficient at the start of parallel flow evaporation, higher due to the major local temperature difference and the resulting higher heat flux density $\dot{q}$.

4.3.3. Heat transfer coefficient of flooded evaporators

The heat transfer coefficients of flooded evaporators in tube bundle design are governed by similar mathematical interrelationships as above.

B. Slipcevic [7] has analyzed the correlations developed in various sources and has translated them in a practical, easy to use relation for the heat transfer coefficients in flooded tube bundle evaporators :

$$h_E = S \cdot \dot{q}^n \cdot R_p^{0,133} \qquad (18)$$

S being a fluid dependent coefficient and

R_p the wall roughness in , um

Table IV lists the parameter S and the pertinent exponents n for some refrigerants. They are valid for an evaporation temperature of about 0°C and for a smooth tube and a low fin tube of 3/4" diameter with 19 fins/inch.

Figure 14 gives a graphic representation of a correlation similar to Equation (18) for a roughness value of $R_p = 1$, um according to [8]. This graph is valid only for a single horizontal tube row. For multiple tube rows, the result will be much better because of the induced turbulence of the fluid. For 5 tube rows the h_E-value will be about 30 % higher. For other roughness values the heat transfer coefficients of Figure 14 are to be multiplied with the factor F given in Table V.

5. OPTIMIZATION OF CONDENSER AND EVAPORATOR

Condenser and evaporator are essential and important components of a heat pump. Their size influences to a large extent the investment cost and the performance. For example, the approach of the working temperature to the temperature of the heat source and the heat sink determines the coefficient of performance COP of the heat pump.

The coefficient of performance results from the cycle of the heat pump as given in Figure 15 :

$$COP = \eta_e \, COP_{Carnot} = \eta_e \cdot \frac{T_C}{T_C - T_O} \qquad (19)$$

with

T_C = condensation temperature

T_0 = evaporation temperature

η_e = exergetic efficiency

At a given heat source temperature T_U and attainable evaporation temperature T_O, the difference required for heat exchange (namely $T_U - T_O = \Delta\theta_U$, also called the "approach") is a factor to be taken into account when sizing the evaporator. The size of the latter varies inversely proportional to the approach. Also the condenser size is influenced by the condenser approach $T_C - T_H = \Delta\theta_H$.

Considering the approach,Equation (19) becomes

$$COP = \eta_e \frac{T_H + \Delta\theta_U}{T_H + \Delta\theta_U - T_U + \Delta\theta_H} \tag{20}$$

Taking into account that the exergetic efficiency decreases with increasing approach, Table VI shows the change in COP of a heat pump generating hot water at $t_W = 45°C$, from a heat source with a temperature $t_U = + 5°C$. The approaches of condenser and evaporator are the independent variables.

With increasing approach of evaporator and condenser, the COP decrease considerably. This necessitates both higher drive power consumption and high-capacity drive units, requiring in turn high investments on the operation side. On the other hand, investment costs of condenser and evaporator decrease abruptly as a result of less exchange surface required with increasing approach. The capitalization of the total investment of drive and exchanger, together with the operating costs give the total cost which has be minimized for given design conditions. From this the optimum approach and optimum unit size can be derived.

Figure 16 shows in the bottom part the variation of COP and specific exchanger performance (U . A) for condensers and evaporators, and in the top part the variation of exchange surfaces A, operation cost B and capitalized variable cost K as a function of the approach, for a domestic heat pump with 18 kW thermal capacity. The optimum approach is about 3 K.

Independent of optimization with respect to approach, suitable in case of considerable temperature approach, optimization of unit size with respect to pressure loss may be useful in case of large temperature differences. With increasing mass flow through the units, the heat transfer coefficients increase with simultaneous decreasing of the mean temperature difference due to the falling evaporation (or condensation) temperature of the working fluid caused by the pressure drop.

It is the product of heat transfer coefficient and mean effective temperature difference which is, however, governing the exchange surface size and it is this quantity which therefore must be optimized :

$$U \cdot \Delta\theta_M = \max \rightarrow A = \min \quad (21)$$

6. FLOODED OR DRY EVAPORATION

Dry evaporation is mainly used in injection evaporators, in which the working fluid is superheated by several degrees after complete evaporation in a single pass. Superheating of the vapor requires major portion of the exchange surface of the evaporator if a sufficient superheating signal for the expansion valve is to be obtained. Superheating however shields the compressor from damage resulting from large quantities of entrained liquid droplets.

The flooded evaporator can save the surface portion required for overheating. It requires, however, an additional droplet separator to guarantee an equal protection against the above effects.

From a thermodynamic point of view there is no major difference between dry or flooded evaporation, so that mainly practical reasons are decisive for choosing either dry or flooded evaporation systems.

When large mass flow rates are required and when it is necessary to regularly clean the heating side, the working fluid must be held in the shell area of a shell and tube evaporator, e.g. a flooded evaporator is to be choosen.

Contrary to past procedure when flooded evaporators were mostly used, presently their number has considerably decreased as they do not offer thermodynamic advantages. Particularly in case of small temperature differences between heating fluids and evaporation temperature, only low heat transfer coefficients are obtained because of the low heat flux density. In-tube evaporation systems can obtain high heat transfer coefficients in simular cases by high mass flow densities notwithstanding the small temperature difference.
In particular with regard to the lower and medium capacity range, the dry evaporator is given preference. The dry evaporator

ensures high heat transfer coefficients and enables the use of thermostatic expansion valves.

7. THROTTLE-TYPE CAPILLARY TUBES OR THERMOSTATIC EXPANSION VALVES TO BE USED AS FEED DEVICE FOR THE EVAPORATOR ?

Method of operation and in field behaviour of the thermostatic expansion valve are well-known, capillary tubes on the other hand are hardly known.

The higher price and higher sensitivity of the thermostatic expansion valve to minor fouling or small particles present in the refrigerant cycle must be taken into account when comparing it to the less sensitive capillary tube. In addition a comparison of the behaviour and efficiency of both systems when applied in heat pumps would be helpful.

As does the expansion valve, the capillary tube expands the liquid refrigerant from condenser- to evaporator-pressure. The undercooled liquid (in some cases after flowing through the intermediat heat exchanger) flows through the only about 1,5 mm wide tube driven by the pressure difference between the two ends. The high speed results in a pressure drop which increases linearly with t.

In spite of the undercooling, the vapor pressure corresponding to the liquid temperature is soon reached and vapor bubbles are formed. The increasing volume of vapor bubbles causes speed augmentation and an abrupt pressure drop which, in return, is the reason for increased evaporation, higher speed and augmented pressure drop.

The capillary tube limits its mass flow automatically when reaching of the velocity of sound of the liquid/vapor mixture and stops to respond to any further pressure drop in the evaporator. For given condenser pressure the flow rate remains nearly constant, independent of the evaporator pressure.

Figure 17 represents the procedure schematically. To the linear pressure drop over the overall length of the capillary tube related to single phase liquid flow is added a sharp pressure drop when vapor bubble formation starts. This stops at a critical pressure difference at which the velocity of sound is reached.

Only the condensation temperature which changes with condenser pressure, influences to a larger extent the throughput capacity of a capillary tube of given length, because of its influence upon the viscosity of the condensate. Rising condensation temperature increases the throughput capacity of a given capillary tube for the same total pressure drop.

For an air/water heat pump with constant condensation temperature, the capillary tube is a simple feed device, almost not affected by the variable evaporation pressure p_0 resulting from the variable ambient temperature. This, however, is valid only for stationary operation of the machine.

Under part load conditions, e.g. in case of repeated machine shut-down, operational difficulties may related to the properties of the capillary tube. The capillary tube forms a continuous connection between condenser and evaporator. As shown in Figure 18, the condenser pressure decreases upon machine shut-down due to overflowing condensate from the liquid catcher through the capillary tube to the evaporator. The evaporator pressure increases simultaneously until the pressure difference between condenser and evaporator has disappeared. It can, however, not be avoided that a certain amount of condensate flows into the evaporator where it is collected in some tubes, possibly blocking these completely, whereas the other evaporator tubes remain almost free from liquid.

Upon starting the machine, low pressure is almost immediately obtained in the evaporator. Due to the low evaporation capacity of the few tubes filled with liquid, this low pressure cannot be countered rapidly. This causes a considerable local undercooling in the area of the tubes filled with liquid. During this period of time, only little liquid or none at all is supplied from the emptied condenser. The pressure in the condenser must first be built up, to be able to condense the delivered vapor.

Subcooling of certain evaporator sections may cause icing on the outside of the evaporator tubes, in particular when undercooling occurs often and when short switching periods occur. Thus complete defrosting of subcooled sections becomes impossible and the formation of thick layers of ice may result. Such uncontrolled local evaporator icing only occurs during part-load operation, e.g. at relatively high ambient temperature, and can be avoided only by complete blocking of the evaporator and condenser connection during standstill, or by a careful determination of the amount of refrigerant in the cycle.

When several evaporator sections are fed in parallel by the same capillary tube as shown in Figure 19, the risk of evaporator icing increases and the use of a thermostatic expansion valve according to Figure 20 should be recommended.

8. APPLICATION EXAMPLE FOR CALCULATION AND OPTIMIZATION

For an air/water domestic heat pump the condenser and evaporator will be calculated and optimized. Refrigerant R 12 is utilized.

The condenser shall be designed as a water-flooded double pipe coaxial unit with low fin tubes and the evaporator as a rectangular element with high fin tubes and forced draught.

8.1. Dimensioning of R 12 Coaxial Condenser

Design data :

Heat duty	$\dot{Q}$	= 18 kW
Condensing temperature	t_c	= 48 °C
Condensing pressure	p_c	= 12 bar
Water inlet temperature	t_{W1}	= 35 °C
Water outlet temperature	t_{W2}	= 45 °C
R 12-Massflow	$\dot{M}$	= 0,104 kg/s

Low fin tube data :

Outside diameter	d_o	= 22,1 mm
Inside diameter	d_i	= 18,6 mm
Reference exchange surface	A_o	= 0,21 m^2/m
Surface ratio	A_o/A_m	= 2,0
	A_o/A_i	= 3,6
Heat transfer resistance	r_w	= $4,2 \cdot 10^{-6}$ m^2 K/W

Shell tube

Outside diameter	D_o	= 30 mm
Inside diameter	D_i	= 27 mm

Physical properties R 12 :

dyn. viscosity, condensate	μ	= $225 \cdot 10^{-6}$ Pas
density, condensate	ρ_L	= 1.218 kg/m^3
density, vapor	ρ_V	= 66,8 kg/m^3

heat conductivity, liquid	k	=	0,065 W/mK
Pr-number	Pr	=	3,5
fouling factor, waterside	r_i	=	$1{,}1 \cdot 10^{-4}\ m^2k/w$

1. Designing on refrigerant side

Double pipe cross section

$$f_R = (D_i^{\ 2} - d_o^{\ 2})\ \frac{\pi}{4} = 1{,}9 \cdot 10^{-4}\ m^2$$

Thermal diameter

$$d_{th} = \frac{4 \cdot f_R}{d_o \cdot \pi} = 1{,}1 \cdot 10^{-2}\ m$$

Re_C-number of condensation (see Equation 7), but as the condensation takes place in the double pipe and on the base tube surface only, it should be modified to :

$$Re_C = \frac{\dot{m} . d_a}{f_R . \mu} = \frac{0{,}104 \cdot 2{,}2 \cdot 10^8}{1{,}9 \cdot 225} = 5{,}3 \cdot 10^4$$

Because $Re_C > 2000$ one has turbulent condensation, shear force oriented (Table I).

Therefore Equation 6a is used with

$C = 0{,}024$

$m = 0{,}8$

$n = 0{,}4$

$$Nu_C = 0{,}024 \cdot (5{,}3 \cdot 10^4)^{0{,}8} \cdot 3{,}5^{0{,}4}$$

$$Nu_C = \underline{238}$$

With equation (10) and (9) one obtains :

$$F_{TP} = \frac{1}{2}\ [(\frac{1218}{66{,}8})^{1/2} + 1] = 2{,}6$$

$$d_{Nu} = \frac{1,1 \cdot 10^{-2}}{2,6 \cdot 0,305} = 1,4 \cdot 10^{-2} \text{ m}$$

$$h_E = \frac{Nu_c \cdot k}{d_{Nu}} = \frac{238 \cdot 0,065}{1,4 \cdot 10^{-2}} = \underline{\underline{1100 \text{ W/m}^2\text{K}}}$$

2. Designing on the water side

Mass flow :

$$\dot{M} = \frac{\dot{Q}}{c \cdot \Delta t_W} = \frac{18,0}{4,19 \cdot 10} = 0,43 \text{ kg/s}$$

Therefore :

$$Re = \frac{4 \cdot 0,43}{18,6 \cdot 10^{-3} \cdot 6 \cdot 10^{-4} \cdot \pi} = 4,9 \cdot 10^4$$

and as

$Pr = 3,9$, Equations(6) and (7) give :

$$Nu = 0,024 \, (4,9 \cdot 10^4)^{0,8} \cdot 3,9^{0,4} = 346,6$$

or $$h_i = \frac{Nu \cdot k}{d_i} = \frac{346,6 \cdot 0,637}{18,6 \cdot 10^{-3}} = \underline{\underline{11870 \text{ W/m}^2\text{K}}}$$

3. Heat transfer coefficient U is determined from Equation (5) :

$$\frac{1}{U} = \frac{1}{h_E} + r_w \cdot \frac{A_o}{A_m} + \frac{A_o}{A_i} \left(\frac{1}{h_i} + r_i\right) ,$$

which gives

$$\frac{1}{U} = \frac{1}{1100} + 2 \cdot 4,2 \cdot 10^{-6} + 3,6 \left(\frac{1}{11870} + 1,1 \cdot 10^{-4}\right)$$

$$= 1,61 \cdot 10^{-3}$$

or $$U = \underline{\underline{620 \text{ W/m}^2\text{K}}}$$

4. Required heat exchange surface

From (1) the heating number is :

$$\Phi = \frac{\Delta T}{\Delta\theta_o} = \frac{10}{13} = 0,77$$

The heat exchanger characteristic according to Equation (2) is :

$$\kappa = -\ln(1-\Phi) = -\ln(1-0,77) = 1,47$$

Specific exchanger performance U A :

$$U \cdot A = \frac{\dot{Q} \cdot \kappa}{\Delta\theta_o \cdot \Phi} = \frac{18000 \cdot 1,47}{13 \cdot 0,77} = \underline{\underline{2637 \text{ W/°C}}}$$

Heat exchanger surface

$$A = \frac{U \cdot A}{U} = \frac{2637}{620} = 4,25 \text{ m}^2$$

5. Total length of the coaxial condenser

$$L = \frac{A}{A_o} = \frac{4,25}{0,21} = \underline{\underline{20,25 \text{ m}}}$$

The condenser can be designed as a flat parallel coil beneath the compressor or as a circular coil around the compressor, depending upon the chosen heat pump lay-out.

6. Checking of speeds and pressure drops

- Water side

a) water velocity

$$W_W = \frac{M}{d_i^2 \frac{\pi}{4} \cdot \rho_W} = \frac{0,43}{(18,6 \cdot 10^{-3})^2 \frac{\pi}{4} \cdot 1000} = 1,58 \text{ m/s}$$

b) pressure drop

$$\Delta p = (n + \xi \cdot \frac{L}{d}) \cdot \frac{\rho}{2} w^2$$

with $\xi = 0{,}316 \cdot Re^{-0,25}$ (for smooth copper tubes)

n = number of tube inlets and -outlets

for this case : n = 2

this gives :

$$\Delta p = [2 + 0{,}316 \ (4{,}9 \cdot 10^4)^{-0,25} \cdot \frac{20{,}25}{0{,}0186}] \frac{1000}{2} \cdot 1{,}58^2$$

$$= [2 + 2{,}12 \cdot 10^{-2} \cdot 1089] \ 1248$$

$$= [2 + 23] \cdot 1248 = \underline{\underline{3{,}13 \cdot 10^4 \ Pa}}$$

- Refrigerant side

a) vapor velocity, entrance

$$w_V = \frac{\dot{M}}{\rho_V \cdot f_R} = \frac{0{,}104}{66{,}8 \cdot 1{,}9 \cdot 10^{-4}} = \underline{\underline{8{,}2 \ m/s}}$$

b) pressure loss

$$\Delta P_R = (1 + a \cdot \xi \cdot \frac{L}{d}) \frac{\rho_V}{2} w_V^2$$

With ξ as before, and

a = constant for mathematical integration for decreasing vapor velocity, a = 0,36 for total condensation

we obtain :

$$\Delta p_R = [1 + 0{,}36 \cdot 0{,}316 \cdot (5{,}3 \cdot 10^4)^{-0,25} \cdot \frac{20{,}2}{0{,}011}] \frac{66{,}8}{2} \cdot 8{,}2^2$$

$$= (1 + 15,1) \cdot 2245 = 36158$$

$$= 0,36 \text{ bar}$$

8.2. Dimensioning of Air Heated Dry Evaporator

Evaporator heat duty $\dot{Q}_v = 12,4$ kW

Ambient temperature $t_U = 6$ °C

Evaporation temperature $t_v = 0$ °C ± 1 °C

Evaporation pressure $p_v = 3,15$ bar

Vapor density $\rho_V = 18$ kg/m^3

Fin tube data

Base tube diameter $d_O = 16$ mm

$d_i = 14,4$ mm

Fin diameter $D_O = 38$ mm

Reference exchange surface $A_O = 0,67$ m^2/m

Tube spacing $s_t = 40$ mm

Surface ratio $A_O/A_i = 14,5$

- (outside/inside)
- (outside/core) $A_O/A_m = 13,6$
- (outside/frontal) $\frac{A_O}{A_F} = 16,8$

Air side heat transfer coefficient at $w_O = 2,8$ m/s :

$$h_O = 63 \text{ W/m}^2\text{K} .$$

Heat transfer resistance $r_w = 3 \cdot 10^{-4} \frac{m^2K}{W}$

(Corresponds with fin efficiency $\eta_F \simeq 0,8$)

Fouling factor, inside : $r_i = 0$

1. Determination of characteristic numbers

Cooling number :

$$\Phi = \frac{3}{6} = 0,5$$

Heat exchanger characteristic

$$\kappa = - \ln (1 - 0,5) = 0,693$$

specific exchanger performance

$$UA = \frac{12400 \; . \; 0,693}{6 \; . \; 0,5} = 2865 \; W/°C$$

Heat exchange surface :

$$A = \frac{(U \; . \; A)}{U}$$

This cannot yet be determined as U is still unknown.

2. Required evaporator front surface

The heat duty, delivered from the air is :

$$\dot{Q}_V = A_F \cdot w_L \cdot \rho_L \cdot C_{p,L} \cdot \Delta t_L$$

This leads to the frontal surface

$$A_F = \frac{\dot{Q}_V}{w_L \cdot \rho_L \cdot C_{p,L} \cdot \Delta t_L} = \frac{12,4}{2,8 \; . \; 1,28 \; . \; 1,0 \; . \; 3} = 1,15 \; m^2$$

3. Dimensioning of the evaporator

a) Heat transfer coefficient of evaporation

$$h_E \simeq 1000 \; W/m^2 \quad \text{(preliminary estimate)}$$

b) Heat transmission coefficient from Equation (5) :

$$\frac{1}{U} = \frac{1}{63} + 3 \; . \; 10^{-4} \; . \; 13,6 + 14,5 \; (\frac{1}{1000} + 0)$$

$$= 1{,}995 \cdot 10^{-2} + 1{,}45 \cdot 10^{-2}$$

$$U = 29 \ W/m^2K$$

c) Heat exchange surface

$$A = \frac{(U \cdot A)}{U} = \frac{2865}{29} = 98{,}8 \ m^2$$

d) Available front surface at n = 4 tube rows

$$A_F = \frac{A}{A_0/A_F \cdot n} = \frac{98{,}8}{16{,}8 \cdot 4} = 1{,}47 \ m^2$$

e) Splitting of front surface into length L and width B

Take L = 1,8 m and B = 0,8 m (corresponds with 20 tubes per tube row)

f) Refrigerant mass flow per tube when split up into 5 tube injections

$$\dot{M}_1 = \frac{\dot{M}}{5} = 0{,}0208 \ kg/s$$

g) Mass flow density :

$$\dot{m} = \frac{0{,}0208}{0{,}0144^2 \ \frac{\pi}{4}} = 128 \ kg/m^2s$$

From Figure 11 one finds : for this value of $\dot{m}$:

$$h_E \geq 1000 \ W/m^2 \ °C.$$

The exact calculation according to Equation (16) shows for the range of convective boiling at $t_0 = 0$ °C :

$$h_E = K \cdot \frac{\dot{m}^{1,4}}{d_i^{0,5}} = 0{,}156 \cdot \frac{128^{1,4}}{0{,}0144^{0,5}} = 1159 \ W/m^2K$$

with K from Table III.

This confirms the evaluation according to item a) above.

Considering

$$\dot{q} = h_E \cdot \Delta t_i = U \cdot \frac{A_0}{A_i} \Delta\theta_M = 29 \cdot 14{,}5 \cdot 4{,}3 = 1820 \text{ W/m}^2$$

it shows that (on the average) evaporation takes place in the range of convective boiling. However also the local heat flux density $\dot{q}$ at the evaporator inlet, with $\Delta\theta_0$ = 6 K and under parallel flow evaporation conditions :

$$\dot{q} = 29 \cdot 14{,}5 \cdot 6 = 2520 \text{ W/m}^2$$

would be well in the range of convectiv boiling so that parallel flow operation would not be advantageous in this case, contrary to what is shown in Figure 12. In order to obtain a sizeable superheat, needed by the expantion valve, cross counter flow operation is choosen here. The now some what oversized frontal and overall surface area ensures good operation and a safe superheating signal.

4. Pressure drop on the refrigerant side

The pressure drop during evaporation is approximately (only friction) :

$$\Delta p_{E,f} = \xi \cdot \frac{\dot{m}^2}{2 \cdot \rho_V} \cdot \frac{n \cdot L}{d_i} \cdot \Psi_{E,f}$$

with

$$\xi = 0{,}316 \cdot Re^{-0{,}25}$$

$$Re = \frac{4 \cdot \dot{M}_1}{d \cdot \pi \cdot \eta} = \frac{4 \cdot 0{,}0208}{0{,}0144 \cdot \pi \cdot 225 \cdot 10^{-6}} = \underline{\underline{8178}} ,$$

this gives for the tube friction factor :

$$\xi = 0{,}316 \cdot 8178^{-0{,}25} = 0{,}033 .$$

Friction function for evaporation $\Psi_{E,f}$

$$\Psi_{E,f} = 0{,}42 + 0{,}01 \cdot \dot{m}^{0{,}7}$$

$$= 0{,}42 \cdot 0{,}01 \cdot 128^{0{,}7} = 0{,}72$$

so that :

$$\Delta p_{E,f} = 0{,}033 \cdot \frac{128^2}{2 \cdot 18} = \frac{16 \cdot 1{,}8}{0{,}0144} \cdot 0{,}72 = 21600 \text{ Pa}$$

$$= 0{,}21 \text{ bar}$$

The pressure drop of 0,21 bar corresponds to a temperature drop during evaporation of 2 K and is of the usual magnitude.

Repeated calculation and designing for different values of $\Delta\theta_U$, $\Delta\theta_H$, t_O air speed, etc. leads to optimization according to Figure 16.

REFERENCES

1. *VDI-Wärmeatlas*, Berechnungsblätter für den Wärmeübergang, Düsseldorf, VDI-Verlag.

2. Boyko, L.D. and Kruzhilin, G.N., Heat Transfer and Hydraulic Resistance During Condensation of Steam in a Horizontal Tube and in Bundle of Tubes, *International Journal Heat Mass Transfer*, 10 (1967).

3. Pierre, B., Wärmeübergang mit siedenden Kältemitteln in horizontalen Rohren, *Kylteteknisk Tidskrift* 28 (1969) nr. 5,

4. Slipcevic, B., Wärmeübergang beim Sieden von R-Kältemitteln in horizontalen Rohren, *Kältetechnik-Klimatisierung* 24 (1972), nr. 12.

5. Schlünder, E.U., Chawla, J.M. and Thomé, E.A., *Wärmeübergangszahlen auf der Kältemittelseite in Verdampferrohren*, Inst. für Therm. Verfahrenstechnik, Universität Karlsruhe (1968).

6. Chawla, J.M., Wärmeübergang und Druckabfall in waagerechten Rohren bei der Strömung von verdampfenden Kältemitteln, *VDI-Forschungsheft 523*, Düsseldorf, VDI-Verlag, (1967).

7. Slipcevic, B., Wärmeübergang bei der Blasenverdampfung von Kältemitteln an glatten und berippten Rohrbündeln, *Klima + Kälte-Ingenieur* 3 (1975) 9.

8. Stephan, K., Berechnung des Wärmeübergangs an siedende Kältemittel, *Kältetechnik*, Bd. 15 (1963), nr. 8.

TABLE I

Constants and exponents of Nu-number of condensing vapors according to Equation 6a for laminar and turbulent film condensation at gravity- and shear forced flow.

	Vertical Tube	Horizontal Tube	
	inside and outside	outside N-tubes in line	inside
$Re_C < 2000$ laminar flow	C = 1,47	$1,47.N^{-0,25}$	1,18
	m = - 1/3	- 1/3	- 1/3
	n = 0	0	0
$Re_C > 2000$ turbulent flow	C = 0,024	$1,47.N^{-0,07}$	-
gravity controlled	m = 0,25	0,47	--
	n = 0,33	0,4	--
$Re_C > 2000$ turbulent flow shear force controlled	vertical and horizontal tubes independent of gravity		
	C = 0,024		
	m = 0,8		
	n = 0,4		

TABLE II

Physical property value B of some refrigerants according to Slipcevic [4] for nucleate boiling at t_0 = - 10 °C, 0 °C and + 10 °C

Refrigerant	B t_0 = 0 °C	B t_0 = - 10 °C	B t_0 = + 10 °C
R 12	0,152	0,146	0,157
R 22	0,169	0,160	0,179
R 114	0,107	0,105	0,109
R 502	0,180	0,175	0,192

TABLE III

Physical property value K of some refrigerants for convection boiling according to Slipcevic [4] at t_0 = - 10 °C, 0 °C and + 10 °C.

Refrigerant	K at - 10 °C	K at 0 °C	K at + 10 °C
R 12	0,200	0,156	0,124
R 22	0,218	0,169	0,140
R 114	0,260	0,208	0,160

TABLE IV

Physical property value S and exponent n of some refrigerants for t_0 = 0 °C evaporation temperature according to [7] .

Refrigerant	Plain tube		Fin tube	
	S	n	S	n
R 12	6,50	0,613	14,10	0,565
R 22	7,95	0,597	16,26	0,554
R 114	7,00	0,544	14,24	0,517
R 502	4,10	0,631	11,00	0,576

TABLE V

Corrective factor F for heat transfer coefficient according to Figure 14 at other surface roughness features.

Surface feature	polished	fine finished			finished		roughed
Rp (μm)	0,1	1	2	3	5	10	15
F	0,74	1,0	1,10	1,16	1,24	1,36	1,43

TABLE VI

Influence of temperature approach of condenser and evaporator on the performance number of a heat pump with T_U = + 5 °C and T_H = 45°C.

$\Delta\theta_U = \Delta\theta_H$	0	1	2	4	6	8
COP_C	8,0	7,6	7,2	6,6	6,2	5,8
COP		4,4	3,6	2,9	2,5	2,3

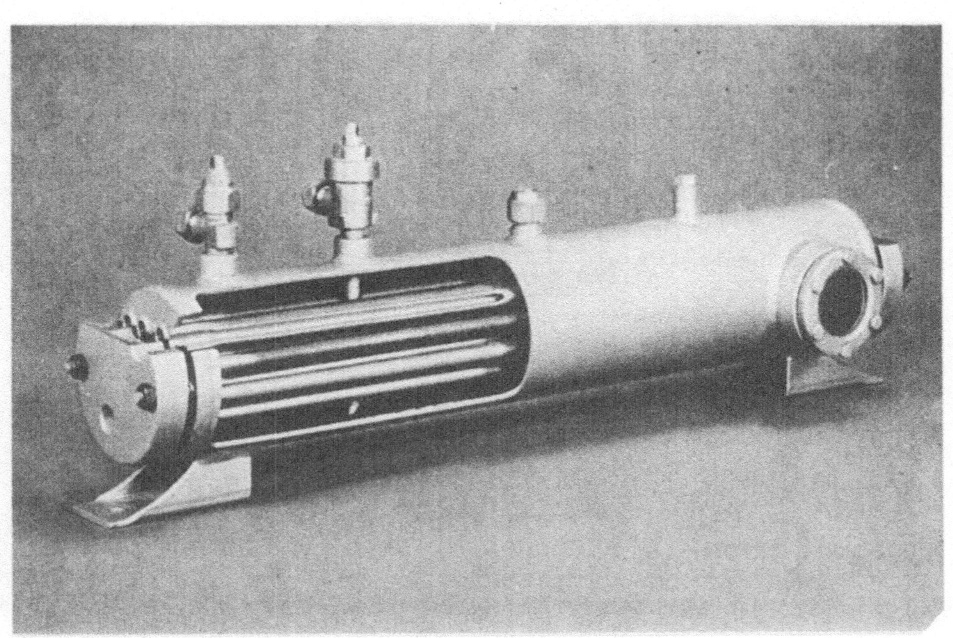

Shell and tube type unit, execution as condenser with replaceable water headers

Figure 1

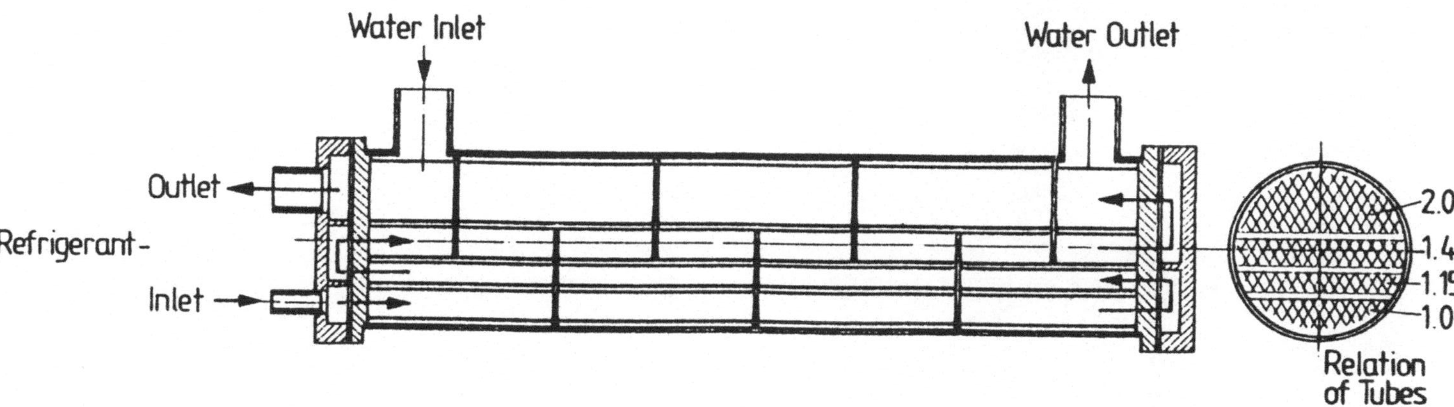

INJECTION-EVAPORATOR. Evaporating of Refrigerants in Horizontal Tubes with Increasing Number of Tubes per Pass.

Figure 2

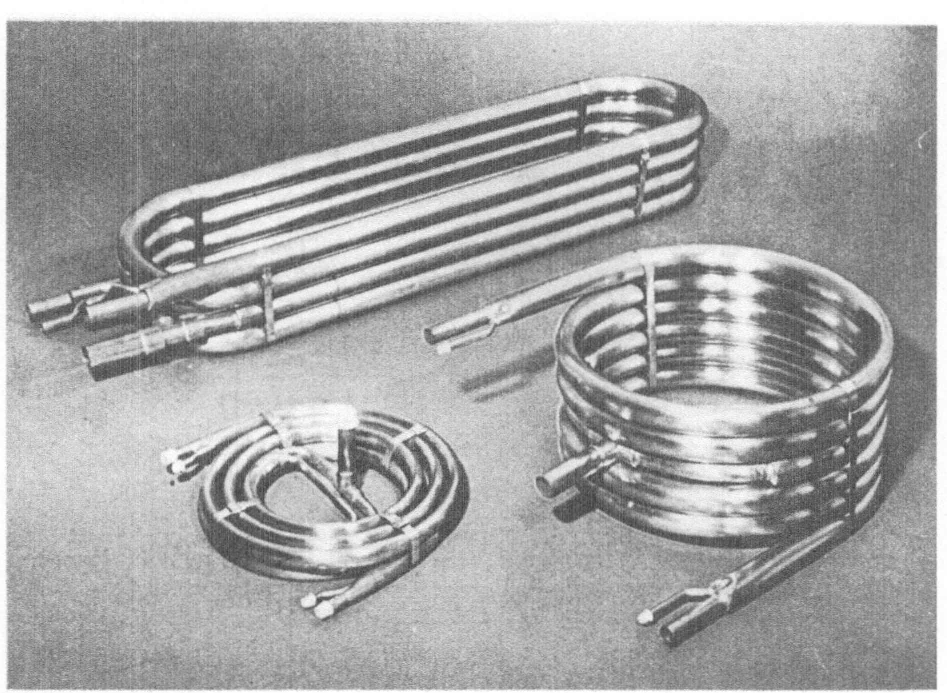

Double pipe coil as condenser or evaporator. Refrigerant in ring space of double-pipe, water in bare tube

Figure 3

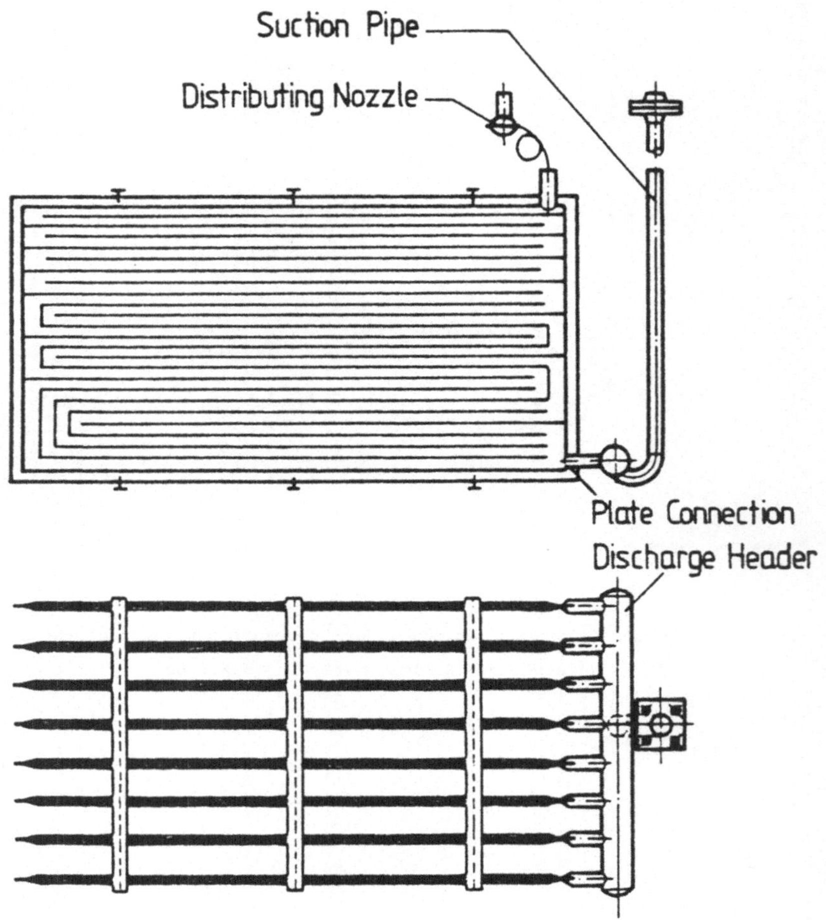

Schematic Design of a Platetype Evaporator

Figure 4

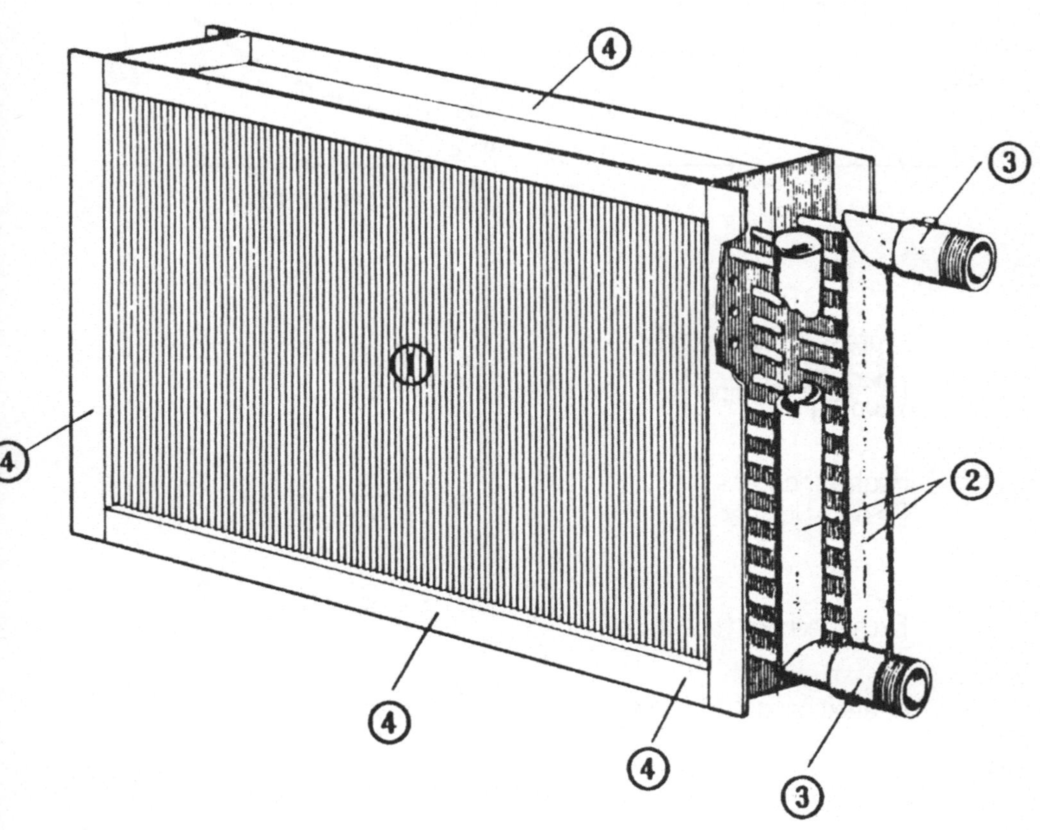

Air cooled condenser in coil execution with fin tubes.
Also commonly used as air heated injection evaporator

Figure 5

Condenser- and Evaporator Characteristics
for Dimensioning

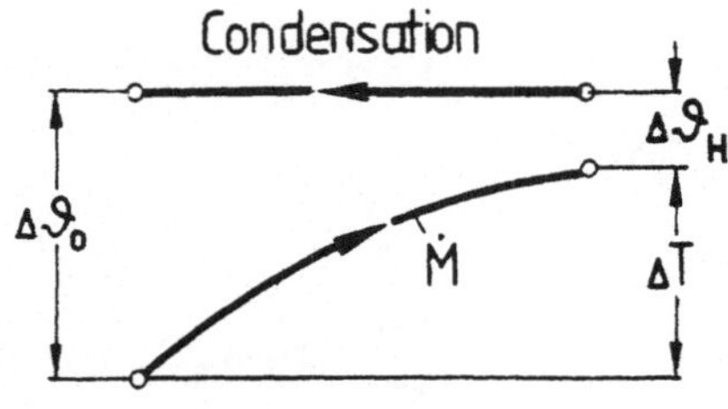

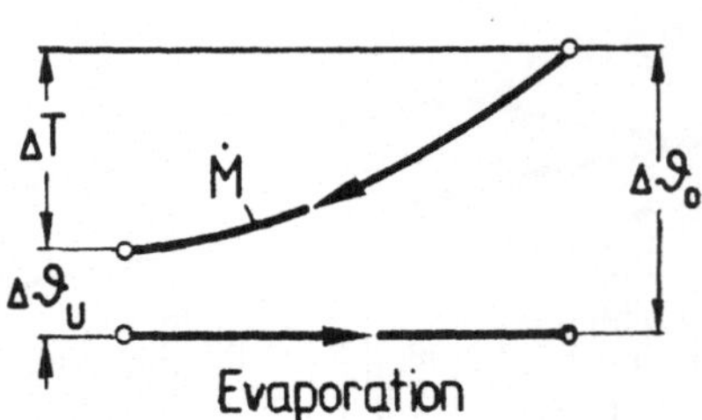

Process Characteristic (Heating- / Cooling- Number)
$$\Phi = \frac{\Delta T}{\Delta\vartheta_0}$$

Heatexchanger-Design Number
$$\varkappa = \frac{U \cdot A}{\dot{M} \cdot cp} = -\left[\ln(1-\phi)\right]$$

Specific Exchanger Performance
$$U \cdot A = \frac{Q}{\Delta\vartheta_0} \cdot \frac{-\ln(1-\phi)}{\Phi}$$

Heatexchanger Surface
$$A = \frac{U \cdot A}{U}$$

Figure 6

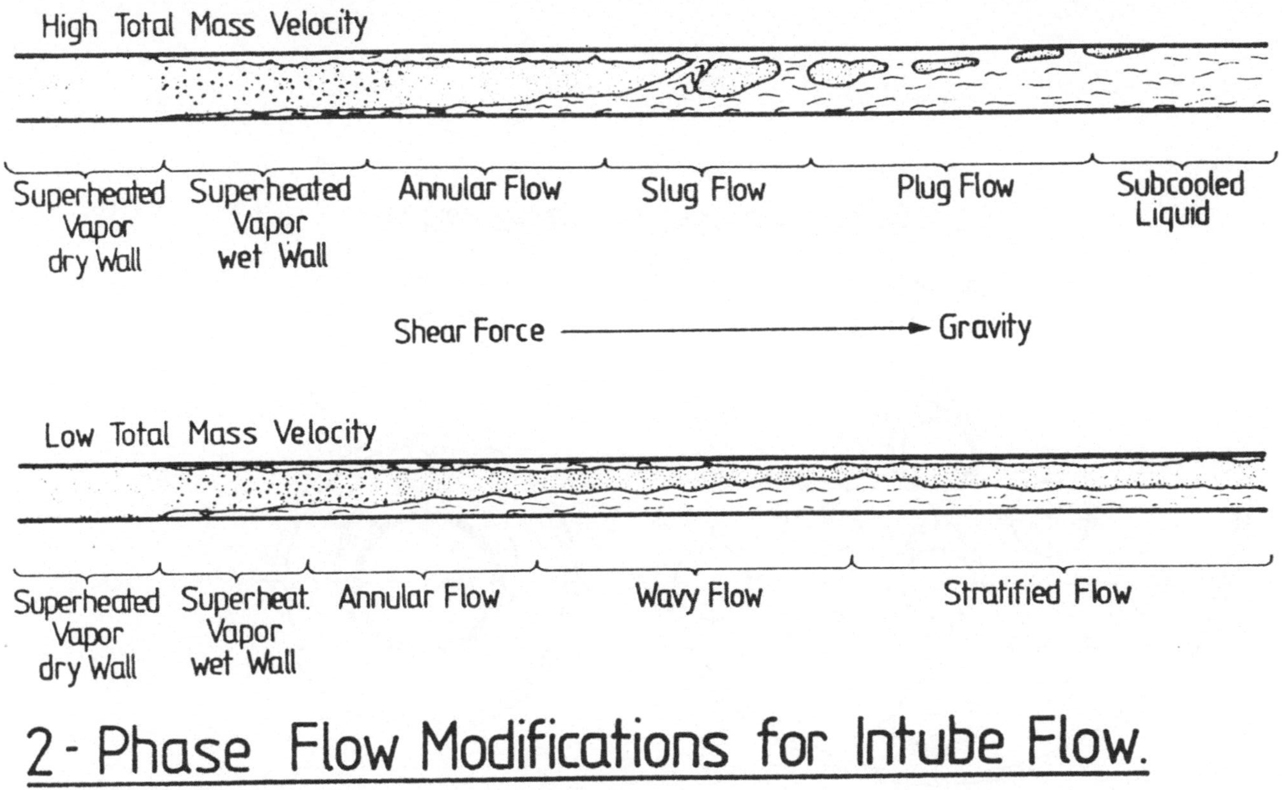

High Total Mass Velocity

Superheated Vapor dry Wall | Superheated Vapor wet Wall | Annular Flow | Slug Flow | Plug Flow | Subcooled Liquid

Shear Force → Gravity

Low Total Mass Velocity

Superheated Vapor dry Wall | Superheat. Vapor wet Wall | Annular Flow | Wavy Flow | Stratified Flow

2-Phase Flow Modifications for Intube Flow.

Figure 7

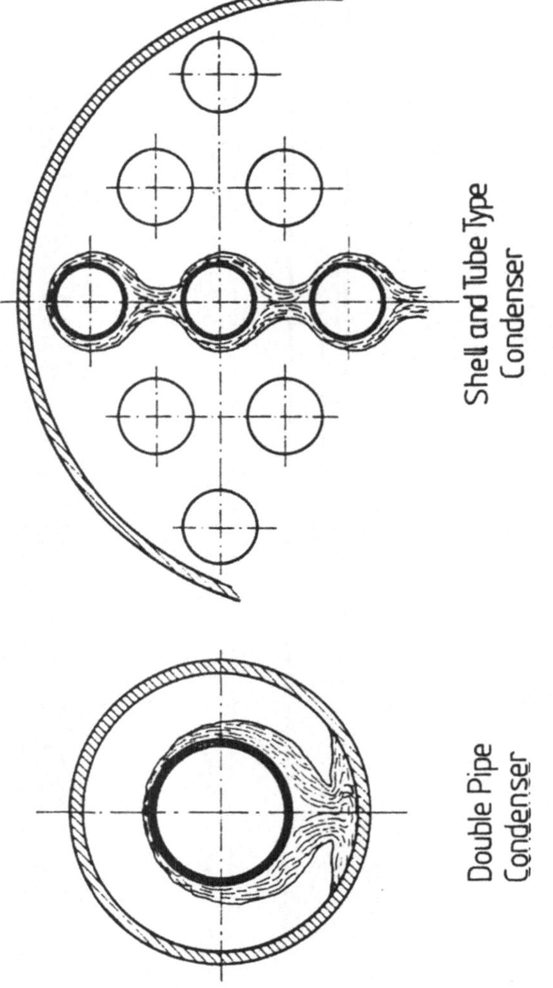

Film Flow Pattern and Condensat Draining

Figure 8

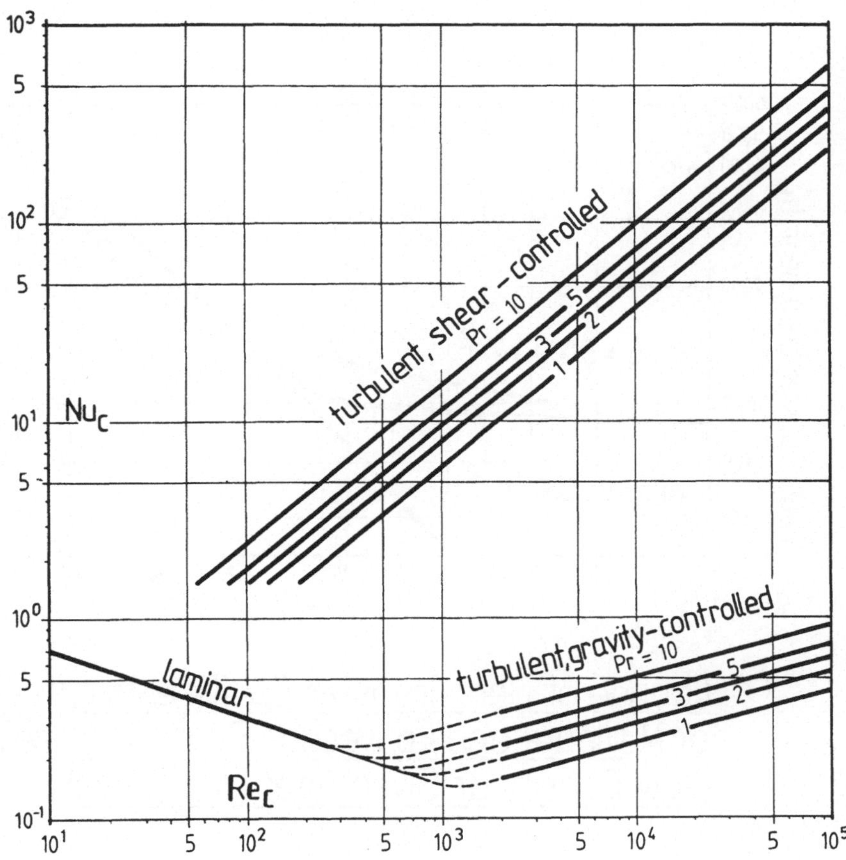

Condensing Pure Vapors

Figure 9

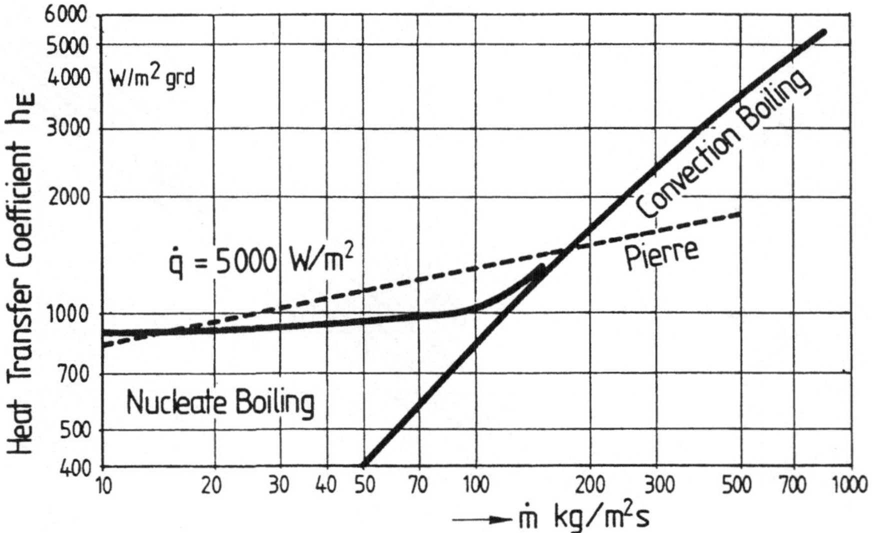

Change of Heat Transfer Coefficients in the Range of Nucleate and Convection Boiling over the Mass Flow Density at Given Heat Flux Density.

Figure 10

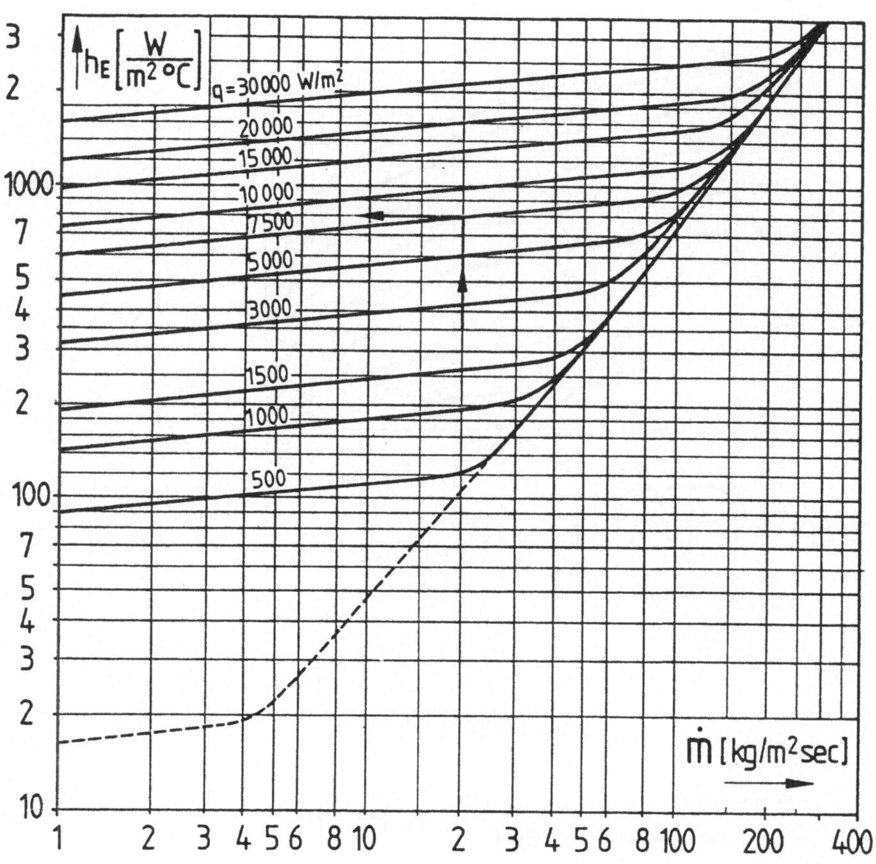

Mean Heat Transfer Coefficient at Complete Evaporation of R 12 in Horizontal Tubes. Tube Diameter D = 13,5 mm.

Figure 11

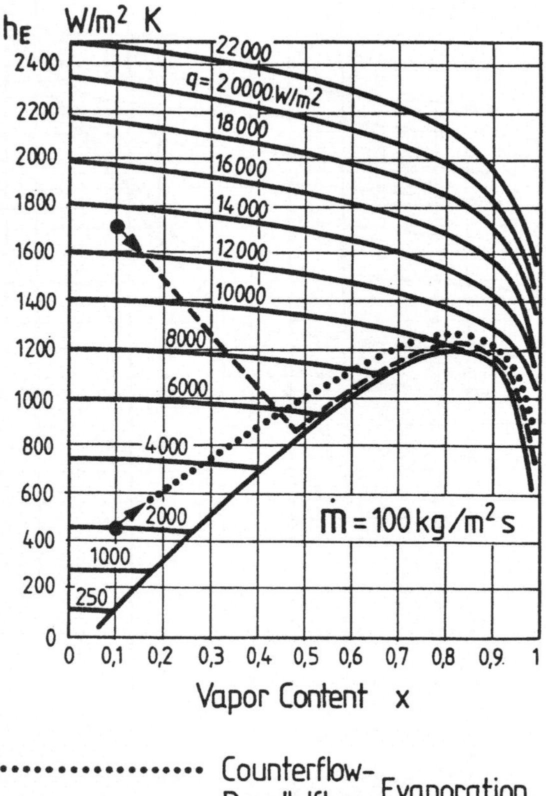

Local Heat Transfer Coefficients for Evaporation of Refrigerants in Horizontal Tubes.

Figure 12

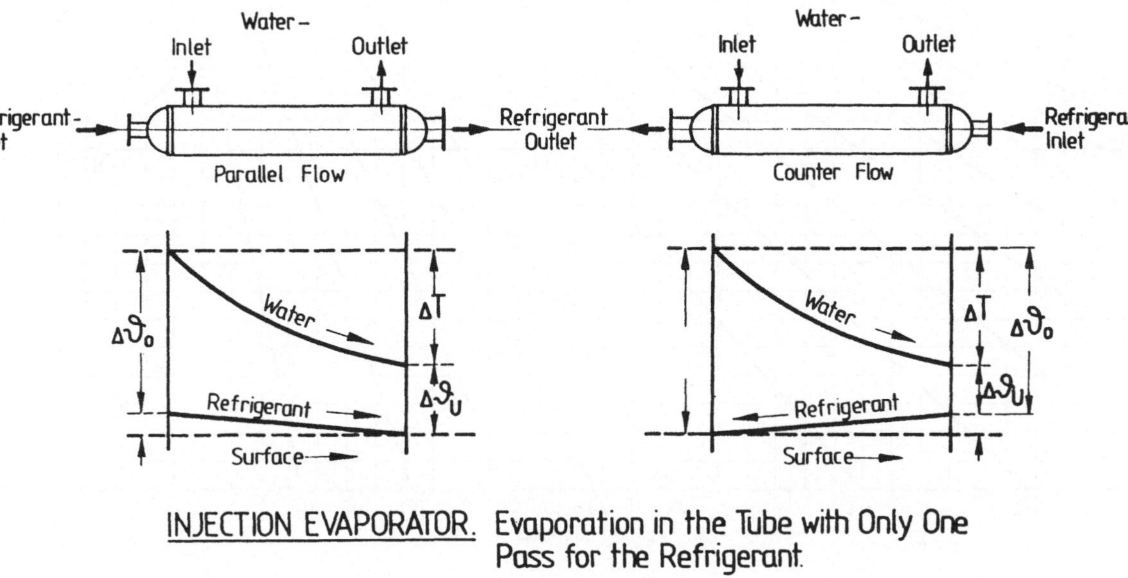

INJECTION EVAPORATOR. Evaporation in the Tube with Only One Pass for the Refrigerant.

Figure 13

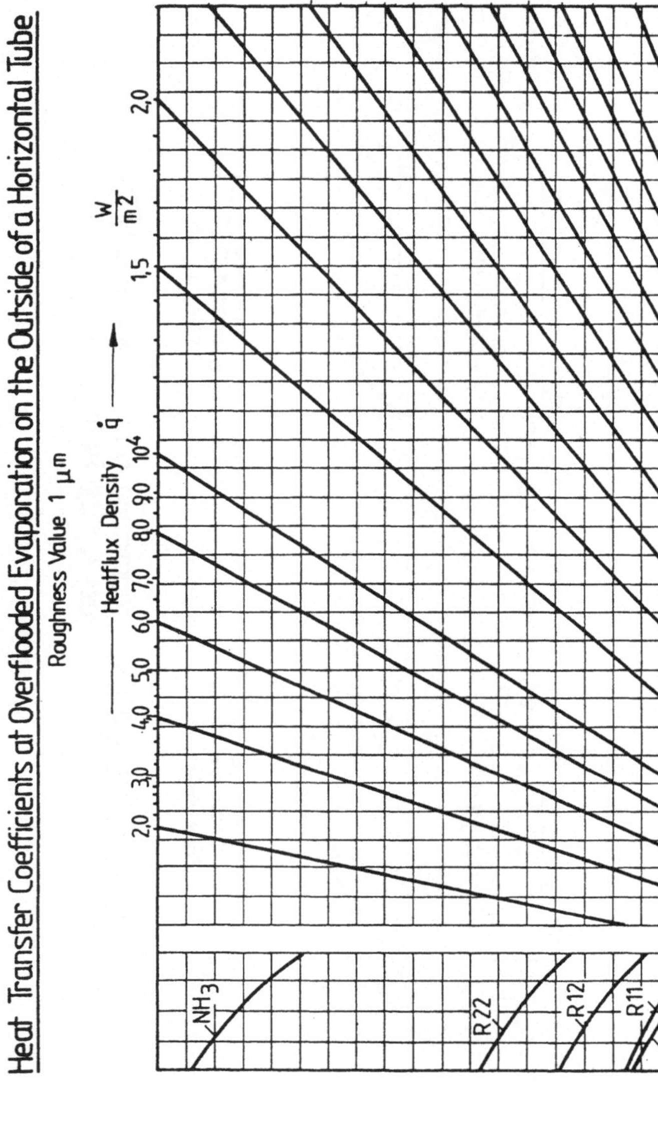

Heat Transfer Coefficients at Overflooded Evaporation on the Outside of a Horizontal Tube
Roughness Value 1 μm
Heatflux Density q̇
W/m2
2,0 3,0 4,0 5,0 6,0 7,0 8,0 9,0 10⁴ 1,5 2,0 2,5 3,0 3,5 4,0 4,5 5,0 5,5 6,0 7,0 8,0 9,0 10⁵
NH3
R 22
R 12
R 11
R 114
3 bar 2 1
Pressure p
300 500 1000 1500 2000 2500 3000 3500
Heat Transfer Coefficient hE
W/m2K

Figure 14

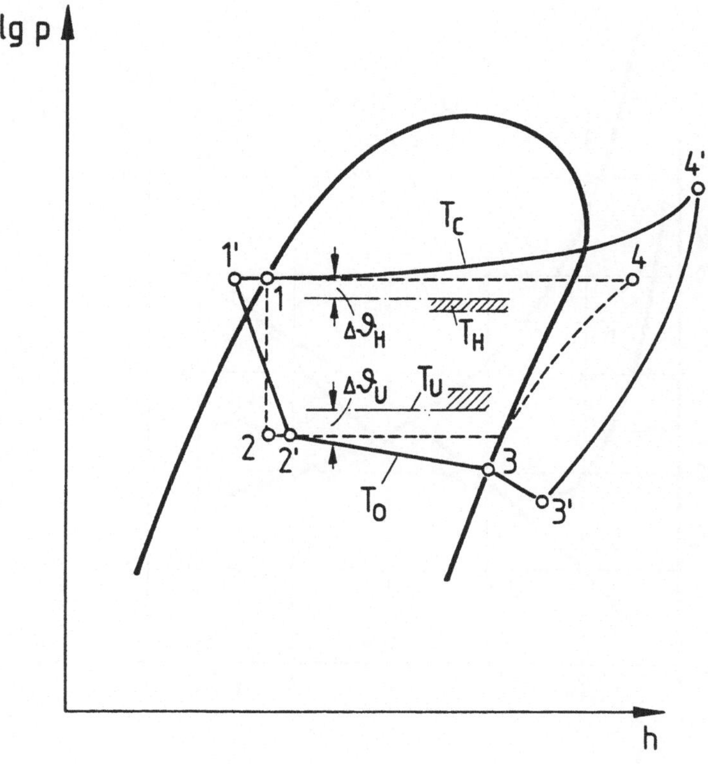

--- ideal
—— real
Heat Pump Cycle Process

T_U Heat Source Temp. (Ambient-)

T_H Heat Absorption Temp. (Heating-)

$\Delta\vartheta_U$, $\Delta\vartheta_H$ Approach - Evaporator, - Condenser

Heat Pump Cycle Process in h, log p-Diagram

Figure 15

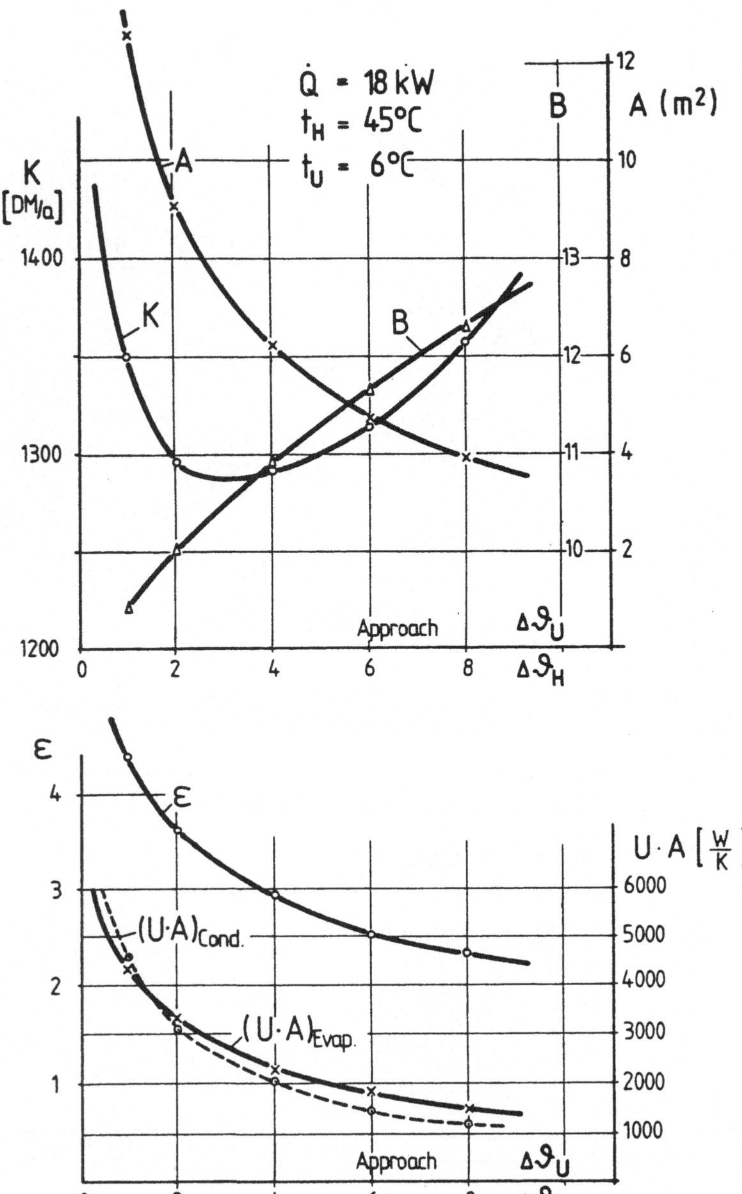

Optimization of Approach and Unitsizes for Condenser and Evaporator

Figure 16

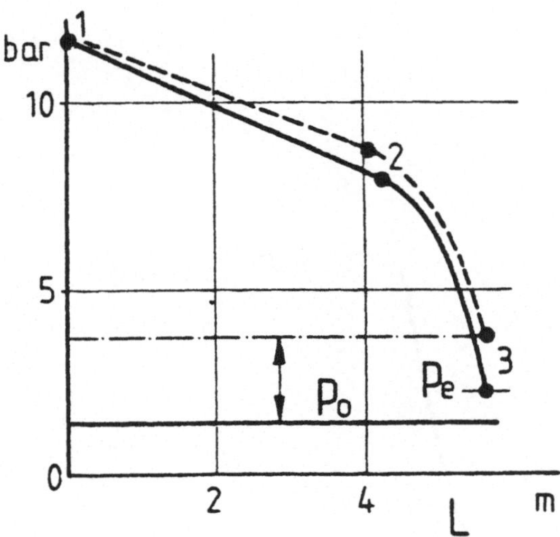

Pressure Change of R 12 at Flow Through a Capillary Tube.

1. Liquid refrigerant at condenser pressure P
2. Starting of vapor formation in the capillary tube
3. Discharge from the capillary tube

Po Evaporator pressure, variable

Pe Lowest final pressure in the capillary tube

Figure 17

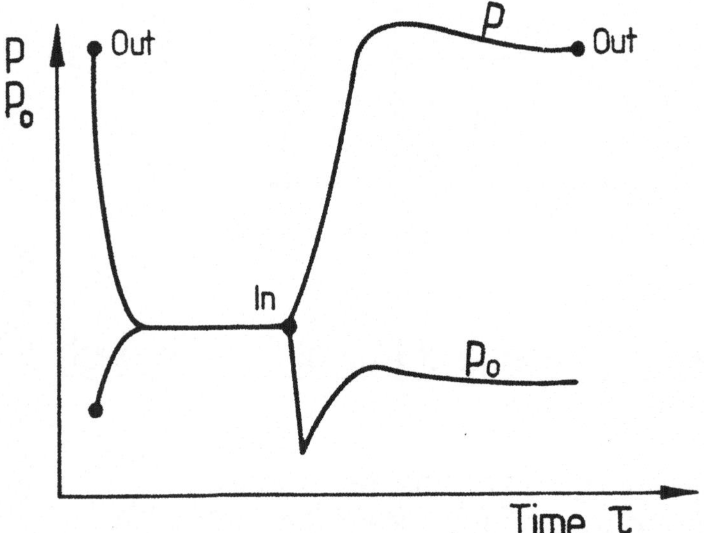

Time Slope of Pressure P in Condenser and P_o in Evaporator of a Heat Pump with Capillary Tube at In- and Out-Operation.

Figure 18

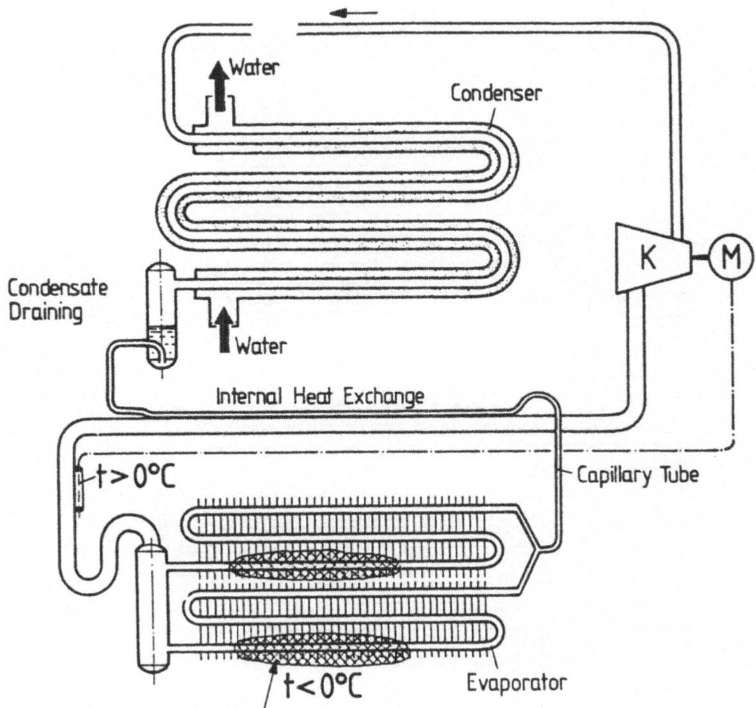

Air to Water Heat Pump

Capillary Tube as Distribution Nozzle

Figure 19

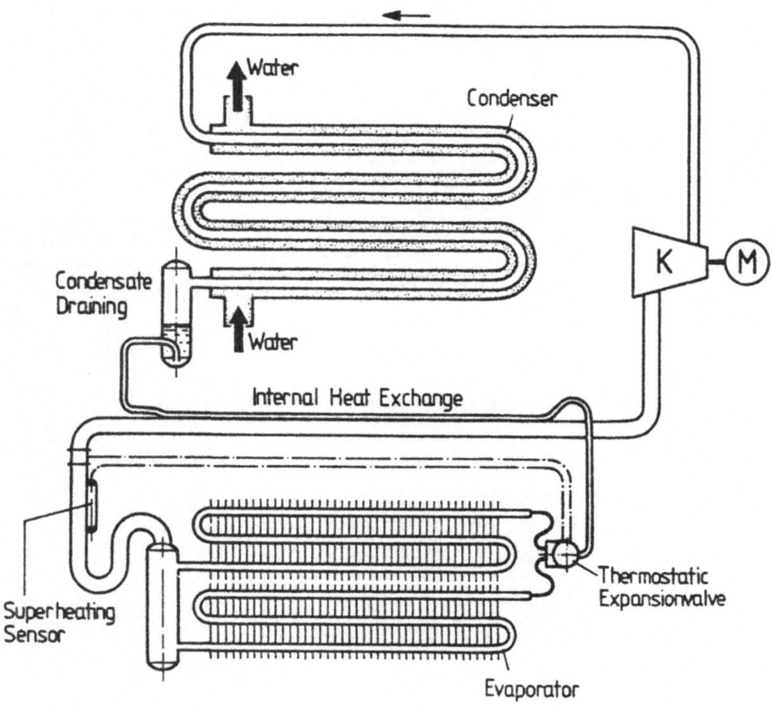

Air to Water Heat Pump

Thermostatic Expansion Valve as Distribution Nozzle

Figure 20

HEAT PUMP COMPRESSORS

John H. Grimm

Copeland Corp., Sidney, Ohio, U.S.A.

INTRODUCTION

The concept of a heat pump is not new, and the basic scientific understanding of the heat pump cycle has long since been thoroughly explored. The first reaction of an unbiased observer would be to wonder why the heat pump has been so slow in development, and why we are discussing heat pump fundamentals decades after the scientific principles were well established.

Obviously the worldwide energy crisis is now a driving force to improve our efficiency in using energy. But the second and major reason for the slow development of heat pump usage has its roots in the practical problems in translating a scientific theoretical principle into a feasible operating system. Primarily those practical problems revolve around compressor failures when applied in heat pump systems.

In the United States, the electric utilities began strongly promoting electric motor driven heat pumps in the 1950's in order to try to make electrical energy more competitive with gas and oil. In the United States defense department a few individuals became convinced of the advantages of the heat pump and decided to use a military housing program to prove its merits.

Thousands of heat pumps were installed in housing units at military bases. This encouraged dozens of manufacturers of air conditioning equipment to rush with heat pumps to the U.S. residential market.

The heat pumps in the military housing program became one of the greatest economic fiascos in history. Compressors failed by the hundred. They failed so fast that replacements couldn't be trucked in fast enough.

It was not only the military program which suffered, the general residential experience was similar. Companies went bankrupt, bitter enmities sprang up between people in the industry, and the heat pump became a controversial topic for years. Most manufacturers withdrew from the heat pump market, and only a few companies, the General Electric Company, Carrier, and Westinghouse, continued with development work. Were it not for the energy crunch the heat pump might never have revived.

The compressor is the heart of any heat pump system, and compressor failures were the heart of that fiasco of 20 years ago. There is no question but that many of the compressor failures of that day originated in poor motor insulation and inadequate mechanical design. But our research indicates far more failures were caused by inadequate unit design.

In concept, the heat pump sounds simple. In practice, it is a sophisticated piece of equipment, requiring a complete understanding of the system operation for proper design. I think we have to pay attention to the lessons of history and experience, or we may be doomed to repeat history. I think the engineers involved in the heat pump fiasco of 20 years ago may not have understood all of the basic problems involved in heat pump applications. My travels around the industry led me to believe that even today a lot of us do not understand everything that goes on in a heat pump.

To fully understand the problems involved in heat pump design and application, it is first necessary to understand the internal working of a compressor, where its vulnerabilities and weak points are, and the unpleasant facts of engineering life that we must face to successfully apply that compressor.

1. COMPRESSOR TYPES

The typical compressor used for heat pump duty is a reciprocating compressor in a welded shell. The reciprocating design with one, two, three, or four cylinders is the predominant design used today in refrigeration and air conditioning usage in the sizes suitable for residential and light commercial duty. It is relatively simple, tolerant of clearances, efficient for high compression ratios, and proven by millions of applications (figure 1). Typical performance curves of hermetic compressors are given in figure 2.

Rotary compressors are also applied for airconditioning and heat pump usage. The rotary has the advantage of being smaller and lighter in weight, and is quite efficient. It is vulnerable to extremes of temperature and pressure, and is relatively intolerant of any foreign material, so its application must be more closely controlled (figure 3).

But the basic heat pump problems affect both types of compressor in quite similar ways so for purposes of our discussion, we can speak generally of positive displacement compressors.

The welded type construction is strictly a matter of economics. It is the lowest cost design which can be produced. Typically the rotor is mounted on the crankshaft which is positioned vertically within the shell. The compressor body is in the lower part of the shell, partially submerged in oil. The entire assembly is internally spring mounted and surrounded by refrigerant vapor.

The welded type of construction is in contrast to the design we call an accessible-hermetic compressor, which is predominantly used for refrigeration and heavy commercial application.

2. HEAT PUMP COMPRESSOR SPECIAL PROBLEMS

In the accessible-hermetic design, the body is a heavy casting with a large heat transfer surface directly exposed to the air. The motor is in direct physical contact with the body, so motor heat can be quickly radiated outward.

Because of its design with the motor suspended in vapor, a welded compressor is a very poor heat radiator, and practically all the heat to be dissipated must be transmitted through the refrigerant vapor. As a result, under high compression ratio conditions when the amount of motor heat input is high compared to the mass flow of the refrigerant vapor, a welded type compressor can easily overheat. Typically these conditions occur on refrigeration applications, and therefore welded compressors in general are used primarily for air conditioning and medium temperature refrigeration usage. For refrigeration applications above 1 H.P. accessible-hermetic compressors are almost universally applied. That problem of preventing a welded compressor from overheating is one of the most basic and least understood heat pump design problems, and I'll return to it in discussing the application.

Many of the U.S. heat pump failures of 20 years ago originated in poor motor insulation. In air-to-air heat pump applications with a defrost cycle, a considerable amount of liquid refrigerant returns to the compressor. Testing after the field failures had occurred uncovered the fact that liquid R-22 refrigerant softened the varnish and insulation used at that time on the motor windings, baring the wire and allowing it to flex, causing a motor failure. Over the years, better modern motor wire insulation has been developed, together with an epoxy dip that has practically eliminated the once common motor problem.

A second area of compressor vulnerability is in its mechanical construction. A compressor built for air conditioning duty must pump against high discharge pressures, but is does not face high compression ratios, which is the ratio of the discharge pressure divided by the suction pressure. One of the peculiar characteristics of a bearing is that it may be able to withstand very high pressure, but may wear badly at much lower pressures and high compression ratios. The difference is really in the lubricating film that can be maintained on the bearing surface. Under high compression ratio conditions, the residual gas in the cylinder continues to expand for much of the suction stroke, maintaining pressure on the head of the piston. This causes the piston to ride on the pin not only on the compression stroke but also for much of the suction stroke, starving the bearing for lubrication. Connecting rod pins which might be entirely suitable for air conditioning duty may wear out in a few months time on high compression ratio applications. We stumbled across this phenomenon quite by accident on commercial applications. Since the basic operating conditions are inherent, the only solution is to increase the bearing size to compensate for the marginal lubrication film.

Another potential problem area in the compressor is the clearance between piston and cylinder. Compressors are built either with plug pistons which depend on minimum clearance and close tolerance, or with piston rings which allow slightly greater tolerance between piston and cylinder.

It is possible that rapid changes in piston and cylinder temperature that could occur with liquid floodback during defrost might cause unequal rates of expansion of the piston and cylinder, and conceivably could close the clearance and cause piston seizing. We have reproduced such conditions in the laboratory, but whether they might be probable in the field is difficult to determine. In any event it appears to be wise in heat pump duty to use ringed pistons with additional clearance.

The reeds used in refrigeration and air conditioning compressors are designed for vapor compression. The reed design, the size of the suction port, and the strength of the reed itself determine the reeds capability to withstand the shock and over-flexing caused by the impact of liquid droplets. Almost any reed can withstand some slight exposure to liquid droplets in the vapor stream, but as the size and number of the liquid droplets increase the stress on the reed is greatly increased. As in the case with bearing design, a reed that might be entirely satisfactory so long as only vapor was being pumped might fail very quickly when exposed to liquid impingement.

Liquid impingement, or <u>slugging</u> as it is commonly called, can result from two sources. One obviously is the flooding of liquid refrigerant to the compressor, a very common occurrence at the initiation and termination of defrost in an air-to-air heat pump. The liquid droplets can be carried along by the vapor stream even despite traps or impingement suction vapor inlets.

Liquid droplets can be oil as well as refrigerant. In any compressor sitting idle, the refrigerant vapor will continuously migrate to the oil and condense in the oil since the vapor pressure of the refrigerant will always be greater than an oil-refrigerant mixture. Given sufficient time, practically all of the refrigerant in a system will migrate to the oil. When the compressor is started under such conditions, the pressure in the crankcase is quickly reduced, falling below the saturation pressure of the oil-refrigerant mixture. Obeying the laws of physics, a portion of the refrigerant in the mixture will instantaneously flash into vapor to reduce its temperature to the new saturation equilibrium. In so doing, depending on the type of oil being used, the mixture may foam violently, the foam being comprised of oil droplets being carried by the refrigerant vapor. As this foamy mixture is pumped through the compressor valves, the droplets of liquid oil and the dense foam can create stresses far in excess of that imposed by vapor alone. There are ways to minimize this threat, and I'll deal with it more directly in discussing compressor application.

The bearing surfaces of a compressor must have a continuous film of oil for proper lubrication, to prevent metal to metal wear. In a conventional automotive engine, this is accomplished by means of an oil pump to create adequate pressure. In a refrigeration compressor, this process is tremendously complicated by the fact we are dealing with a closed system with no vents, with only refrigerant and oil within the system.
Further, the refrigerant is constantly circulating through the crankcase, the compressor must pump some oil in order to lubricate the pistons and cylinders and the only way this oil can be returned is to circulate it through the complete system, the refrigerant acting as a transport medium. Add to this the threat of the refrigerant in its liquid form entering the crankcase, either due to floodback or migration, and it becomes obvious that <u>lubrication</u> is a very complex process.

In order to withstand the extremes of temperature which are inherent in a refrigeration system, and at the same time retain the capability of flowing through a cold evaporator transported only by refrigerant vapor, refrigeration oils are highly specialized, highly refined oils, usually obtained from a napthenic base. The viscosity of the oil must be low in order to allow it to flow, and the danger exists that either at elevated temperature or when mixed with excessive amounts of liquid refrigerant, the viscosity may be reduced so low it loses its basic lubricity characteristics.

Typical refrigeration oils start to carbonize and break down at approximately 350°F (175°C), the lubricity drops to a danger point at about 200°F (93°C), and although there is no firm laboratory data, it is probable that lubricity becomes marginal when diluted with 8 to 10 % of liquid refrigerant. It is critical in system design that compressor operation be maintained within these parameters, and for long life and reliability the compressor must operate well away from the critical limits.

One other area of compressor design that is critical from a heat pump standpoint is motor protection. However this is primarily a problem only with single phase motors on air-to-air heat pumps.

At low outdoor ambient conditions, an air-to-air heat pump is very lightly loaded, and draws very little current. Since the conventional line break motor protector usually employed on air conditioning compressors responds primarily to running current, the protector may actually lose its ability to protect against excessive temperatures under low outdoor ambient conditions. To compensate for this condition, a new three lead protector has been developed which senses both start winding and run winding current. Under light load conditions, the motor tends to run faster with less slip, and in so doing develops a greater generating effect in the start winding, increasing the current flowing through the start winding. This increased current makes the protector far more sensitive to temperature and provided good protection against overheating due to loss of charge or other threatening conditions.

3. APPLICATION OF COMPRESSORS IN HEAT PUMPS

As a compressor manufacturer, we recognize our responsibility to provide a product suitable for this demanding usage which a heat pump application is. Based upon intensive laboratory testing, and on an intensive study of the critical stresses involved in heat pump operation, together with years of background experience in commercial refrigeration applications we designed a new line of heat pump compressors developed specifically for this demanding application.

It was mentioned before that motor insulation which deteriorated from liquid refrigerant was one of the past triggers to compressor failure. Today we use mylar slot liner and ester-amide wire insulation; none of which are vulnerable to liquid refrigerant.

The bearing surfaces were dramatically increased to reduce the loading so that even under marginal lubrication conditions we had an ample safety factor. New compressors were designed with ringed pistons so that we could allow a greater clearance between the piston and the cylinder. We developed a laboratory torture test for suction reeds, and modified our reed design and our valve porting until the valving could withstand stresses far beyond those involved with pumping vapor.

After extensive laboratory testing we changed to a white mineral oil in our heat pump compressors rather than the conventional naphthenic oils. We found that the solubility characteristics of the conventional oils contributed to excessive foaming, causing valve slugging and loss of oil from the compressor, while the white oil had low foaming characteristics much more suitable for heat pump usage. On single phase electrical motors in heat pump compressors, we went to the new three lead protector to provide better light load motor protection.

But as a compressor manufacturer we would be the first to admit that no matter what the design of the compressor, if the system is not designed properly the compressor is doomed to failure the day it is installed. So the second stage of our heat pump program was to develop sufficient expertise in heat pump design and operation so we could intelligently advise our customers on compressor application.

3.1. Specific Heat Pump related Problems

As mentioned earlier, once we have some appreciation of the compressors areas of vulnerability, we can consider the unpleasant facts of engineering life involved in heat pump application. These are :

1. An air-to-air heat pump operating in the heating cycle is more like a low temperature refrigeration unit than an air conditioner.

   The evaporating temperatures required to extract heat from the cold outdoor air under low outdoor ambient conditions are so low the system in effect becomes a low temperature refrigeration system, and as a result we have to concern ourselves about such things as excessive discharge temperatures, oil return, and hot gas defrost.

2. Welded compressor design is not well adapted to low temperature refrigeration.

   As mentioned previously, welded compressor design is such that heat is not easily dissipated from the compressor body and the motor, and the result is that a welded compressor in the size range suitable for heat pump usage if applied at low evaporating temperatures with conventional low temperature design criteria will develop temperatures so hot they will literally destroy the compressor.

   We can build a much more suitable compressor for low temperature application in the accessible-hermetic design, and it is working every day at no more demanding conditions in refrigeration usage. But its construction costs approximately twice as much as a welded compressor, and so long as it is possible to apply a welded compressor without an excessive failure rate, first cost alone will dictate the choice of a welded compressor.

   In order to adequately cool a welded compressor on an air-to air heat pump operating at low outdoor temperatures in the - 15°C range, we have to deliberately flood it with liquid refrigerant, an operating condition which is directly contrary to the beliefs, basic training, and experience of the thousands of servicemen upon whom we must rely to maintain the equipment. Laboratory tests have proven that some liquid refrigerant dilution of the crankcase oil - possibly from 5 % to 10 % - can be tolerated at heat pump operating conditions for reasonably prolonged periods with no apparent adverse effects on compressor lubrication.

   It appears that such desired flooding can be easily obtained with capillary tube control due to the inherent flooding characteristics of capillary tubes, while the good control characteristics of conventional thermal expansion valves may actually rule against their use.

   In flooding refrigerant to the compressor we expose it to potential hazards, both of a lubrication and mechanical nature, that we would rather avoid from a reliability standpoint. It can be done, but it means we must maintain a delicate balance between sufficient flooding to cool the compressor, and excessive flooding which could adversely affect lubrication with disaster looming if we let the system get out of control.

Happily the suction accumulator provides us with a reliable means of accomplishing that balance - more on that later.

3. R-22 is not a good refrigerant for high compression ratios.

   Because of its physical characteristics, R-22 which is almost universally used for air conditioning is not a good refrigerant for low temperature refrigeration or for high compression ratios. It's relatively low density as a vapor and its specific heat characteristics result in a much greater temperature response to heat than is true of refrigerants such as R-12 or R-502.

   In the air-to-air heat pump we can compensate for this by controlled liquid refrigerant flooding since the number of hours at critical conditions are relatively small. But in the air-to-water heat pump so popular in Europe, the compressor must operate continuously with a very high condensing temperature and a resulting high compression ratio. For long term reliability from a lubrication standpoint it is not desirable to reduce the lubricity of the oil under high loads for long periods by diluting the oil with liquid refrigerant.

   Typically water temperatures in the range of 60°C have been desired. Condensing temperatures to reach that water temperature level are so extreme they pose major reliability questions, and most manufacturers have been able to compromise with water temperatures of 50°C to 55°C. Even at that level, it is reasonable to anticipate that condensing temperatures of 63°C will be required.

   Even assuming a bivalent system with a secondary source of heat for lower ambient temperature, it is typical with a frosting evaporator that evaporator temperatures of - 10°C might be encountered.

   A convenient way of comparing the expected discharge temperatures at a given operating condition is by means of a pressure enthalpy diagram. Assume that we are evaporating at - 10°C. The vapor on entering the compressor comes in contact with the warm motor and compressor body, and becomes much warmer before entering the cylinder on the suction stroke. Laboratory data indicates a 60°C increase is typical, so the gas entering the cylinder for the compressor stroke is approximately 50°C. The peak discharge temperature occurs at the end of the compression stroke. For R-12, it is 132°C, for R-502 it is 127°C, but for R-22 it is 157°C (figures 4,5 and 6).

   A compressor will tolerate discharge temperatures in excess

of 150°C for short periods of time, but for long term reliability peak temperatures should be kept below that level.

Therefore we strongly recommend R-502 for air-to-water heat pump applications. Too many engineers look only at the conditions they wish to achieve, ignoring the practical realities of compressor life. We sometimes tend to forget that a compressor is essentially designed as a cooling and refrigeration device, not a heating mechanism.

4. Hot gas defrost causes liquid refrigerant floodback.

   The potential hazard of liquid refrigerant flooding back to the compressor at the initiation and termination of a hot gas defrost cycle is well known in commercial refrigeration work, but is a problem that is relatively new to engineers primarily involved in air conditioning. We must not ignore the fact that a compressor is not physically capable of digesting large amounts of liquid refrigerant without encountering stresses that if continued long enough, are almost certain to destroy the compressor.

3.2. Recommendations

With that background of general problem areas, what then are the specific application recommendations that we feel are critical to a long and happy compressor life in a heat pump.

1. Use a compressor designed for heat pump duty.

   There are differences between a compressor designed for air conditioning duty and one designed for heat pump usage. One should use a compressor designed for heat pump duty.

2. For air-to-air heat pumps, use a capillary tube to allow liquid refrigerant flooding at low ambient conditions.

   The operating characteristics of a capillary tube are such that essentially it is constant feed device. When applied to a heat pump, it typically results in liquid control with 5° to 8°C superheat at mild outdoor weather conditions, the superheat diminishing as the outdoor ambient drops, with liquid flooding occurring as the outdoor ambient approaches - 12°C. Thus the capillary tube has ideal characteristics from a compressor cooling standpoint for an air-to-air heat pump.

3. On an air-to-air heat pump, or any heat pump with a defrost

cycle, use a properly designed suction accumulator.

While the capillary tube can prevent overheating by liquid flooding, it tends to flood excessively at low ambient temperatures, and without some further means of control, the compressor crankcase may be literally swimming in liquid refrigerant. On units with a defrost cycle, at both the initiation and termination of the cycle, liquid refrigerant that has accumulated in the condensing coil will come roaring down the suction line like a river. Under extreme conditions the oil can be literally blown out of the compressor, and the resultant slugging may take the suction reeds along.

Obviously we need some device both to control liquid flooding to a rate the compressor can safely digest, and to prevent the surges of liquid during defrost from damaging the compressor.

A suction accumulator was the obvious candidate but laboratory testing indicated that most commercially available accumulators failed to provide adequate protection. Laboratory testing has proven that sizing of the oil return orifice is the critical key to suction accumulator design for heat pumps, and an orifice of approximately 1 mm size will provide excellent compressor protection during all phases of heat pump operation. An accumulator with an orifice in this size range will act as a receiver for excess liquid in the system, storing the excess in the accumulator when not required.

On an air-to-air heat pump, or any heat pump with a defrost cycle, use a low foaming oil.

To avoid the threat of loss of oil from the compressor or liquid slugging, we have found that a highly refined mineral oil, commonly called "white oil" or "water white oil" which has excellent low foaming characteristics can measurably improve the compressor's ability to withstand potential stresses from liquid refrigerant. We insist this be used in place of conventional oils in all of our heat pump compressors.

Use a high pressure control.

Studies of failure patterns on air-to-air heat pumps indicate that one of the major sources of unit malfunction and compressor failure are dirty filters on the indoor coil. As the filter becomes increasingly plugged, the discharge pressure increases on the heating cycle, increasing the operating compression ratio.

One of the less recognized operational pitfalls is that welded compressors used for this type application commonly have an internal pressure relief valve. With a dirty filter the compressor may generate enough pressure to pop the valve open, but since there is still some airflow through the dirty filter, the compressor may stabilize operation with the valve open, creating an almost continuous overheating condition.

On air-to-water systems, a similar problem may develop after a period of operation if the heating coil starts to accumulate scale, reducing its heat transfer capabilities. This becomes increasingly critical as the desired water temperature is increased.

6. Use a crankcase heater to prevent liquid refrigerant migration.

   Since the compressor on a heat pump must operate year round during all sorts of weather variations and for uncertain operating periods, it is going to be subject to repeated long off-cycles, when liquid migration to the crankcase can occur. A crankcase heater will keep the sump warm, preventing migration in mild conditions, and easing the starting burden on the compressor motor.

7. On air-to-air heat pumps, use continuous compressor operation on the heating cycle.

   In air-to-air heat pumps operating in geographical areas where low outdoor ambient temperatures are encountered, design engineers for years have felt they were enhancing the reliability of a system by using a low ambient lockout. Studies of failure patterns have convinced us that this is a mistaken belief, and greater reliability can be obtained with continuous operation.

   Admittedly the coefficient of performance at low ambient temperatures is not significantly better than resistance heat. However, if the compressor were to be cycled off as the outside temperature dropped, becoming thoroughly chilled during long hours exposed to the cold, the oil in the crankcase would become the consistency of molasses, the electrical controls would be exposed to ice and snow, and the probability exists when the temperature rose to the cut-in point.

   The available evidence would appear to indicate that there is less hazard to the compressor, and greater reliability can be obtained utilizing a heat pump compressor designed for high compression ratio operation, capillary tube control to avoid excessive temperatures, and no low ambient cut-out.

4. CONCLUSIONS

If the simple design precautions above are followed in any reasonably decent system design, you will have a reliable heat pump whether it is located in Norway or Italy. Ignore them, and I can almost guarantee you field problems. To illustrate this, let me give two specific examples.

Our standard SR line of welded compressors are designed for high volume production in the 2 to 2-3/4 H.P. range. In 1976 we modified the design on certain models to introduce a parallel heat pump model. In 1977 and 1978 we had high volume production of comparable air conditioning and heat pump models. Based on the increased hours of operation and potential hazards inherent in heat pump operation, it was the expectation of our engineering department that heat pump compressor failure rates would be at least twice that of air conditioning models. We keep fairly detailed records of in-warranty failure rates, and our actual experience contrary to our expectations has been that the heat pump compressors experienced a failure rate 1/2 that of the air conditioning compressors.

In the summer of 1977 we introduced our new CR line of compressors, designed from the beginning as a heat pump compressor. As a field trial, we purchased twenty-five 8,75 kW air-to-air heat pumps from major manufacturer, installed CR compressors, and installed these heat pumps in employees homes in the Sidney, Ohio area. The winters of 1977 and 1978 were some of the bitterest in Ohio history. We had prolonged periods of sub-zero temperatures, months of snow-covered ground, and temperatures to - 25°C and below.

The units were designed in accordance with our recommendations, and it turned out that we had a good heat pump test. Over the three years the units have been installed, we have experienced mass failures of defrost controls, fan motors, fan blades, fan motor brackets, and heating monitor controls, but we have not had one compressor failure. Despite the fact that there appears to be a crying need for more reliable accessory equipment, I think we have proven our point that heat pumps are a viable product in the most severe northern weather.

There are two dangers as I see them. First all of the emphasis today is on efficiency and there are areas where maxium efficiency and reliability are in conflict. If the industry is forced into a design concentrating solely on efficiency - then it is possible that reliability may be compromised.

The second danger as I see it is that traditionally the air conditioning industry has been highly price competitive, with possibly too much emphasis on first cost and not enough on long term costs of ownership. As the heat pump grows in volume, there will be those who may be tempted to trade system design features which insure reliability for a lower first cost, and if there is doubt and confusion in the minds of those who might be participating in such decisions then that conscientious design engineer who is trying to do a responsible job may come under increasing pressure to cheapen his system design.

The only really unhappy heat pump customers I have encountered in the past two years are those who required service on the air-to air refrigeration system in the winter months. The servicemen, with good reason, will not work on the refrigeration system until outside ambient temperatures are above 40°F (5 °C). I am convinced that if the air-to-air heat pump is to be a long term acceptable consumer product, the compressor must be located indoors so that it is serviceable under any weather conditions.

In general, the majority of the operating hours for heat pumps are in a fairly comparable operating range with air conditioning duty. Therefore compressors developed with higher efficiency characteristics are in general applicable to both types of duty. In the United States we face governmental regulation setting minimum efficiency standards for the end use product that are going to force us to the extreme limits of compressor design. Therefore any heat pump compressors being produced in the U.S. represent today the best present state of the art.

In reality the overall heat pump efficiency is much more dependent on system design than compressor design. Compressor operating curves demonstrate clearly the tremendous effect of changes in evaporating and condensing temperatures. Since both the compressor's capacity and efficiency decrease with a decrease in suction pressure, it is important to maintain as high an operating evaporating temperature as possible. Any pressure drop in the suction line from reversing valves or control devices can drastically effect the operating efficiency, and the designer must keep any unnecessary losses to a minimum.

It is doubtful if electric heat pumps are as yet economically competitive with natural gas, especially considering maintenance expense. But if the goal is to reduce the usage of oil and gas, regardless of economics, then the basic competitive comparison is between electric resistance heat which by definition has a coefficient of performance of 1.0, and a heat pump. Assuming there were no losses of heat to the atmosphere, a heat pump should always have a coefficient of performance greater than 1.0, since the heat rejected must be the sum of the power input and any refrigeration effect. As a practical matter, there are always some losses, and as the compressor refrigeration capacity drops rapidly with a decrease in suction pressure, the improvement over a coefficient of performance of 1.0 in an air-to-air heat pump at outside ambient temperatures of 0°F (- 18°C) and below is so minimum that differences in efficiency at that level are relatively meaningless.

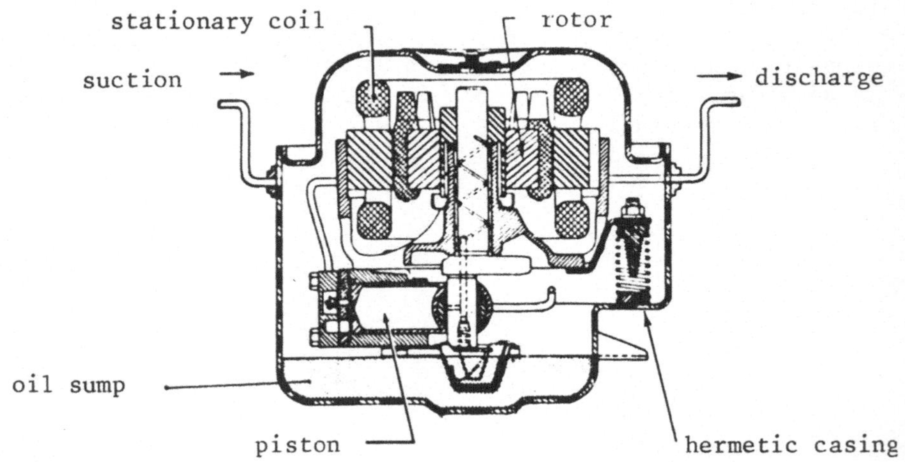

FIGURE 1

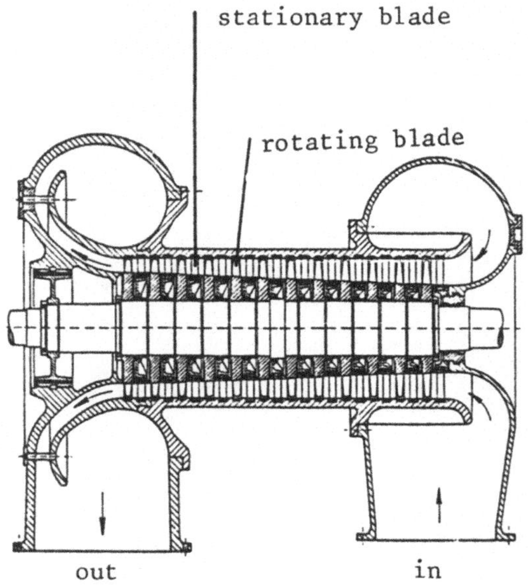

FIGURE 3

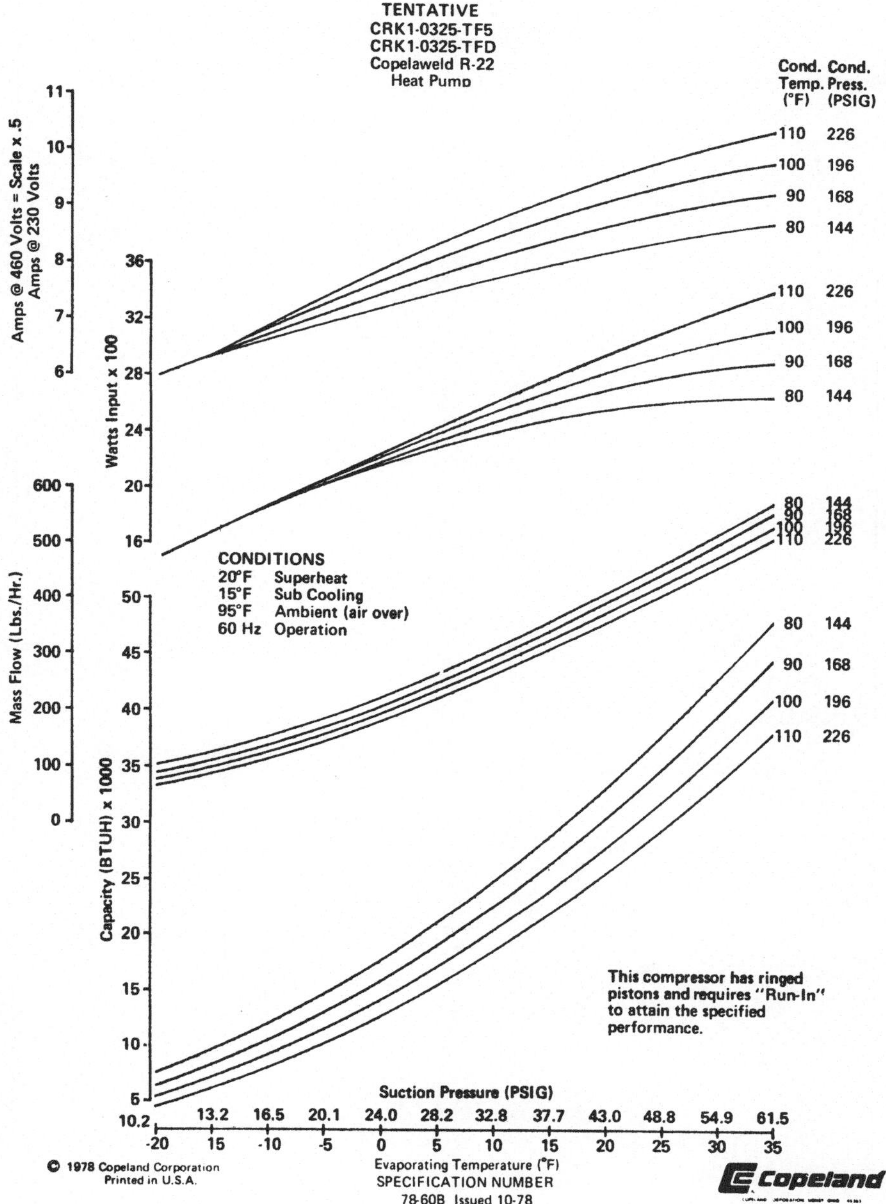
TENTATIVE
CRK1-0325-TF5
CRK1-0325-TFD
Copelaweld R-22
Heat Pump
Cond. Temp. (°F)
Cond. Press. (PSIG)
110 226
100 196
90 168
80 144
Amps @ 460 Volts = Scale x .5
Amps @ 230 Volts
Watts Input x 100
Mass Flow (Lbs./Hr.)
Capacity (BTUH) x 1000
CONDITIONS
20°F Superheat
15°F Sub Cooling
95°F Ambient (air over)
60 Hz Operation
This compressor has ringed pistons and requires "Run-In" to attain the specified performance.
Suction Pressure (PSIG)
10.2 13.2 16.5 20.1 24.0 28.2 32.8 37.7 43.0 48.8 54.9 61.5
-20 15 -10 -5 0 5 10 15 20 25 30 35
Evaporating Temperature (°F)
SPECIFICATION NUMBER
78-60B Issued 10-78
© 1978 Copeland Corporation
Printed in U.S.A.
Copeland

FIGURE 2

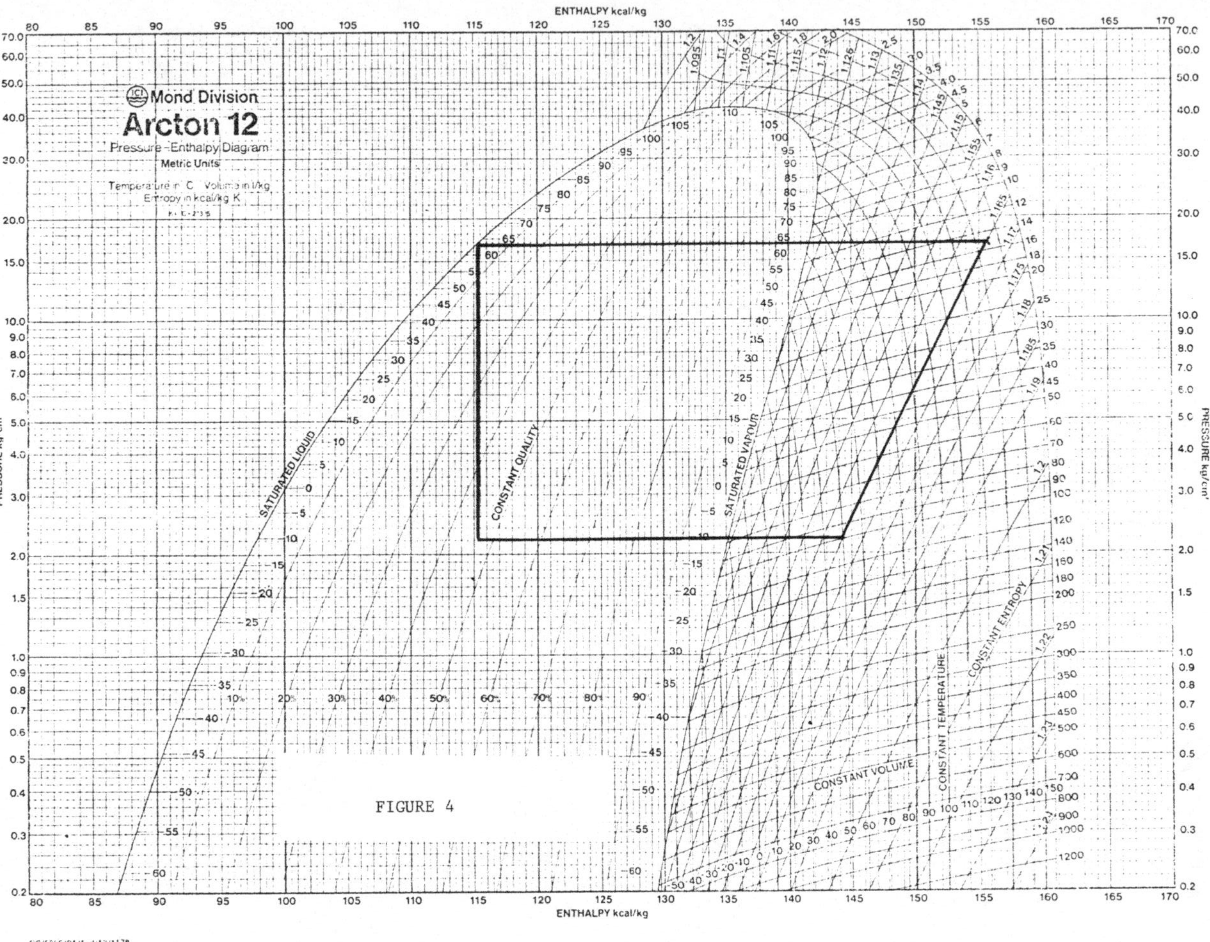
Mond Division
Arcton 12
Pressure-Enthalpy Diagram
Metric Units
Temperature in °C Volume in l/kg
Entropy in kcal/kg K
ENTHALPY kcal/kg
PRESSURE kg/cm²
SATURATED LIQUID
SATURATED VAPOUR
CONSTANT QUALITY
CONSTANT VOLUME
CONSTANT TEMPERATURE
CONSTANT ENTROPY

FIGURE 4

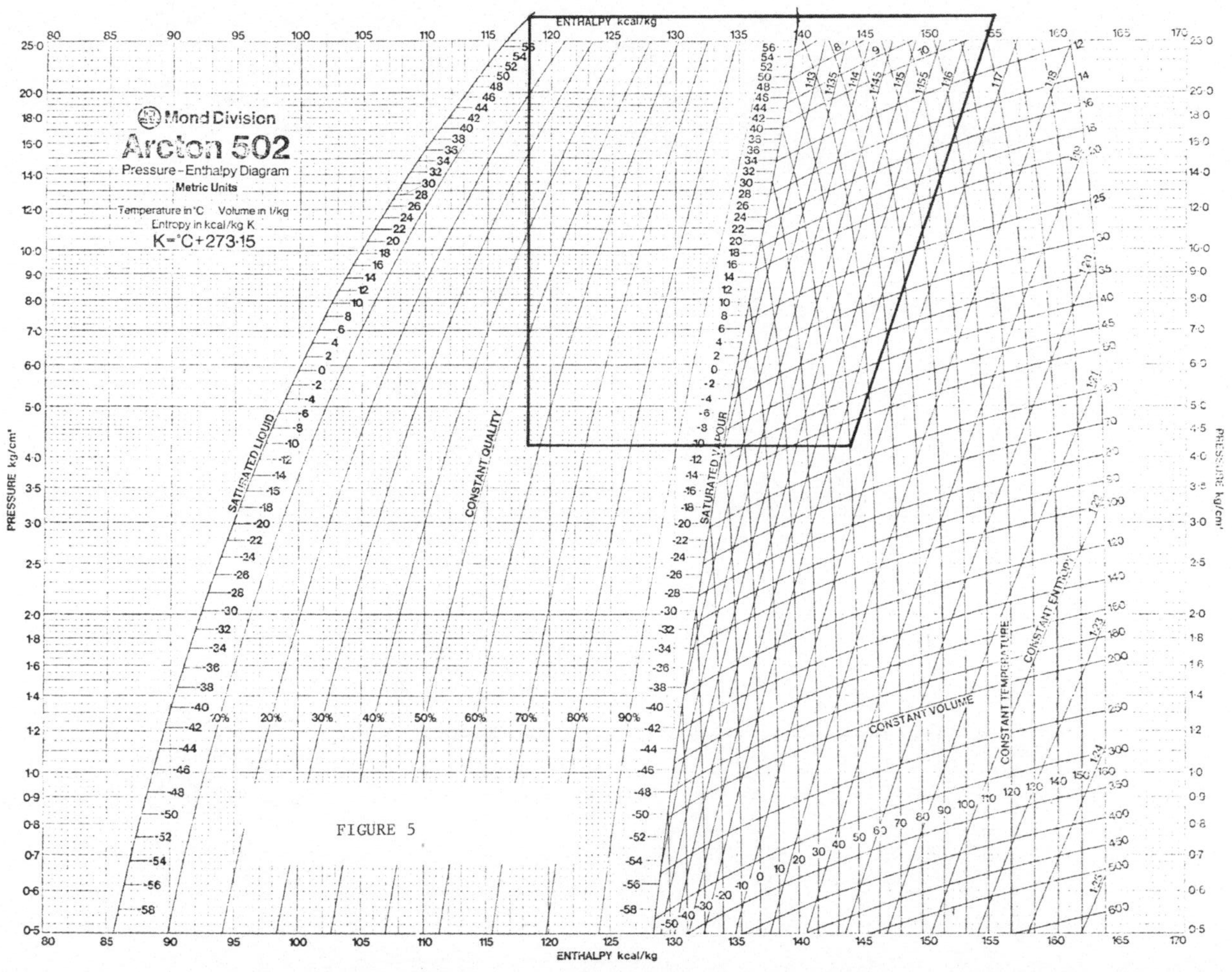

Mond Division
Arcton 502
Pressure - Enthalpy Diagram
Metric Units
Temperature in °C Volume in l/kg
Entropy in kcal/kg K
K=°C+273·15
PRESSURE kg/cm²
ENTHALPY kcal/kg
SATURATED LIQUID
SATURATED VAPOUR
CONSTANT QUALITY
CONSTANT VOLUME
CONSTANT TEMPERATURE
CONSTANT ENTROPY

FIGURE 5

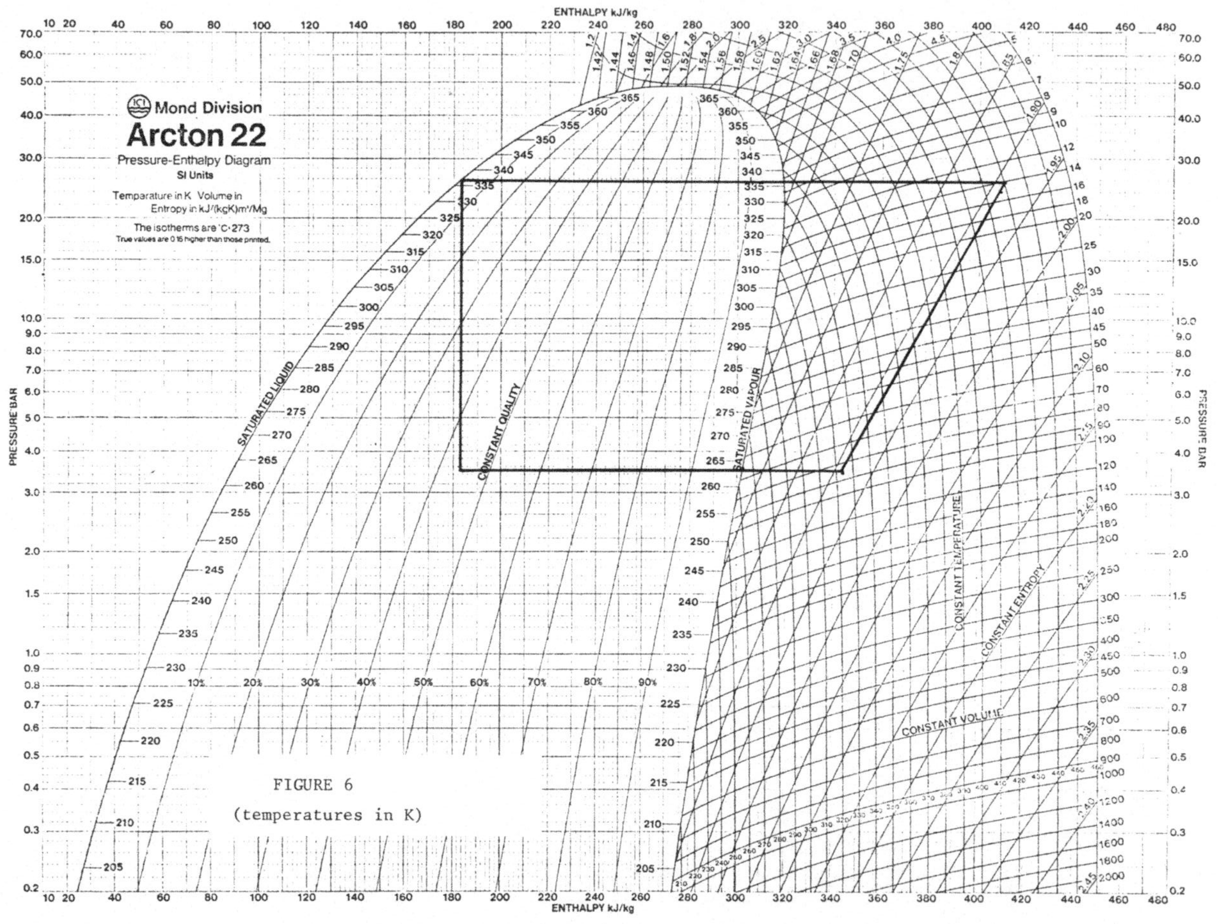

FIGURE 6
(temperatures in K)

HEAT PUMP MODELING, SIMULATION AND DESIGN

F.W. Ahrens

The Institute of Paper Chemistry
Appleton, Wisconsin, U.S.A.

This chapter begins with some introductory comments on the elements of heat pump modeling and simulation. The major portion of the chapter then deals with specific details of steady state modeling and simulation; a brief discussion of transient modeling is also presented. Finally, the chapter concludes with a section on heat pump design.

1. PRELIMINARY REMARKS ON HEAT PUMP MODELING AND SIMULATION

1.1. Definitions of Modeling and Simulation :

The goal of heat pump modeling is to describe the important aspects of the heat pump operation in terms of mathematical relationships. In particular, heat pump modeling refers to the development of models (i.e. assumptions, equations and data) which characterize the performance of all the discrete system components (*) (compressor, condenser, expansion device, evaporator). In general, these component models include mass, momentum and

---

(*) Most heat pump modeling is confined to the refrigeration system components, but an extended model might also include details of the heat source and sink (e.g. air of water circulation) systems.

energy balance equations, thermophysical property functions and appropriate (i.e. dependent on the specific component configuration) heat or work transfer relationships. Thus, a component model describes the relationship between the working fluid conditions within the component (e.g. flow rate, inlet and outlet thermodynamic states) and its design parameters and operating environment. In principle, the simultaneous solution of the model equations for all the components (subject to the constraint that the refrigerant system as a whole is closed) would yield a prediction of the heat pump operating behavior.

In turn, heat pump simulation refers to the implementation of the mathematical model relationships, and a numerical solution method, into a computer program which predicts the operating cycle and related information of interest, such as heating or cooling capacity, power consumption, COP, refrigerant flow rate, etc. An extended meaning of simulation is the process of using this computer model in predicting the effect of changes in design parameters or operating conditions (e.g. outdoor ambient temperature) on heat pump performance.

1.2. Approaches to Modeling

The approach often employed in modeling individual processes of heat transfer or fluid flow involves direct application of the partial differential equations of mass, momentum and energy with appropriate boundary conditions. In contrast, the usual approach to modeling heat pumps, or other complex thermal systems, attempts to characterize the numerous and varied processes by means of simpler algebraic equations (steady state analysis) or possibly ordinary differential equations (transient analysis). Even within this framework, however, an extremely broad range of models is possible, depending on the generality and accuracy demanded. Some factors influencing the size and complexity of the model include the level of detail with which each component is modeled, the degree of empiricism employed and the simplifying assumptions considered. For example, a very simple condenser model might neglect refrigerant pressure drop and might describe the heat transfer rate as some empirical constant times the difference between refrigerant saturation temperature and environment temperature. A very detailed condenser model, however, might utilize generalized correlations for inside and outside heat transfer coefficients, incorporate explicit fin efficiency equations, separately represent the heat transfer in the vapor, two-phase and liquid zones, include frictional and momentum pressure drops (which influence both the refrigerant temperature distribution and the outlet pressure condition), etc. Obviously the latter model would allow many condenser design variables to be studied and would be expected to do a better job of predicting off-design performance, but the former one would be far easier to implement into a computer simulation program and might be satisfactory for some limited purposes. Another point worth noting is that the latter model requires much more fundamental information (e.g. heat transfer correlations). If such information is not available for the particular heat exchanger configuration being modeled, there may be little value in retaining the highly detailed approach. Similar comments would apply for the other heat pump components.

1.3. Benefits of Modeling and Simulation.

Due to the quantitative, analytical process of modeling, its initial benefit is improved understanding of the fundamental technical details and interactions important to the performance of the heat pump. A related benefit is the identification of critical processes for which more or better technical information is needed. This can help to define research needs and priorities.

The use of heat pump simulation is beneficial both in research/development and in design. In the former area, variations in system or component configurations, working fluid, etc., can be evaluated in order to identify significant opportunities for improvement. When a heat pump simulation program is sufficiently accurate and detailed, its use greatly reduces the time and cost in designing a heat pump for a given application, offers the possibility for achieving a more nearly optimal design, via the consideration of more alternatives, and enables the off-design performance to be investigated prior to actually building the unit. Although most heat pump simulation programs deal with steady state operation, transient simulations, based on solution of unsteady forms of the model equations, are also of interest; these would be useful in analyzing the startup process, the defrost cycle, rapidly-varying operating conditions, etc.

2. STEADY STATE MODELING AND SIMULATION.

One might define a steady state heat pump model as a set of mathematical relationships and data sufficient to uniquely locate (for specified constant indoor and outdoor conditions) the time-independent refrigerant cycle on a thermodynamic (e.g. pressure-enthalpy) diagram and to determine the corresponding refrigerant flow rate (or heating capacity). As discussed earlier, models describing various levels of physical detail may meet this requirement, but certainly these different models of a particular heat pump will yield quantitatively different predictions of the operating cycle. Ideally, if enough detail is included in the model, and the parametric data required is accurate, the predicted behavior would closely match that of the actual heat pump.

The choice of level of detail in modeling/simulation thus depends on required accuracy. For some purposes, only broad trends are of interest (e.g. an estimate of the change in heating capacity with outdoor temperature or with nominal compressor size) and a relatively simple model may suffice. On the other hand, a manufacturer of mass-produced residential heating and cooling equipment might strive for a highly-detailed simulation, with a high degree of accuracy, so that the effects of many design parameters and alternatives can be evaluated with confidence.

In this section, a very simplified heat pump model is first presented. Then, the structure and component model characteristics for a reasonably general and detailed heat pump model/ simulation are discussed. Possible extensions or modifications are also indicated.

2.1. A Simplified Model.

The following heat pump model illustrates some of the ingredients of more general models, while being simple enough that the equations can be solved explicitly, avoiding the need for a numerical solution procedure. It is presented primarily for the purpose of introducing the concept of heat pump modeling, but should also be of value in indicating the effect of the overall component "size" parameters on heat pump performance. Since it is a linear model (with respect to the refrigerant), it can also be viewed as a model which predicts the effects of small changes in design or operating variables, relative to a nominal condition which is known. For this viewpoint to be valid, one must assume that a nominal set of parameter values and nominal operating cycle are known and compatible with the model equations.

Referring to the idealized refrigerant cycle in Figure 1, the primary objective of the heat pump model is to predict the evaporation temperature (T_E) and condensation temperature (T_C) in terms of heat pump component performance parameters and of the indoor and outdoor environmental conditions. The subcooling of the liquid leaving the condenser (SC) and the superheat of the vapor entering the compressor (SH) are assumed to be constant (and small) and are not explicitly included in the model.

An air-to-air electric heat pump in the heating mode is assumed in this analysis, with the following air conditions :

Ta_E = outdoor air temperature entering evaporator

$\dot{m}_E$ = air mass flow rate through evaporator

Ta_C = indoor air temperature entering condenser

$\dot{m}_C$ = air mass flow rate through condenser

c_p = specific heat of air at constant pressure

The air flowing through the evaporator is assumed to be relatively dry, so latent heat transfer (moisture condensation) is neglected.

The compressor is modeled in terms of parameters of its performance curves, which are assumed to be available from the manufacturer. Two performance maps are usually available, one giving the "evaporation capacity" (or, equivalently, the mass flow rate) and the other giving the electrical power consumption, both as functions of T_E and T_C. If it is assumed that the actual operating point (T_E, T_C) is not vastly different from a nominal point (T_{E_o}, T_{C_o}) at which the evaporation capacity and power consumption have known values, C_{E_o} and P_o, the performance curves may be approximated by the linear relationships :

$$C_E = C_{E_o} \left[1 + \alpha_c \left(T_E - T_{E_o}\right) - \beta_C \left(T_C - T_{C_o}\right)\right] \tag{1}$$

and

$$P = P_o \left[1 + \alpha_p \left(T_E - T_{E_o}\right) + \beta_p \left(T_C - T_{C_o}\right)\right] \tag{2}$$

The above equations have been written so that the slope parameters (α's and β's) are positive. It is worth noting that the α's and β's, as well as the ratio $\zeta \equiv P_o/C_{E_o}$, tend to remain relatively constant over a range of compressor sizes within a given family (similar design) of compressors. Thus, the single parameter C_{E_o}, representing the compressor size, might be considered as the fundamental compressor design variable of the present heat pump model.

For completeness, the relationship between the compressor "evaporation capacity" (C_E) and a more fundamental compressor quantity, the refrigerant mass flow rate ($\dot{m}_R$) is also presented. If the compressor is inserted into a heat pump to yield the idealized cycle shown in Figure 1, the rate of energy absorption in the evaporator (C_E) is related to $\dot{m}_R$ by :

$$C_E = (\Delta h)_E \, \dot{m}_R \tag{3}$$

where : $(\Delta h)_E$ = refrigerant enthalpy rise from evaporator inlet to compressor inlet.

Considering as isenthalpic expansion between the condenser outlet and the evaporator inlet, neglecting pressure drops in the evaporator and condenser, and assuming that SC and SH remain fixed for all values of T_E end T_C, $(\Delta h)_E$ is seen to be a function of T_E and T_C only. Similarly, since SH is fixed, $\dot{m}_R$ is primarily a function of T_E and T_C, since these reflect the refrigerant pressure levels at the suction and discharge sides of the compressor, both of which influence physical processes within the compressor. The mass flow rate tends to increase with increasing suction vapor density (and thus with T_E). It tends to decrease with increasing discharge pressure (and thus with T_C), due to back-leakage from the high pressure side toward the low pressure side. Combining the various points in the above discussion, one can verify the qualitative correctness of Equation 1 and realize that the "compressor capacity" data presented by some manufacturers is really a means of characterizing the mass flow rate.

The evaporator model is based on simple heat exchanger theory [1]. It is assumed that there is no refrigerant pressure drop and that the possibility of a region of superheated vapor near the outlet has a negligible effect. Finally, it is assumed that the overall heat transfer coefficient between the refrigerant and the air is uniform. Under these conditions, the evaporator heat transfer rate can be written [1] :

$$Q_E = \dot{m}_E \, c_p \, \varepsilon_E \, (T_{a_E} - T_E) \tag{4}$$

where : $\varepsilon_E = 1 - \exp\,(- \overline{UA}_E / \dot{m}_E \, c_p)$ = evaporator effectiveness

$\overline{UA}_E$ = product of overall heat transfer coefficient and evaporator heat transfer area

Thus, the quantities $\dot{m}_E$ and $\overline{UA}_E$ are the fundamental design parameters in the present evaporator model. A simple energy balance shows, from Equations (3) and (4) :

$$Q_E = C_E \tag{5}$$

Applying a similar analysis to the condenser yields the condenser heat transfer rate to the air :

$$Q_C = \dot{m}_C \, c_p \, \varepsilon_C \, (T_C - T_{a_C}) \tag{6}$$

where : $\varepsilon_C = 1 - \exp\,(-\,\overline{UA}_C/\dot{m}_C \, c_p)$ = condenser effectiveness

$\overline{UA}_C$ = product of overall heat transfer coefficient and condenser heat transfer area.

This heat is transferred from the refrigerant, giving a condenser capacity as :

$$C_C = \dot{m}_R \, (\Delta h)_C = Q_C \tag{7}$$

where : $(\Delta h)_C$ = magnitude of refrigerant enthalpy change between compressor outlet and condenser outlet.

Finally, a refrigerant energy balance for the complete heat pump cycle yields :

$$C_C = C_E + \delta P \tag{8}$$

where : δ = fraction of P actually delivered to the refrigerant.

It is assumed that an estimate of δ is available for the particular compressor type being considered. If not, $\delta = 1$ can be used. Manipulation of Equations (1), (2) and (4-8) leads to two linear equations in the two unknowns, T_E and T_C. Solving these yields :

$$T_C - T_{C_o} = \frac{A_1}{A_3} \tag{9}$$

and

$$T_E - T_{E_o} = \frac{A_2}{A_3} \tag{10}$$

with : $$A_1 = (C_{E_o} \, \alpha_C + \delta \, P_o \, \alpha_p) \, \dot{m}_E \, c_p \, \varepsilon_E \, (T_{a_E} - T_{E_o})$$

$$- (C_{E_o} \, \alpha_C + \dot{m}_E \, c_p \, \varepsilon_E) \, \dot{m}_C \, c_p \, \varepsilon_C \, (T_{C_o} - T_{a_C})$$

$$+ C_{E_o} \, \delta \, P_o \, (\alpha_C - \alpha_p) + (C_{E_o} + \delta \, P_o) \, \dot{m}_E \, c_p \, \varepsilon_E$$

$$A_2 = C_{E_o} \, (\delta \, P_o \, (\beta_C + \beta_p) - \dot{m}_C \, c_p \, \varepsilon_C)$$

$$- (\delta \, P_o \, \beta_p - C_{E_o} \, \beta_C - \dot{m}_C \, c_p \, \varepsilon_C) \, \dot{m}_E \, c_p \, \varepsilon_E \, (T_{a_E} - T_{E_o})$$

$$- C_{E_o} \, \beta_C \, \dot{m}_C \, c_p \, \varepsilon_C \, (T_{C_o} - T_{a_C})$$

$$A_3 = C_{E_o} \alpha_C (\dot{m}_C c_p \varepsilon_C - \delta P_o \beta_p)$$

$$- \dot{m}_E c_p \varepsilon_E (\delta P_o \beta_p - \dot{m}_C c_p \varepsilon_C - C_{E_o} \beta_C)$$

$$- C_{E_o} \beta_C \delta P_o \alpha_p$$

Once T_C and T_E are calculated, all the other quantities of interest (Q_C, P, COP, etc.) may be determined from the equations presented.

The simplified model results which have been derived, although linear in the refrigerant temperatures, do explicitly exhibit the nonlinear effects of equipment design parameters like heat exchanger sizes, air flow rates and compressor characteristics on the heat pump performance. It could thus be used to investigate tradeoffs between different design parameters in achieving desired heating capacity and COP targets. One aspect not considered, which could be readily incorporated, is the power consumption of the fans which produce the air flows (typically about 10-15 % of the compressor power). In most cases, one would base the actual COP on the total power consumption of compressor <u>and</u> fans.

2.2. A General Structure for Heat Pump Simulation

The purpose of the following material is to describe a relatively general approach to heat pump simulation (steady state). Subsequently, a more detailed discussion of specific component models which can be incorporated into the simulation will be given.

A schematic diagram of the heat pump considered and the corresponding 'real' refrigerant cycle are shown in Figures 2 and 3. It is noted that an air-to-air configuration, having an adiabatic, fixed-restriction expansion device (capillary tube) and no accumulator, is assumed. This choice is made for convenience in illustrating various ideas of simulation and modeling, and does not imply that such a configuration is particularly well-suited for general applications. After considering the modeling and simulation of this system, however, the modifications or extensions necessary for handling other configurations should become apparent.

In developing a heat pump simulation, it is useful to distinguish between operation at a nominal design point condition(*), for

(*) The design point refers to operation under 'standard' indoor and outdoor conditions.

which certain system parameters are selected to provide a refrigerant cycle with desired attributes, and off-design operation (with a fixed heat pump). For discussion purposes, it is assumed that, in either of these operating modes, the type (configuration) and size have been selected (tentatively, perhaps) for the evaporator, condenser and compressor, and that the refrigerant has been selected. It is further assumed that, at the design point, desired values of the refrigerant superheat ($\Delta T_{SH} = T_1 - T_A$) and subcooling ($\Delta T_{SC} = T_C - T_3$) have also been selected (again tentatively, perhaps). These latter two refrigerant cycle constraints, it may be observed, replace the specification of the capillary tube configuration and the total mass of refrigerant in the system. In the "design mode", the task of the simulation is to identify the operating cycle for which these constraints and the appropriate evaporator, condenser and compressor model equations are satisfied. After the simulation results are obtained, they can be used in selecting an appropriate capillary tube size and refrigerant 'charge', thus completing the specification of a system which, when operated at the design point, has desired values of superheat and subcooling. The ability to avoid specifying the capillary tube configuration prior to the design point simulation results from the fact that the required capillary tube model is simply the statement that the expansion process is isenthalpic.

With the heat pump now totally specified, its performance under off-design conditions may be investigated. In this situation, the values of ΔT_{SH} and ΔT_{SC} are no longer free to be specified. Instead, the simulation must be given the task of finding that operating cycle (including values for ΔT_{SH} and ΔT_{SC}) for which not only the evaporator, condenser and compressor equations are satisfied, but also for which the mass flow rates through the capillary tube and compressor are equal and for which the total mass of refrigerant in the system is unchanged from the value determined at the design point.

A general (but somewhat simplified) simulation program structure for handling both design and off-design conditions in the manner just discussed is indicated in Figure 4. The air system simulation section is optional (i.e. one can either view the evaporator and condenser air flow rates as input parameters to be specified or as quantities to be predicted via a separate simulation of the air system).

The iterative method for convergence to the steady state solution at the design point is based on the recognition that two cycle variables may be independently adjusted in seeking to satisfy the constraints on ΔT_{SH} and ΔT_{SC}. The compressor inlet and discharge pressures(*) (p_1 and p_2) are chosen, since they have a direct influence on the compressor mass flow rate, as well as on the evaporating and condensing temperatures (and thus heat transfer rates).

Because of the iterative nature of the solution process, it is necessary to provide an initial guess for the variables p_1 and p_2. Initial values for any other cycle thermodynamic variables may be based on ideal cycle values. At any stage in the iterations, the compressor, evaporator and condenser models are called to calculate the information necessary to test the energy balance constraints :

$$C_E \equiv \dot{m}_R (h_1 - h_4) = Q_E \qquad (11)$$

and

$$C_C \equiv \dot{m}_R (h_2 - h_3) = Q_C \qquad (12)$$

where $\dot{m}_R$ is the refrigerant mass flow rate and Q_E, Q_C are evaporator and condenser heat transfer rates.
Throughout the calculations, the capillary tube constraint

$$h_4 = h_3 \qquad (13)$$

and the constraints

$$T_1 - T_A = \Delta T_{SH} \qquad (14)$$

and

$$T_C - T_3 = \Delta T_{SC} \qquad (15)$$

must be imposed. The dashed lines shown between the component models in Figure 4 reflect the fact that some "internal iterations" may be needed in order to achieve self-consistent flow rate, heat transfer rate and thermodynamic state point predictions.

If the energy balances are not both met within some desired level of accuracy, it is necessary to revise p_1 and p_2 (or ET end CT) and try again. The revisions are best obtained by use of a numerical derivative version of the Newton-Raphson method [2].

(*) In practice, the saturation temperatures (ET and CT) corresponding to these pressures have been used as the independent variables, for convenience, because the compressor performance characteristics are usually given in terms of ET and CT.

If residuals F_1 and F_2 are defined by :

$$F_1 = C_E - Q_E = F_1\ (ET,\ CT) \tag{16}$$

and

$$F_2 = C_C - Q_C = F_2\ (ET,\ CT) \tag{17}$$

then, at the solution,

$$F_1 = F_2 = 0 \tag{18}$$

Assuming that the "old" values of ET and CT are close to the correct values, so that F_1 and F_2 are linear functions of ET and CT, new values of ET and CT which cause equation (18) to be satisfied can be developed. For example,

$$ET_{new} = ET_{old} - \left[\frac{F_{1_{old}} - \dfrac{f_{12}\ F_{2_{old}}}{f_{22}}}{f_{11} - \dfrac{f_{12}\ f_{21}}{f_{22}}} \right] \tag{19}$$

where, $f_{11} = \frac{\partial F_1}{\partial ET}$, $f_{12} = \frac{\partial F_1}{\partial CT}$, $f_{21} = \frac{\partial F_2}{\partial ET}$, $f_{22} = \frac{\partial F_2}{\partial CT}$

These partial derivatives are evaluated at the point (ET_{old}, CT_{old}) and may be calculated numerically. For example :

$$f_{11} \approx \frac{F_1\ (ET + \Delta,\ CT) - F_1\ (ET,\ CT)}{\Delta} \tag{20}$$

where Δ is a small increment in the temperature ET.

The evaluations, of course, require appropriate calls to the component models.

Upon achieving the design point solution (both evaporator and condenser energy balance criteria satisfied), a capillary tube model may be used to design a tube which provides the flow rate needed by the system at the known inlet and outlet conditions (points 3 and 4 in Figure 3). Furthermore, since the distribution of thermodynamic states around the system is known, the local densities are readily calculated. So, if the internal volumes of the components are also known, the total mass of refrigerant required by the system can be predicted. The capillary tube and refrigerant "charge" information is now available for use in the off-design part of the simulation.

An off-design solution is obtained (for given new values of indoor and outdoor conditions or, perhaps, for different air flow rates) by an iterative series of "design point" simulations. In these, the independent variables are now ΔT_{SH} and ΔT_{SC}, which are adjusted (by the numerical Newton-Raphson technique) until the following constraints are approximately satisfied :

$$\dot{m}_{R_{C.T.}} = \dot{m}_{R_{COMP}} \qquad (21)$$

and

$$M_{TOT} = M_{TOT}^{*} \qquad (22)$$

where : $\dot{m}_{R_{C.T.}}$ = mass flow rate predicted by a capillary tube model

$\dot{m}_{R_{COMP}}$ = mass flow rate predicted by the compressor model

M_{TOT} = total mass of refrigerant required by system

M_{TOT}^{*} = mass of refrigerant required at original design point.

The implementation of the simulation strategy which has been outlined requires that a number of computer subroutines be prepared. The major subroutines needed, together with a brief description of their purposes and requirements, are indicated in Table 1. The primary purpose of the MAIN program is to execute the solution strategy shown in Figure 4. Equations and subroutines describing thermodynamic properties are available from refrigerant manufacturers and in the literature [3]. The subroutine DIMEN, as defined in Table 1, assumes that a rather detailed mathematical description of the heat exchangers is being employed. Less detailed coil models may not require a separate DIMEN subroutine.

2.3. Component Models

The principal calculations to be performed by the component model subroutines needed in simulating the heat pump depicted in Figure 2 are indicated in Table 1. Now, some of the more important details to be considered in developing these component models are discussed.

Compressor :

The compressor is by far the most complex component in the heat pump. Rotary and reciprocating electric motor-driven compressors predominate for most small (domestic) and intermediate heat pump sizes; the present discussion is restricted to these types.

In principle, a highly-detailed compressor model could be imbedded in the heat pump simulation. Many such models are available [4, 5], accounting for the various flow, heat transfer and mechanical processes which occur within a compressor. Unfortunately, the size and complexity (i.e. numerous design parameters, detailed predictions) of such a compressor simulation model would tend to overshadow the rest of the heat pump simulation, suggesting that a less-detailed model would be more prudent for most purposes. Of course, a simplified model which relates the overall performance quantities of interest (see Table 1) to the operating conditions(*), for a fixed set of compressor design parameters, could be derived from the results of a detailed compressor simulation.

The simplest model for the class of positive-displacement compressors considered here would predict the mass flow rate, discharge temperature and power requirement in terms of the corresponding ideal values and appropriate volumetric, isentropic, mechanical and motor efficiency parameters. However, the efficiencies required in such a model are not constants, but depend on the operating conditions, thus requiring that empirical data be incorporated into the model. The source of this data might be the results of a detailed compressor simulation, but in most cases it would be the manufacturer's performance data.

As a practical matter, there seems to be little advantage to attempting to extract efficiency values from empirical performance data. Rather, it appears to be simpler and generally adequate to use curve-fitting or interpolation techniques to employ the data more directly in the compressor model. The subsequent discussion will consider this approach.

The completeness and method of presentation of performance data appears to vary, depending on the manufacturer, the compressor size/application, etc. The mass flow rate data (needed in predicting the evaporator and condenser performance) might be available directly or it might have to be deduced from "evaporation capacity curves". Assuming the latter situation, the capacity if often defined as the refrigeration rate associated with operating the compressor in an ideal cycle, having certain "standard" inlet temperature (T_I), condenser subcooling and ambient temperature values. The capacity (C_{std}) is essentially just a function

(*) Note that the compressor operation is a function of the three independent variables p_1, T_1 and p_2 (see Figure 3).

of ET and CT in this case, and is readily related to the corresponding mass flow rate at the "standard" conditions :

$$\dot{m}_{R_{std}} = C_{std}/\Delta h_{std} \qquad (23)$$

where Δh_{std} is the evaporator enthalpy change for the standard, ideal cycle at the actual ET, CT conditions. The mass flow rate given by equation (23) would likely be within a few percent of the actual value ($\dot{m}_R$) over a reasonable range of inlet temperatures. Some manufacturers are able to furnish data to correct for this inlet temperature (suction side vapor density) effect :

$$\dot{m}_R = F\ \dot{m}_{R_{std}} \qquad (24)$$

$$F = \text{correction factor} = F\ (ET,\ T_1)$$

To summarize, the mass flow rate "model" consists of interpolating between data points corresponding to selected combinations of ET, CT, in order to find the value of C_{std} (or $\dot{m}_{R_{std}}$, if this data is directly available) at the actual ET, CT (p_1, p_2) operating point being considered, and then applying equations (23) (if needed) and (24). The value of F could also be found by interpolation.

The compressor electrical power input and the electric current are also useful quantities to be predicted by the compressor model, as a function of operating conditions. The former can be used in determining the actual COP of the heat pump, while the latter can be checked against any limiting (constraint) value which may be in effect for the installation site. In both cases, a simple interpolation or curve-fit of the available manufacturer data is recommended.

The next compressor quantity required is the discharge temperature (T_2), which influences the condenser heat transfer calculations. If manufacturer data on "heating capacity" (C_H) is available for the "standard" cycle conditions, one could use the definition :

$$C_H = \dot{m}_{R_{std}}\ (h_2 - h_3)_{std} \qquad (25)$$

to solve for $h_{2_{std}}$, since the other quantities would all be known. The discharge enthalpy ($h_{2_{std}}$), together with the CT value, specify the discharge temperature ($T2_{std}$) which would occur if the compressor were at the standard inlet temperature. As an approximation, this value could be corrected by the equation :

$$T_2 \approx T_{2_{std}} + (T_1 - T_{1_{std}}) \qquad (26)$$

The appropriate C_H value for use in equation (25) would be found by interpolation.

If data for C_H is unavailable, an alternative way of estimating T_2 would be to use the following polytropic "process" relation :

$$T_{2_{abs}} = T_{1_{abs}} (p_2/p_1)^{\frac{n-1}{n}} \qquad (27)$$

where :

$T_{2_{abs}}$ = absolute temperature of refrigerant at compressor discharge

$T_{1_{abs}}$ = absolute temperature of refrigerant at compressor inlet

p_2 = absolute pressure at compressor discharge

p_1 = absolute pressure at compressor inlet

n = an empirically determined "polytropic exponent".

It can be assumed that n is proportional to the mean isentropic exponent :

$$n = c_1 \bar{\delta} = c_1 \left(\frac{\delta_1 + \delta_2}{2}\right) \qquad (28)$$

where :

$\delta_1 = c_p/c_v$, evaluated at inlet conditions

$\delta_2 = c_p/c_v$, evaluated at discharge conditions

Use of c_1 = 0.9 to 1.0 is often adequate.

The final aspect of the compressor model is the calculation of the mass of refrigerant inside the compressor at the operating point. This requires that the internal free volumes on the suction and discharge sides be known and that the corresponding vapor densities be estimated. A more subtle contribution to be included is the mass of refrigerant dissolved in the <u>oil</u> inside the compressor. Curve fits of solubility data [6] can be used to predict this contribution, as a function of estimated oil temperature and pressure and mass of oil present.

Evaporator :

In contrast to the compressor model, the evaporator model to be described is relatively fundamental in character. An idealized diagram of the system considered is shown in Figure 5. Note that the possibility of a superheated vapor section in the evaporator proper and a vapor "suction line" between the actual evaporator and the compressor inlet are included, and that a cross-flow fin-and-tube heat exchanger is assumed. In the following discussion, no specific correlations or equations for air-side and refrigerant-side heat transfer coefficients, refrigerant pressure drop, or fin efficiency will be given. It is assumed that these relationships have been determined, from the literature [e.g. 7-13] or other sources, for the specific configuration being considered. It is also assumed that, within each zone (i.e. air side, superheated vapor section, two-phase section) a correlation will be used which predicts the "average" heat transfer coefficient for that zone based on the appropriate flow rate and "average" fluid state (including quality, in the two-phase region). In effect, this approach requires that an iterative solution, involving heat transfer and refrigerant pressur drop calculations, and the compressor mass flow rate model, must be performed so that self-consistent values of flow rate and fluid properties are used in the correlations. Consistent with this approach, a mean refrigerant temperature (denoted T_{EVP} in Figure 3) is used in computing the total heat transfer rate. For simplicity,

$$T_{EVP} \approx \frac{1}{2} (T_A + T_4) \tag{29}$$

is recommended, since only small pressure (saturation temperature) drops would be allowed in a practical system.

With reference to Figure 5, the key task of the evaporator model, preliminary to the calculation of the heat transfer rate, is the prediction of the evaporating fraction, f_E. This quantity can be interpreted as the fraction of the total evaporator heat transfer area, and as the fraction of the total evaporator air flow rate, which is actually utilized in the evaporation heat transfer process. With respect to pressure drop calculations, it can also be viewed as the fraction of the evaporator tube length causing a two-phase flow pressure drop. The determination of f_E is done by a process of elimination. That is, the fraction (f_{SH}) of the heat exchanger required to provide the specified vapor superheat (ΔT_{SH}) is found first. Then, by definition :

$$f_E = 1 - f_{SH} \tag{30}$$

The first step in calculating f_{SH} is to find $T_{1'}$ (see Figure 5).

Assuming the ambient air seen by the suction line is at constant temperature ($T_{A_{SL}}$), an elementary heat transfer analysis yields :

$$T_{1'} = T_1 \, e^{NTU_{SL}} - T_{A_{SL}} \, (e^{NTU_{SL}} - 1) \qquad (31)$$

where :

$NTU_{SL} = (UA)_{SL}/\dot{m}_R \, c_{p_{RV}}$

$(UA)_{SL}$ = overall heat transfer coefficient-surface area product for suction line

$c_{p_{RV}}$ = specific heat of the vapor

The amount of superheating occurring in the evaporator is thus :

$$(T_{1'} - T_A) = \Delta T_{SH} - (T_1 - T_{1'}) \qquad (32)$$

It is assumed that the fins on the vapor section tubes of the evaporator are exposed only to air at the entering temperature (T_{a_E}). With this model, it can be shown that :

$$f_{SH} = \frac{\dot{m}_R c_{p_{RV}}}{(UA)_{SH}} \, \ell n \, \left(\frac{T_{a_E} - T_A}{T_{a_E} - T_{1'}}\right) \qquad (33)$$

where :

$(UA)_{SH}$ = overall heat transfer coefficient - surface area product based on <u>total</u> evaporator operating with superheated vapor

Having established a value for f_E (= $1 - f_{SH}$), the heat transfer rate in the two-phase (evaporating) section can be calculated. It can be shown that, neglecting refrigerant temperature variation in the two-phase zone, the heat transfer effectiveness for the evaporating section is independent of f_E. Therefore, it can be calculated from :

$$\varepsilon_E = 1 - e^{-NTU_E} \qquad (34)$$

where :

$NTU_E = (UA)_E/\dot{m}_{a_E} \, c_{p_a}$

$(UA)_E$ = overall heat transfer coefficient - surface area based on <u>total</u> evaporator operating with evaporating refrigerant

$\dot{m}_{a_e}$ = total air mass flow rate through evaporator

c_{p_a} = specific heat of air

The effect of $f_E < 1$ enters the analysis only by influencing the effective air flow rate in the two-phase section heat transfer rate equation :

$$Q_{E_{TP}} = f_E \, \dot{m}_{a_E} \, c_{p_a} \, \varepsilon_E \, (T_{a_E} - T_{EVP}) \qquad (35)$$

Equation 35 assumes that air dehumidification effects are negligible.

Finally, combining an energy balance on the vapor section with the above expression yields the total evaporator heat transfer rate :

$$Q_E = f_E \, \dot{m}_{a_E} \, c_{p_a} \, \varepsilon_E \, (T_{a_E} - T_{EVP}) + \dot{m}_R \, c_{p_{RV}} \, \Delta T_{SH} \qquad (36)$$

It is this value which must be compared with the evaporation capacity ($\dot{m}_R$ $(h_1 - h_4)$) in testing for convergence of the heat pump simulation iterations.

The final prediction to be made by the evaporator model is the mass of refrigerant contained in the evaporator as a function of the operating conditions. The portion contained in the suction line and superheated section of the evaporator are readily calculated from the appropriate volumes (using f_{SH}) and vapor densities. The portion contained in the two-phase section is more involved because, as suggested in Figure 5, the liquid and vapor volume fractions do not necessarily vary linearly with position. A model of the two-phase flow process in the evaporating zone must be used in finding the actual liquid distribution. A very appropriate model is available [15], which predicts the equivalent of f_{VE} (and the corresponding refrigerant mass) as a function of vapor/liquid density ratio and vapor/liquid velocity ratio. It appears that the steam/water data of Thom [16] would be adequate for relating velocity ratio to density ratio, thus facilitating the mass calculation.

Condenser :

The condenser model follows the same general pattern as the evaporator model, and will not be discussed in detail. The zones to be considered are shown in Figure 6. The temperatures $T_{2'}$ and $T_{3'}$ at the inlet and outlet of the condenser proper can be determined from single phase heat transfer analyses of the discharge line and liquid line, respectively, if it is assumed that, at any particular stage in the condenser iterations, both T_2 and T_3 are known. Subsequently, the fractions of the condenser iterations, both T_2 and T_3 are known. Subsequently, the fractions of the condenser needed to complete the desuperheating and subcooling processes, f_{DSH} and f_{SC}, can be found from equations analogous to equation (33) :

$$f_{DSH} = \frac{\dot{m}_R \, c_{p_v}}{(UA)_{DSH}} \ln \left(\frac{T_{2'} - T_{a_c}}{T_B - T_{a_c}} \right) \qquad (37)$$

and

$$f_{SC} = \frac{\dot{m}_R \, c_L}{(UA)_{SC}} \ln \left(\frac{T_c - T_{a_c}}{T_{3'} - T_{a_c}} \right) \qquad (38)$$

where :

c_{p_v} = average specific heat of refrigerant vapor

c_L = average specific heat of refrigerant liquid

T_{a_c} = air temperature entering condenser

$(UA)_{DSH}$ = overal heat transfer coefficient - surface area product based on <u>total</u>condenser operating with superheated vapor

$(UA)_{SC}$ = overall heat transfer coefficient - surface area product based on total condenser operating with subcooled liquid

The portion of the condenser actually used for condensation is then :

$$f_c = 1 - f_{DSH} - f_{SC} \qquad (39)$$

It is worth emphasizing that, for given values of T_2 and ΔT_{SC}, the values for the temperatures T_B, T_C, $T_{2'}$, $T_{3'}$ and T_3 appearing in the condenser model will depend on the amount of refrigerant pressure drop in the condenser side of the system, thus requiring some iteration processes within the condenser subroutine.

Analogous to equation (36) in the evaporator model, the total condenser heat transfer rate is :

$$Q_C = f_c \, \dot{m}_{a_c} \, c_{p_a} \, \varepsilon_C \, (T_{CND} - T_{a_c}) + \dot{m}_R [c_{p_v} (T_2 - T_B) + c_L (T_c - T_3)] \qquad (40)$$

where :

$$\varepsilon_C = 1 - e^{-NTU_C}$$

$$NTU_C = (UA)_C / \dot{m}_{a_c} c_{p_a}$$

$(UA)_C$ = overall heat transfer coefficient - surface area product based on total condenser operating with condensing refrigerant

$$T_{CND} \approx \frac{1}{2} (T_B + T_C)$$

The value of Q_C must be compared with the heating capacity ($\dot{m}_R$ $(h_2 - h_3)$) in testing for convergence of the heat pump simulation iterations.

With regard to the refrigerant mass occupying the condenser, the contributions in the single-phase zones are readily evaluated once the operating conditions and internal volume fractions, f_{DSH} and f_{SC}, are known. The model recommended for the two-phase zone of the evaporator, based on information in References [15] and [16], can also be applied to the condensing region of the condenser.

Capillary Tube :

The required capillary tube models for design and off-design operation (Table I) must characterize the interaction between flow rate, pressure drop, length and diameter, and upstream conditions. A fundamental model would require integration of rather elaborate two-phase flow equations; it is probably not worthwhile to use this approach. Instead, curve-fits of available design charts [17] are recommended.

In the design mode (subroutine CAPTUB), the known flow rate and operating conditions would be used in solving for various suitable length-diameter combinations. One of these pairs would then be selected for use in the off-design mode.

In the off-design mode (subroutine CPLRY), the dimensions and operating conditions would be used in solving for the mass flow rate. This would then be compared with the compressor flow rate in testing for convergence of the heat pump simulation.

It might be mentioned that the usual capillary design charts assume a straight tube geometry. If the capillary is to be coiled, it may have an appreciably larger flow resistance, requiring a correction factor in the model.

2.4. Discussion

The presentation on steady-state modeling/simulation has focused on the particular configuration shown in Figure 2. No details were given on modeling the air supply system; this was considered too specialized and beyond the scope of the present discussion. It was not possible to give all the details of the heat pump model, but the aspects considered most important or not available in the literature were covered. The basic goal was to suggest a way of approaching the modeling and simulation of an example heat pump system. Many of the ideas should be useful in simulation of somewhat different configurations. However, those planning to develop a simulation should recognize that their purposes may require significant variations on the strategy suggested here. In all cases, it is recommended that a basic simulation structure be developed first, with the aid of relatively simple component models, with subsequent evolution toward a more detailed representation, as needed, after a working program has been achieved.

Some comments about possible extensions or modifications to the heat pump model, to represent situations of potential importance, are in order. The following examples appear to be significant.

First, an accumulator in the suction line between the evaporator and compressor could be added. Its function is to prevent liquid entering the compressor under certain off-design operating conditions and to avoid excessive superheating of the vapor under other conditions. Assuming the accumulator to contain some liquid at all times, it could be modeled by equating conditions at point 1' (see Figures 2, 3 and 5) to those at point A, enforcing $f_{SH} = 0$ ($F_E = 1$) at all operating conditions, and eliminating ΔT_{SH} as a variable in the simulation procedure. The calculation of the maximum and minimum refrigerant mass in the other system components at the extremes of anticipated off-design conditions would then allow the needed refrigerant charge and accumulator volume to be determined.

Another system change of interest would be to allow for a condenser liquid line/compressor suction line heat exchanger (intercooler). This addition would provide the desired superheat to the suction vapor entering the compressor (minimizing the use of the evaporator for this purpose) while providing subcooling to the liquid entering the capillary tube. One would probably simulate this heat exchanger by selecting its size parameters to be compatible with the specified ΔT_{SH} value at the design point, and then using these parameters to predict the amount of superheating/subcooling occuring in the intercooler under off-design conditions. It appears that the simulation structure outlined in Figure 4 may need some revison to accommodate this device, due to the thermal interaction it causes between condenser and evaporator sides of the system.

The replacement of the capillary tube by a thermostatic expansion valve is another option of interest. These valves tend to maintain a somewhat constant level of superheat at the evaporator outlet (over a range of operating conditions), due to the control action effected by a sensing bulb located there. The valve's effect could be approximated by retaining the specified superheat value in the off-design mode, while eliminating the capillary tube models and the mass flow rate constraint from the off-design simulation procedure. The predicted compressor flow rates and operating conditions at the expansion valve over the range of off-design conditions of interest could then be used in selecting an appropriate valve size for the system.

A few other areas of possible heat pump configuration change or useful component model refinement are also suggested, without detailed comment. These are : modeling of reversed heat pump operation (cooling mode), modeling of non-electric compressors, inclusion of a compressor thermal model to more accurately predict the discharge temperature, modeling of other heat sources and sinks (and corresponding evaporator and condenser configurations) and allowance for dehumidification effects in the evaporator model.

The emphasis of this chapter has been on simulation and modeling of steady state heat pump operation. This approach should be adequate for many purposes. Certainly a steady state heat pump simulation can be a useful tool in the design process (to be discussed in the final section of this chapter). It also appears that such a model can predict the trends in heating capacity and in COP as a function of operating conditions (and thus provide the capacity to relate heat pump design parameters to seasonal COP), even if cycling (on-off operation) effects are not explicitly included. This statement becomes most valid for conditions where the heat pump "on-time" is long compared to the duration of "startup effects" (typically on the order of a few minutes). It must be acknowledged, of course, that there are situations of great importance for which a transient heat pump model/simulation is essential. These are discussed briefly in the following section.

3. TRANSIENT MODELING AND SIMULATION

A transient heat pump simulation model would be necessary in predicting the response of the system and its components to various kinds of changes. Important examples would include the startup and shutdown processes, rapid changes in environmental (indoor or outdoor) conditions and the effects of different flow control/ expansion devices on the system response to transient conditions. Results of particular interest from simulation of these processes would include response times and "worst case" compressor operating conditions. Certainly the ability to predict the "instantaneous COP" would enable a more accurate estimate of seasonal COP. The physical insight to be derived from knowledge of the redistribution of refrigerant in the system under transient conditions would be of value, also.

It is worth noting that, if one had a transient simulation program available, it could be used in the analysis of steady state operation, at the design point or off-design conditions, as well. In essence, the numerical integration technique of the transient simulation would become an alternative to the purely iterative solution technique (exemplified by Figure 4) described earlier. If this method were used, however, it appears that selection of the "proper" capillary tube and refrigerant charge (to satisfy desired values of ΔT_{SH} and ΔT_{SC} at the design point) would become an iterative process.

The essential goal of a transient model is to follow the evolution of the refrigerant mass and thermodynamic state distributions within the system components (in response to changes in operating conditions, etc.) from known or assumed initial conditions. This requires that transient versions of the component models are needed in order to allow for refrigerant mass and energy storage and solid material energy storage effects. These terms cause the mass and energy balance equations to become ordinary differential equations rather than algebraic equations, as they were in the steady state case. This also results in the introduction of some additional component parameters (thermal capacities) into the model. Although a detailed description of transient modeling and simulation will nog be given here, it should be noted that such a discussion is available [18].

4. HEAT PUMP DESIGN

A natural application of heat pump simulation is in the area of design. This <u>does not imply</u> that all aspects of design can be dealt with by modeling techniques. However, the emphasis of this section will be on the analytical aspects of heat pump design. Many of the practical design considerations, as well as discussions of heat pump applications, are discussed elsewhere [6, 17 19] and in other chapters of this book. The book by Stoecker [2] is a good reference to the ideas of analytical design.

The section will attempt to convey some of the concepts of (analytical) heat pump design through the presentation of comments on the following topics : component balances, use of steady state simulation in design, and heat pump optimization. In so doing, attention will be focused mainly on the primary design criterion - that the heating capacity meets a selected target value when the heat pump is operating at the design point. Some discussion of possible secondary design criteria will be given in the material on optimization.

4.1. Component Balances

The traditional, most elementary, approach to analytical heat pump design is the method of component balancing [6, 17]. Typically, this involves the use of manufacturers' performance curves for the compressor, evaporator and condenser in establishing, usually via a graphical method, the evaporation and condensation temperatures (ET, CT) and associated heating or cooling capacity at which the components will be in balance (i.e. satisfy steady state energy balance criteria). Example component curves are schematically indicated in Figure 7. The numerical values on the evaporator and condenser cruves would actually be derived from more funamental manufacturer data, for the type of coil being considered, such as heat transfer rate per unit face area as a function of air velocity and number of rows of refrigerant tubes.

It might be observed that the component balance method emulates the traditional, but often costly and inefficient, "build it and try it" approach. That is, if the results of the component balance, for the particular component curves (i.e. component sizes and configurations) that were employed, do not meet the design capacity target, it is necessary to adjust one or more of the components and try again. Typical changes would be to adjust air flow rates, coil sizes or compressor size. Obviously, experience in heat pump design will decrease the number of tries required to find a workable design by this method.

For clarity, one recipe for performing a component balance, assuming the availability of curves similar to those in Figure 7, is a follows :

1. Combine the compressor evaporation capacity and power curves to yield a new family of curves of $C_C \equiv (C_E + P_C)$ vs. CT with ET as the parameter. If the compressor manufacturer's curves on heating capacity are already available, this step can be omitted.
2. Combine the evaporator heat transfer rate and the compressor power curves to form a family of curves of $C_C \equiv Q_E + P_C$ vs. CT with ET as the parameter.

3. Noting that at the steady state solution $Q_C = C_C$, and also $C_C = Q_E + P_C = C_E + P_C$ (because $Q_E = C_E$), superpose the original condenser heat transfer curve (Figure 7 (d)) and the two families of curves developed in steps 1 and 2. The operating point is that unique combination of ET and CT for which Q_C (or C_C) is the same on all three figures.

It is interesting to note that the "simplified model" presented earlier in this chapter is essentially the algebraic-solution equivalent of the component balance method just discussed, for the special case in which all the component performance curves are linear and in which the heat transfer curves are defined in terms of "fundamental" heat transfer parameters.

It is also useful to note that, if the curves in Figure 7 were all represented by curve-fit equations (e.g. quadratic functions of ET and CT), the Newton-Raphson method described in connection with the general simulation structure presentation could be used in finding the component balance solution via computer. This would constitute a simplified heat pump simulation, with the curve-fits being the "model".

4.2. Role of Steady State Simulation in Design

In essence, a highly detailed heat pump simulation program (steady state) represents a very refined and automated form of the component balance technique. Of course, the simulation greatly facilitates the basic goal of identifying heat pump component size combinations, air flow rates, etc., which meet the design point performance target. Because of the ease of making predictions of the effects of the major size anf flow rate variables via computer, one can very readily perform parametric studies (the equivalent of repeated application of the graphical method) that allow a suitable design to be found. In fact, one can then proceed to investigate equipment tradeoffs and identify alternative workable designs, allowing secondary factors such as equipment cost or COP to be used in making a final selection of the design parameters. This ability to conveniently consider more options is certainly one of the major values of simulation in heat pump design. If the level of detail of the component models is sufficient, there may easily be several dozen parameters available to the designer for use in studying design options. The availability of these parameters can allow effects to be studied which are often overlooked in applying the much more time-consuming graphical method or even more simplified simulation models. For example, the effect of various evaporator or condenser tube circuiting options (which influence the pressure drop), or the effects of varying the superheat and subcooling, on system performance can be investigated. This could allow a fine-tuning of the design or an improvement in COP to be made.

It should be observed that, although the capacilities, accuracy and potential value of the simulation increase with the level of detail of the component models, the requirements for more detailed component data and for model development time increase also. For example, the model may require data or correlations related to the variation of various heat transfer coefficients with refrigerant or air flow rate or with geometrical parameters. Some of this information may not be readily available and may have to be developed experimentally.

4.3. Heat Pump Optimization

The ultimate in design is <u>optimal</u> design - finding <u>the best</u> of all possible designs. Of course, to achieve this goal, it is necessary to define "the best". In general, a specific measure of the optimum, such as maximum COP, minimum life cycle cost, etc., must be selected as the criterion of excellence. In addition, constraints that must be satisfied by the optimum design must also be specified. For example, if minimum cost was the objective, some constraints might be : the COP must exceed some minimum allowable value, the electrical current must not exceed a certain value, the physical dimensions of the system must not exceed certain values, the noise level (air velocity through the coils) must not exceed a certain value, etc. The primary constraint, of course, is that the heat pump capacity at the design point meet its target value. Furthermore, those design parameters which are free to be altered in seeking the optimum must be stipulated, together with a specification of any limitations on their allowable values.

Having formulated a definition of the "optimization problem" for a particular heat pump design situation, a method for finding the set of design variables corresponding to the optimum system must be found. The traditional parametric study approach can really only yield an approximation to the optimum, due to the practical limits on the number of parameter combinations that can be investigated.

The most promising approach to heat pump optimization is to employ a generalized constrained optimization computer procedure in solving the problem. It is beyond the scope of this chapter to discuss these "nonlinear programming" methods here. Let it be noted, however, that many powerful and general computer codes are available, commercially or in the literature, which could be interfaced with the heat pump simulation and statements of the optimization criterion and constraints. The optimization program adjusts the simulation model parameters, in accordance with its particular algorithm, until the optimum is found. The simulation results developed from each of the trial combinations of model (design) parameters are used in testing contraints and evaluating the optimization criterion. While published details of heat pump optimization techniques or studies appear to be unavailable, these techniques have been successfully applied in recent years, on a limited basis. Simulation/optimization will undoubtedly play an important role in the future analytical design of heat pumps.

REFERENCES

1. Kays, W.M. and London, A.L. *Compact Heat Exchangers* 2nd Ed. (New York, McGraw-Hill, 1964).

2. Stoecker, W.F. *Design of Thermal Systems* 2nd Ed. (New York, McGraw-Hill, 1980).

3. Kartsounes, G.T. and Erth, R.A.,"Computer Calculation of the Thermodynamic Properties of Refrigerants 12, 22, and 502" *ASHRAE Trans.* 78(2) (1971).

4. Qvale, E.B., et al.,"Problem Areas in Mathematical Modeling and Simulation of Refrigerating Compressors" *ASHRAE Trans.* No. 2215 (1972).

5. Soedel, W. *Introduction to Computer Simulation of Positive Displacement Type Compressors* Short Course Notes, Ray W. Herrick Laboratories (U.S.A., Purdue University, 1972).

6. American Society of Heating, Refrigerating and Air Conditioning Engineers *ASHRAE Systems Handbook* (New York, 1976).

7. American Society of Heating, Refrigerating and Air Conditioning Engineers *ASHRAE Handbook of Fundamentals* (New york, 1972).

8. Traviss, D.P., Rohsenow, W.M. and Baron, A.B. "Forced Convection Condensation Inside Tubes : A Heat Transfer Equation for Condenser Design" *ASHRAE Trans.* No. 2272 RP-63 (1973).

9. Rich, Donald, G. "The effect of the Number of Tube Rows on Heat Transfer Performance of Smooth Plate Fin-and-Tube Heat Exchangers *ASHRAE Trans.* No. 2345 (1975).

10. McQuiston, Faye, G. "Correlation of Heat,Mass and Momentum Transport Coefficients for Plate-Fin-Tube Heat Transfer Surfaces with Staggered Tubes" *ASHRAE Trans.* No.2487 PR-155 (1978).

11. Rich, Donald, G. "The effect of Fin Spacing on the Heat Transfer and Friction Performance of Multi-row, Smooth Plate Fin-and-Tube Heat exchangers" *ASHRAE Trans.* No. 2288 (1973).

12. Trelkeld, J.L. *Thermal Environmental Engineering* (Englewood Cliffs, Prentice-Hall, Inc., NJ (1962)

13. Wallis, G.B. *One-Dimensional Two-Phase Flow* (New York, McGraw-Hill, 1969).

14. Paikert, P. "Chapter on Heat Pump Evaporators and Condensers *NATO/ASI Heat Pump Fundamentals* (Alphen a.d. Rijn, Noordhoff & Sijthoff, 1982).

15. Rigot, G. "Contenance en Fluide d'un Evaporateur en Detente Directe *Chaud-Froid-Plomberie* No. 328, pp. 133-144 (1973).

16. Thom, J.R.S. "Prediction of Pressure Drop During Forced Circulation Boiling of Water" *International Journal of Heat and Mass Transfer* Vol. 7, pp. 709-724 (1964).

17. American Society of Heating, Refrigerating and Air Conditioning Engineers *ASHRAE Guide and Data Book - Equipment* (New York, 1972).

18. Dhar, M. and Soedel, W. "Transient Analysis of a Vapor Compression Refrigeration System : Part I - The Mathematical Model; Part II - Computer Simulation and Results" *Proc. XV International Congress of Refrigeration* (Venice, 23-29 Sept. 1979)

19. Reay, D.A. and Macmichael, D.B.A. *Heat Pumps - Design and Application* (Oxford, Pergamon Press, 1979).

TABLE 1

Subroutines for steady state heat pump simulation

Subroutines	Purpose
MAIN program	- read input - initialization of variables - iteration procedure - print output
COND	- discharge line and condenser heat transfer calculations - discharge line and condenser refrigerant pressure drop and average condensing temperature - amount of refrigerant charge stored in high pressure side of system
EVAP	- evaporator and suction line heat transfer calculations - refrigerant pressure drop and average evaporating temperature - amount of refrigerant charge stored in low pressure side
COMP	- compressor mass flowrate - compressor discharge temperature - power input - current (optional) - amount of refrigerant charge stored in compressor
THERMO	- therodynamic properties of refrigerants
BLOCK DATA	- default values of model parameters
CPLRY	- mass flow rate through capillary tube
CAPTUB	- capillary length and diameter (design)

Table 1 (continued)

DIMEN	- coil heat transfer areas - fin and surface efficiency - coil heat transfer coefficients - coil heat transfer effectiveness - coil internal volumes
AIRSYS (Optional)	- fan efficiency, power, speed, flowrate - coil flow resistance - motor torque, efficiency, power, current

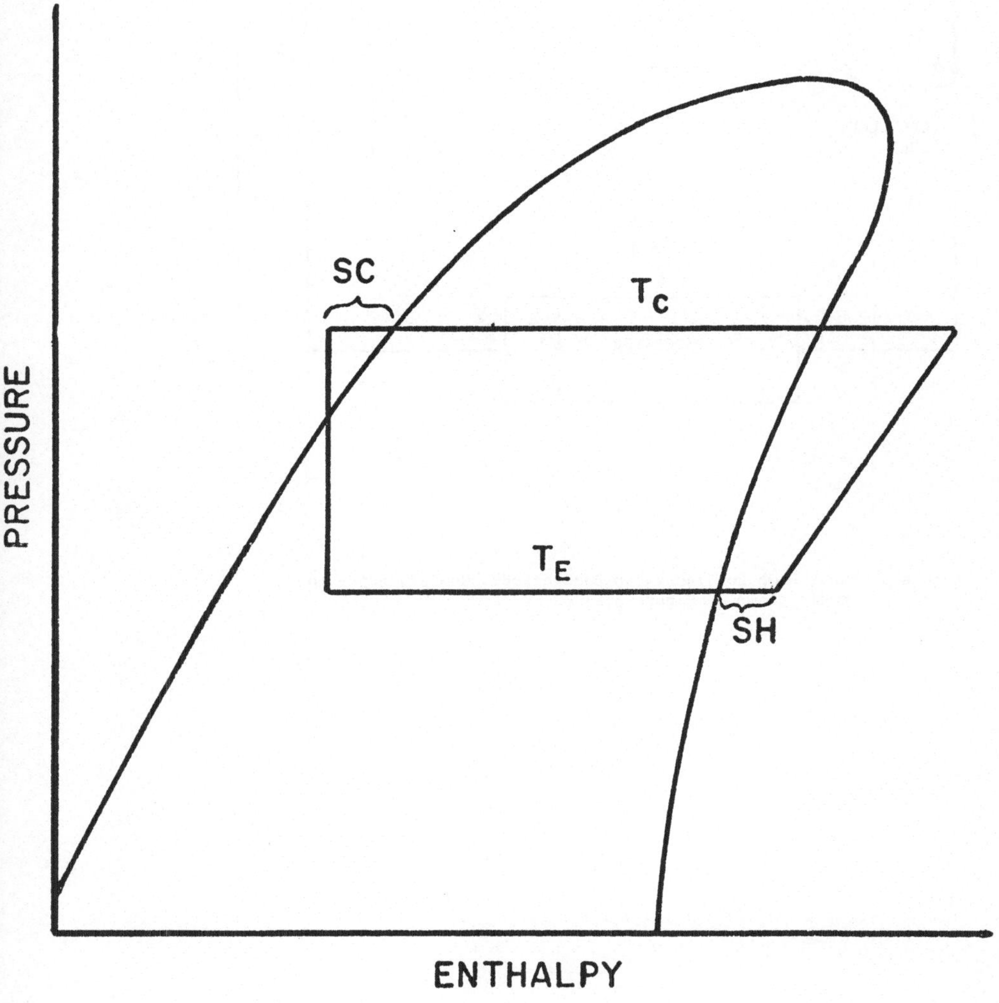

Figure 1. Idealized Refrigerant Cycle for Vapor Compression Heat Pump.

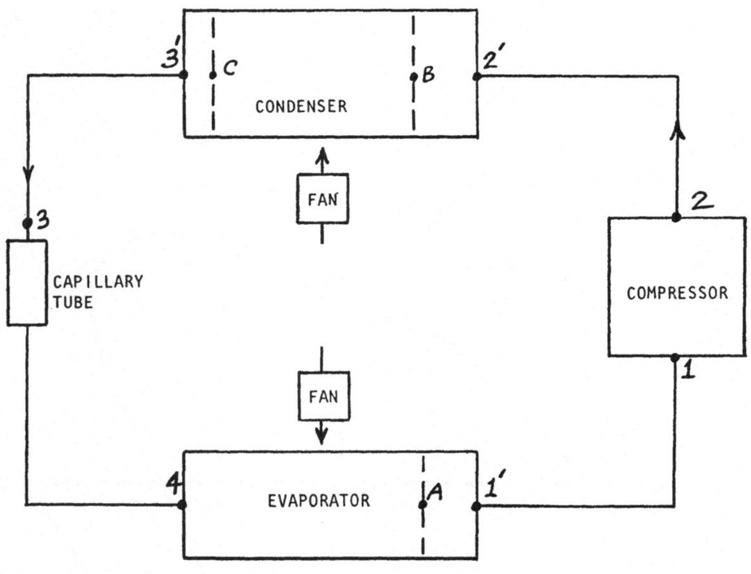

Figure 2. Heat Pump Configuration Considered. Numbers refer to cycle shown in Figure 3.

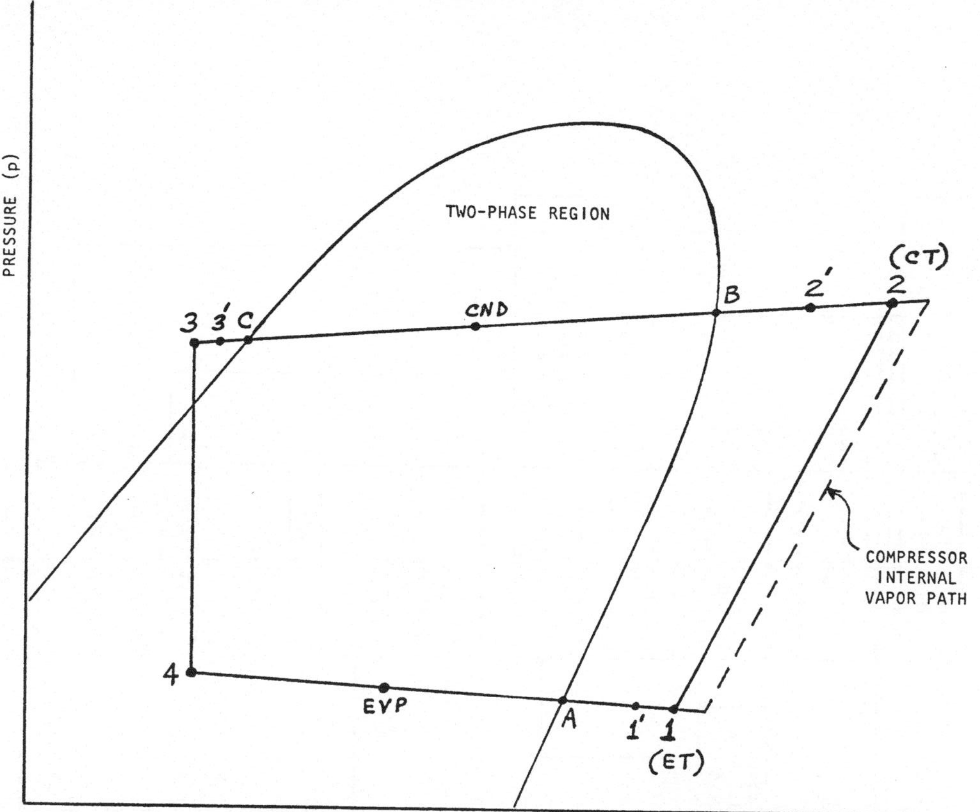

Figure 3. Typical Refrigerant Cycle. ET and CT are Saturation Temperatures Corresponding to p_1 and p_2.

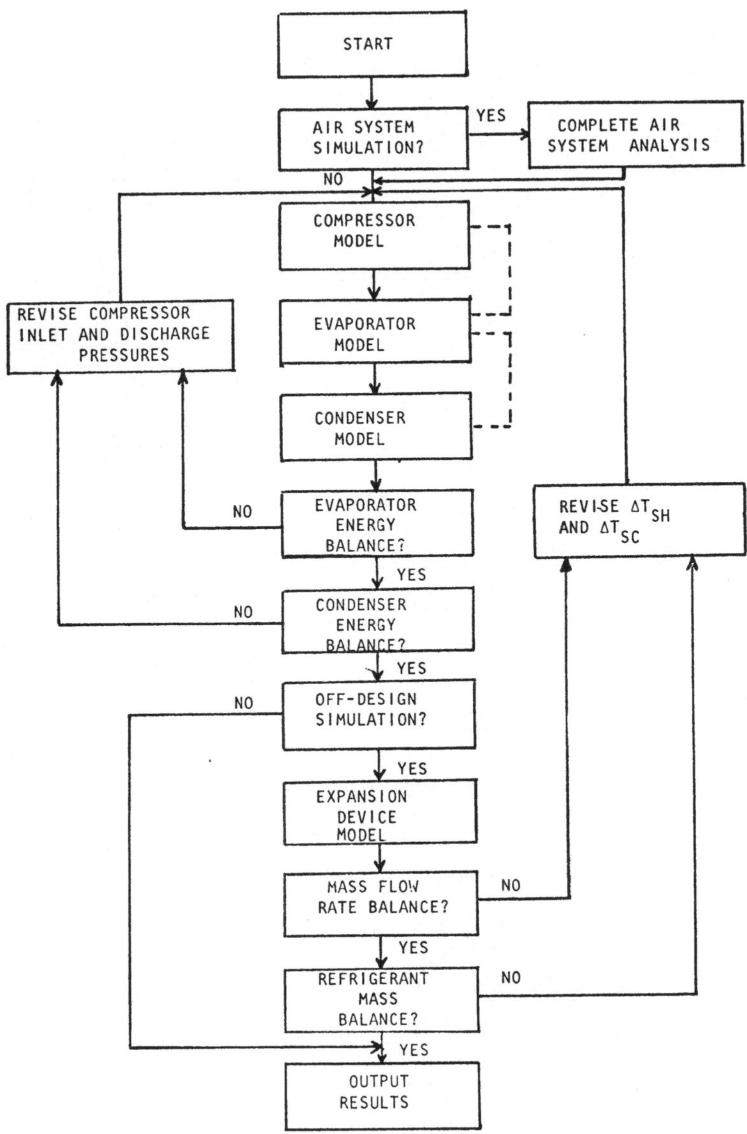

Figure 4. General Steady State Simulation Structure for Design and Off-Design Operation.

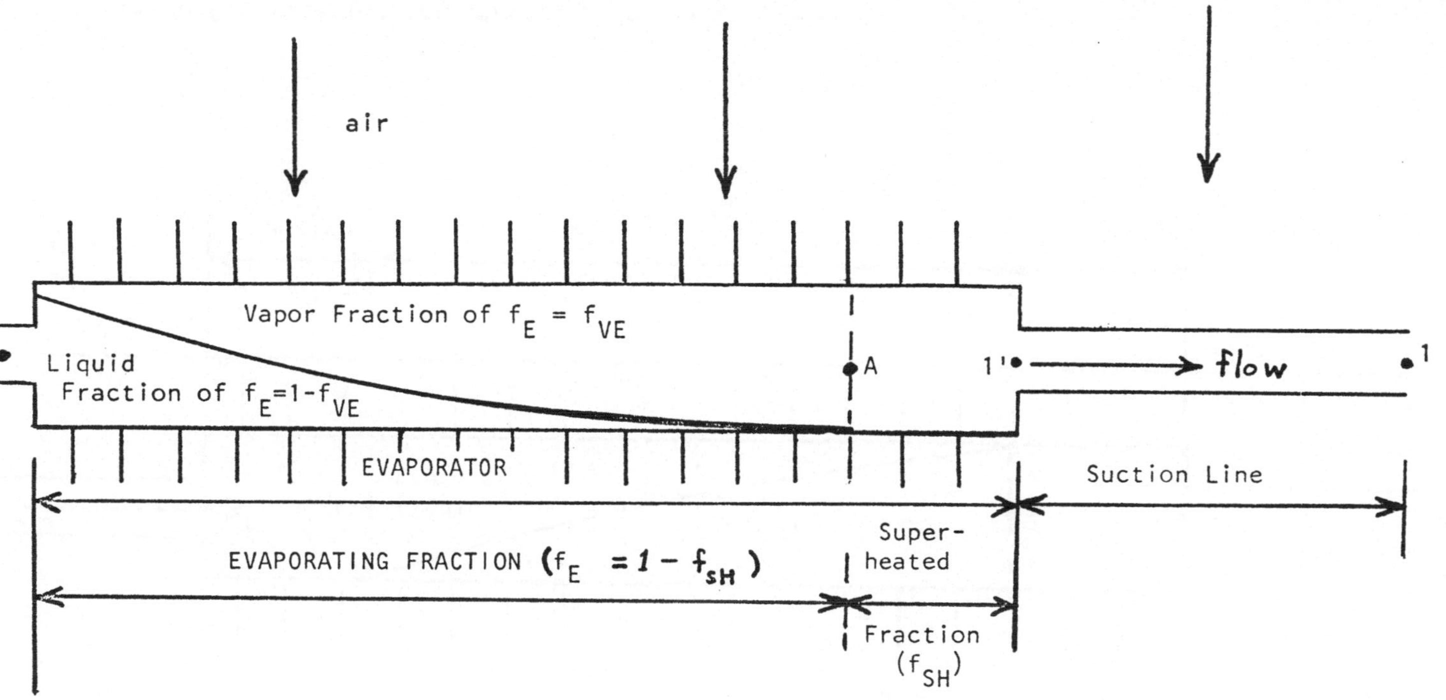

Figure 5. Notation for Evaporator Model.

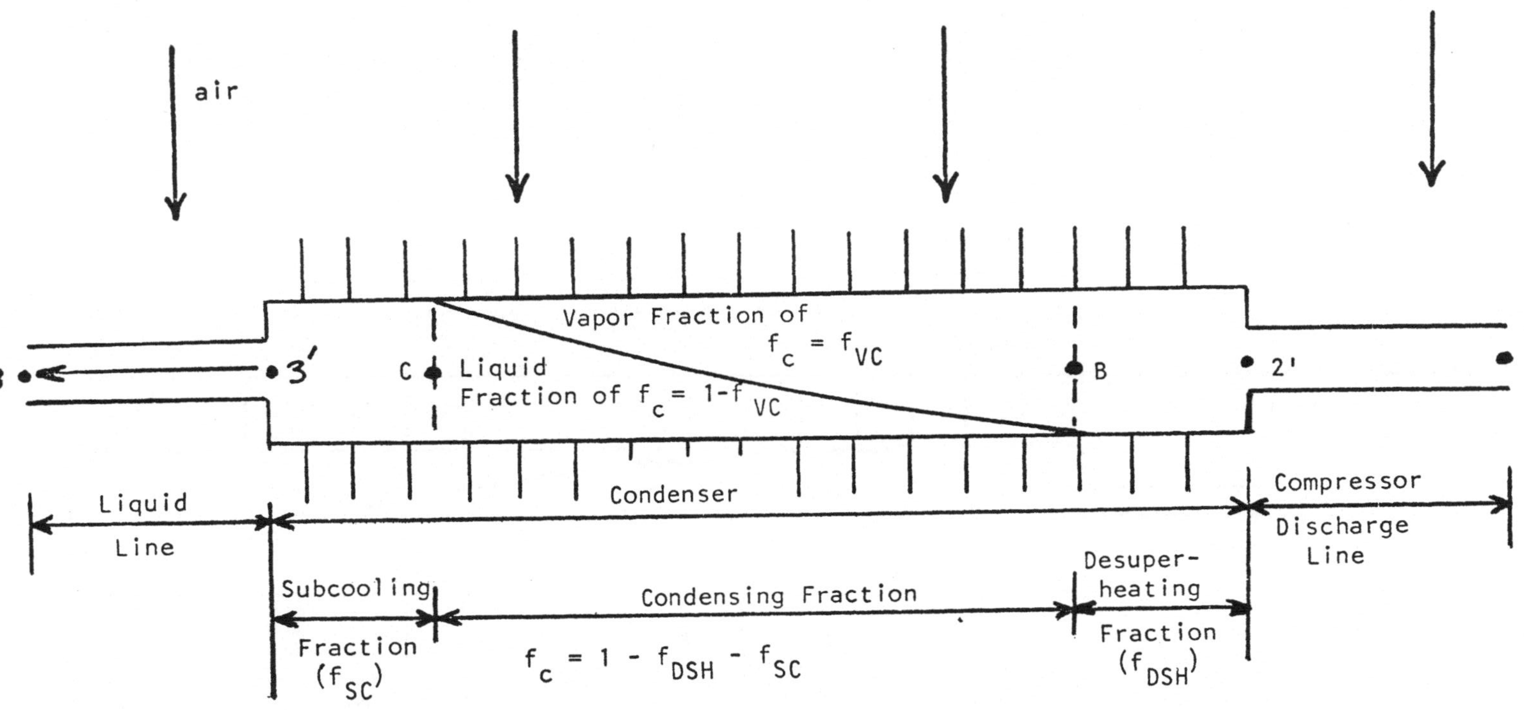

Figure 6. Notation for Condenser Model.

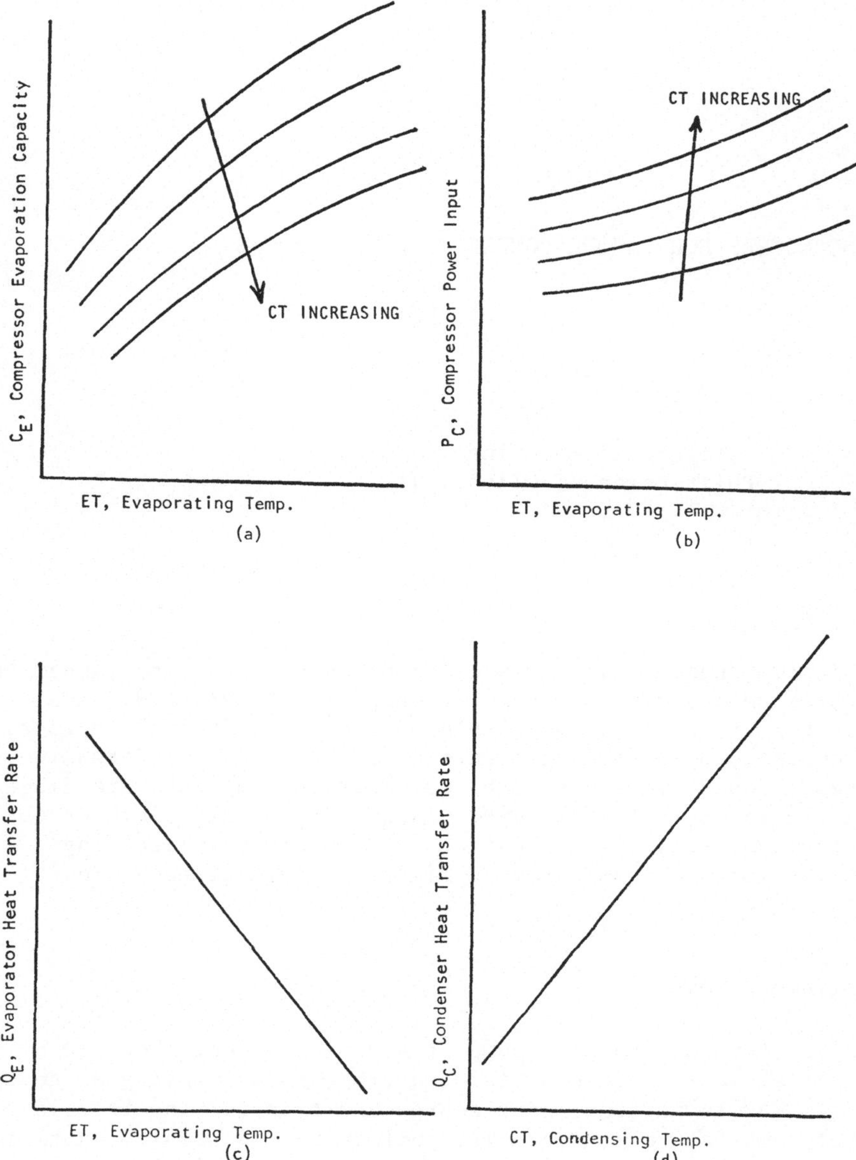

Figure 7. Schematic Representation of Component Performance Curves.

PRIME MOVERS FOR VAPOUR COMPRESSION HEAT PUMPS

E. Macchi

Istituto di Macchine
Politecnico di Milano, It.

ABSTRACT

The thermodynamic advantages of adopting heat engines rather than electric motors for driving heat pump are illustrated. The characteristics of the available heat engines (Diesel engines, gas engines, open cycle gas turbines, steam turbines, Rankine cycles, closed cycle gas turbines, Stirling engines) are discussed. It is shown that several classes of engines, both of the internal and external combustion type, exhibit interesting features for the considered application, even if they require some development.

1. INTRODUCTION

A vapour compression heat pump is a device which, when fed by mechanical work L, is capable of extracting a quantity of heat Q_3 from low temperature heat source and transferring it to a higher temperature source. Its performance is usually expressed by the so-called coefficient of performance, defined by

$$COP = \frac{Q_3 + L}{L} \qquad (1)$$

Obviously the COP does not consider how efficiently the mechanical work is generated and cannot be by itself an index of the fuel economy of the whole system. Since the mechanical work is in most cases obtained by burning fuel in a thermal power station, a quantity ζ which better accounts for fuel economy can be defined as the ratio :

$$\zeta = \frac{\text{heat to the low temperature user}}{\text{heat generated by ideal fuel combustion}} \qquad (2)$$

For conventional heating systems not using heat pumps, ζ has actual values in the range 0.7 to 0.92. The ideal ζ value, achievable by following fully reversible thermodynamic paths (see [1]), is many times larger. Intermediate values can be achieved in systems using heat pumps. The process can be schematized in the following phases (see Figure 1) :

1. a heat engine, which transforms part of the heat generated by the ideal fuel combustion Q_1 in the mechanical work W'_m; an other fraction Q'_2 is made available at a useful temperature level; the difference $L' = Q_1 - (W'_m + Q'_2)$ is lost;

2. a transmission system, which transfers the mechanical work from the prime mover shaft to the compressor shaft of the heat pump; at the end of this process, only a portion of the mechanical work W'_m is still available as mechanical work W''_m, since a portion is degraded into useful heat Q''_2 and eventually $L'' = W'_m - Q''_2$ is lost;

3. a heat pump which extracts $Q_3 = (COP - 1)\, W''_m$ from the cold heat source and transfers to the user $Q'''_2 = Q_3 + W''_m = COP.W''_m$.

If the combined efficiency of the heat engine and transmission system :

$$\eta = \frac{W''_m}{Q_1} \qquad (3)$$

and the combined heat recovery factor :

$$f = \frac{Q'_2 + Q''_2}{Q_1} \qquad (4)$$

are defined, the obvious relation holds :

$$\zeta = \frac{Q'_2 + Q''_2 + Q'''_2}{Q_1} = f + \eta\, COP \qquad (5)$$

Actual values of η and f achievable by the various prime movers will be discussed in the next chapter. For the present discussion, average values of three typical systems are indicated in Figure 2 : for electrical drivers, η must account for the power station losses, for distribution losses and for electric motor losses (see the Sankey diagram of Figure 3a), a small contribution of f can be achieved, only if the cooling system of the electric engine is working at useful temperatures; for heat engines, directly driving the heat pump compressor, negligible transmission losses take place, and most of the heat not converted into work can be recovered, the sum (η + f) can be as high as 80 - 90 % (Figure 3b); finally, for conventional heaters (Figure 3c), the produced mechanical power is zero (it is actually slightly negative, due to power absorption of fans and pumps), and the coefficient f is equal to the heater efficiency.

Consider the results represented in Figure 4, where equation (5) is solved for three different values of the COP. For a low - but not unrealistic - coefficient of performance (COP = 2), the electric heat pump doesn't allow any fuel economy with respect to the conventional heaters; far better results can be obtained by using "total energy" engines : for example, with the gas engine described in [2], even with COP = 2, the equation (5) yields ζ = 1.2, and therefore a fuel savings of about 30 % with respect to conventional heaters.

For larger coefficients of performance (COP = 4 and 6), where also the electric driven heat pumps can yield better fuel economy than conventional systems, again the heat engine performance is quite superior. The quoted gas engine for example, yields fuel consumptions 50 % lower than the conventional heaters for COP values larger than 4.

As discussed in [1], the COP can be related to the heat source and sink temperatures, T_{min} and T_{max} respectively, by the relation :

$$COP = \beta \frac{T_{min}}{T_{max} - T_{min}} + 1 \tag{6}$$

where β is a coefficient which accounts for all the irreversible processes occuring in the heat pump components (heat exchangers, compressors, valves and piping). Values of β around 0.5 are representative of current, industrial type equipment. Values given in Figure 5 are computed for β = 0.5. Quite lower values are found for small domestic heating units [5].

The preceding elementary analysis and the results of Figures 4 and 5 show that heat generation by means of heat pumps driven by electric motors can yield significant energy savings only for low temperature differences between the available heat sink (either ambient air, water, ground, or a process fluid in the case of industrial applications) and the user. The adoption as heat pump drivers of thermal engines with heat recovery can greatly improve the potential of these interesting machines.

Of course, the above reasoning is oversimplified : the fuel consumption at the design point is not the only parameter to look for in the selection of a system. Other points which are related to the energy savings are the plant behaviour at various operating conditions, and its compatibility with various fuels. Moreover the plant first cost, its availability and durability, the capability of unattended operation, the maintenance requirements and the polluting characteristics are important.

2. THERMAL ENGINES ANALYSIS

In this section a brief review of the available thermal engines, suitable for heat pump driving is performed. In principle, all types of thermal engines could be considered for this application. However, since heat pumps work under relatively small temperature differences and the transfer of the generated heat to various remote utilizers can be impractical and not economic, the range of power outputs of interest is limited to values unusually low for some class of engines. In particular for domestic heating, the most interesting applications require a few kW shaft power of the engines. The relationship between shaft power and thermal power output is given in Figure 6 as a function of the above defined parameters f and η. As an example, the gas engine [2], with $\eta = 0.285$ and $f = 0.634$ and $W_m = 16.5$ kW, driving a heat pump working between 0°C and 45°C with a COP of 4.03 (from Figure 5) generates $6.25 \times 16.5 \doteq 103.2$ kW.

The thermal engines which will be considered in this section are listed in Table I where a subdivision is made according to the mode of heat generation : in internal combustion engines the combustion occurs in the working fluid, which is therefore necessarily a mixture of air and fuel while in external combustion engines the heat is generated outside and given to the engine by means of an heat exchanger. All heat engines reject some heat, in the mode indicated in the table.

A comprehensive description of all prime movers quoted in Table I is outside the purpose of this chapter. In what follows some comments on the potential of various engines for the considered application are made. Reference will be made to a few specific existing engines particularly suitable for heat pump driving, rather than to the average characteristics of the various classes of engines.

2.1. Internal Combustion Engines

2.1.1. Reciprocating engines

Diesel engines are very well known, reliable engines and are often used in total energy applications. Like most volumetric engines, they are available in a large power range : from few kilowatts to several tenths of megawatts. A typical heat balance for a Diesel engine specifically designed for cogeneration is given in Figure 7 [3].

The heat is recovered by three main sources : the cylinder cooling system, using either water or low pressure steam; the lubrifying oil and the exhaust gases. The cooling of exhaust gases down to low temperatures brings about severe corrosion problems, due to the presence of sulphur in the fuel. While diesel engines are probably today the best choice for industrial application for which their durability can be adequate, the adoption of small units for domestic heating raises several problems related to maintenance, unattended operation, noise, etc.

An interesting solution for small power output engines is proposed by an Italian firm [2]. The engine, directly derived from a gasoline engine for automotive propulsion, is represented in Figure 8, while its heat balanceis given in Figure 3b.

Designed for cogeneration, it drives, an electric generator at constant speed. It can use several gaseous fuels, and is designed for continuous operation at a power load about 50 % lower than the maximum power in the automotive version (see Figure 9).

Its maintenance routine assumes a TBO (Time Between Overhauls) of 1000 hours, and it lasts without requirement of major operations for 3000 hours, that is to say for one heating season. The complete substitution of the engine every year doesn't penalize the considered application, thanks to the low prime cost of the engine, achieved by the mass production techniques typical of automotive industry. This cogeneration package has not yet achieved a commercial success.

One difficulty is the contemporary production of two outputs, electric power and heat,which are not balanced. For domestic application the electric power is by far overabundant compared to the heat supplied.

The development of a heat pump package using this engine, which produces low temperature heat with much lower fuel consumption than a boiler, would be, in the author's opinion, a technically feasible operation with great potential energy savings.

An other problem, related to the domestic applications of reciprocating engines is their noise. The results of Figure 10 demonstrate that this problem can be solved, provided that proper insulation is applied.

The experience gained during a study sponsored by the Italian CNR (National Research Center) [6], has shown that a succesful combined heat pump + reciprocating engine + heat recovery unit cannot come about by simply coupling these three machines. Its development should be preceded by a systems approach, to provide an efficient and reliable package working in a fully automatized way.

2.1.2. Gas Turbines

Gas turbines probably are the most reliable and safe thermal prime movers. Many engines commonly experience 20.000 to 30.000 hours of operation prior to overhaul, and maintenance requirements are minimum (see for example data of Table II, from [7]). The engine's small size and weight provides for simple installation. Moreover, if mass produced, their cost could be very low (see the 1974 prediction, given in Table III, after NREC [8]).

A great obstacle to the diffusion of gas turbines as heat pump prime movers is their poor efficiency, particularly in the low power range, which is the most attractive for heat pump applications (see for example the data of Figures 11 and 12, where performance of small gas turbines are compared to Diesel engines). Efficiencies of about 20 % are achievable either with simple cycle gas turbines in the range of 500 kW power output [10], or with regenerative gas turbine engines in the 50 kW range [11].

Due to their low efficiency, an important contribution to fuel economy is given by the heat recovered from the exhaust gases of gas turbines. This requires the adoption of specially designed heat exchangers (see for example [12]).

Even if the outstanding performance of the above mentioned reciprocating engines are not reached by the gas turbines, still, with $\eta = 0.20$ and $f = 0.50$, a combined gas turbine - heat pump system has better fuel consumption than an electrically driven unit, up to a COP of about5.

2.1.3. Combined Cycles

Better performance could be achieved by using combined cycles, according to the scheme presented in Figure 13. In this case, the combined gas turbine-organic Rankine recovery cycle has a performance similar to a diesel engine. However, most of the advantages of the gas turbine (simplicity, reliability, etc.) are lost, and thermodynamic advantages are justified only in relatively large power stations.

2.2. External Combustion Engines

An important advantage of closed cycles versus open cycles is the possibility of a better control of the combustion process, and therefore the achievement of lower emissions of polluting products, a feature very important for domestic heating systems. A discussion of this problem can be found in [1] and will not be repeated here. This advantage is common to all the engines discussed in the below.

2.2.1. Steam Turbines

Although the use of steam turbines is the most conventional approach to the total energy system, their use as heat pump drivers is in general not attractive since their performance is rather poor at low power outputs. For example a cycle having a maximum working pressure of about 25 bars and a 100°C superheat, condensing at atmospheric pressure, hardly reaches an efficiency of 10 % for power outputs in the 500 kW class. Moreover, the steam turbine plant is inherently costly and complicated.

2.2.2. Closed Cycle Gas Turbines

This relatively old engine has not yet found a commercial success. It is however currently used in total-energy applications for large power output plants [14]. A particularly interesting class of closed cycle gas turbines was developed for space power production [15]. Some of these engines like the one shown in Figure 14 , would be ideal for driving a heat pump in a domestic application. The major obstacle is of course the very high cost of these units. A detailed economic analysis probably could demonstrate that an interesting cost reduction is achievable by making use of mass production techniques and redesigning the units with some simplifications.

It is difficult to anticipate if this cost reduction would be large enough to make these engines economically attractive for the considered application.

2.2.3. Organic Rankine Cycles

The substitution of steam by other working fluids brings about important advantages in many fields of application.

Organic Rankine cycles (ORC) have been proposed for industrial heat recovery, for solar energy and geothermic applications and for automotive propulsion. The author took part in the design of a number of ORCs developed in Italy for various applications, described in Table IV. It can be seen that some of these engines operate at low maximum temperatures, since they were designed for low temperature heat sources. For heat pump driving, the ORC is generally directly fed by a fuel combustor, and best performance can be obtained only for high temperatures of the working fluid. This circumstance limits the choice of the working fluid to few candidates, since most organic media are not chemically stable at high temperatures. A comprehensive description of the available technical data on thermal stability of organic compounds can be found in [1]. Recently, experimental data regarding thermal stability of fluorocarbons has been published by the author [16], which demonstrate the capability of some of these fluids to withstand operation in the temperature range of 350 - 400°C. This data, obtained in the sealed loops shown in Figure 15, are summarized in Figure 16. It was shown that the vapour pressure curve of the three tested fluids did not show any alteration under the test conditions described in the following table :

FLUID	Maximum Temperature °C	Maximum Pressure bar	Duration hours
pp-2	360 (± 10)	10 (± 1)	300
pp-3	320 (± 10)	10 (± 1)	425
pp-5	320 (± 10)	10 (± 1)	245

Two of the three fluids tested (commercially available under the name FLUTEC PP2 and PP5) were selected as the working fluids for the prime movers of two Rankine-Rankine heat pump systems which are being developed.

The main advantage of ORC engines is the possibility of obtaining high efficiencies at relatively low temperatures. As shown in Figure 17, at 380°C, a C_8F_{16} cycle, condensing at 50°C, has a better efficiency (32 %) than the highly sophisticated closed cycle gas turbine of Figure 14, which operates with T_{max} = 840°C. The other fundamental advantage is the possibility of selecting a working fluid such that a low stress, efficient turbine can be designed for each power level. For low power outputs the turbine diameters which seem preferable from the manufacturing point of view are in the 100 to 300 mm range. These dimensions can be obtained for various power outputs, say from few kilowatts to some Megawatts, just by changing the fluid thermodynamic properties. The most important parameter is the condensing pressure, which affects the volumetric flow rate at the turbine discharge (see Figure 19). With peripheral speeds of about 100 m/s, speeds of revolution of about 7.000 to 20.000 rpm result.

At these speeds of revolution efficient centrifugal compressors can be designed so that the rotating assembly of a Rankine-Rankine heat pump plant can be a compact, low stress, potentially low cost machine. A critical element will be the intermediate system between the power and the heat pump cycle. This problem can be avoided by using the same working fluid in the two loops, as is done in the Glynwed heat pump (see Figure 18 [18]), or in the CNPM single fluid heat pump.

Another problem to be solved is the heat recovery from the exhaust gases of the primary heater : the ORC is a highly regenerative cycle and therefore requires heat in a relatively narrow temperature range. The heat recovered by the exhaust gases can be used either direclty by the thermal user, or to preheat the combustion air.

2.2.4. Stirling Engines

These engines have been intensively studied in the last years, mostly for automotive propulsion [19]. The thermodynamic performance is excellent (see Figure 20). Small power units are being developed for solar energy applications (Figure 21 [20]). For the considered application, there are still doubts about their cost when mass produced. Their reliability and use of maintenance has still to be proven.

3. CONCLUSIONS

The use of heat engines with heat recovery greatly improves the fuel economy of heat pumps. In principle several types of engines, both with internal and external combustion, could be used for heat pump driving. While for industrial applications, having relatively large power output, diesel engines seem to be the most promising heat pump drivers several options are available for smaller heat pumps for domestic heating. All of them require some development either to improve their reliability and/or to reduce the costs.

The importance of considering the prime mover and the heat pump as a whole cannot be overemphasized. A significant advantage of the heat engine versus the electric motor is the possibility of a continuous variation of the speed of revolution, to adapt the plant to lead variations.

REFERENCES

1. Angelino, G. "Development of Thermal Prime Movers for Heat Pump Drive" in *Heat Pump Applications* Camatini and Kester eds. (Alphen aan de Rijn, Sijthoff and Noordhoff, 1976).

2. Anon. "Totem Total Energy Module *Fiat Auto S.p.A. Technical Report* (1977).

3. Anon. "Modulo per Servizi Energetici (Module for Energy Production)" *Franco Tosi Technical Report* (1979).

4. Angelino, G. et al. "The CNPM Thermal Heat Pump - Part 1 General Description and Thermodynamic Analysis" *Revue Internationale du Froid* 3, 1 (1980).

5. ANON. *Heat Pumps for Domestic Heating* Aerimpianti, Technical Report for the Italian National Research Council (CNR), within the "Progetto Finalizzato Energetica".

6. Anon. "CNR Finalized Project on Heat Pumps - Demonstration Plant with an Idependent power Generation" *Fiat Research Center Technical Report* (Torino 1976).

7. Mackay, R. "Cogeneration to the Rescue" *Electric Power Research Institute Dual Energy Use Systems Workshop* (September 1977).

8. Heitman, A.M. and Rizika, J.N. "Low cost small gas turbine" *ASME paper 74-GT-133* (1974).

9. Macchi, E. and Gigliolo, G. "On the use of combined gas-organic vapour cycles for climatization plants with low pollution and high efficiency" (in Italian) *La Thermotecnica* n. 4 (1972).

10. Anon. "The Industrial Gas Turbine Engine" *Garrett Technical Bulletin* 1977.

11. Walzer, P. et al. "Passenger Performance with the Experimental Gas Turbine VW-GT 70" *ASME paper 74-GT-108* (March 1974).

12. Boyen, J.L. *Practical Heat Recovery* (New York, John Wiley, 1975).

13. Macchi, E. and Sacchi, E. "Thermal Energy Production by Means of large Heat Pumps" (in Italian) *AICARR Journal* (1973) 551-570.

14. McDonald, C. *Large Closed Cycle Gas Turbine Plant, Closed-Cycle Gas Turbines* (Von Karman Institute for Fluid-dynamics, Bruxelles, 1977).

15. Mock, E.A. "Closed Cycle Gas Turbine Optimization -- Procedures and Examples", in Lecture Series 100 on *Closed-Cycle Gas Turbines* (Brussels, Von Karman Institute for Fluid-dynamics, 1977).

16. Morini, A., Macchi, E. and Giglioli, G. "Experimental Results on the Thermal Stability of Some Fluorocarbons" *Proceedings of 15th IECEC* (August 1980).

17. Angelino, G. et al. "Organic Rankine Cycles for Geothermal Hot Waters" (in Italian) *Proceedings of the First Seminar on Geothermic Energy* (Roma, CNR, June 1979).

18. Strong, D.T.G. "Directly Fired Heat Pump for Domestic and Light Commercial Applications" *New Ways to Save Energy* CEE International Seminar Proceedings (Brussels, October 1979).

19. Van Wittenveen, R.A.J.O. "The Stirling-cycle Engines" *EPA Technical Report of the Conference on Low Pollution Power Systems Development* (Eindhoven, February 1971).

20. Beale, W.T. and Rankin, L.F. Jr. "A 100 Watt Stirling Electric Generator for Solar or Solid Fuel Heat Sources" *Proceedings of 10th IECEC* (August 1975).

21. Angelino, G., Ferrari, P., Giglioli, G., Macchi, E. "Combinated thermal engine - heat pump systems for low-temperature heat generation" *The Institute of Mechanical Engineering Proceedings 19027* (1976) 255.

22. Macchi, E. "Design Criteria for Turbines Operating with Fluids Having a Low Speed of Sound" *Closed Cycle Gas Turbines* (Brussels, Von Karman Institute for Fluid-dynamics, 1977).

23. Vavra, M.H. *Axial Flow Turbines* (Brussels, Von Karman Institute for Fluid-dynamics, 1969).

24. Deich, M.E. et. al. *Atlas of Axial Turbine Blade Characteristics* (Mashinostroenie Publishing House, 1965).

25. Macchi, E., Osnaghi, C. "Flow Angles, Profile and Mixing Losses Estimation for Axial Flow Turbine Cascades by Means of Blade-to-Blade and Boundary Layer Calculation" *La Termotecnica* 8 (1975) 25-34.

26. Osnaghi, C. and Macchi, E. "A Methode of Characteristics for Calculating Unsteady Transonic Flow in Turbomachinery Cascades" *Proceeding of XXXIII ATI National Meeting* Vol I, 1978 445-477.

27. Bassi, F. "Calcolo non isentropico di flussi supersonici per lo studio di schiere di pale di turbomacchine" (Non- isentropic Calculation of Supersonic Flows for the Study of Turbomachine Blades) *Proceedings of XXIII ATI National Meeting* Vol. II (1978) 1335-1392.

28. Macchi, E. and Perdichizzi, A. "Theoretical Prediction of the Off-design Performance of Axial-flow Single Stage Turbines" *Proceedings of XXXII ATI National Meeting* Vol. II (1977) 1867-1896.

TABLE I

Thermal Prime Movers

ENGINE	HEAT REJECTION MODE
A) Internal combustion engines	
1. Reciprocating engines a) Diesel engines b) gas engines	- Cooling system (cylinder jacket) (water or steam) + exhaust gases
2. Open cycle gas turbine a) regenerative b) non regenerative	- exhaust gases
3. Combined cycles	
a) reciprocating engine + bottoming cycle	- Reciprocating engine cooling system + exhaust gases + bottoming cycle condenser
b) gas turbine + bottoming cycle	- exhaust gases + bottoming cycle condenser
B) External combustion engines	
1. Steam turbines	- condenser
2. Closed cycle gas turbine	- precooler + intermediate refrigerator
3. Organic Rankine cycles	- condenser
4. Stirling engines	- cooling system
5. Combined cycles	- bottoming cycle condenser + precooler

TABLE II

	Number of Turbines	Total hours	Average hours/turbine
Southern California Gas Co	2	101.534	50.767
Northern Illinois Gas Co	2	102.793	51.397
Michigan Consolidated Gas Co	2	97.599	48.800
United Airlines	4	180.628	45.157
Elizabethtown Gas Co	4	199.181	49.795
Union Gas Co	2	93.920	46.460
Canyon Crest Apartments	3	156.062	52.021
Mountain Fuel Supply	2	126.206	63.103
Brooklyn Union Gas Co	1	72.526	72.526
Mountain Fuel Supply	5	255.296	51.059
Western Airlines	2	161.516	80.758
Carswell Airforce Base	2	143.204	71.677
Shell Oil	4	276.204	69.051
Northern Illinois Gas Co	6	402.029	67.005
Phillips/Arpet	4	247.176	61.794
TOTAL	45	2.616.024	58.134

TABLE III

Small Gas Turbine Costs, Predicted by NREC [8] in 1979 (for a production of about 15.000 units/year, with shaft power of about 100 kW)

Compressor components	20.4 $
Combustor components	36.3 $
High pressure turbine components	159.5 $
Low pressure turbine components	151.1 $
Gear box components	46.7 $
Controls	72.5 $
Others	4.7 $
Assembling	48.5 $
Total	539.7 $

TABLE IV

Organic Rankine Cycle Heat Pumps Development in Italy

Working Fluid	Heat Source	Power kW	Max Temp. C	Min Temp. C	Efficiency	Z	Turbine RPM	R_M	End of Constr.
C_2Cl_4	Solar Coll.	4	75	30	0.094	1	13000	110	1977
C_2Cl_4	Geothermal	4	75	30	0.094	1	13000	110	1978
C_2Cl_4	Solar Coll.	3	75	40	0.067	1	13000	110	1979
C_2Cl_4	Geothermal	3	75	40	0.067	1	13000	110	1979
C_2Cl_4	Exhaust gases	40	110	40	0.115	1	6700	240	1979
C H Cl_3	Geothermal	50	70	40	0.061	1	6500	200	1980 april
Flutec PP3	Solar Coll.	45	280	40	0.245	4	6700	120	1980 august
Flutec pp5	Fuel	25	340	45	0.295	3	8500	130	1980 december
C_6H_5Cl	Solar Coll.	8	177	33	0.210	2	24000	90	1981
$C_2Cl_2F_4$	Geothermal	500	102	40	0.098	1	5500	300	1981

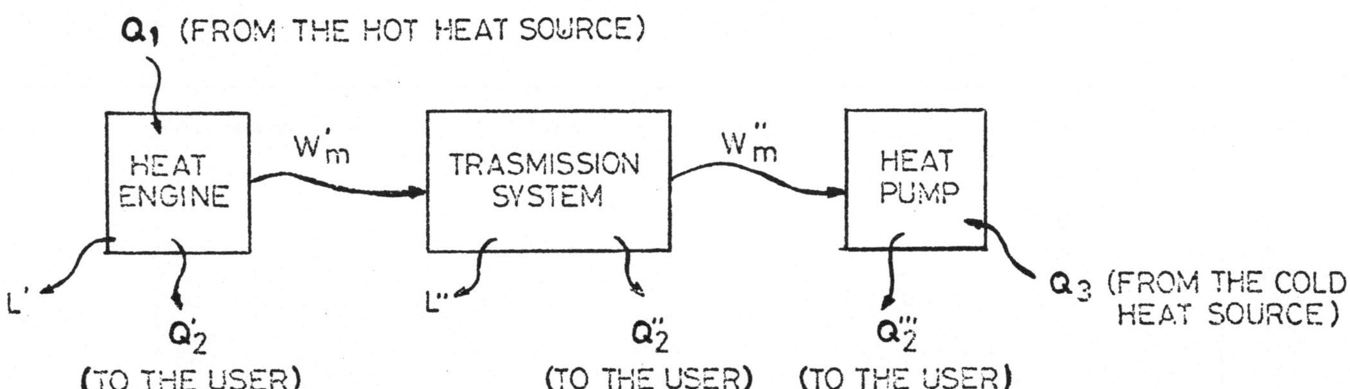

$$\varepsilon = \frac{Q'_2 + Q''_2 + Q'''_2}{Q_1}$$

Figure 1 : Scheme of the energy process for a heat pump

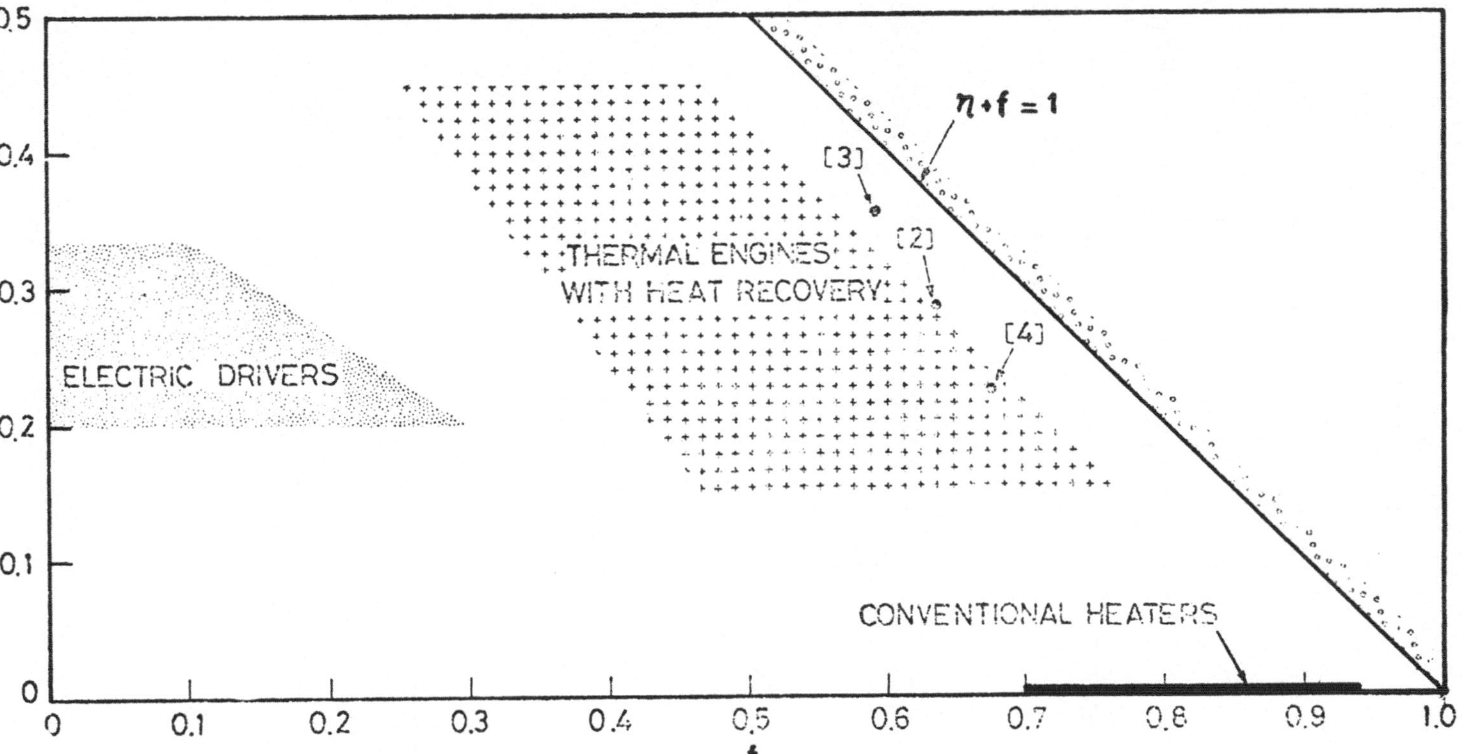

Figure 2 : Performance of various systems in the η - f chart

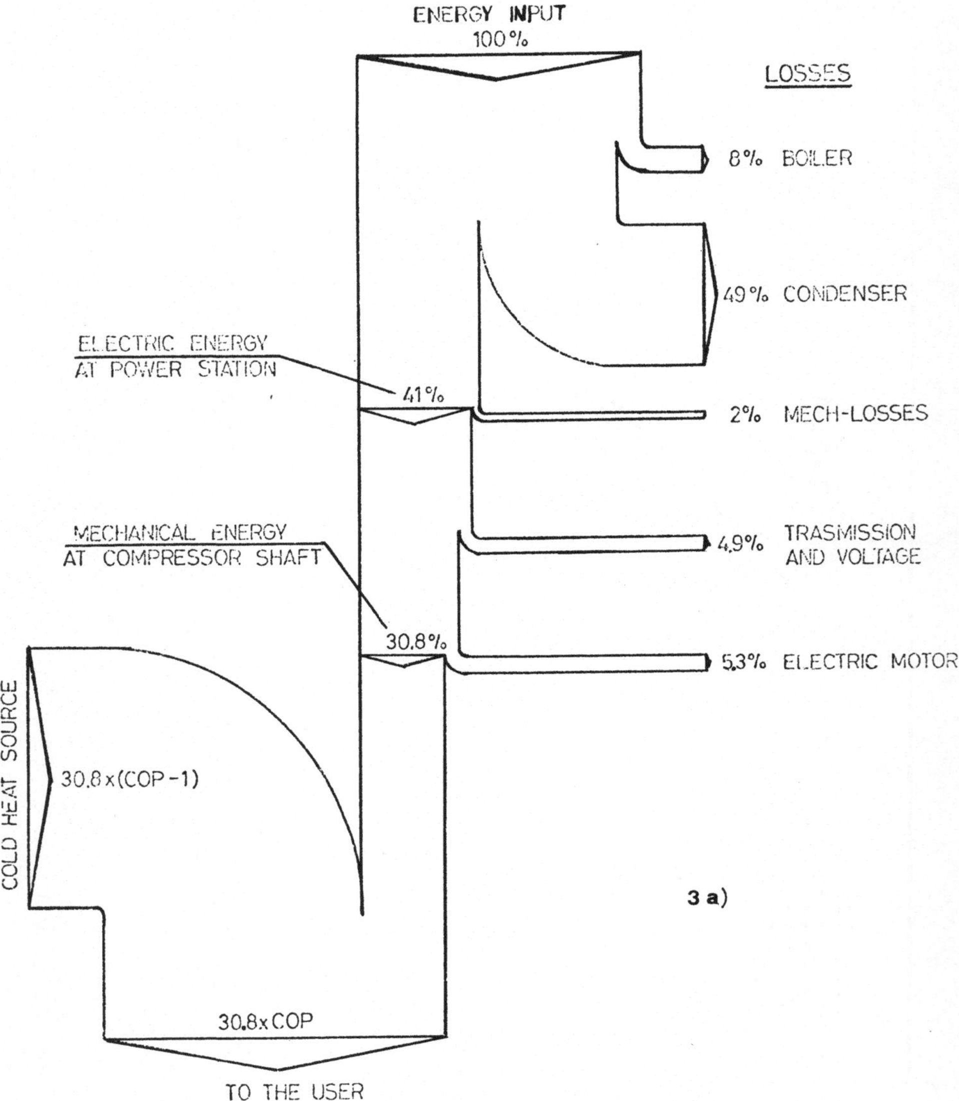

Fig. 3 Sankey diagrams for various heating plants

a) Electrically driven heat pump
b) Thermal engine + heat pump [2]
c) Conventional heater

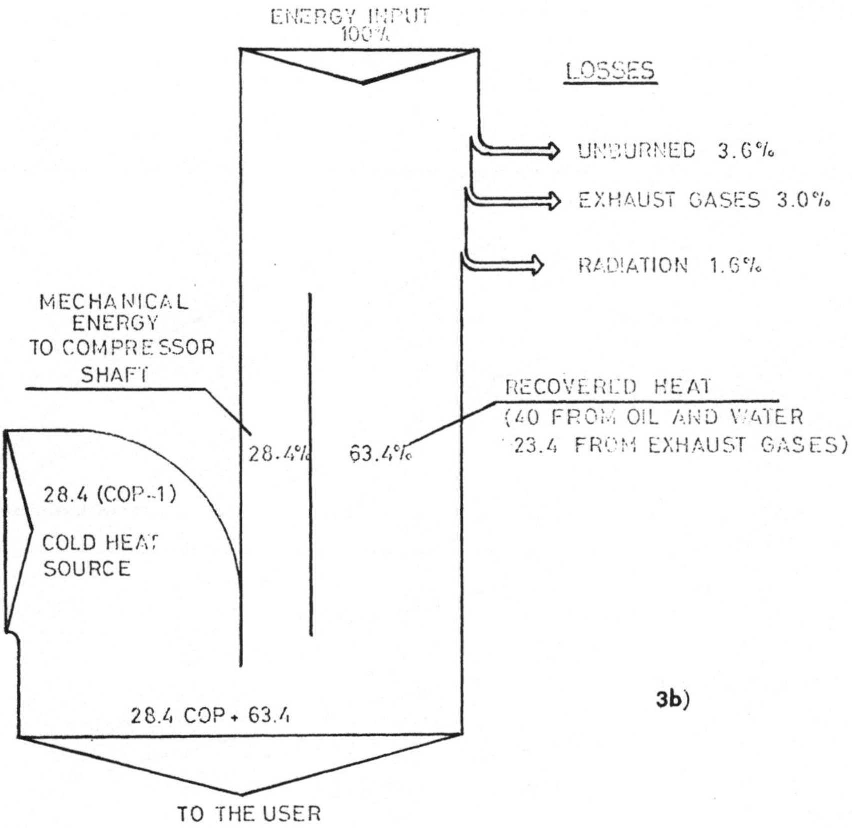

ENERGY INPUT
100%
LOSSES
UNBURNED 3.6%
EXHAUST GASES 3.0%
RADIATION 1.6%
MECHANICAL
ENERGY
TO COMPRESSOR
SHAFT
RECOVERED HEAT
(40 FROM OIL AND WATER
23.4 FROM EXHAUST GASES)
28.4%
63.4%
28.4 (COP-1)
COLD HEAT
SOURCE
28.4 COP + 63.4
TO THE USER

3b)

ENERGY INPUT
100%
15%
VARIOUS LOSSES
(RADIATION, EXHAUST GASES)
85%
TO THE USER

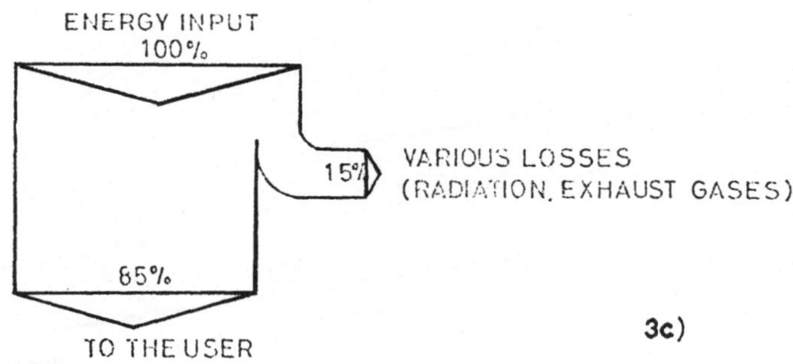

3c)

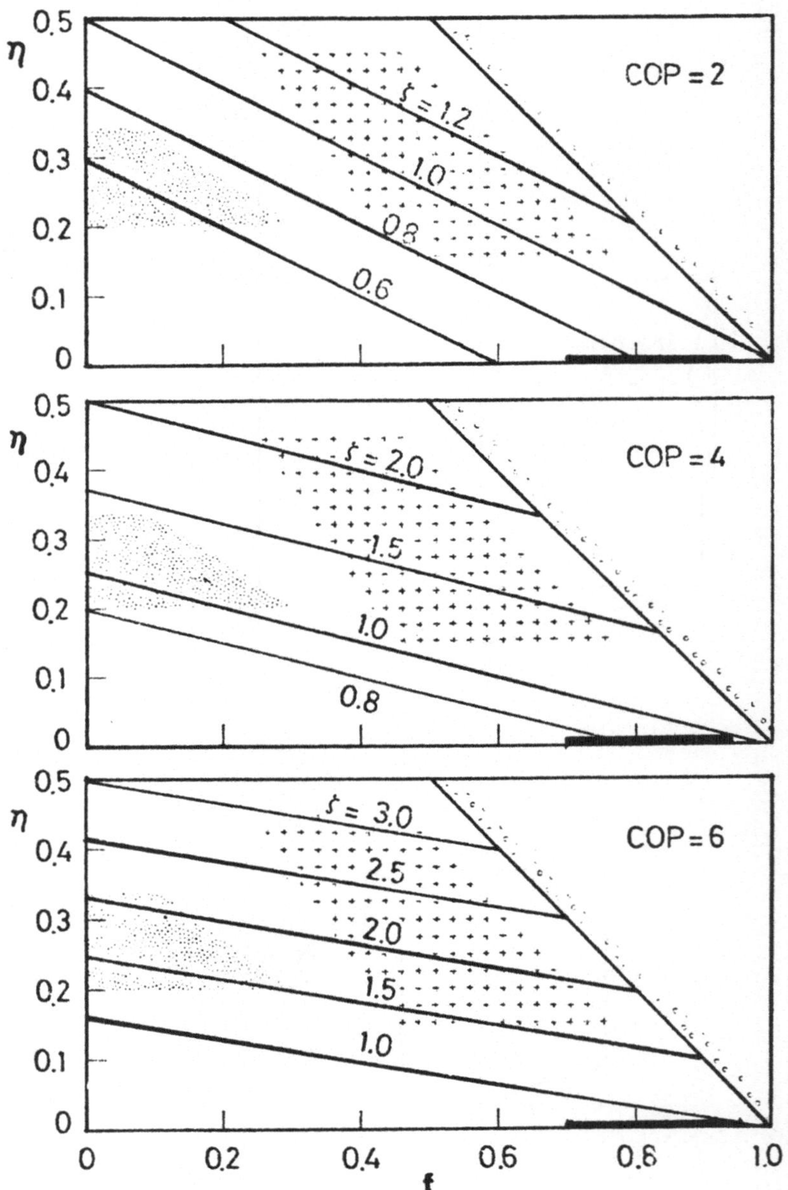

Fig. 4 Values of ζ (see Eq.(2)) for various COP, η and f. For the meaning of the shaded areas refer to Fig.

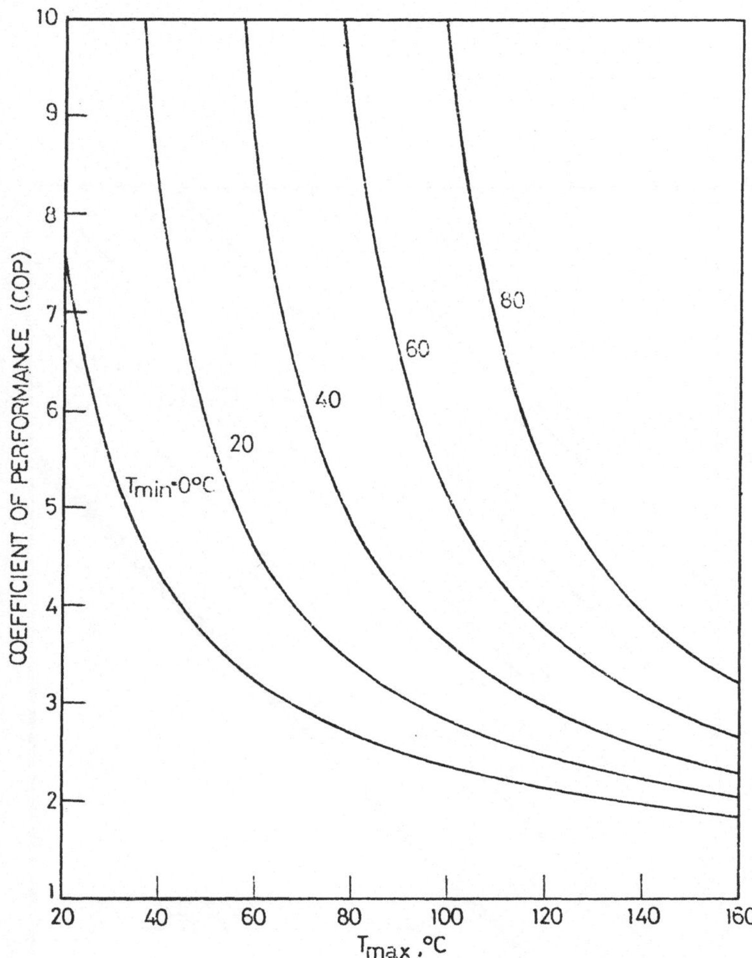

Figure 5 : Coefficient of performance of industrial-type heat pumps (from Eq. (6), with β = 0.5), as a function of the temperatures of the heat sink and of the user.

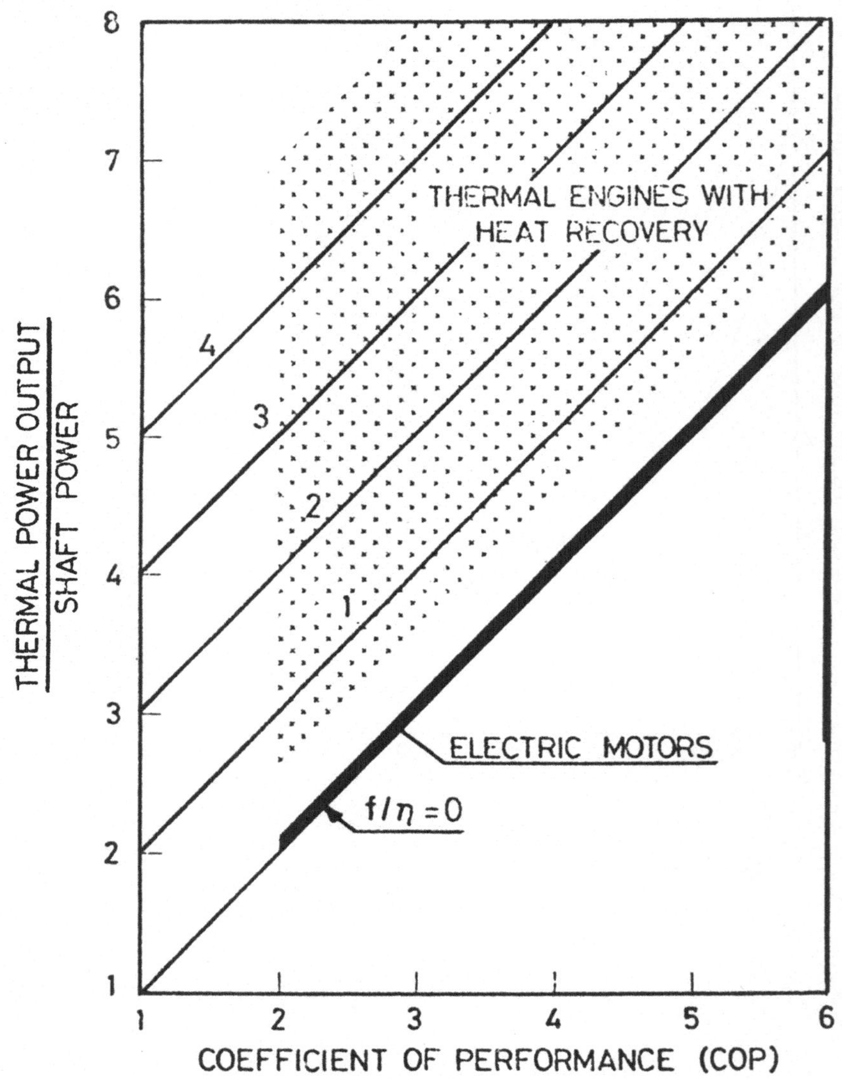

ẓure 6 : Dependance of the thermal power output mechanical shaft power as a function of f, η and COP.

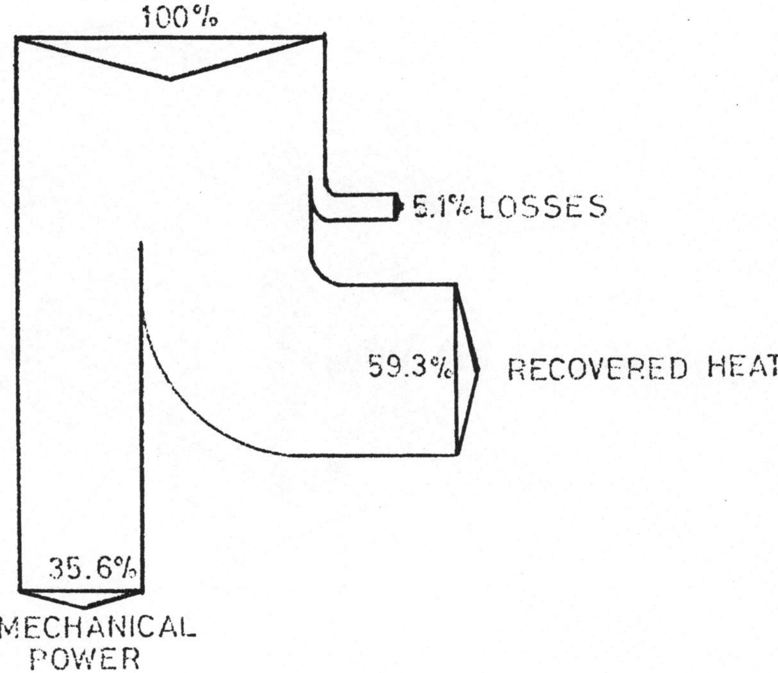

Figure 7 : Heat balance for a Diesel engine with heat recovery

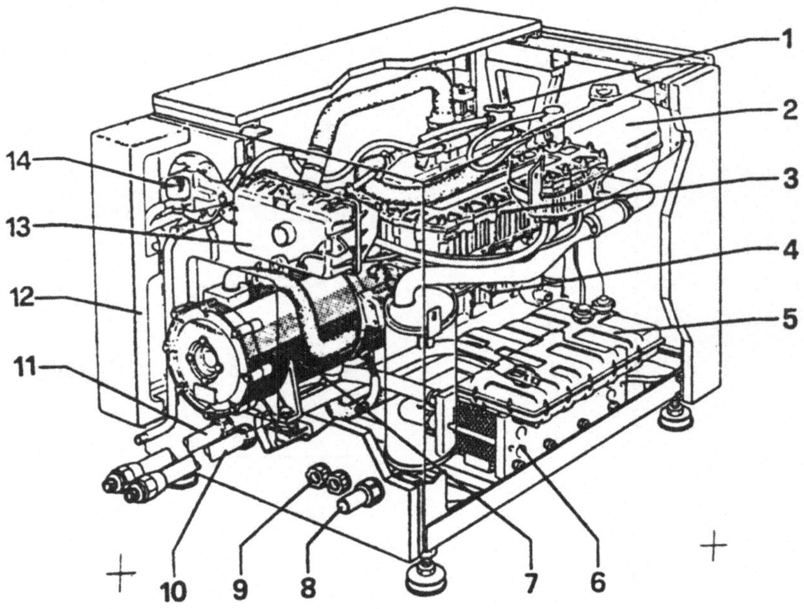

Figure 8 : The TOTEM module; the gas engine is derived from a popular automotive gasoline engine.

1. engine 127
2. water reservoir
3. gas/water heat exchanger
4. oil/water heat exchanger
5. oil reservoir
6. water/water heat exchanger
7. electric generator
8. exhaust
9. electrical connections
10. hot water outlet
11. cold water inlet
12. insulation
13. air inlet
14. gas inlet

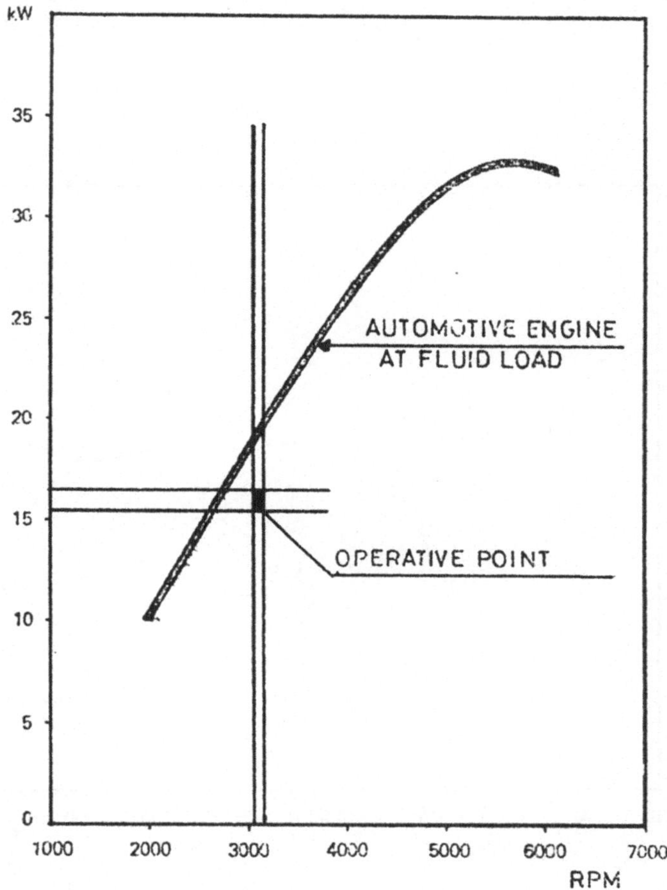

Fig. 9 : Power curve of the automotive engine and operation point of the derived gas engine [2].

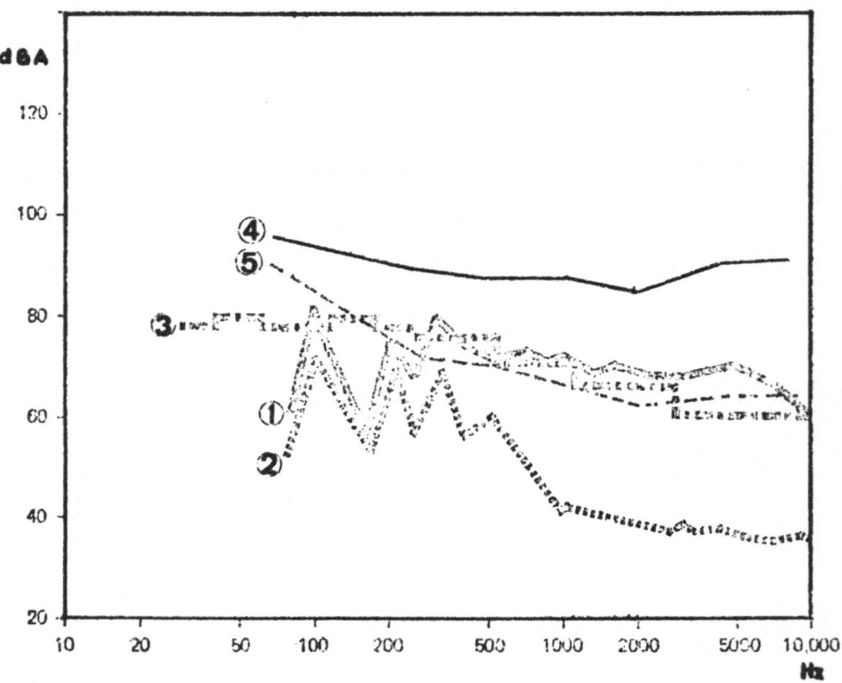

Fig. 10 Noise characteristics of reciprocating engines

1 - original automotive engine [2]
2 - insulated gas engine [2]
3 - conventional pressurized boiler (200.000 kcal/h
4 - Diesel engine without insulation [3]
5 - Diesel engine with insulation [3]

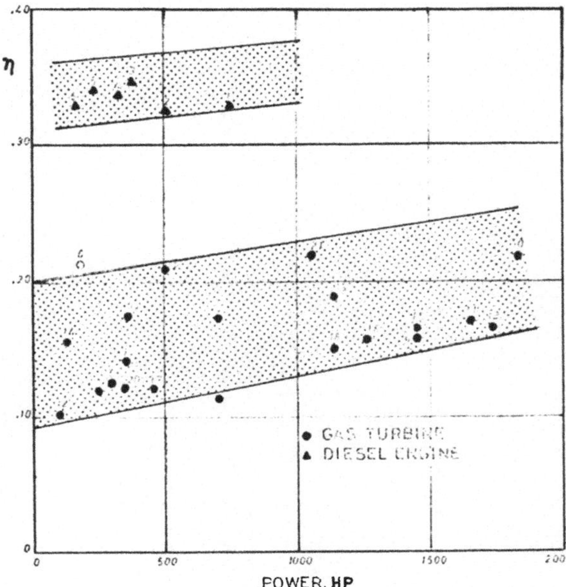

Fig. 11 Efficiency of small commercial gas turbines and Diesel engines (from[9]).

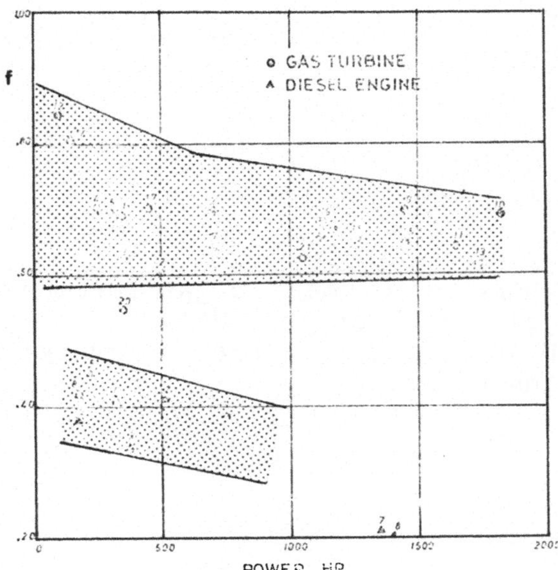

Fig. 12 Recoverable heat from small commercial gas turbines and Diesel engines (from[9]).

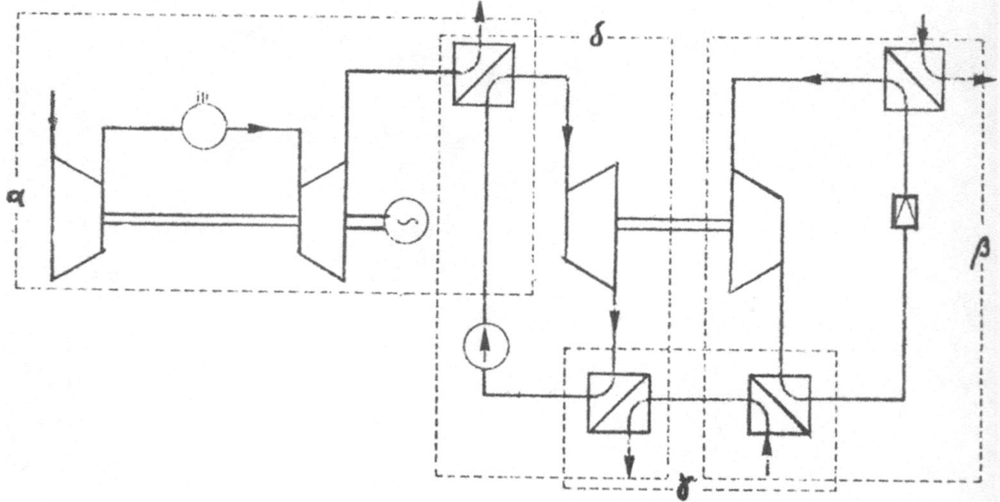

Fig. 13 Scheme of a combined gas turbine-organic Rankine Cycle for heat pump driving; α -gas turbine; β -heat pump; γ -heat recovery; δ -Organic Rankine cycle.

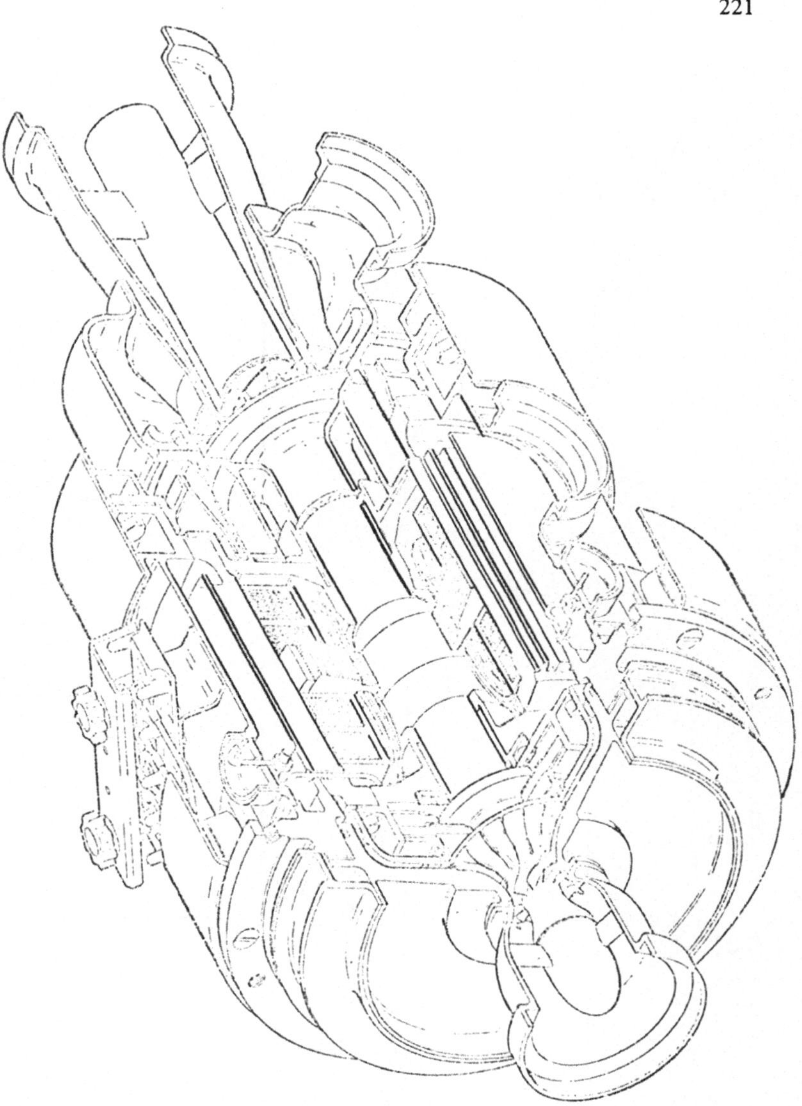

Fig. 14 View of a small closed cycle gas turbine engine, called MINI BRU (Brayton Rotating Unit) [15]. The unit has a power output of 2 kW; it was designed for 10 years of continuos, unattended operation. The measured efficiency was 0.27. A larger unit, called BRU, with power output of 10-15 kW, reaches an efficiency of 0.36.

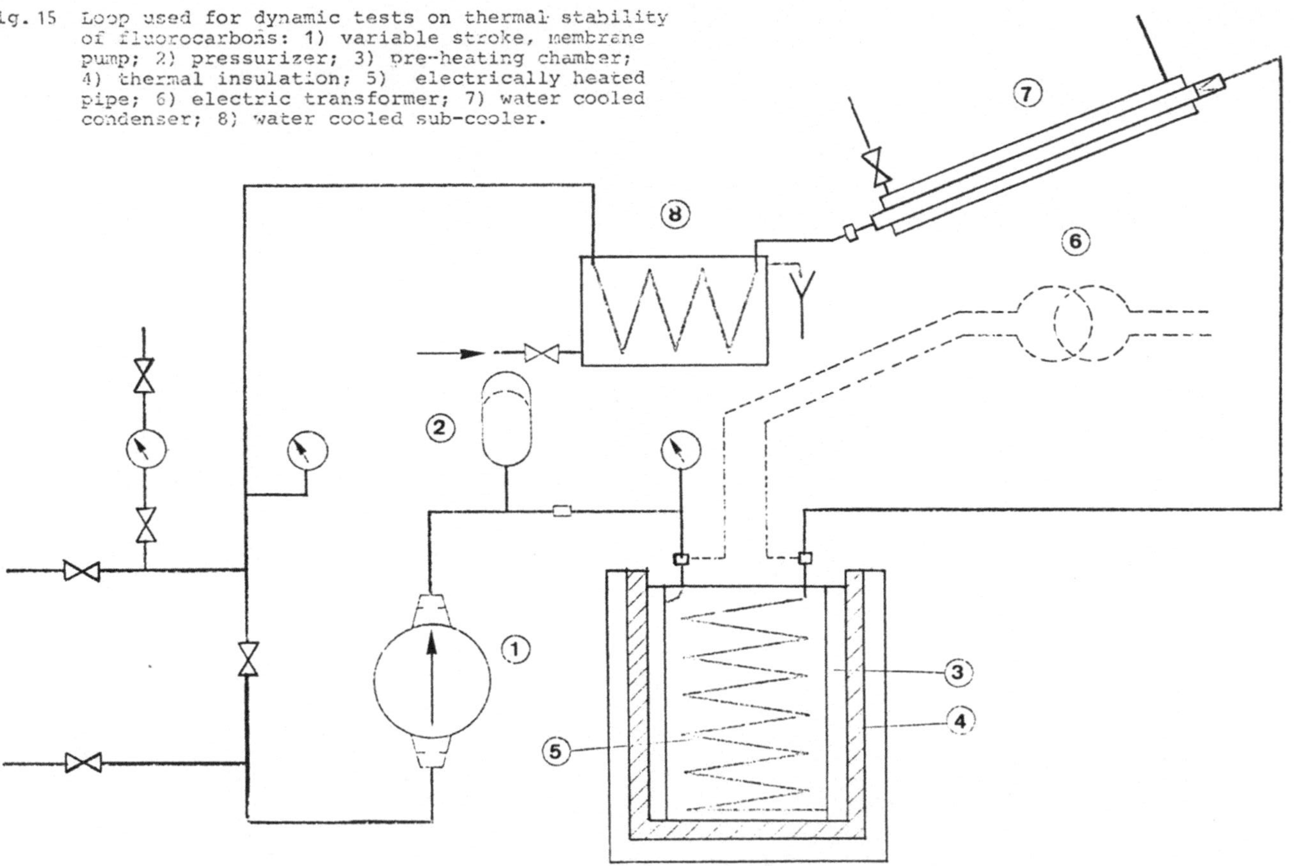

Fig. 15 Loop used for dynamic tests on thermal stability of fluorocarbons: 1) variable stroke, membrane pump; 2) pressurizer; 3) pre-heating chamber; 4) thermal insulation; 5) electrically heated pipe; 6) electric transformer; 7) water cooled condenser; 8) water cooled sub-cooler.

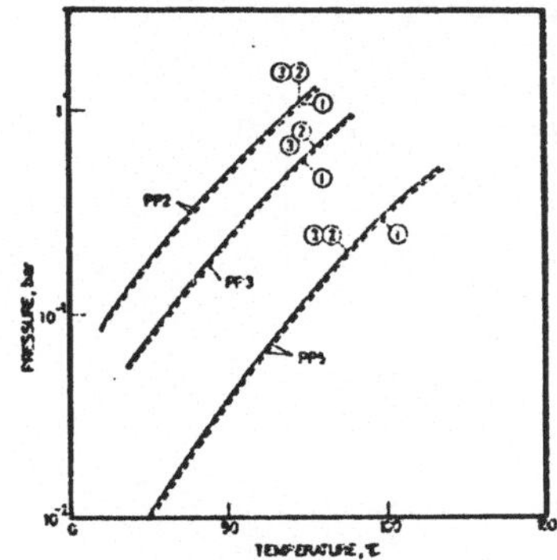

Fig. 16 Results of thermal stability on some fluorocarbons (from [16]).

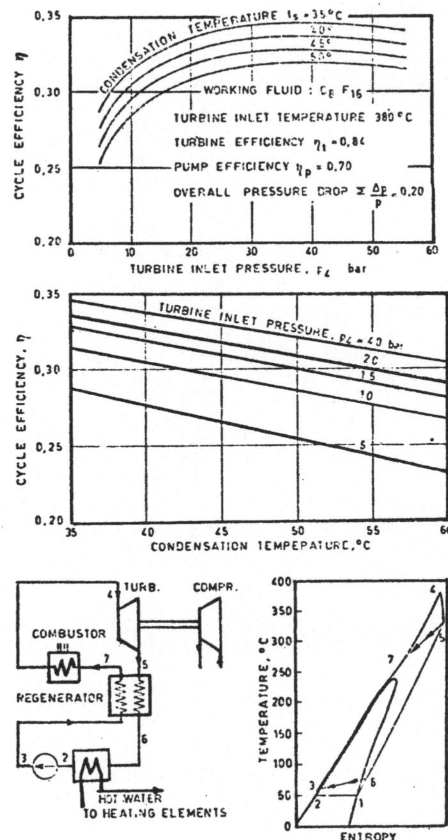

Fig. 17 Configuration and performance of C_8F_{16} power cycles (from [1]).

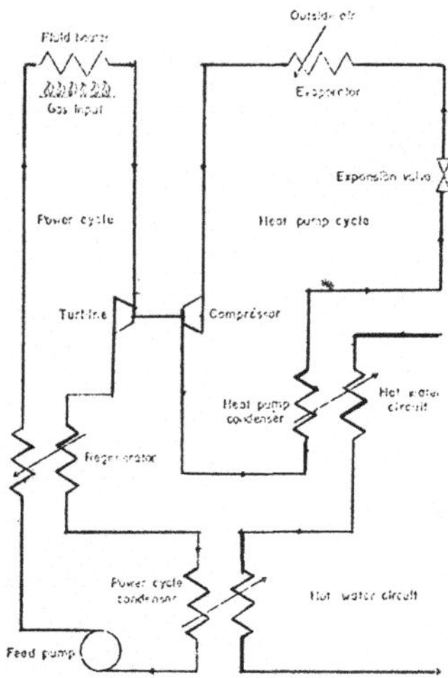

Fig. 18 The single fluid Rankine-Rankine thermal heat pump developed by Glynwed.

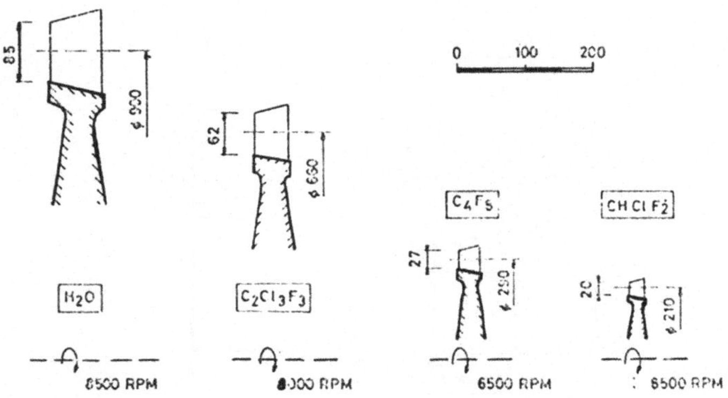

Fig. 19 Influence of the working fluid on the size of the turbine of a Rankine cycle [17] ; output = 500 kW.

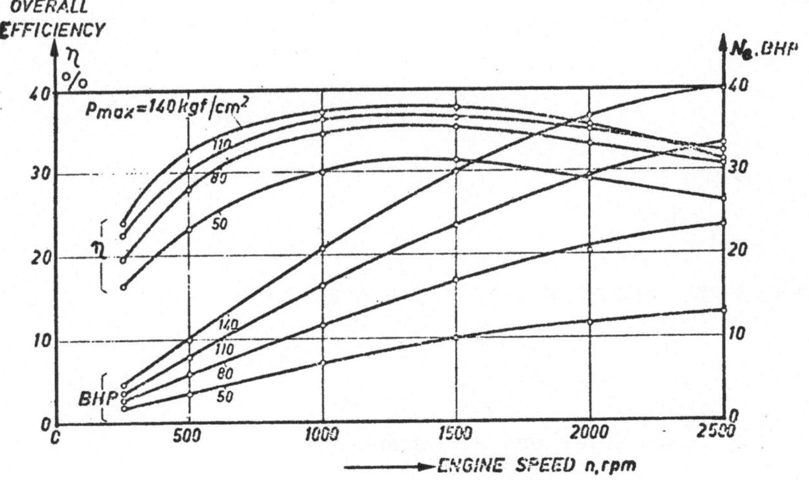

Fig. 20 Efficiency and power curve of a Stirling engine [19].

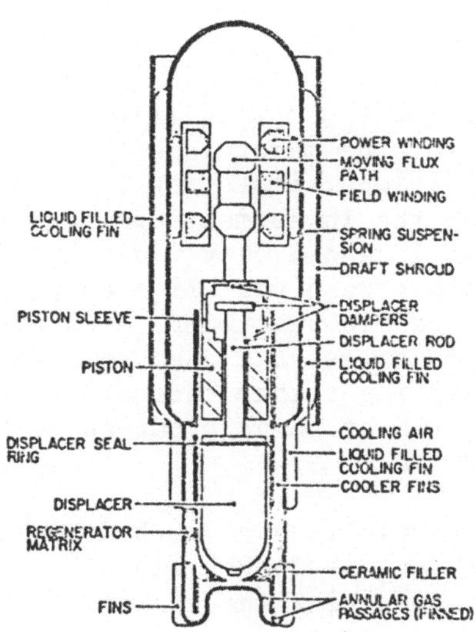

Fig. 21 A small (100 kW) Stirling engine prototype [10], with predicted efficiency of 15%.

HEAT PUMP OPERATING WITH A NON-AZEOTROPIC FLUID MIXTURE[1)]

A. Rojey and J. Durandet

Institut Français du Pétrole, Rueil-Malmaison, France

1. GENERAL BACKGROUND

The development of heat pumps in industry and for home heating appears to be one of the major aspects of energy conservation. Under present conditions, improving the coefficient of performance and lowering the investment are necessary conditions before heat pumps will become very wide-spread. In industry, the use of heat pumps has been very limited up to now. Considering home heating, reversible heat pumps are very common in the United States. This is much less the case in Europe where they are primarily used for heating rather than air conditioning. Nevertheless, the number of heat pumps produced in Europe is growing rapidly (see Table I).

New improvements should lead to much more rapid growth. This is why the Institut Français du Pétrole has undertaken a broad research program on improving the various technologies available and on developing new processes using either absorption or compression cycles.

In the case of compression heat pumps, great improvements can generally be obtained by more reversible heat exchanges at the evaporator and the condenser.

---

1) This research has been undertaken with the support and financial assistance of the European Economic Community.

When heat is exchanged, the outside fluid temperatures generally vary whereas in conventional heat pumps the temperature of the working fluid remains practically constant during the vaporization and condensation stages. If this is the case, the use of a mixture, selected to have a temperature variation during the vaporization and condensation stages parallel to that of the outside fluid with which the heat exchange takes place, enables the coefficient of performance to be greatly increased.

The use of a non-azeotropic fluid mixture is known from refrigeration (References [1], and [2]). In the case of heat pumps, this solution is comparatively new and is covered by several Institut Français du Pétrole patents.

2. THEORETICAL AND EXPERIMENTAL INVESTIGATION

2.1. Basic Design

The flowsheet of the heat pump which has been investigated is shown in Figure 1. The main elements in the circuit are the evaporator E1, the condenser E2, the compressor K and the pressure reducer V1. Heat is recovered from an external fluid 1 and supplied to an external fluid 2. It is essential that exchangers E1 and E2 operate in counterflow. Undercooling the liquid phase in exchanger E3 reduces vapor flashing after expansion through the pressure reducer V1 and therefore enables a greater vaporization temperature range to be obtained. Superheating the vapor phase in exchanger E4 is possible only if the final condensing temperature in exchanger E2 is higher than the final evaporating temperature in exchanger E1. Superheating is useful to increase the coefficient of performance and/or to avoid liquid slugging during compression. Exchangers E3 and E4 are generally optional and are not used when the pressure ratio is low.

2.2. Fluid Mixture Cycle Modeling

When selecting the fluid mixture, various properties have to be considered. First, it is necessary to determine the vaporization and condensation temperature ranges which are needed. It is also necessary to take into account the specific volume at the compressor intake, the thermal transport properties, the thermal stability, the price of the products and their safety of use (flammability, toxicity). These various properties were examined and different possible mixtures of hydrocarbons and halogenated fluids of the Freon type were selected.

For optimizing the fluid mixture composition, a thermodynamic model based on the Soave equation of state was used. This model represents all the fluid mixture properties, namely vapor-liquid equilibrium and enthalpy properties. Various modeling programs based on the previously mentioned thermodynamic model were devised for comparing a fluid mixture cycle with a pure component cycle.

The PACSOA1 (pure component) and PACSOAN (mixture) programs enable such a comparison to be made when the minimum temperature difference in exchangers E1 and E2 is fixed. For a pure component, the ideal cycle is a Carnot cycle based on the exit temperature of external fluid 1 and the exit temperature of external fluid 2. For a fluid mixture it is a Lorenz cycle based on the entry and exit temperatures of the external fluids. It was shown that the real cycle coefficient of performance is close to the ideal cycle coefficient of performance. Therefore, for a fixed minimum temperature difference in exchangers E1 end E2, the energy savings which can be obtained when using a mixture may easily be estimated. For instance, when the difference between the entry and exit temperatures of the external fluids is close to the difference between the mean temperatures of the external fluids, energy consumption in the case of a fluid mixture cycle is about half of the consumption in the case of a pure component cycle.

Predicting the behavior of either a pure component or a fluid mixture heat pump for various external conditions requires modeling of the heat pump cycle for given heat exchange surfaces. This task is performed by the POMP1 and POMPN programs. In these programs it is assumed that only sensible heat is exchanged with external fluids 1 and 2 and that the specific heat of these external fluids is constant. The evaporator and condenser surfaces are then specified through a given surface coefficient σ, which is the product hS of the heat transfer coefficient h and the heat exchanger surfaces.

2.3. Experimental study

An experimental loop equipped with a membrane compressor was used for testing a fluid mixture heat pump as compared to a pure component heat pump. The photograph of the experimental set-up is shown in Figure 2.

Two kinds of exchangers were used : plate exchangers and double-pipe exchangers. Better results were obtained with double-pipe exchangers, due to their higher through-put velocity. In this experimental loop, pressures, flowrates and temperatures (15 temperature sensors) are continuously recorded.

In Table II, the results obtained for a pure component (R-12) and a mixture are compared. The experimental results are in good agreement with the results obtained by the computer model.

Heat transfer coefficients were measured, and for high enough throughput velocities (double-pipe exchangers), no significant difference was found between the pure component and the mixture heat transfer coefficients during the vaporization and condensation stages.

The computer model calculates the temperature evolution during the vaporization and condensation stages by assuming equilibrium between the liquid and vapor phases. For testing this assumption and comparing the experimental and computed temperature profile during the vaporization stage, an experimental set-up was used, consisting of a vertical tube heated by a counter-current flow of water. A moving temperature sensor measured the temperature profile all along the inner tube. Temperature sensors were placed on the tube to measure the heat transfer coefficient during evaporation. The photograph of this experimental set-up is shown in Figure 3.

Measurements which were performed for different hydrocarbon mixtures have shown that, for mixture compositions which are to be used in heat pumps, the vapor-liquid equilibrium assumption is valid. Also, with through-put liquid velocities higher than 1 cm/s, heat transfer coefficients obtained for mixtures are of the same order of magnitude as those of pure components.

3. APPLICATIONS

3.1. Home Heating

Different cases have been considered :

- Exhaust air/external air
- External air + exhaust air/external air + internal recycled air
- Water/external air + internal recycled air.

In these three cases, the fluid mixture heat pump was compared to a pure component heat pump, heating a 250 m^3 house under climatic conditions typical of the Paris region (zone B, reference [3]). A heat loss coefficient of 1.25 $W/m^3 \times {}^\circ C$ in accordance with French regulations for electric heating was considered (Reference [4]).

The average coefficient of performance and the performance factor, taking into account a resistance heating backup system, were calculated for a heating season.

Figure 4 shows a comparison between the average coefficients of performance obtained during a heating season with a pure component and a fluid mixture for different values of the evaporator surface coefficient and with the condenser surface varying proportionally.

The coefficient of performance of a commercial water-water heat pump when the pure component was replaced by a mixture was estimated. Without changing the evaporator and condenser surface areas, an average coefficient of performance of 4.43 is obtained when working with a mixture as compared to an average coefficient of performance of 3.16 when working with a pure component. The capital investment is expected to be about the same or slightly lower when a mixture is used, due to the smaller power needed for the compressor and the motor. This means that the heat pump becomes economically much more attractive and, if the investment can be further reduced by mass production, the heat pump should find a large market, not only in the field of new buildings but also in the field of existing housing which corresponds to a much larger market.

3.2. Industrial applications

Industrial applications may be very numerous and varied. A practical case corresponding to a warm air drying process was examined. The warm air drying process is represented in Figure 5.

An economic evaluation was made for a dry air flowrate of 6000 Kg/h and a wet air saturation temperature varying between 40° and 60°C. Three possible solutions were considered :

A. Pure component heat pump (normal butane)

B. Fluid mixture heat pump (85 % butane - 15 % hexane)

C. Pure component heat pump + heat exchange circulating loop

The results obtained are shown in Table III. They show that the fluid mixture heat pump is the most attractive from the view-points of both energy consumption and investment.

4. CONCLUSION

In the present study, the fluid mixture heat pump concept has been proven to be sound. Attractive prospects have been found for both home heating and industrial applications. Further work is needed for full scale testing such fluid mixture heat pumps with commercial equipment.

REFERENCES

1. Bourguet, J. Garnaud, R. and Grenier, M. *The Oil and Gas Journal* (1971) August 30, p. 71-75.

2. Kinard, G.E., and Gaumer, L.S. *Chem. Eng. Progress* 69, (1973) nr. 1, p. 55-61.

3. CSTB - Document Technique Unifié - Règles Th-G77 *Cahiers du Centre Scientifique et Technique du Bâtiment* (Paris, Nov. 1977) nr. 184.

4. Ministère de l'Equipement et de l'Aménagement du Territoire *Journal Officiel de la République Française* Paris, 23 July 1977.

TABLE I

Number of heat pumps produced in France

Residential heat pumps	Water-Water	Air-Water	Air-Air
Year :			
1975	116	52	1722
1976	336	204	3228
1977	467	280	4423
1978	712	952	9103

TABLE II

Experimental Results

		Pure component R-12	70 % R-22/ 30 % R-11
Evaporator	Ext. fluid t_{in} (°C)	44.9	44.7
	t_{out} (°C)	15.8	15.1
	Int. fluid t_{in} (°C)	10.3	7.4
	t_{out} (°C)	44.6	39.7
Condenser	Int. fluid t_{in} (°C)	91	66.1
	t_{out} (°C)	25.8	25.8
	Ext. fluid t_{in} (°C)	25.6	25.6
	t_{out} (°C)	56.7	53.2
External fluid 1 flowrate (1/h)		13.4	13.4
External fluid 2 flowrate (1/h)		14.7	14.7
Internal fluid flowrate (1/h)		7.92	6.53
Heating power (W)		531.72	473.34
Compression power (W)		77.67	35.19
COP		6.84	13.45

TABLE III

Warm air drying process

Comparison between solutions A, B and C

	A PC heat pump	B MIX heat pump	C PC heat pump + exchange
Investment FF (Installed unit)	1814153	1631718	1849656
Utilities Electricity kW Cooling water m^3/h	 90 7	 44 5	 76 6
Annual expense (0.15 FF/kWh 0.08 FF/m^3 c.w.)	112084	62065	96651

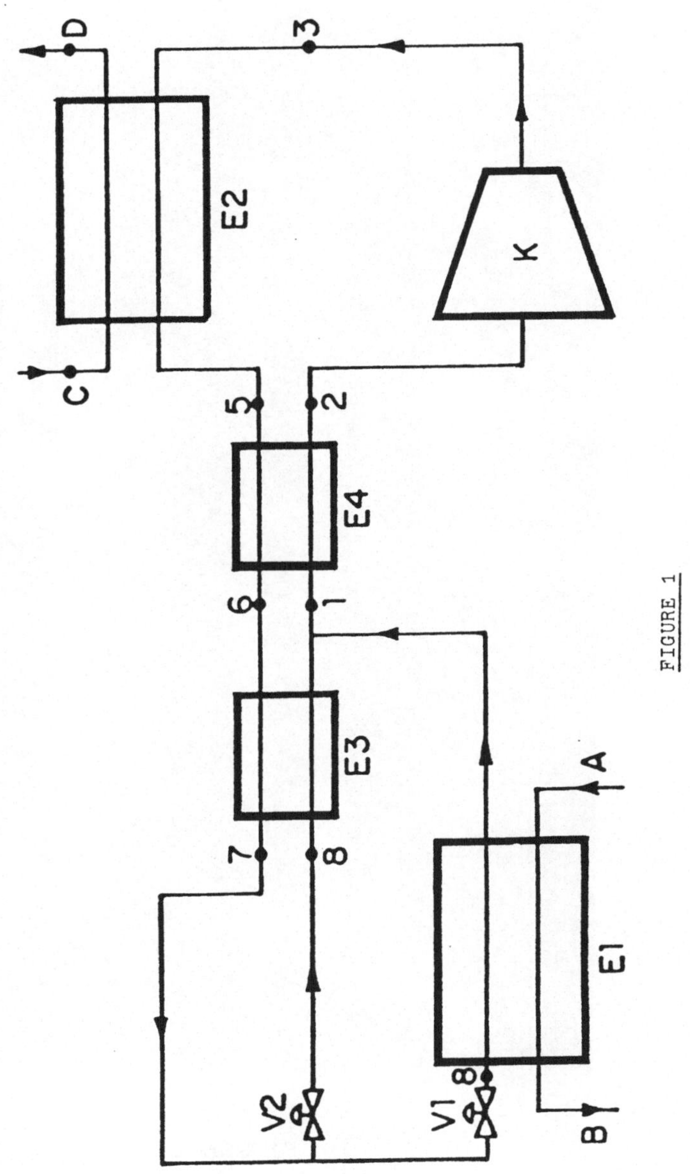

FIGURE 1

FIGURE 2

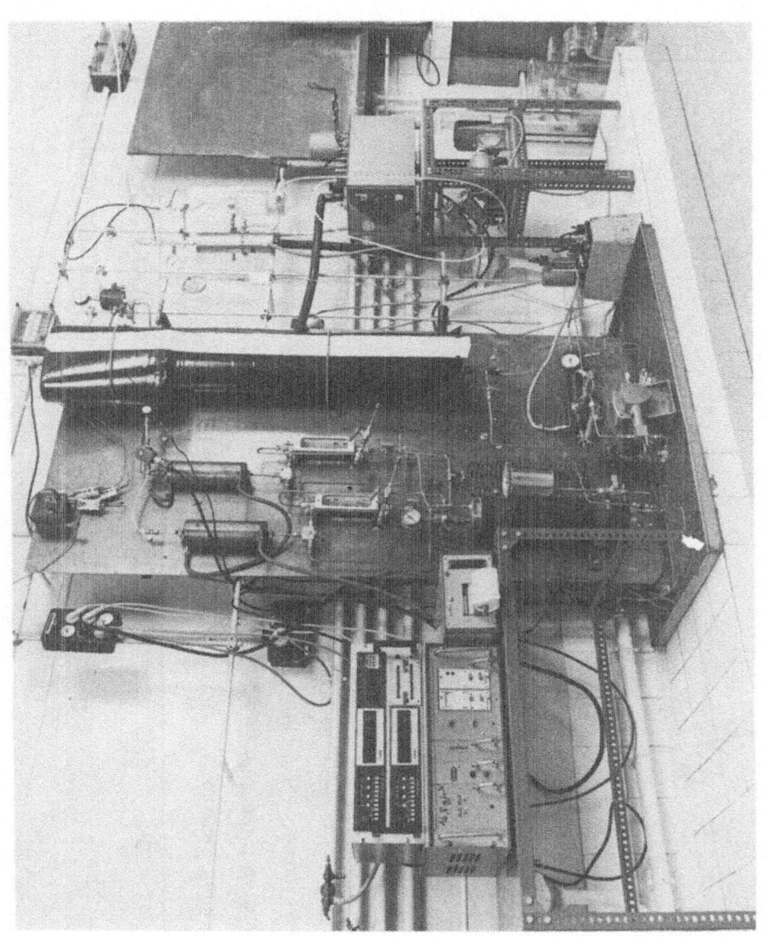

FIGURE 3

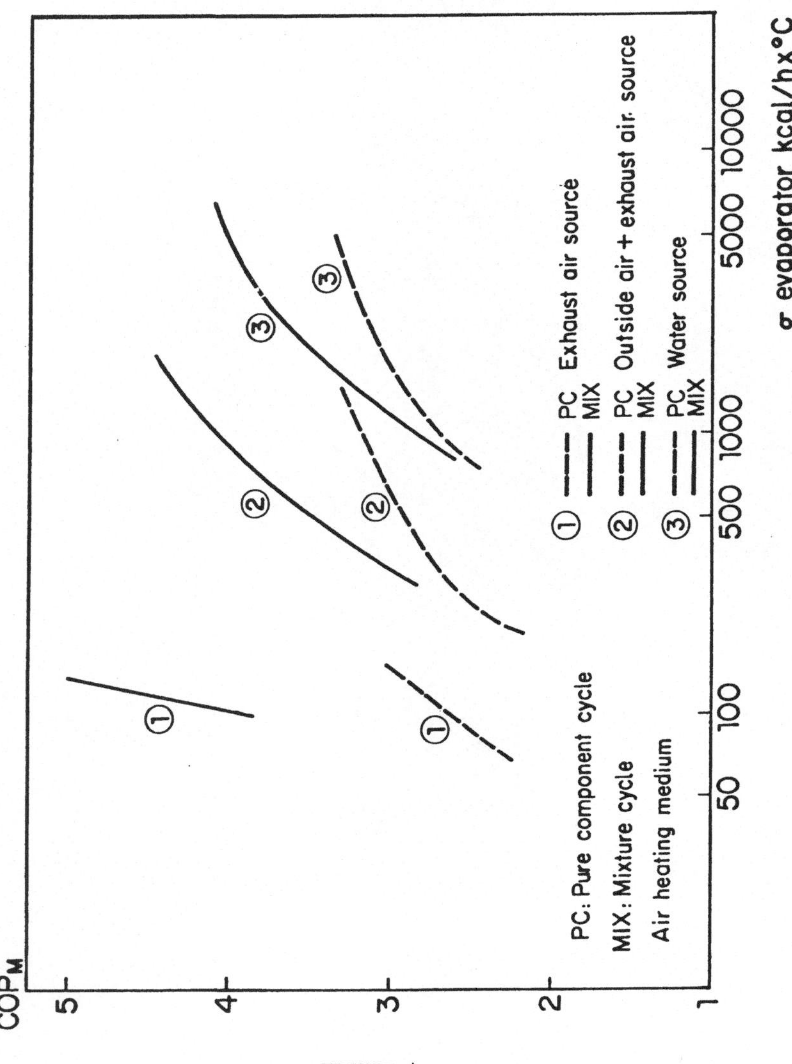
COP_M
5
4
3
2
1
50
100
500
1000
5000
10000
σ evaporator kcal/hx°C
PC: Pure component cycle
MIX: Mixture cycle
Air heating medium
① PC MIX Exhaust air source
② PC MIX Outside air + exhaust air source
③ PC MIX Water source

FIGURE 4

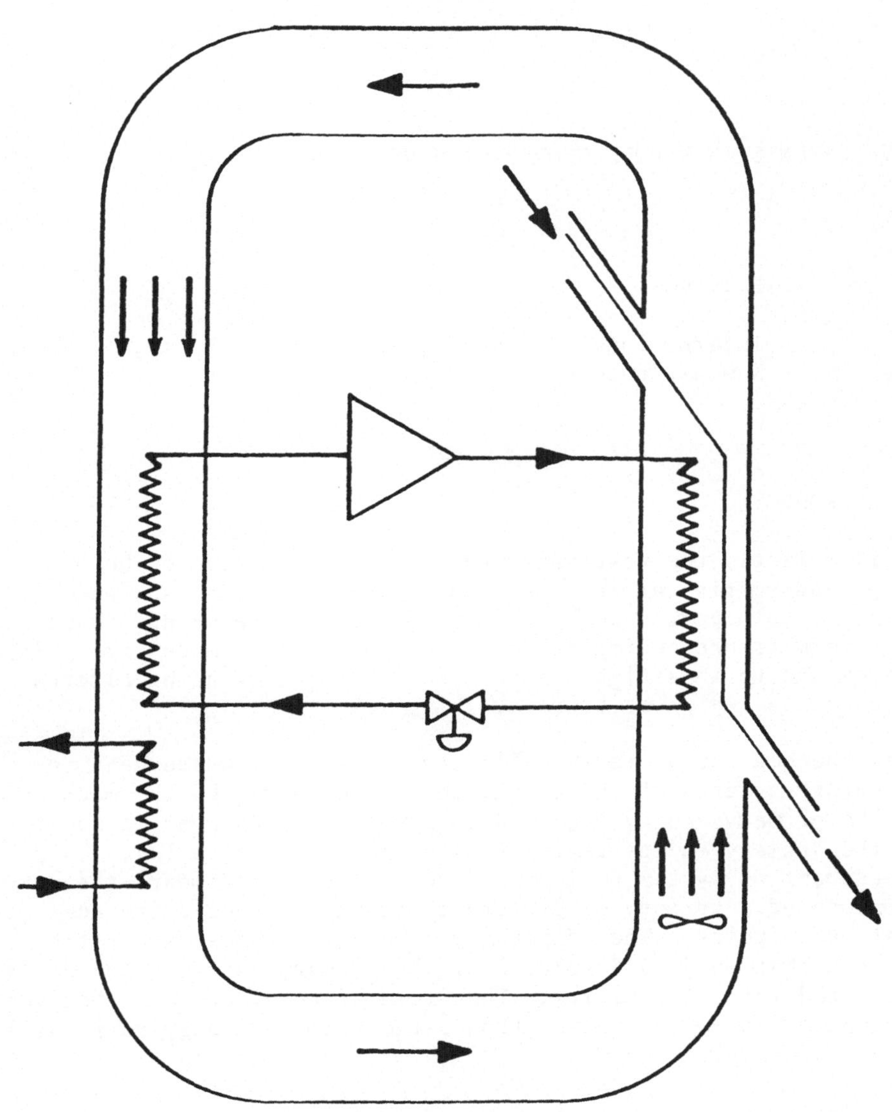

FIGURE 5

DEVELOPMENTS IN VAPOUR COMPRESSION HEAT PUMPS

P.A. Kew

International Research & Development Co Ltd,
Newcastle upon Tyne, U.K.

1. INTRODUCTION

Until recently the vast majority of vapour compression heat pumps and refrigerators worked on the Rankine Cycle and were powered by an electric motor drive. However research and development work is proceeding to enable advances to be made with this system and to investigate and exploit the potential of alternative cycles and power sources.

When the heat pump is installed for heating or waste heat recovery the ratio of the useful energy delivered by the machine to the primary energy required to power the heat pump is known as the primary energy ratio (PER) or the heat ratio (ζ). This assessment of system performance can be used to compare the performance of a variety of different arrangements of drive and heat pump cycles. The efficiency of electric power generation is such that only 25 % to 30 % of the primary energy consumed is converted into electricity. Therefore if an electrically driven heat pump has a COP of less than 3 the primary energy ratio will be below unity.

If however the heat pump is powered by a locally situated internal or external combustion engine it is possible to recover as useful heat much of the energy supplied to the engine while the shaft power is used to drive the compressor.

The development of a gas engine driven, high temperature heat pump will be described later, illustrating the problems in the design of an internal combustion engine driven heat pump and the extension of the Rankine Cycle heat pump to operate at higher temperatures than those conventionally used.

Firstly there is a brief discussion of some of the alternative cycles and drives which are being considered for use in a heat pump context. Any thermodynamic cycle which may be used for producing power can in principle be reversed and used in a heat pump or refrigerator. A number of possible cycles may be used either for the drive or the heat pump circuit. Certain cycle combinations are favoured because of the ease of matching the heat pump and power cycle characteristics. The selection of heat pump and drive cycle is influenced not only by the theoretical performance of the thermodynamic cycle but also by the characteristics of the type of machinery which operates using the cycle.

The discussion here will mainly be concerned with the use of internal combustion engines to power Rankine cycle heat pumps. Mention should be made of the Joule or Brayton cycle which is used at present in certain air conditioning applications and is the cycle on which the gas turbine operates. Another system worthy of consideration is the organic Rankine powered heat pump, this can operate in two modes, either using two working fluids or with a single fluid for both the power and heat pump circuits. Stirling cycle machines are also being developed for use as both the prime mover and the refrigerant circuits in heat pumps.

In an industrial application where there are perhaps tens of megawatts to be recovered one can envisage a gas turbine being used to drive a centrifugal compressor driving a Rankine heat pump cycle. Virtually all the waste heat from the gas turbine is rejected via the exhaust and an exhaust gas heat exchanger could be used to recover much of this to add to the heat from the condenser. Although no new technology would be required for this type of machine the first cost and running cost would be very great so despite large potential energy savings it is unlikely that such an installation will be seen in the near future.

2. ALTERNATIVE CYCLES

2.1. Brayton Cycle

A simple reversal of the open cycle Brayton power cycle as given in Figure 1 gives a type of heat pump which may be used in two forms as shown in Figure 2.

The arrangement of Figure 2(b) is used in vehicle air conditioning. Ambient temperature air is compressed and then cooled back to the ambient temperature. It is then expanded through the turbine and thereby cooled, this air is then passed into the vehicle interior. Brayton cycle cooling is especially appropriate for the air conditioning of passenger aircraft where a supply of compressed air is available from the engines and it must simply be cooled and expanded through a turbine before being passed into the passenger compartment.

2.2. Rankine-Rankine Cycle

Figure 3 shows an arrangement of the Rankine/Rankine Cycle using a single working fluid together with the thermodynamic cycle shown on a pressure enthalpy diagram.

The cycle illustrated here is using a low pressure refrigerant which is suitable for use with rotary machines which, it is assumed, are used here. In practice the cycle would work equally well with a reciprocating compressor/expansion engine but it is thought that it may be most useful in a domestic heating application where the quiet running rotary machinery is more widely used.

The points ABCD follow the pattern of a mechanical vapour compression cycle. ABEF gives the power cycle. The refrigerant is compressed adiabatically from B to E by a small liquid pump and the work added is negligible since the liquid is virtually incompressible. Between states E and F heat is added by direct external combustion of the fuel in what can be called the boiler. The upper temperature limit reached at F is dictated by thermal stability of the refirgerant and oil rather than by thermodynamic limits and it is this limit which restricts the COP obtainable by this cycle.

This heat pump consists really two separate cycles, but they are combined for simplicity. There is only one condenser, which is common, and the rotary machines can be virtually identical and combined on a single shaft.

Without worrying here about the details of controlling this type of system, the cycles will be briefly analysed to give a crude indication of the likely primary energy ratio. For simplicity the combined cycles will be considered.

There are three mass flows corresponding to the three pressure levels of the system, Me, Mc and Mb in the evaporator, condenser and boiler respectively.

$$\zeta = \frac{(h_A - h_B) \times M_c}{(h_F - h_E) \times M_b}$$

Because the expansion engine is used to drive the compressor, one can equate the shaft work for the two machines.

$$\text{Work} = (h_F - h_A) \times M_b = (h_A - h_D) \times M_e$$

And finally one can say

$$M_c = M_e + M_b$$

Solving these three equations together gives the following expression for the primary energy ratio :

$$\zeta = \left[\frac{h_F - h_A}{h_A - h_D} + 1\right] \frac{h_A - h_B}{h_F - h_E}$$

One practical embodiment of this type of combined cycle is in the use of rotary vane compressors and expansion engines. This has been proposed in the USA by Battelle (Columbus) who adopted a novel type of vane with pivoting tips, to allow oil-free operation.

In the UK, Denco-Miller Ltd have proposed a system in which the expansion and compression units are even more tightly integrated, in a "two-lobe" rotary vane engine. This is designed with the minimum of moving parts to reduce first costs and keep maintenance to a minimum. This system is intened to be a gas fired heat pump, whereas the Battelle design is to be solar powered. Note that similar studies have been undertaken at General Electric.

The ability of Rankine powered heat pump to operate with a low boiler temperature has made them particularly interesting to workers seeking to provide useful air conditioning or water chilling or refrigeration from solar energy. These developments are especially attractive for use in countries where sufficient solar power is available, the cooling effect is most needed, and it is needed during the heat of the day when the power is strongest.

If the power and heat pump circuits are separate and two working fluids are used the thermodynamic analysis of the system is identical to that for an electric driven heat pump, together with the Rankine cycle used to generate the electricity. However by situating the power source close to the heat pump the waste heat from the power cycle condenser may be added to the energy from the heat pump. This is advantageous even if the primary energy used is of a high grade.

2.3. Stirling Cycle

The development of Sterling engine driven systems is at a very early stage. Kinematic Stirling engines suitable for driving heat pumps are not commercially available. Several firms are currently working on Stirling engines for various uses, some of which may be useful as heat pump prime movers. Much of the work carried out on kinematic Stirling engines is based upon the early work by Philips in the Netherlands. Free piston systems are primarily being developed in the USA. The free piston machines are unsuitable for use with conventional compressors so it is necessary to develop inertia piston compressors for use with the engine. The Stirling Cycle which is illustrated in Figure 4 operates as follows.

As gas is compressed isothermally from 1 to 2, the gas is cooled externally. It is passed through the generator and heated internally. Between 3 and 4 the gas is externally heated and allowed to expand, in doing so it delivers useful work then finally the gas is passed back through the regenerator, which internally cools the gas back to state 1.

The Stirling cycle is interesting because all the processes described in the ideal cycle are reversible, and the external heat exchangers are isothermal which means that an ideal Stirling heat engine could equal the efficiency of the Carnot cycle. It is of interest because, in reverse, the same applies to a Stirling refrigeration machine or heat pump.

In practice there are several idealizations which make the Stirling machine difficult to design, and bulky. These are listed below :

1) The pistons must move intermittently, not sinusoidally
2) The regenerator is frictionless and 100 per cent effective
3) Ideal heat exchangers are assumed for external heat transfer.

Heat exchangers always pose problems with external combustion engines. In effect, an extra two temperature differences are being introduced between heat source and sink.

A free piston system incorporating both heat engine and heat pump is illustrated in Figure 5. Performance energy ratios for this type of system compare reasonably favourably with those for Rankine/Rankine systems working between the same temperature, however this comparison should be confirmed by experiment.

3. ALTERNATIVE HEAT PUMP DRIVES

Internal combustion engine driven heat pumps were first utilised during the last decade for swimming pool and leisure centre heating. Often combined heating and cooling is necessary making a heat pump system the natural choice. As well as increasing the primary energy ratio by heat recovery from the gas engine it is possible to raise the output temperature of the system above the condensing temperature of the refrigerant by passing the process water through the engine and exhaust heat exchangers. Thus a high COP may be retained for comparatively high water outlet temperatures.

Currently most commercial heat pump installations use four stroke gas engines of the type used for standby power generation. Though Diesel engines have higher efficiencies than Other engines the general preference has been for gas engine driven system. This is partly because of the lower fuel cost and partly because of the greater reliability and less corrosive exhaust of gas engines. The development of automotive derived power units for heat pump use is being carried out by the industrial divisions of several motor manufacturers. These engines will be suited to duties requiring a shaft power of the order of 10-1000 kW. There are several major problems to be overcome in adapting an automobile engine for use as a heat pump drive. A typical automobile is used for perhaps 500-1000 hours per annum and regular maintenance is regarded as the rule rather than the exception. These engines are typically designed to produce power at 2000-5000 RPM whilst the majority of compressors are designed for 1500 RPM drive, thus a gearbox or belt drive is required to reduce speed.

The automobile engine is designed for a relatively short life and is less reliable in general than a purpose built stationary engine. However a converted automotive engine for compressor drive may have a capital cost of less than one third of that of the equivalent purpose built engine.

Engine maintenance costs are not as easily estimated as fuel consumption, recoverable heat, or initial cost; they are, however, not entirely elusive. The increasing use of guaranteed maintenance and service contracts most common in the United States has eliminated much of the estimating formerly required in feasibility studies. Maintenance contracts vary from complete maintenance and service, including all parts, supplies and labour, to contracts that provide only a guaranteed cost for engine rebuild. For this reason, a maintenance cost figure is meaningless unless well defined. Complete maintenance costs are composed of three basic items :

a) The miscellaneous maintenance and service cost, including service manual recommendations plus make-up oil (excluding labour to perform this routine duty).

b) The overhaul maintenance cost, usually expressed in terms of cost per engine operating hour. This item should cover all labour and parts necessary to perform major and minor overhauls at the recommended intervals.

c) The third item is the labour cost necessary to perform the miscellaneous service for item (a).

Items (a) and (b) will vary considerably with the severity of service the engine must perform. For on-site power installations the conditions under which the engines operate, the quality of fuel, the routine maintenance the engine receives, are usually considered to be good. Item (c) will vary with labour costs and the location of the engine plant with respect to the point from which service personnel must be despatched. Because of the many variables, it is difficult to provide realistic figures that would be useful for all applications. Maintenance costs should be based on past experience.

In many applications of the gas engine, this past experience may be somewhat limited. In the United Kingdom, British Gas have installed several units comprising of refrigeration compressors for air conditioning duties driven by Ford, Perkings and Pelapone automotive engines. Their experiences are very encouraging to potential users of these systems.

4. A GAS ENGINE DRIVEN HIGH TEMPERATURE HEAT PUMP

The IRD high temperature gas engine driven heat pump is at present under test at the IRD laboratories in Newcastle upon Tyne and is described below, together with the problems encountered in the design, construction and operation of the machine.

The heat pump has been designed to produce low pressure steam at 110 °C, however it would be a comparatively simple operation to replace the condenser so that pressurized hot water at the same temperature could be produced.

Figure 6 shows the heat flows and temperatures in the various parts of the cycle. The heat pump will recover about 170 kW from a process effluent stream in the temperature range 70-90 °C. This high temperature is necessary to maintain a refrigerant evaporating temperature of approximately 60 °C. A lower evaporating temperature would result in an unacceptably low COP.

Heat is delivered from the heat pump via the refrigerant condensing at 120 °C. The shell side of the condenser is fed with water at 80 °C this is heated to 110 °C and boiled to produce steam at this temperature. Water from the condenser at 110 °C is supplied to the gas engine and the exhaust gas boiler which also produce steam.

The combined heat output of the condenser, engine and exhaust boiler is 356 kW.

The high condensing temperature severly limits the number of refrigerants which can be used in the heat pump. Table I lists those which are available, together with their critical temperatures. The refrigerant must also be chemically stable at the compressor discharge temperature which is highest temperature reached in the cycle, for this reason it is necessary to choose a refrigerant with a low discharge temperature relative to the condensing temperature and good thermal stability at this temperature.

Decomposition of the refrigerant reduces the performance and lifetime of the machine by the formation of corrosive sludges and vapours which can attack the components of the compressor and refrigerant circuit.

A high saturation pressure at the condensing temperature results in an excessively heavy and costly condenser and associated pipework and easily available compressors are unable to cope with unduly high pressures. Therefore a moderate discharge pressure is of prime importance.

As with any heat pump or refrigeration system latent heat, flammability, toxicity and viscosity play an important role in the choice of refrigerant.

After consultation with compressor manufacturers the limit on discharge pressure was set at 2.1 MPa. Thus several of the refrigerants listed on Table 1 were eliminated. By eliminating those refrigerants which could present a fire risk the choice was limited to the halogenated hydrocarbons or to water. Of the halogenated hydrocarbons four were studied in detail, R11, R21, R113 and R114 together with water. Thermodynamic analysis of the proposed cycle showed that none of these fluids was entirely suitable, water was eliminated for several reasons and the choice was considered to be a critical factor in the successful operation of the system.

While breakdown can be tolerated, refrigerants are expensive and thus regular replacement is undesirable. Also, acids can be formed by breakdown, leading to corrosion in the compressor and pipework. Breakdown is accelerated at high temperature, and is also a function of the materials which the refrigerant is brought into contact.

Refrigerant 114 appeared to be consistently much more stable than other candidates, probably because of the high proportion of fluorine contained in the molecule. As fluorine is a strongly electronegative element, the bonds within the refrigerant molecule becomes polarised and therefore much stronger, thus making the molecule more stable.

R113 appeared to be the next most stable refrigerant, followed by R11, and R21 was the least stable of the four.

Consideration of the calculated discharge temperatures revealed that in all cases they were above the ASHRAE recommended maximum continuous operating temperatures, but R114 would be operating closest to its maximum recommended temperature at 125 °C - only 4 °C above it. R113, the next best fluid with a discharge temperature of 133 °C, would be operating at 26 °C above its maximum continuous operating temperature of 107 °C.

Narrowing the selection down to R114 and R113, an examination of volumetric flow rate shows that for R113 the value is too high for reciprocating compressors which we believe are most appropriate for the duty here, and also the pressure ratio is high. While R114 has a lower COP and higher discharge and suction pressures, with implications for thrust bearing design, its stability, lower discharge temperature, and greater suitability for reciprocating compressors made it the fluid to be used in the prototype.

Solubility of the refrigerant in the compressor oil is related to temperature, and of course can have a marked effect of compressor operation. The most critical factor is the oil viscosity, which decreases with increasing refrigerant concentration, leading to bearing lubrication problems. At high temperatures the oil and refrigerant can react with one another, forming acids and sludges. IRD have been liaising with several oil companies and one has supplied a high viscosity synthetic oil which is expected to be adequate in this application. Even with such an oil, it is necessary to maintain compressor temperature in excess of 100 °C, or the decrease in viscosity would lead to lubrication problems.

Figure 7 shows how oil viscosity varies with temperature and pressure when combined with refrigerant in the compressor sump.

The full range of compressors was studied prior to selecting a wet reciprocating compressor for the prototype. For compressors which can meet the technical requirements, the final selection criterion must be cost. From this study the choice must be the wet reciprocating compressor for the sizes of drive which are studied here. The study also showed that centrifugal compressors are prohibitively expensive at these ratings but it is thought they will become more competitive at substantially higher unit ratings.

The technical problems associated with wet reciprocating compressors can be overcome by reasonably conventional means. Problems associated with the discharge valves, for instance, are greatest, as they are at the hottest part of the cycle (125 °C), although it should be emphasised that discharge temperature as high as this can occur with more conventional refrigerants condensing at only 60 °C or 70 °C. The condition is therefore not unusually severe.

The economic success of the gas engine heat pump is dependent upon recovery of waste heat from the engine. If a useful amount of heat is not recovered then the primary energy ratio or fuel utilization of the complete unit will be too low.

The first, and most important consideration in this design is the thermal efficiency of the engine. This will vary between different types of prime mover and will also vary with speed and load conditions.

Typical values of the thermal efficiency of a variety of prime movers is given below :

Small steam turbine (Rankine cycle)	23 %
Gas turbine (Brayton cycle)	28 %
4 stroke non-turbocharged gas engine (Otto Cycle)	31 %
4 stroke turbocharged gas engine (Otto Cycle)	38 %
Diesel engine (Diesel cycle)	40 %

Even with good waste heat recovery from the prime mover it is important that the brake thermal efficiency should be maximised since power delivered as shaft work is multiplied by the COP before being delivered at the condenser.

The prime mover considered for this application is a four stroke, naturally aspirated reciprocating piston engine, and a thermal efficiency of 31 % is anticipated.

The remaining 69 % of input heat emerges from the engine in the form of hot exhaust gas, engine cooling water, oil cooling water and stray losses. The relative proportions of these are as follows :

Exhaust gas	28 %
Engine water jacket	30 %
Oil cooler	3 %
Stray losses	8 %

The exhaust from a 75 kW engine will emerge at about 650 °C at a rate of 0.1 kg/s. Useful heat (approximately 44 kW) can be recovered by cooling this down to 180 °C. The figure of 180 °C was chosen for two reasons : to avoid condensation and to avoid the need for an excessively large heat exchanger.

The shape of the heat exchanger will depend on the use to which it is put, i.e. for water heating or for steam generation. In this case low pressure steam production is required and the heat exchanger is therefore designed as a boiler. The most convenient configuration is a shell and tube unit using extended surface tubes with the exhaust gas on the shell side. It is proposed that the water feed should be taken from the heat pump condenser.

About 80 kW of heat will be released in the water jacket and the lubricating oil of the engine and this must all be removed to prevent the engine from overheating.

If the heat removal can be achieved at 110 °C or 120 °C then it can be added to the heat delivered by the condenser. On the other hand, if the engine is cooled at a more conventional 80 °C or 90 °C then this heat is considerably less useful. The actual temperature which can be used in practice depends on the recommendations made by the engine manufacturers.

Summarizing, because the efficiency of the prime mover is fundamental to the overall cycle efficiency it is important to know the distribution of energy from the engine. This is classified as :

Useful work	31 %
Useful waste heat	49 %
Non-recoverable heat	20 %

The high quantity of useful waste heat can only be achieved by using ebullient cooling, which also requires a more expensive type of engine than does conventional cooling.

The selection of the Waukesha gas engine, which utilises ebullient (or 2-phase) cooling to produce steam, is necessary for a project where high condensing temperatures are required, but a conventional water-cooled engine would be more attractive, both technically and economically, for heat pumps operating at lower temperatures.

The thermodynamic analysis for refrigerant 114 shows that the heat fluxes will be

Condenser output	239 kW
Evaporator input	164 kW
Intercooling heat transfer	31 kW
Compressor input power	75 kW
COP	3.2

An intercooler (also known as an interchanger or superheater) is used in the heat pump cycle to provide the necessary superheating of 13 °C by subcooling condensed liquid at high pressure. This superheat is required to ensure that there is no chance of liquid forming in the compressor at any stage during the compression. The effect of this is to relieve the evaporator of any need to superheat the refrigerant which means that the size can be minimised. If there was a superheating region in the evaporator it would be substantially larger. The intercooler is not an essential part of the circuit and it is quite possible that if the evaporator performance is better than anticipated or if the pressure loss is more than expected then the intercooler could be removed.

Initial tests have shown that this may be the case, however better performance may be obtained by adjustment or modification of the expansion valve to allow a greater mass of refrigerant through the evaporator.

The primary energy ratio of the system is given by the total heat output of the condenser engine jacket and exhaust boiler divided by the energy value of the gas used by the engine, this is 356/242 = 1.47. Comparing the heat pump with a boiler of 75 % economic efficiency the actual energy saving over a given period can be calculated.

Heat pump input (fuel)	= 242 kW
Heat pump output (steam)	= 356 kW
Equivalent boiler input (fuel)	= 475 kW
Thus fuel saving = 475 - 242	= 233 kW

If the heat pump is applied to a continuous process and operates for 8000 hours a year then the annual saving E, is given by

$$E = 233 \times 8000 \text{ kWh}$$

The expression for the payback period of the system is

$$P = \frac{I}{E \times A}$$

Where I is initial investment and A is the cost of the fuel saved. Using this expression the payback period for the heat pump is slightly less than three years.

At present the heat pump is in prototype form but could be incorporated into any process which was considered suitable with comparatively little modification.

The engine and compressor are mounted in line on a bed plate, drive from the engine being transmitted through a flexible coupling direct to the compressor. Smaller units powered by an automotive derived engine which would rotate at 3000 rpm may need a reduction either by belt or gearbox between engine and compressor. When connecting two reciprocating machines care must be taken to ensure that a suitable coupling is chosen. If a resonant frequency of the engine, coupling, compressor system coincides with an existing frequency within the system the effect on engine and compressor bearings can be catastrophic.

The condenser is mounted on top of the evaporator in order to reduce the floor space required by the unit. Space is one of the resources which may be at a premium when installing a heat pump, in particular if the system is being added to an existing process. Piping is routed to allow easy access to the engine for maintenance and to the water and steam valves which must be regularly adjusted on the prototype under test. A more compact unit could be constructed.

The use of ebullient (2-phase) cooling in an engine is unusual although not unique, however integrating the engine cooling circuit into the heat pump system has presented some problems, these are being overcome with the help of the engine manufacturer. In many applications where gas or diesel engine heat pumps may find a use the heat may be recovered from the engine by a conventional water cooling system. This is less expensive and technically simpler. The range of engines for which ebullient cooling is provided as an option is very limited.

The final configuration of the heat pump which will be put into service will depend upon the process in which it is used.

REFERENCES

1. Reay, D.A. and Macmichael, D.B.A. *Heat Pumps Theory, Design and Applications* (Oxford, Pergamon Press, 1979).

2. Macmichael, D.B.A. and Reay, D.A. "Feasibility and Design Study of a Gas Engine Driven High Temperature Heat Pump" *IRD Report 78/79* (1979).

3. Heap, R.D. *Heat Pumps* (London, E and F.N. Spon, 1979).

4. Reay, D.A. *Waste Heat Recovery, A Directory of Equipment and Techniques* (London, E and F.N. Spon, 1979).

5. Henry, J.A.R. "Heat Pumps, Part I, Compression Cycle Applications" *HTFS DR 49* (AERE Harwell).

TABLE I

Suitable refrigerants with respect to critical temperature

ASHRAE Ref. No.	Refrigerant Name	Chemical formula	Critical temp., °C	Critical pressure, MPa
717	Ammonia	NH_3	133	11.3
600a	Isobutane	C_4H_{10}	134	3.5
40	Methyl chloride	CH_3Cl	143	6.6
114	Dichlorotetrafluoroethane	$CClF_2CClF_2$	146	3.3
600	Butane	C_4H_{10}	152	3.8
764	Sulphur dioxide	SO_2	157	8.0
630	Methylamine	CH_3NH_2	157	7.8
21	Dichlorofluoromethane	$CHCl_2F$	178	5.1
631	Ethyl amine	$C_2H_5NH_2$	183	5.6
160	Ethyl chloride	C_2H_5Cl	187	5.2
11	Trichlorfluoromethane	CCl_3F	198	4.4
611	Methyl formate	$C_2H_4O_2$	214	4.2
113	Trichlorotrifluoroethane	CCl_2FCClF_2	214	3.4
1130	Dichloroethylene	$CHCl{=}CHCl$	243	5.5
30	Methylene chloride	CH_2Cl_2	249	4.6
1120	Trichlorethylene	$CHCl{=}CCl_2$	271	5.0
610	Ethyl ether	$C_4H_{10}O$	272	2.6
718	Water	H_2O	374	22.0

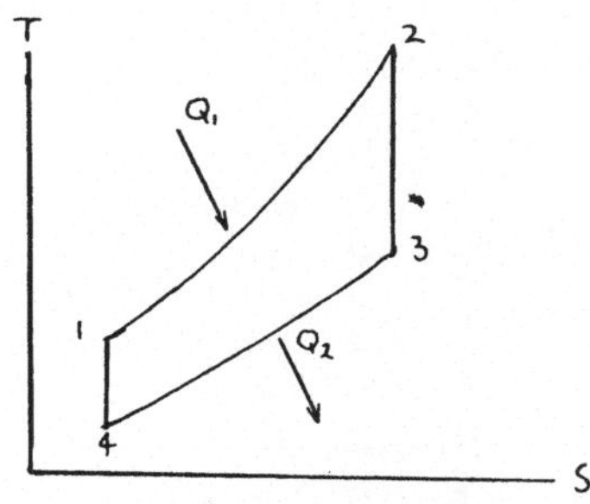

Fig. 1 Brayton Cycle

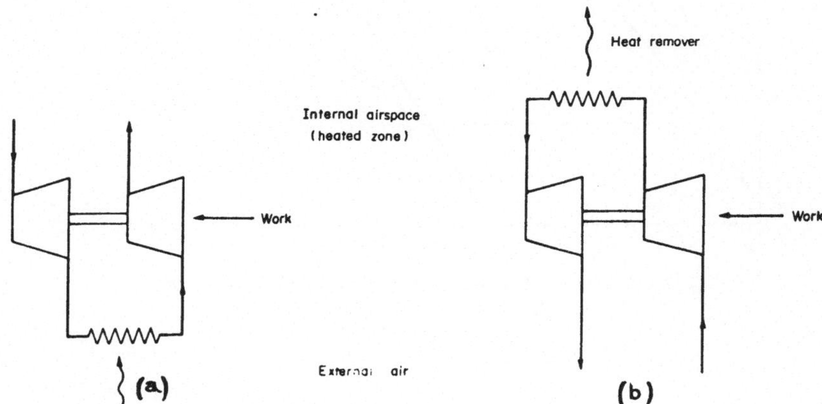

Fig. 2 : Open Cycle Brayton heat pumps

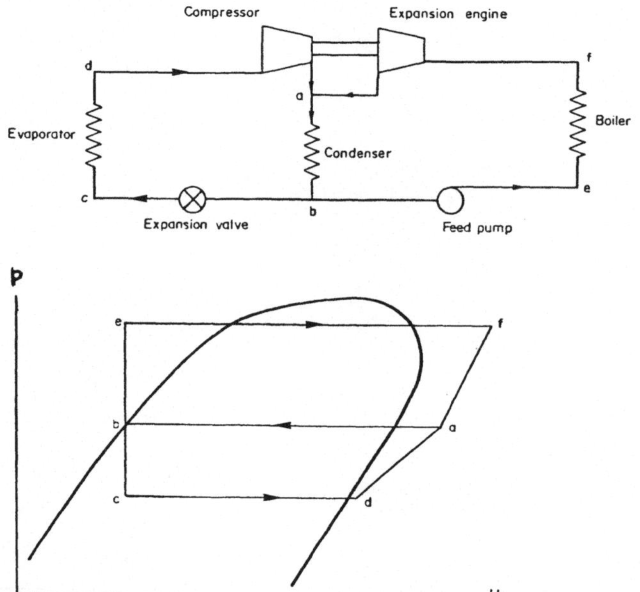

Fig. 3 Rankine-Rankine cycle, schematic and on a pressure-enthalpy diagram.

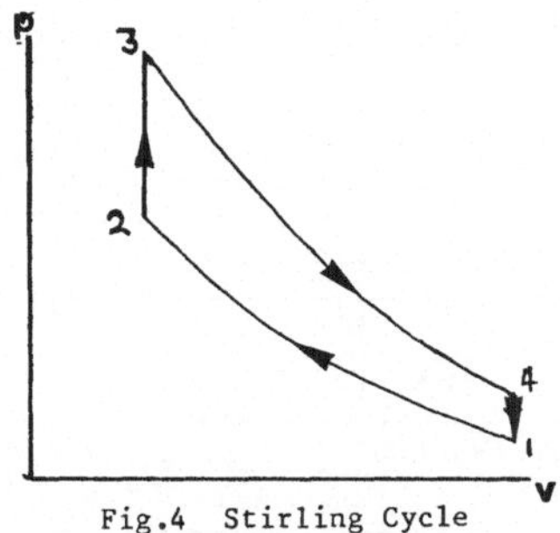

Fig.4 Stirling Cycle

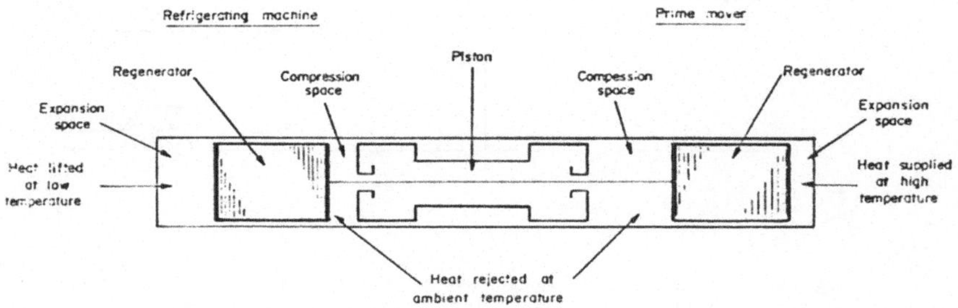

Fig. 5 The 'Duplex' Stirling cycle.

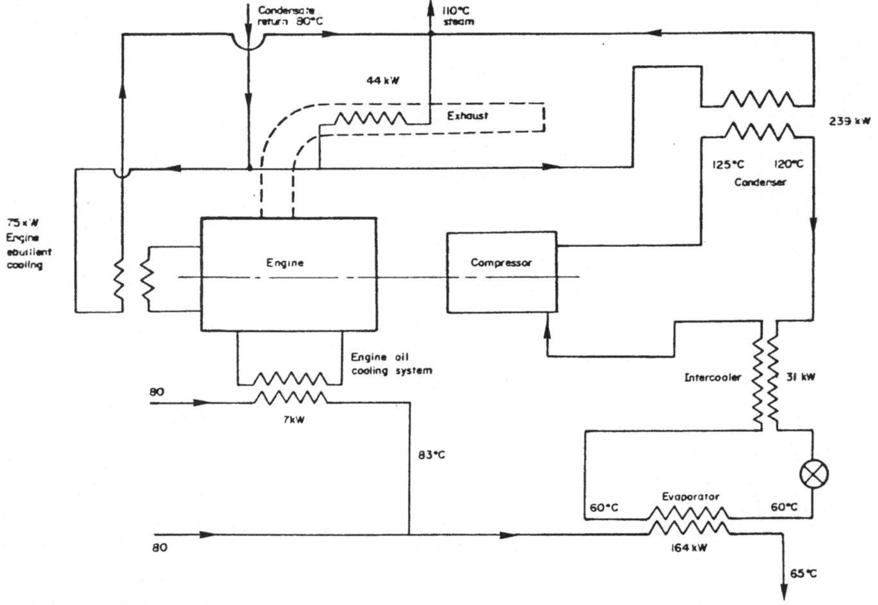

Fig. 6 Circuit diagram of a gas engine-driven heat pump with an ebullient cooling system on the prime mover.

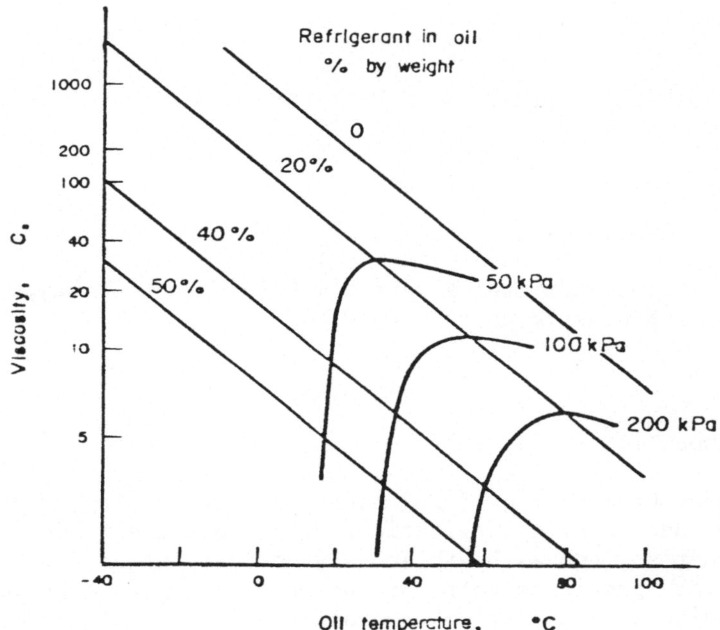

Fig. 7 Oil viscosity showing constant pressure curves.

THE INDUSTRIAL APPLICATION OF HEAT PUMPS

P.A. Kew

International Research & Development Co Ltd, Newcastle upon Tyne, U.K.

1. INTRODUCTION

Heat pumps have been used for many years in heating, ventilating and air conditioning applications where they are commonly regarded not so much as heat recovery devices but as a means for effectively providing both heating and cooling in a building by functioning as a heater in one section of a building and as a refrigerator in another. Application of heat pumps to industrial and process heat recovery has proceeded much more slowly.

The largest potential market for industrial heat pumps has, until recently, been identified as heat recovery from refrigeration or cooling plant for space or water heating and drying. In most of these cases the heat pump is driven by an electric motor. There is now a growing interest in the application of heat pumps to process heat recovery.

The distinguishing characteristics of most industrial applications, in contrast to commercial and domestic applications, are the longer operating hours required (up to 8000 hours per annum), the necessity for high reliability and low maintenance and often the ability to operate satisfactorily in a dirty environment. Under these conditions the internal combustion engine driven Rankine cycle shows poor characteristics. This system would otherwise often be the most efficient. The present constraint on condensing temperature limits the heat supplied to 120 °C with conventional halocarbon working fluids which also restricts the industrial application of heat pumps. In most cases industrial heat pumps must be custom designed to suit the needs of a particular application.

2. HEAT RECOVERY FROM REFRIGERATION PLANT

It is obviously economically desirable to combine heating and cooling duties in the same works with a heat pump acting as both refrigerator and heater. When the refrigeration system is already installed the replacement of the conventional condenser by one suitable for heat recovery is comparatively inexpensive. It must be borne in mind that the heat pump coefficient of performance will not be optimised for the heating mode, however, because of the need to install the plant for refrigeration duty. Producing some heating at a relatively low cost makes this point less significant.

One of the classic applications of refrigeration heat recovery is in the field of plastics injection moulding, typified by the installation at the Link 51 factory in the United Kingdom. The annual energy savings due to this installation amount to 15000 pounds, and the layout is illustrated in Figure 1.

Cooling water from the injection moulding machines is continuously recycled via a hot-well and a Prestcold Central heat pump. The heat pump, which replaces a conventional cooling tower, reduces the temperature of the cooling water from about 11 °C as it leaves the injection moulders to a steady 7.2 °C at the rate of up to 1140 l/min. The heat extracted from the cooling water is transferred to the space heating system, consisting of fourteen horizontal fan-type air heaters.

A major benefit of using the heat pump system has been a 5 per cent reduction in cycle time and a corresponding increase in productivity. This is because the heat pump provides cooling water to the injection moulding machines at the correct low temperature all the year round, whereas temperature variations are common with a cooling tower.

There is also a significant saving in fuel costs because oil or gas is not needed to heat the 3160 m^2 factory. When the machines are working the heat saved from the process, plus the heat generated by the compressor plant, amount to about 325 kW - which provides adequate space heating under the coldest conditions. Provision has been made for space heating when the machines are off-line or just prior to start-up after weekends. Two 140 kW electric heaters have been installed in the hot-well and can be switched on prior to work recommencing or during injection moulding machine maintenance so as to keep the temperature of the hot-well adequate for the space heating requirement. This makes optimum use of energy since any low grade heat in the hot-well is not wasted but brought up to operating temperature by the immersion heater.

A second example of an installation of this type is at the Revell factory, where plastic hobby kits are made. Although the system is somewhat smaller than that at the Link 51 factory, a 22 kW drive to the compressor being used, the system cost of 6000 pounds has been recovered in less than one year (This includes a 1500 pounds saving due to the fact that the cooling water can be directly recirculated, leading to a reduction in both water supply and treatment costs). As in the other example, space heating is via air-cooled condensers.

Typical of the equipment available for the above type of duty is the range of units supplied by Heat-Frig Ltd. These are sold as water chillers for the plastics and allied industries, but are available with a heat reclaim facility, in the form of an air cooled condenser. Heat rejection is achieved by dual centrifugal fans, independently operated and designed for external ducting application as appropriate to waste heat recovery.

3. HEAT RECOVERY FROM EFFLUENT

3.1. General

The recovery of heat from liquid effluents is a major growth area for industrial heat pumps using either electric or internal combustion engine drive. Figure 2 shows the range of industries which require a heat supply within the limitations of a conventional vapour compression heat pump. The range of industries which produce effluent at 10 °C to 60 °C is large. Preheating of boiler feedwater is general to many industries.

Figure 3 shows the layout of a Westinghouse Templifier heat pump which is a liquid, liquid heat pump. Figure 4 gives typical temperatures found in the various parts of the cycle. The Templifier has been developed over a number of years and is based on a range of standard heat pumps using centrifugal compressors and capable of generating useful heat at temperatures of up to 110 °C and having heating capacities from 200 kW to 3 MW. Units employing reciprocating compressors are available for use at lower duties, up to 300 kW. The units are designed to extract heat from a low grade heat source at between 27 °C and 77 °C and upgrading this heat to the required delivery temperature.

About 15 Templifier units utilising the centrifugal compressor are in use in the USA. In Europe the first unit incorporating the centrifugal compressor is being installed at the Unigate Dairy in Walsall, England. The heat pump installed by Northern Engineering Industries (NEI) is part of a new dairy installation.

The system has a COP of 5.4 and uses 170 kW of electric power to recover heat from a Hydrolock milk sterilization unit and delivers 910 kW of heat as hot water for use as boiler feed water and for process and space heating. This arrangement also gives a considerable water saving as the water from which the heat is extracted in the evaporator is cooled down to a temperature which allows it to be re-used as cooling water in the sterilizer.

A typical application of the Templifier incorporating a reciprocating compressor is shown in Figure 5.
The three processes in this case are arc welding machines in continuous operation for 10,5 hours throughout the day and occasionally during the night shift. Before the Templifier unit was installed, heat from the processes was rejected through a cooling tower to atmosphere, representing a waste of up to 90 kW of low grade heat. During normal operation of the welding machines, the Templifier is supplied with 3 l/sec of process water at 38 °C and returns the water to the arc welding machines at 30 °C (the same conditions as prevailed when the cooling tower was in operation). Heat recovered from the process water is delivered at the condenser, raising 2.6 l/sec from 70 °C to 82 °C. Since the delivery water is heated to 82 °C it can be supplied directly to the heating system, via an insulated hot water storage tank if necessary, without any further heating.

The capacity of the system is balanced to the office heating load by cycling the compressor in response to variations in return hot water temperature (The cooling tower has been retained in this particular system to handle process water not required by the heat pump during periods of low heating load or shut-down).

3.2. Bottle Washing Processes

There is substantial scope for energy recovery in bottle washers which are to be found in the dairy brewery and soft drink industries. Machines operating on a similar principle are used in a wide variety of industries for cleaning components and for dishwashing. Both heat and water conservation play an important role in the economic operation of these machines.

A bottle washing machine which can process several tens of thousands of bottles an hour conveys the bottles through a number of stations in order that they may be effectively cleaned. The process must be designed to remove both visible and bacterial contamination. In order to do this the bottles must be raised to comparatively high temperatures, up to 80 °C, and in the case of bottles for use with pasteurized milk they must then be cooled to ambient temperature. Several stages are involved in this process which is shown diagramatically in Figure 6.

Considerable quantities of heat, typically in excess of 500 kg/hr of steam are needed to maintain the water and detergent solution temperatures in the machine. In addition the water used in the final rinse flows to waste at a rate of up to 10,000 litres/hr at 20 °C - 30 °C.

Heat transfer within the machine itself is complex as heat is carried from one section to another by the bottles. The heat absorbed by the bottles in the hot section is given up to the cooling water in the final rinses. Some heat recovery and water conservation are employed within the bottle washer and a considerable amount of heat is lost to the atmosphere. However with the cost of steam being approximately 3.50 pounds per 100 kilogram and the total charge for water supply and effluent treatment may be over 0.20 pounds per 1000 litres it can be seen that recovery of heat and water is, in principele at least, very attractive.

In addition to the incorporation of conventional heat exchangers in the bottle washer the presence of a number of low grade heat sources and higher grade heat sinks makes the heat pump particularly attractive in this application. An integrated circuit incorporating water treatment facilities, conventional heat exchange and a heat pump has been proposed by Milpro NV. The layout of this system is shown in Figure 7. The temperature profile within the machine is similar to that shown in Figure 6. The process works as follows.

A conventional heat exchanger is used to recover heat from the final detergent tank, this heat being used to raise the temperature of the pre-rinse spray water. The input to the pre-rinse sprays is supplemented from the final rinse tank (30 °C) and also by overflow from the final detergent tank (45 °C). A filtration system, which rejects a proportion of the pre-rinse water, is used to maintain pre-rinse water in a comparatively clean condition. Once some of the heat has been moved from the water passing out of the final detergent tank for transfer into the pre-rinse water, the final detergent tank water is then passed through an evaporative heat exchanger, serving as one of the heat sources for the heat pump, before being returned at 30 °C to the final warm rinse and final detergent tanks. Prior to passing through the evaporator, a proportion of this flow is taken for filter treatment and cooled to 12 to 15 °C in a second evaporator. This clean water is then used as the final cold rinse water supply.

The heat recovered in these evaporators is taken up in the form of refrigerant vapour, which is then compressed, raising its temperature by the addition of the work of compression. This heat is then rejected at the higher temperature in a condenser to water from the first detergent tank (raised from 55 to 65 °C) and also in a second condenser, where water for the high temperature section is heated from 65 °C to 80 °C.

Any make-up water required is added to the pre-rinse water circuit. Based on a machine washing 30.000 bottles per hour, consuming with conventional heating and water usage, 13.600 litres and 600 kW heat energy per hour, Milpro claim that the heat pump and filter system can reduce consumption to 70 kW and 2600 litres per hour. Payback periods of two to three years are claimed for the unit, which can run either on natural gas or electricity.

This is one of the few cases seen to date where a heat pump is offered as a standard energy-conservation device or a process package.

The washing cycle of conventional large dishwashing machines is similar to that of bottle washers and the heat pump may find application here also. The highest temperature required is somewhat higher at approximately 90 °C and much of the waste heat is discharged as hot vapour, and this is used as the heat source. As well as recovering energy the cooling of this vapour improves comfort in the vicinity of the machine.

3.3. Heat Recovery from Drying Processes

The field of drying, evaporating, boiling and dehumidifying is a growth area for heat pumps and the heat pump in its conventional form is in competition with other heat recovery systems and a variety of other compression cycles.

Commercial systems are on the market for the purpose of drying and dehumidification. Figure 8 shows diagramatically the layout of a heat pump dehumidifier. One of the first applications of heat pumps in a dehumidification role was as a result of work by Sulzer in 1943 for the dehumidification of underground caverns. The heat accrueing from condensing the water vapour in the air was used to heat the incoming air. An early study was carried out in the 1950's in the USA associated with the application of heat pumps to grain drying and an experimental rig was constructed. At the time the project showed that the energy savings were significant but capital costs proved prohibitive. Commercial systems introduced to the market now should show rapid return on investment.

One of the first applications of heat pumps to drying processes on a commercial scale, and one where control of temperature and humidity are of prime importance, is in timber kilning. Pioneers in this field are Westair Systems Ltd, and their range of equipment has been developed over a period of ten years, in about one thousand installations throughout the world. A typical installation, showing the air path, is illustrated in the sketch in Figure 9. The unit operates generally as described earlier, but the air is passed over the electrical components of the system, where it both picks up some heat and ensures that the fan motor remains cool. It finally passes over a heating coil before reentering the chamber.

The Westair units include a humidistat and a thermostat which control the switching of the refrigeration and heater circuits to suit the particular timber being dried. Periodic adjustments to controls may be required as the timber is dried, depending upon the moisture content. In some cases it is not necessary to adjust the controls at all during the drying process. All necessary information required to control the drying of specific woods is supplied by the manufacturer.

Conditions can be spot checked by means of a whirling hygrometer or recorded on a thermohygrograph chart to ensure that the recommended conditions are being achieved. Throughout the process the moisture content of the timber should be checked, using strategically placed test pieces. Once the desired level has been reached the controls should be adjusted to ensure an equilibrium condition for a short time before the timber is removed from the chamber.

The chamber itself can be constructed from ordinary building materials such as timber, brick or concrete, or indeed, an existing building can be converted quite easily. The important requirement is that it must be well insulated and as near air-tight as practical. The floor, usually concrete, and other internal surfaces should be waterproof to prevent absorption of moisture. This can be achieved easily by bitumastic paint, alternatively, for the walls and ceiling, insulation board with a plastic coating can be used.

In addition to the fans incorporated in the Westair units it is necessary to install fans for primary air circulation in the chamber. These are normally mounted above the timber stack and, in order to ensure the most efficient air flow through the stack, it is important to partition off open spaces around and above it. This can be achieved by means of flexible polythene or canvas baffles.

The flexibility of the system ensures that an efficient drying chamber can be provided to suit a client's individual requirements and facilities.

The drying capacity of the installation is determined entirely by the layout and construction of the chamber and by the client's requirements in respect of the following factors :

1. wood types to be dried
2. starting and finishing moisture content
3. quantity of timber to be dried per week/per month etc.
4. dimensions of timber and standard stacks if applicable

Another drying application of the heat pump which has recently been exploited is the drying of ceramic filters for water purification. The installation of a heat pump drying system at Portacel Ltd in Kent, England, has resulted in more effective use of factory space, manpower, capital, energy and materials - in fact all the resources employed in the process.

A drier employing superheated steam as the drying agent is being developed by, amongst others, the electricity council in the UK. In this system the superheated steam is passed over the articles to be dried. Water evaporates from these articles and reduces the degree of superheat of the steam, which after passing through the drying chamber is divided into two streams. One of these streams is compressed to above the saturation temperature and pressure of the second stream. This second stream is then heated in a heat exchanger by the first stream which will condense in the exchanger and the condensate, representing the removed water, is drained off.

4. MECHANICAL VAPOUR RECOMPRESSION

An open cycle heat pump operating on the mechanical vapour recompression (MVR) principle can produce very significant energy savings in evaporating applications. MVR can be used for three types of processes, namely product concentration, reduction in the volume of liquid effluents and the recovery of water for re-use, as for example, in a desalination plant. Figure 10 shows four different layouts used in distillation apparatus, three of which involve energy recovery.

In a single effect evaporator the evaporate is rejected, but by adding a further stage of evaporation this rejected vapour may now be used in a closed system to effect further evaporation. If the second vessel is to boil at atmospheric pressure, the steam supplied to it from the first vessel must be under pressure.

Assuming that vessel 2, as shown in Figure 10, is open to the atmosphere when steam enters vessel 1, boiling will occur there, but not in the vessel 2. However, as the pressure in the first vessel increases, the temperature difference between the input steam and output vapour will decrease, reducing heat exchange. At the same time, a temperature and pressure difference is building up between the vapour and liquid in the second vessel, initially resulting in condensation of the vapour from the first effect vessel.

The latent heat given up will eventually provide sufficient energy in the second vessel to effect evaporation. The number of stages used in evaporation typically may be up to six. In the dairy industry, for example, as plant replacement is implemented, the number of effects used in evaporators is being gradually increased from the current two or three, to four or five. This will be logically followed by the use of vapour recompression, particularly in larger plants.

Referring again to Figure 10, the third and fourth diagrams show indirect and direct vapour recompression systems. In the context of distillation column applications of vapour recompression, as opposed to simple evaporator duties, the vapour emitted by the former rarely is water. This generally means that direct steam recompression cannot be used, but compressors capable of handling the product of the distillation column, or an indirect system using a conventional refrigerant, must be incorporated.

The indirect system is shown in the third diagram. Here a conventional heat pump circuit is employed, using heat from the distillation column condenser (at the top of the column) to evaporate the refigerant, which is then compressed before proceeding to the heat pump condenser section, heating the reboiler at the base of the distillation column, replacing in whole or in part, the steam supply to this heat exchanger.

In addition to the energy benefits of this system when compared with multiple effect units, the main advantage of the indirect system is the fact that it is a combination of two conventional and proven systems, thus control and reliability data are well-known. On the debit side, it is often difficult to select a suitable working fluid for the heat pump circuit because of the temperature range (150 to 200 °C) within which most distillations of organic compounds take place. Secondly, in most cases it has been found that the compression ratio for the heat pump working fluid is high, necessitating a high energy input to the compressor.

In general, the direct system is preferred, as illustrated in the fourth diagram in Figure 1. Applicable to both evaporation and distillation processes, in this case the vapour reaching the top of the column is transported directly to the compressor and thence to the reboiler where it is used instead of steam. The principal advantages of this system are that the working fluid for the "heat pump" is always available, and a lower compression ratio is required than that needed for the indirect method. One disadvantage pointed out by Sulzer is that when a rotating machine is installed in the vapour stream, as is the case here, it may produce a psychological resistance on the part of the plant operatives.

The type of compressor used for mechanical vapour recompression depends upon the pressure ratio required and the properties of the vapour to be compressed. In many evaporator duties, typified by units such as the APV Rosco module illustrated in Figure 11, the normal procedure is to use a single stage high speed centrifugal compressor, constructed, of course, from materials selected to avoid any risk of corrosion from the vapour handled. In these well-established application areas, machines have been shown to have a high reliability and require only a minimum amount of maintenance. (As with the conventional heat pumps, the drive for the compressor, incorporated in an MVR system may take one of several forms. APV cite the use of pass-out steam turbines to drive the compressor, this being particularly attractive where there is a demand for low pressure steam elsewhere in the process or on the site. The more common electric motor drive may also find competition from gas or diesel-fuelled reciprocating engines, or gas turbines).

Other companies have selected the Roots blower as the compressor type. Aiton, who use vapour recompression systems based on the Roots Blower in seawater desalination plant, where of course steam is the fluid being compressed, employ either electric motors or internal combustion engine drives for the system, the layout of which is shown in Figure 12.

On a heat supplied basis an MVR system can reduce energy consumption by almost 80 % over a four effect evaporator and even when the conversion losses associated with electricity generation are associated for MVR still represents a saving in primary energy of over 40 per cent.

5. CONCLUSIONS

As can be seen from the above the potential applications of heat pumps in industry are many and diverse. The economic viability of using a heat pump for a given duty depends predominantly upon the capital cost of the heat pump system, the COP, or the Primary Energy Ratio of the system, the output of the system and the hours per year for which it will run. The last three of the above delivering the amount of energy saved per annum. The cost of the fuel saved is also of importance.

Often the criterion used in choosing a system is the simple payback period defined as the capital cost of the sytem divided by the saving in running costs per year. Other factors which will influence the selection of a type of heat pump in favour of conventional energy recovery are the maintenance required and reliability of the system, although generally the heat pump will add complication to a machine it may, as in the Portacell and Link 51 application increase productivity by improving the control of the system.

Figure 13 shows how COP varies with the temperature lift for a practical heat pump. Obviously the application of a heat pump becomes less attractive as the temperature lift increases and the COP decreases. However, by use of the internal combustion engine driven heat pump rather than an electric drive primary energy ratios greater than unity may be achieved with low COP, this may offset the extra capital cost incurred when calculating the payback period.

REFERENCES

1. Reay, D.A. and Macmichael D.B.A. *Heat Pumps Theory, Design and Applications* (Oxford, Pergamon Press, 1979).

2. Macmichael D.B.A. and Reay, D.A. "Feasibility and Design Study of a Gas Engine Driven High Temperature Heat Pump" *IRD Report 78/79* (1979).

3. Heap, R.D. *Heat Pumps* (London, E. and F.N. Spon, 1979).

4. Reay, D.A. *Waste Heat Recovery. A Directory of Equipment and Techniques* (London, E. and F.N. Spon, 1979).

5. Henry J.A.R. "Heat Pumps, Part 1, Compression Cycle Applications" *HTFS DR 49* (AERE Harwell).

6. Hodgett, D.L. "Efficient Drying Using Heat Pumps" *The Chemical Engineer* (July/August 1976).

7. Anon. "Waste Heat Recovery Saves 15.000 pounds a year" *The Engineer* (5 Feb. 1976).

8. Davis, C.P. *A Study of the Adaptability of the Heat Pump to*

Drying Shelled Corn. (M. Sc. Thesis Purdue Univ., USA).

9. Miller, W. "Energy Conservation in Timber Drying Kilns by Vapor Recompression" 27 *Forest Prod. Jn.*, 9 (1977).

10. UK Patent Spec. 1494780 *Improvements in or Relating to a Method and Apparatus for Washing Articles* (UK Patent Office, 14th Dec 1977).

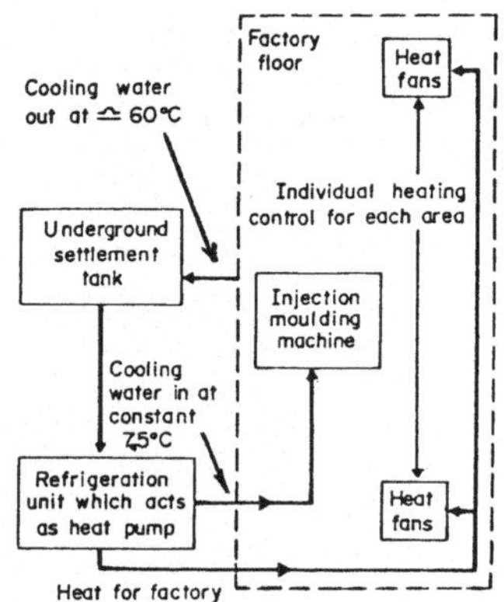

Fig. 1 A heat pump functioning as a water chiller and space heater in a plastics factory.

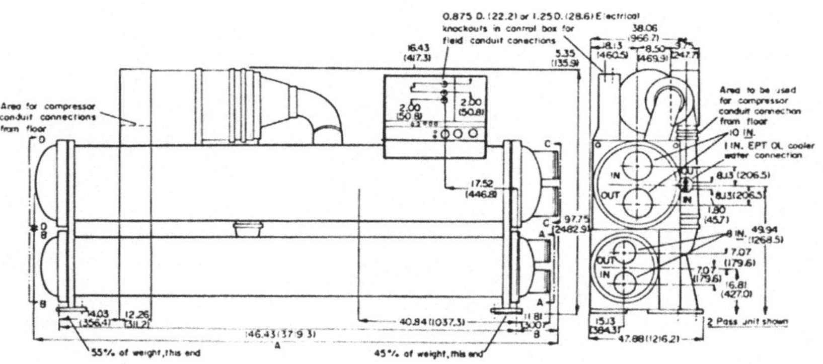

Fig. 3 Layout of the TP Templifier heat pump, showing location of heat exchangers.

Major process use (Temperature requirements are indicated in °C) \ Industry	Food products																			
	Meat products	Dairy products	Canned, cured and frozen foods	Grain mill products	Bakery products	Sugar	Confectionary products	Beverages	Miscellaneous	Textile mill products	Apparel and other textile products	Lumber and wood products	Pulp and paper mills	Chemical and allied products	Petroleum refining	Fabricated rubber products	Fabricated metal products	Machinery, except electrical	Electrical equipment and supplies	Transportation equipment
Washing/sanitizing per cleanup (60)	X	X	X	X	X	X	X	X	X											
Cooking (100 – 115)	X		X				X	X												
Pasteurization (65)		X																		
Blanching (85)			X																	
Dye heating (88)										X										
Pressing (100)											X									
Log soaking (90)												X								
Metal cleaning and plating (60 - 90)																	X	X	X	X
Paint drying (80 - 120)																	X	X	X	X
Drying (105)										X										
Mold release solution (85)																X				
Vessel heating														X	X					
Heat tracing															X					
Evaporators		X											X							
Process heat (hot water)														X						
Paper machine													X							
Pulp heating													X							
Sterilization (110)		X	X																	
Scalding (50 - 60)	X																			

* Courtesy westinghouse electric corporation

Fig. 2 Potential applications and temperature requirements of a Templifier heat pump.

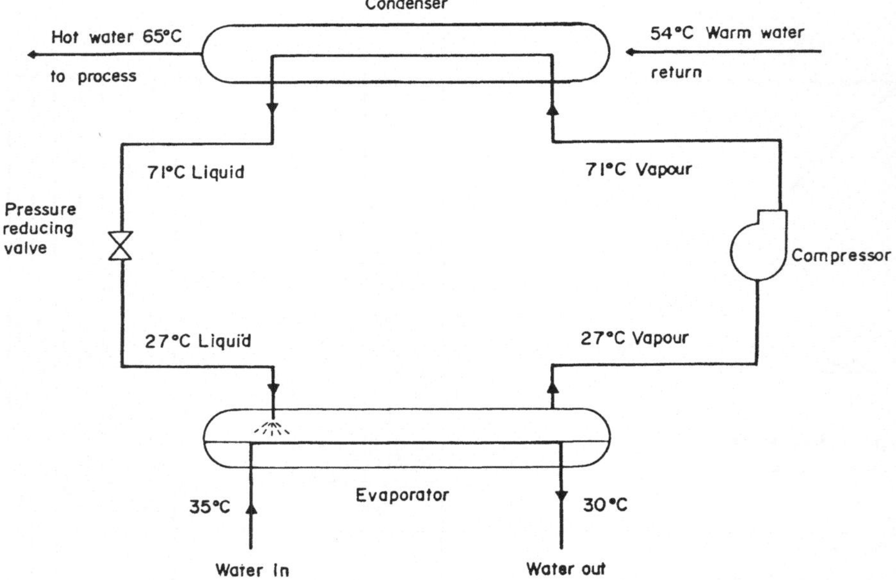

Fig. 4 Templifier schematic diagram.

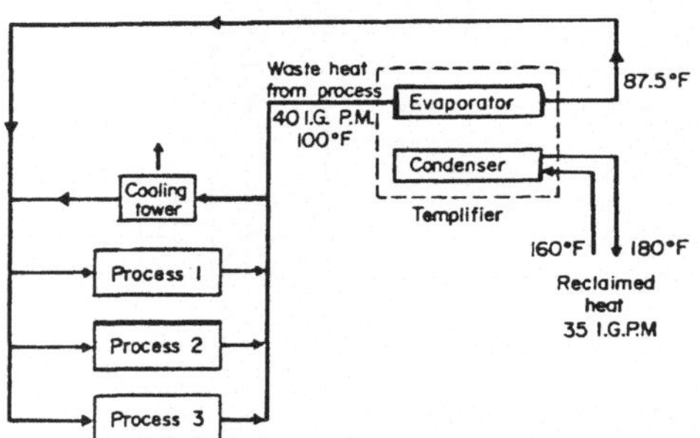

Fig. 5 Reciprocating compressor-driven Templifier application, using heat rejected by welding sets.

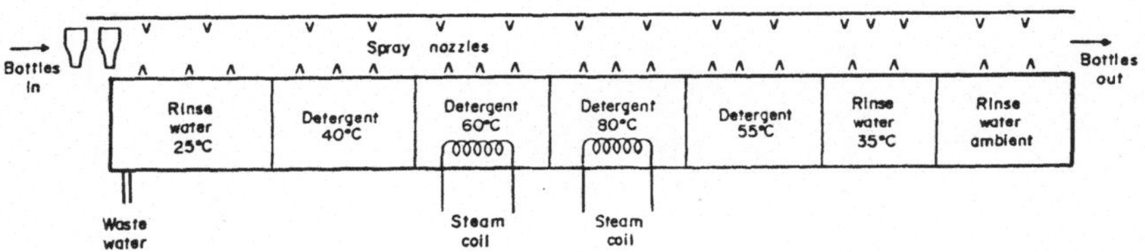

Fig. 6 Sectional view of a jetting type bottle washing machine.

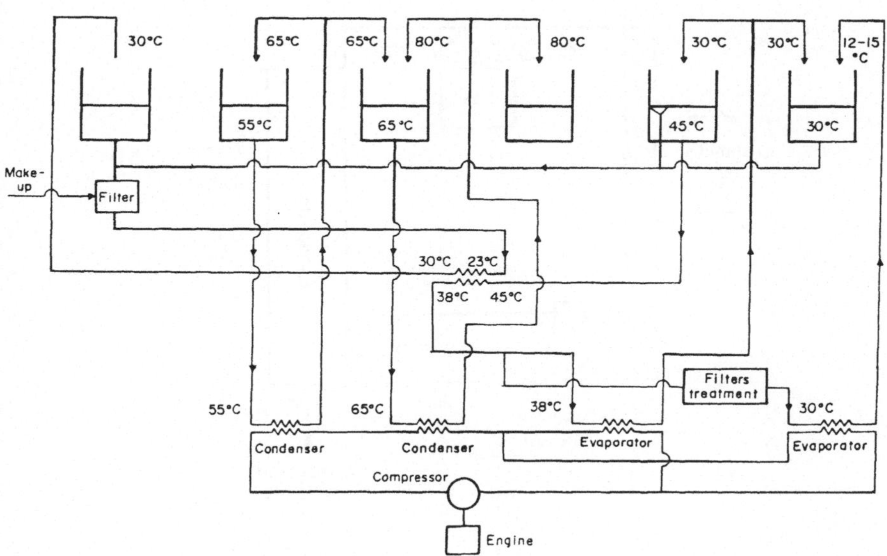

Fig. 7 A heat pump and water recovery system developed by Milpro NV for bottle washing machines.

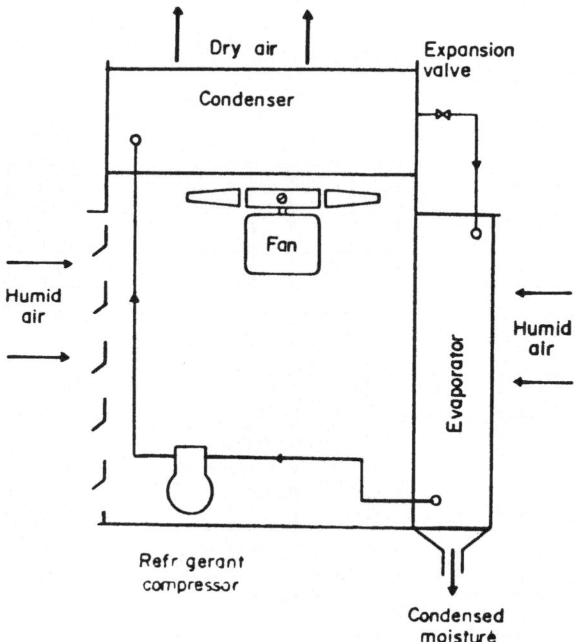

Fig. 8 Heat pump dehumidifier

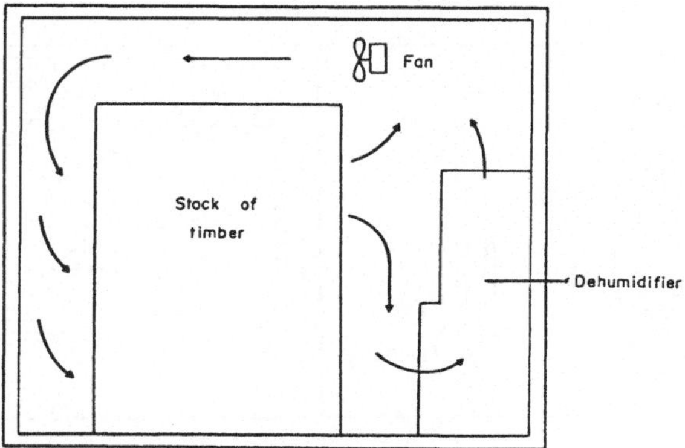

Fig. 9 A heat pump dehumidifier used for timber kilning.

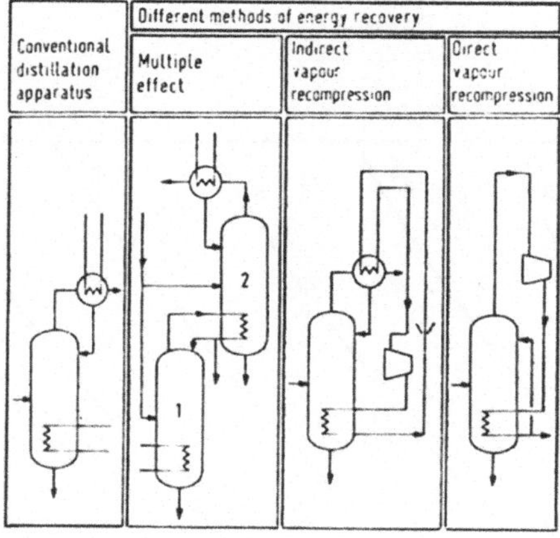

Fig.10 Three different methods of energy recovery compared with conventional distillation apparatus.

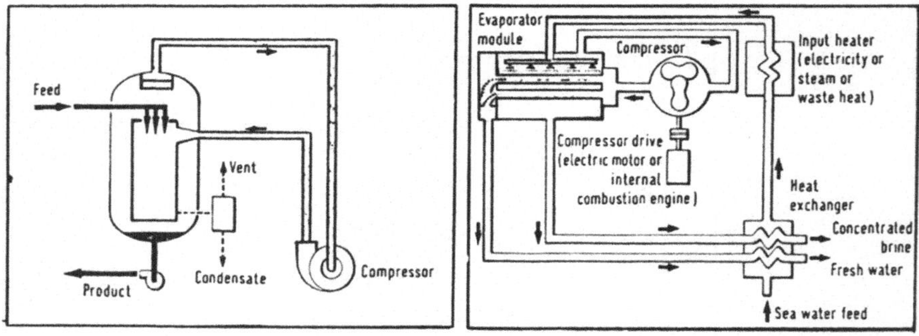

Fig.11 APV Rosco Module

Fig.12 Aiton MVR seawater desalination plant, using a Roots blower compressor.

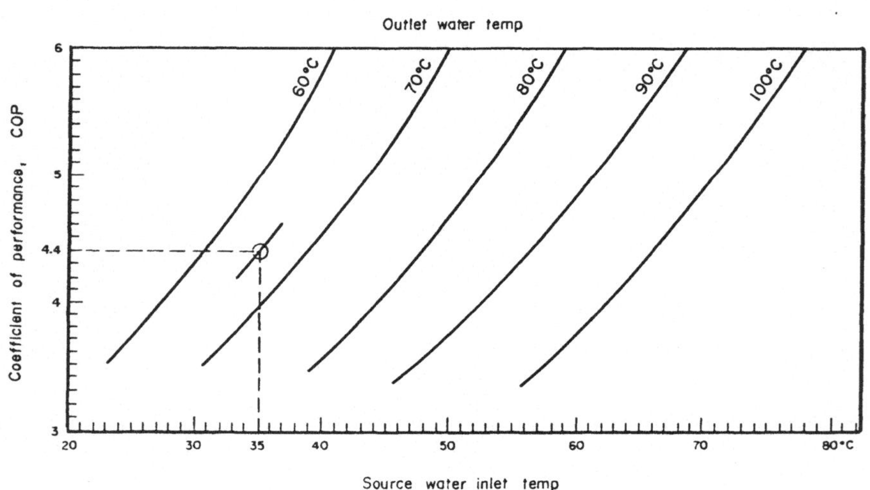

Fig. 13 Variation of Templifier heat pump COP with source and sink water temperatures.

DOMESTIC HEAT PUMP APPLICATIONS

J. Berghmans

Katholieke Universiteit Leuven, Leuven, Belgium

The most important heat pump applications in the domestic sector are : hot water production, space heating and cooling and swimming pool heating. These applications will be discussed here in a general way, the energetic performance of these systems strongly depending upon the local conditions (climate, heat source availability...). The same can be said about the economic aspects of these applications.

Most of the heat pumps applied in the domestic sector are of the vapour compression type with electrically driven compressor. Heat pumps driven by internal combustion engines (gas, oil) have only recently been able to penetrate the domestic sector in a significant way.

1. HOT WATER PRODUCTION

An example of a heat pump developed for the production of hot sanitary water is shown in figure 1.

The evaporator (1) withdraws heat from the room air by means of a ventilator (2). The heat pump further consists of the hermetic compressor (5), the condenser (4) and the expansion valve (3). The average coefficient of performance of this system is between 2 and 3, depending largely upon the amount of water consumed daily (50 to 300 l) and the temperature to which it is heated (see ref. [1]). In most cases hot water is produced at 50-55 °C. Systems of this type with storage capacities up to 200 l are presently available. It is claimed that pay back periods of such systems are of the order of 3 years if comparison is made with direct electric heating boilers [2].

2. SPACE HEATING AND COOLING

The heating and cooling of single and multi-family houses has become the most succesful application of heat pumps thus far. A large variety of systems exists depending upon whether they are intended for both heating and cooling or only heating, the nature of the low temperature source and the medium distributing the heat (cold) to the dwelling (air, water, ...).

2.1. Heat Sources

As far as low temperature sources is concerned, ground or surface water, air and soil are most commonly used.

a) Ground and Surface Water

Most ground water at depths more than 10 m is available throughout the year at temperatures high enough (10 °C) to be used as low temperature source for heat pumps. Its temperature remains practically constant over the year and makes it possible to achieve high seasonal heating COP's (3 and more). For single family heating purposes ground water flow rates of several m^3/h are necessary in W. European climatic conditions. The pump energy necessary to pump up this water has a considerable effect upon COP (10 % reduction per 20 m pumping height). It is necessary to pump the evaporator water back into the ground to avoid depletion of ground water layers. The ground water has to be of a purity almost up to the level of drinking water to be able to be used directly in the evaporator (flocullation). The rather large consumption of water of high purity limits the number of heat pump systems which can make use of this source. Also surface waters constitute a heat source which can be used only for a limited number of applications.

Ground water at considerable depth (aquifers) may offer interesting possibilities for direct heating or for heating with heat pump systems. The drilling and operating costs involved require large scale applications of this heat source. The quality of these waters often presents serious limitations to their use (corrosiveness salt content...).

b) Soil

The ground consitutes a suitable heat source for a heat pump in many countries. At small depth temperatures remain above freezing. Furthermore the seasonal temperature fluctuations are much smaller than these of the air. This is demonstrated by Table I which is valid for Belgium. Heat is extracted from the soil by means of a glycol solution flowing through tubing embedded in the ground. If a horizontal grid of tubing is utilized, several hundreds of square meters surface area are needed to heat a single family dwelling. In urban areas such space is rarely available. In addition considerable costs are involved. For these reasons vertical ground heat exchangers are studied presently.

c) Air

Outside air is the most interesting heat source as far as availability is concerned. Unfortunately when the space heating demand is highest the air temperature is lowest. The coefficient of performance of vapour compression heat pumps decreases with decreasing cold source temperature. In addition at evaporator temperatures below 5 °C air humidity is deposited on the evaporator surface in the form of ice. This does not improve the heat transfer and leads to lower working fluid temperatures and therefore lower COP values. This is demonstrated in figure 2 which shows the dependence of the COP upon the temperature of the air flowing over the evaporator. If ice formation occurs periodic de-icing of the evaporator surface has to be applied. This invariably leads to decreased values of the overall system coefficient of performance (5 to 10 %).

d) Each of the above mentioned heat sources for heat pumps present some drawbacks. Presently considerable research is devoted to the problems involved. Also alternative heat sources are studied. Geothermal energy has been mentioned already. Also solar energy may provide a suitable heat source. Unfortunately solar systems presently are very costly. Furthermore the intermittend character of solar energy requires the use of large and costly storage volumes.

The presently available compression heat pumps are limited in condensation temperature (55 °C maximum). This poses rather stringent requirements on the high temperature heat sink. If the heat is transferred to air, the size of the condenser has to take into account the small temperature difference between air and condensing freon (less than 35 °C). If the heat is transferred to water (hydronic heat distribution systems are predominant in Europe) then a low temperature heat distribution system has to be used as floor heating, forced draft convectors... This has a considerable effect on the applicability of heat pumps for space heating as in Europe most existing houses use radiators which require high water temperatures (70 to 90 °C). Fortunately these radiator sytems traditionally have been largely oversized such that for limited heating loads (not too low outside temperatures) the heat pump can still be used in a number of cases.

2.2. Systems

First it should be pointed out that in principle the vapour compression heat pumps utilized for heating purposes also can be used for cooling purposes by incorporation of a four-way value (figure 3 shows such a system in the cooling mode). The latter allows for the inversion of the role of the heat exchangers (evaporator - condenser). In practice proper sizing of the heat exchangers is necessary taking into account heating and cooling loads. It is evident that the heat exchangers have to be of the same type (air or water). Hereafter special attention is given to the heating function of domestic heat pump systems while the above remark concerning cooling operation should be kept in mind.

Depending upon the capacity of the heat pump compared to the peak heating load of the dwelling, distinction can be made between monovalent and bivalent heat pump systems.

a) Monovalent systems

In a monovalent system the heat pump is designed to cover the peak heating losses of the dwelling. The heat pump is the only device delivering heat during the whole heating season and therefore has to be sized at the maximum heating load of the dwelling. This leads to considerable overcapacity of the heat pump as shown in figure 4. In this figure the heating capacity of a heat pump withdrawing heat from the outside air (full line) is compared to the heat losses (dashed line). For a heat pump operating on ground water (dotted line) the overcapacity is less pronounced.

Overcapacity of the heat pump leads to more intermittent use and therefore to less efficient heat pump operation and lower average COP values. Therefore monovalent systems are applied only when heat sources like ground water and ground heat are available. Such systems are characterized by large investments. On the other hand the seasonal COP values usually are higher than those which can be obtained with air systems.

b) Bivalent systems

Bivalent systems are used whenever the performance of the heat pump deteriorates due to low source temperatures. Air heat pumps are of this type. The heat pump capacity is compared to the heating load in figure 5. In these systems heat pumps of smaller capacity are utilized. They can provide the heat to the dwelling down to the outside temperature t_d (design temperature). Below this temperature an auxiliary system has to be called in operation. Depending upon the operation of the system one may distinguish between parallel bivalent systems and alternative bivalent systems.

In the first case the heat pump remains in operation, also at temperatures below t_d. In alternative bivalent systems the heat pump does not operate below t_d at these temperatures. The head load is totally met by the auxiliary sytem.

In figures 6 and 7 the heat delivered by the auxiliary system relative to the total heat consumption is represented by the shaded portions. This can be shown as follows. The heat H needed to maintain a temperature level t_r in a house during a heating season is given by :

$$H = \sum_{j=1}^{N} K\,(t_r - t_j)\,\Delta n_j \qquad (1)$$

in which : K = the heat loss coefficient
t_j = average outside air temperature
Δn_j = number of hours at which t_j occurs

The above formula is based upon a temperature bin method i.e. a method in which the temperature interval occurring during a heating season is subdivided in N-bins. Each bin is characterized by an average temperature t_j which occurs during Δn_j hours per heating season.

In each bin the required heating power Q_j is given by :

$$Q_j = K\,(t_r - t_j) \qquad (2)$$

From which it follows that :

$$H = \sum_{j=1}^{N} Q_j\,\Delta n_j \qquad (3)$$

Now the relation between t_j and Δn_j is known for a given climate. From this one can also derive the relationship between t_j and the number of hours during which the outside air temperature is below t_j (cumulative number of heating hours). By means of equation (2) one can then also plot the required heat load as a function of cumulative number of heating hours. Such a curve is represented in figures 6 and 7 (solid line). From equation (3) it follows that the area between this curve and the horizontal coordinate axis is proportional to the total seasonal heat demand (in dimensionless numbers for the case of figures 6 and 7).

In figure 6 the case of parallel operation is shown. The dashed curved represents the heat pump power. The shaded area represents the heat which has to be provided by the auxiliary system. The heating power of the auxiliary system (Q_a) is also indicated. Figure 7 represents the case of alternative heat pump operation.

From these figures it can be seen that parallel operation requires much less auxiliary heat than alternative operation. The operation of the heat pump in the latter system occurs in a more efficient manner. If the temperature t_d is well chosen ice formation on the evaporator can be completely avoided. On the other hand the heat provided by the auxiliary system, which is less efficiently produced, increases with increasing t_d values. It is evident that an optimum value of t_d exists.

During parallel bivalent operation considerable defrosting has to occur. The frequency of defrost cycles depends upon the local climate. In most cases the energy required for defrosting is limited to less than 3 % of the energy produced.

The auxiliary heat to be provided during alternative operation may amount up to 30 - 40 % of the total amount of heat consumed during the heating season. In addition the power of the auxiliary heater has to be equal to the peak heating load. This makes it necessary to use conventional heating systems as auxiliary systems (oil, gas). In the case of parallel operation the amount of auxiliary heat typically amounts to only 10 % of the total consumption. Here one may rely on direct electric resistance heating as auxiliary heating. The power of the latter is considerably lower than the peak heating load.

Figure 8 shows the contribution of the heat produced by the heat pump relative to the total seasonal heat demand as a function of heat pump power relative to the peak heating load. This curve is valid for parallel heat pump operation in Belgium. It is found that a heat pump with a capacity of only 50 % of the peak load already produces more than 90 % of the heating demand.

One group of bivalent heat pump systems is characterized by heat recuperation from the air extracted from the dwelling. It is necessary to renew dwelling air continuously. Typically 1 tot 1,5 times the dwelling volume of fresh air has to be provided hourly. In dwellings with little uncontrolled ventilation of inside air this can be achieved by controlled extraction of air from the house. The heat in this air can be recuperated by means of a heat pump (see figure 9). The incoming air is led over the condenser and further heated by an auxiliary system. The extracted air flows over the evaporator. It has to be pointed out that the coefficient of performance of the heat pump increases with decreasing outside temperature. This is due to the fact that the average condenser temperature decreases when the temperature of the fresh air decreases. These systems are of the parallel-bivalent type. Heat pump systems relying upon extracted air but with a hydronic heat distribution system are also used.

2.3. Performance

As a measure for the energetic performance of space heating heat pumps one may take the average COP over a heating season. The value of this COP will depend upon the characteristics of the heating season. Table 2 lists typical seasonal COP for a number of systems valid for the average belgian heating season (lowest average temperature being -10°C; number of degree-days 2165)[6].

Table 2 also lists the type and capacity of the auxiliary system and the percentage of the heat load provided by the heat pump. Of course only bivalent heat pump systems require auxiliary heat, the parallel systems requiring less of this heat than the alternative ones.

The seasonal COP values listed in Table 2 take into account the electricity consumed by fans, pumps, etc. The water-water monovalent heat pump shows the highest COP values due to the high temperature of the heat source available throughout the whole heating season. The lower temperatures of ground sources considerably reduce the COP values.

For the bivalent systems the heat pump COP looses a lot of its meaning since it does not take into account auxiliary heat. Because of this the seasonal system COP, including auxiliary heat as defined by :

$$\text{Seasonal COP} = \frac{\text{heat produced per season}}{\text{heat pump + auxiliary heat per season}}$$

is used for the bivalent systems.

Finally table 2 also lists the seasonal primary energy efficiency which is the ratio of heat produced to the primary energy consumed to produce this heat. Here one has taken into account that all heat pumps are electrically driven and that the efficiency of electricity production is about 33 %. The last column shows that only the monovalent heat pump systems using the soil or ground water as heat source may give rise to primary energy savings compared to conventional oil or natural gas heaters.

Systems with very high primary energy efficiency which are presently available for heating of single and multi family houses are the heat pumps driven by internal combustion engines (gas or oil motor). The schematic representation of these heat pumps is given in figure 10. They allow for the recuperation of waste heat from the cooling of the engine and from the exhaust gases. Thus with ground water heat pumps of this type primary energy efficiencies of 150 % can be reached. The disadvantages of these systems are : high initial and maintenance cost and a limited lifetime.

More detailed discussions regarding special heating applications of heat pumps can be found in [5] and [7], to [10].

3. SWIMMING POOL HEATING

A succesfull application of heat pumps in the domestic sector is the heating of swimming pools, in particular of the indoor type. First of all pool water has to be kept at a temperature of the order of 28 °C. Furthermore for comfort reasons the relative humidity of the pool air has to be limited to about 60 %. If no measures were taken the relative humidity would tend to values of 90 %. Both problems can be solved by means of a heat pump which has extracted humid air going over the evaporator. Heat can be extracted from this air by lowering the temperature of the air (sensible heat) as well as by condensing part of the water content of the air (latent heat). This heat can be discharged to dry, fresh air flowing over the condenser and entering the pool area. It can also be transmitted to the pool water and even as preheat to hot water reservoirs for warm showers. It is also possible to utilize the heat pump only as air dryer by sending the same air over evaporator and condenser. A detailed discussion of this heat pump application can be found in [8]. Considering the temperatures involved, the heat pump functions at high COP values (3 ... 4) and gives rise to considerable energy and financial savings.

4. COMMERCIAL AND PUBLIC SECTOR

Heat pumps are applied more and more for the heating and cooling of commercial and public buildings. The need of both heating and cooling gives rise to considerable energy savings when heat pumps are used. This makes it possible to arrive at short payback periods to repay for the additional investment which heat pump systems always represent. Simultaneous heating and cooling loads are encountered in large office buildings where peripheral zones have to be heated during the winter while central zones have to be cooled. Also commercial buildings (supermarkets) often require heating and cooling at the same time.

A large variety of systems exists. This is due to the peculiarities of each building. An example of a system using ground water for the heating and cooling of an office building including shops is shown in figure 11 (reference [3]). Ground water can be sent either to the condenser or to the evaporator of the heat pump depending on wheter cooling or heating is needed. The water is then dumped in a river. Two water distribution systems are provided, one for hot and one for cold water. In this system the temperatures in the offices can be in dvidually controlled while the temperature of the shops are globally controlled. It is possible to provide cooling to one part of the building while heating is provided to other parts.

A system providing heating and cooling to a supermarktet is shown in figure 12 (reference [3]). The heat source consists of outside air and extracted air. In figure 12a the heating mode is shown. Heat is added to fresh incoming air in heat pump 3. Fresh air is added which was heated by recuperation of heat rejected by the condensers of the supermarket refrigerators. This air is further heated in the condenser of heat pump 2. Air extracted from the building is recirculated and added to the fresh air. This air is heated in the condenser of heat pump 1 to which auxiliary heat can be added to meet the required heating load. Air is mainly extracted in the lower portion of the building. The heat contained in the ventilated air is recuperated by means of heat pumps 1 and 2. Heat pump 3 operates on outside air only. At increasing heating loads heat pumps 1, 2 and 3 are started consecutively.

The cooling mode of operation of this heat pump system is shown in figure 12b. Heat from the refrigerator condensers has to be discarded to the atmosphere and cannot be recuperated. The role of the heat exchangers is inverted (evaporator becomes condenser and vice-versa).

The two systems described above rely on a large central heat pump to provide heating (and cooling). In recent years a considerable number of systems have been installed in office buildings in which small heat pumps (1 - 5 kW) are placed in the offices themselves. The heat pumps are all connected to a water distribution system. Depending upon the needs in each separate office the heat pump functions as heating or as cooling device. The heat is withdrawn or rejected to the water lines. This system has, for office buildings, a number of advantages such as better serviceability and reliability which render it preferable to the central heat pump system.

Important areas for application of heat pumps are schools and hospitals. The former are characterized by heavy ventilation losses : 2 to 4 air renewals per hour are applied. It is obvious that a heat pump recuperating the heat from the extracted air leads to considerable energy savings. In hospitals very often totally fresh air (no recirculation) is required. Here also considerable ventilation losses occur and heat pumps can be applied succesfully. More information regarding systems in this sector are found in the references [2], [4] and [5].

5. CONCLUDING REMARKS

From this overview of heat pump systems applied in the domestic sector it appears that the most important problem which has to be overcome is that of the temperature level of the cold source, if the function of the heat pump is to heat only. Presently it is difficult to reach high seasonal heating COP values and so it is difficult to achieve primary energy conservation with heat pumps. The widespread application of heat pumps in single family dwellings is also limited by the temperature which present heat pumps can attain. This is also true for systems which can be used also for cooling purposes. Here the heat pump investment can be recuperated more rapidly however. This explains why in areas where both heating and cooling are required the presently available heat pumps (i.e. electrically driven) are able to penetrate the market to a considerable extent. If only heating is required it seems that the existing heat pumps are able to penetrate only because of local economic factors (e.g. low off-peak electricity rates...).

For large office and commercial buildings, schools and hospitals it can be expected that heat pumps will be applied more frequently in the future.

REFERENCES

1. Ruff, A.N., The Heat Pump Water Heater, Refr. Eng. 59, 153-154 (1951).

2. Von Cube, L., Steimle, F., *Wärmepumpen*, Grundlagen und Praxis, VDI Verlag (Düsseldorf, 1978).

3. Anders, P., Kalischer, P., Betriebserfahrungen mit Grosswärmepumpenanlagen und Schlussfolgerungen für die Weiterentwicklung, Elektrowärme International, 30, A4, 171-179 (1972).

4. *Handbuch der Wärmerückgewinnung*, Vulkan Verlag, 2nd Edition.

5. Elektrowärme International, 1960 - present.

6. Geeraert, B., Technico-economic study of the heat pump, Report B4, Ministry of Scientific Policies (Brussels, 1978).

7. Kirn, H. and Hadenfelt, A., *Wärmepumpen, Vol. 1 : Einführung und Grundlage*, Müller Verlag (Karlsruhe, 1981).

8. Kirn, H. and Hadenfelt, A., *Wärmepumpen, Vol. 2 : Anwendung der Elektrowärmepumpe*, Müller Verlag (Karlsruhe, 1980).

9. Jüttemann, H., *Wärmepumpen, Vol. 3, Anwendung der Gas- und Dieselwärmepumpe in der Haustechnik*, C.F. Müller Verlag (Karlsruhe, 1981).

10. Eickenhorst, H., *Wärmepumpen, Vol. 4, Installation, Betrief und Wartung der Elektrowärmepumpen*, Müller Verlag (Karlsruhe, 1982).

TABLE I

	J	F	M	A	M	J	J	A	S	O	N	D
Average soil temp. at 60 cm (°C)	4,7	4,1	5,7	7,8	11,7	15,0	16,6	17,2	15,5	12,3	8,7	6,3
Minimum soil temp. at 60 cm (°C)	1,9	1,0	1,3	4,0	7,0	11,9	13,0	15,0	12,1	7,7	5,0	2,9

TABLE 2

Heat Pump System	Heat Source	Heat Source Temperature (°C)	Auxiliary Heating System Type	Auxiliary Heating System Capacity (%)	Heat Pump delivered Heat (%)	Seasonal COP	Primary Energy Efficiency (%)
monovalent soil/water	soil	- 2 to + 10	-	-	100	2 to 3	66 to 100
monovalent water/water	ground or surface water	10	-	-	100	3 to 4	100 to 130
bivalent parallel air/water	outside air	- 10 to 15	oil or gas boiler	40	90	2,5 to 3,0	70
bivalent alternative air/water	outside air	- 10 to 15	oil or gas boiler	100	60 - 70	1,6 ... 1,8	75
bivalent extracted air/air or water	extracted air	18 to 20	direct electric resistance heating	75 to 85	50 - 60	1,6	50

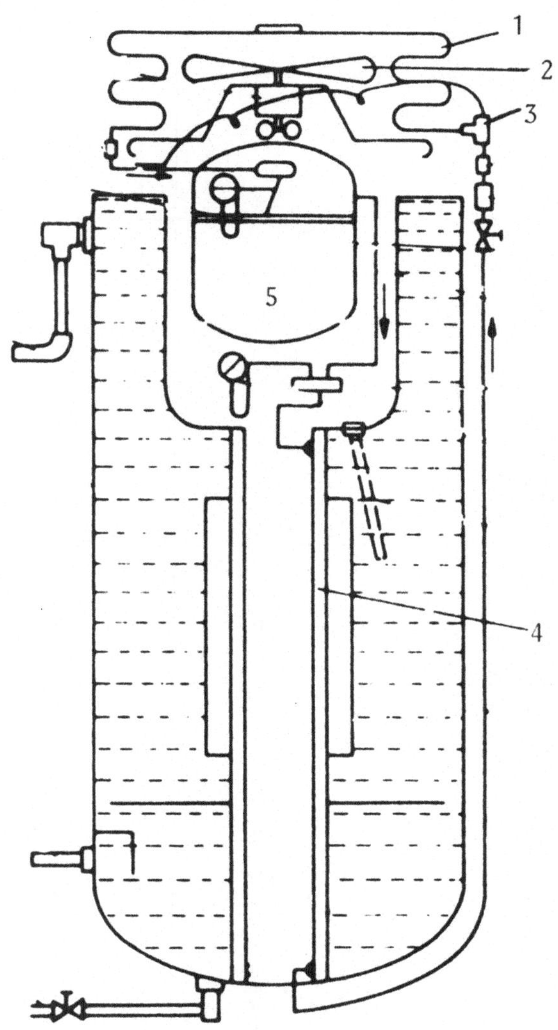

FIGURE 1

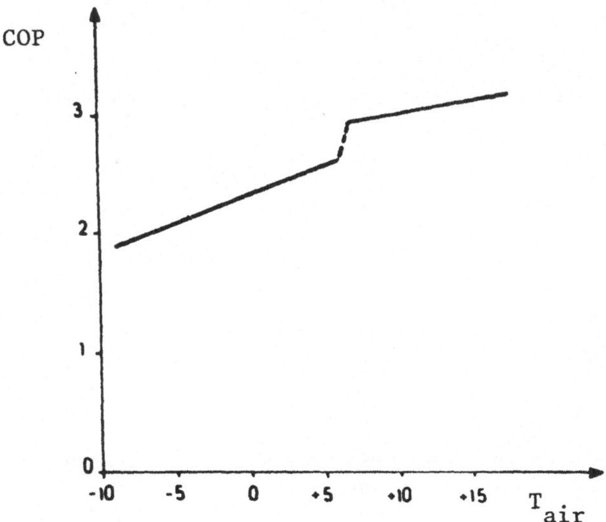

FIGURE 2

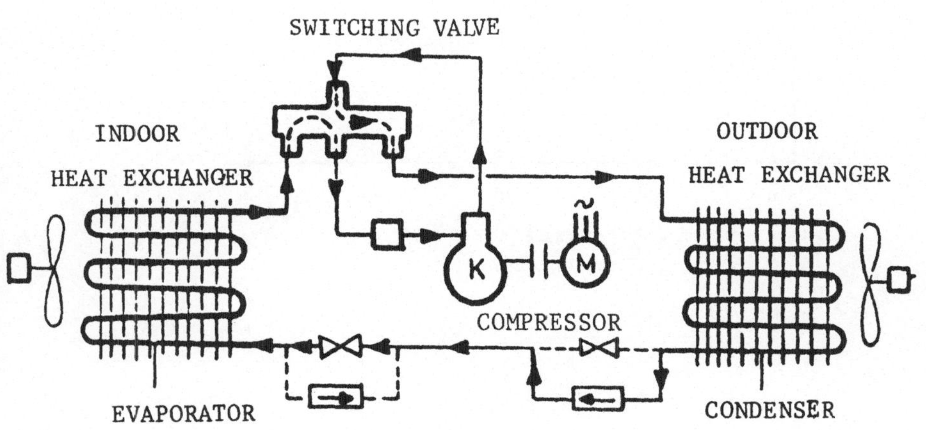

FIGURE 3

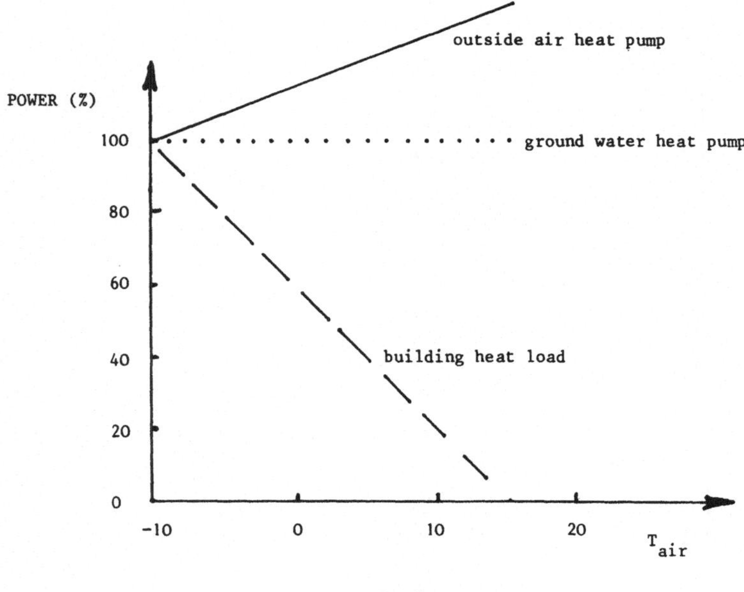

FIGURE 4

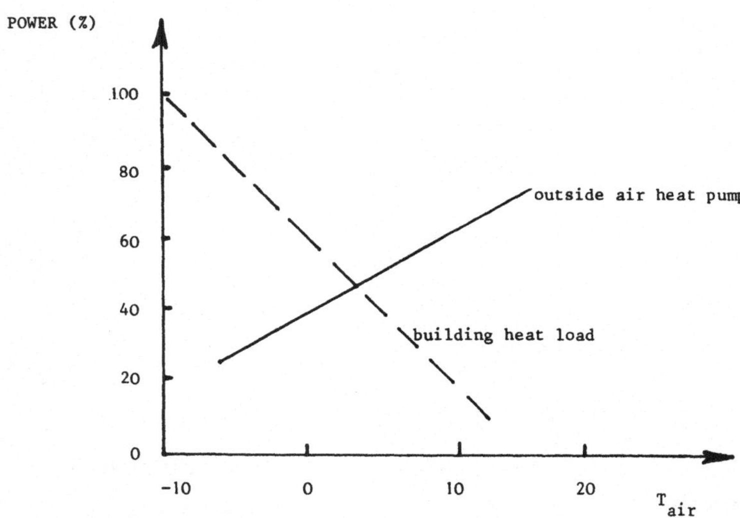

FIGURE 5

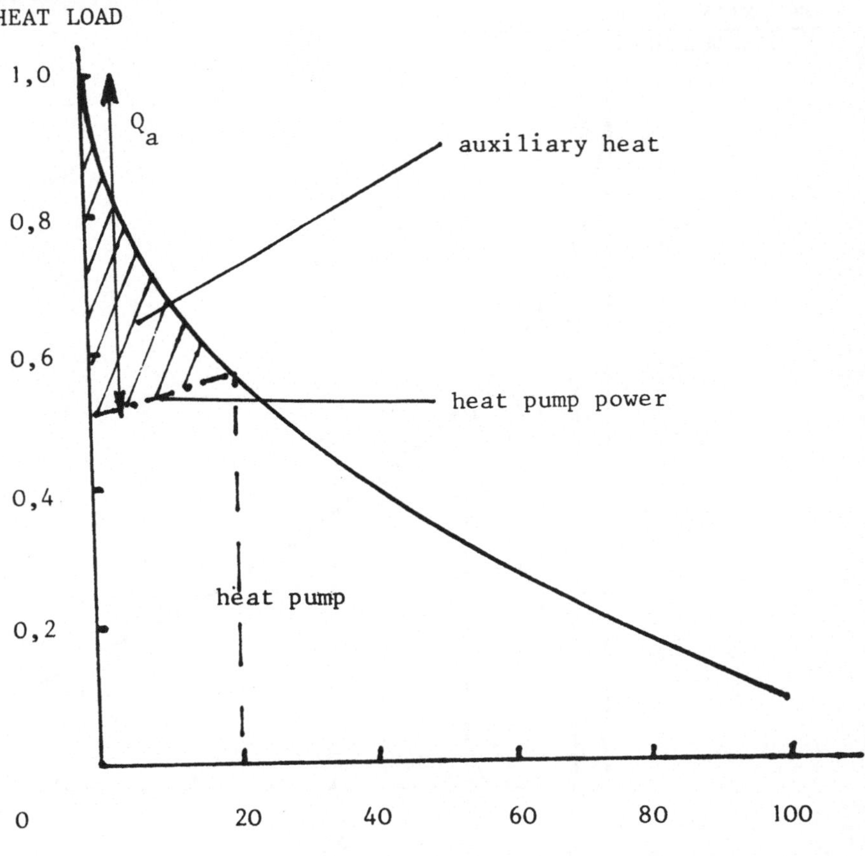
INSTANTANEOUS HEAT LOAD
MAXIMUM HEAT LOAD
1,0
0,8
0,6
0,4
0,2
0
Q_a
auxiliary heat
heat pump power
heat pump
20
40
60
80
100
NUMBER OF HEATING HOURS (cumulative)
TOTAL NUMBER OF HEATING HOURS
(%)

FIGURE 6

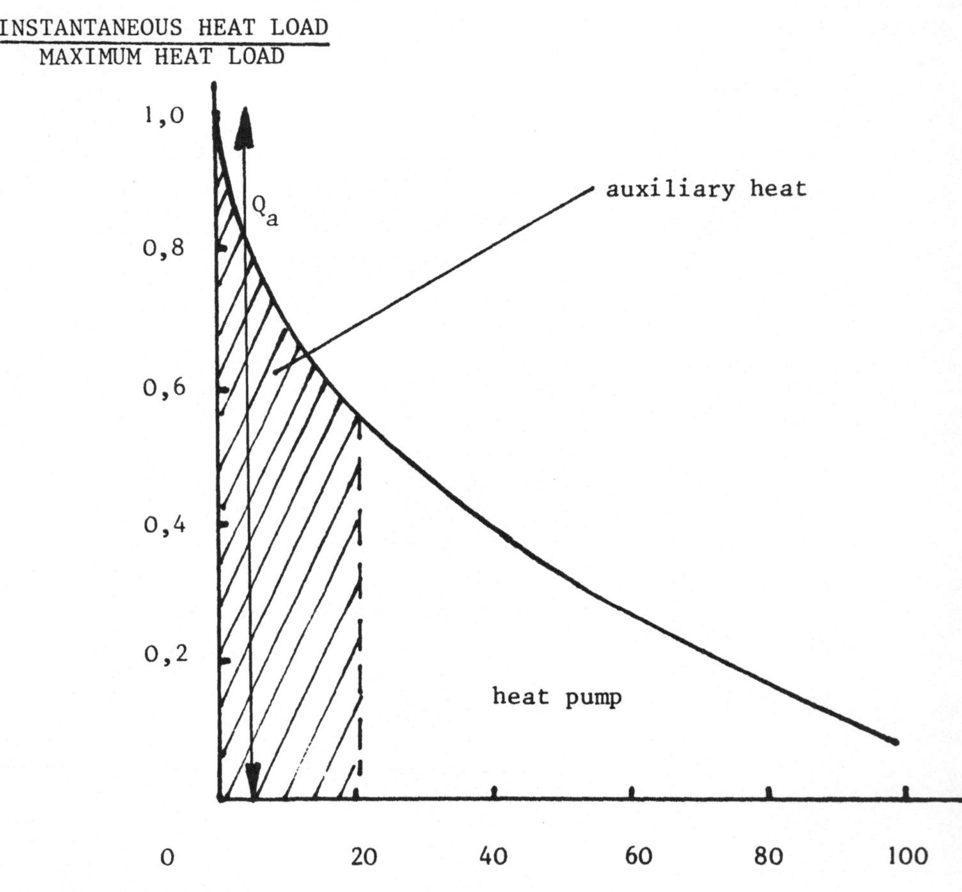

INSTANTANEOUS HEAT LOAD / MAXIMUM HEAT LOAD
1,0
0,8
0,6
0,4
0,2
Q_a
auxiliary heat
heat pump
0
20
40
60
80
100
NUMBER OF HEATING HOURS / TOTAL NUMBER OF HEATING HOURS (%)

FIGURE 7

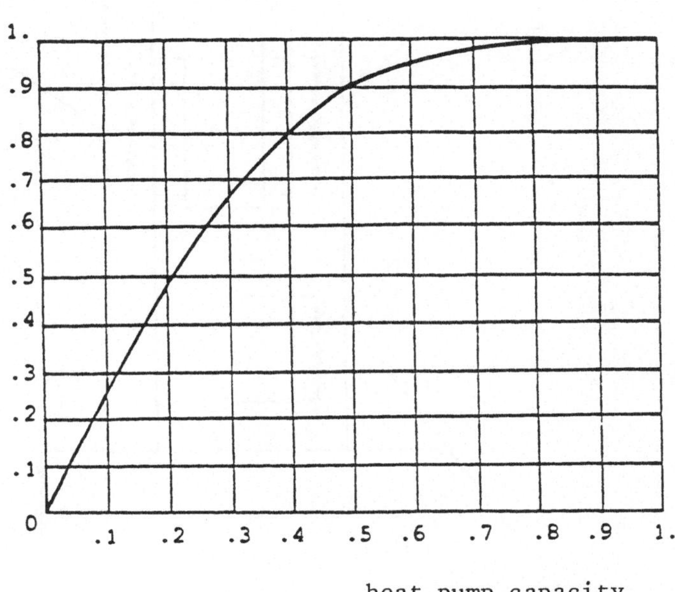
heat pump delivered heat / total heating needs
1.
.9
.8
.7
.6
.5
.4
.3
.2
.1
0
.1
.2
.3
.4
.5
.6
.7
.8
.9
1.
heat pump capacity / peak heating load

FIGURE 8

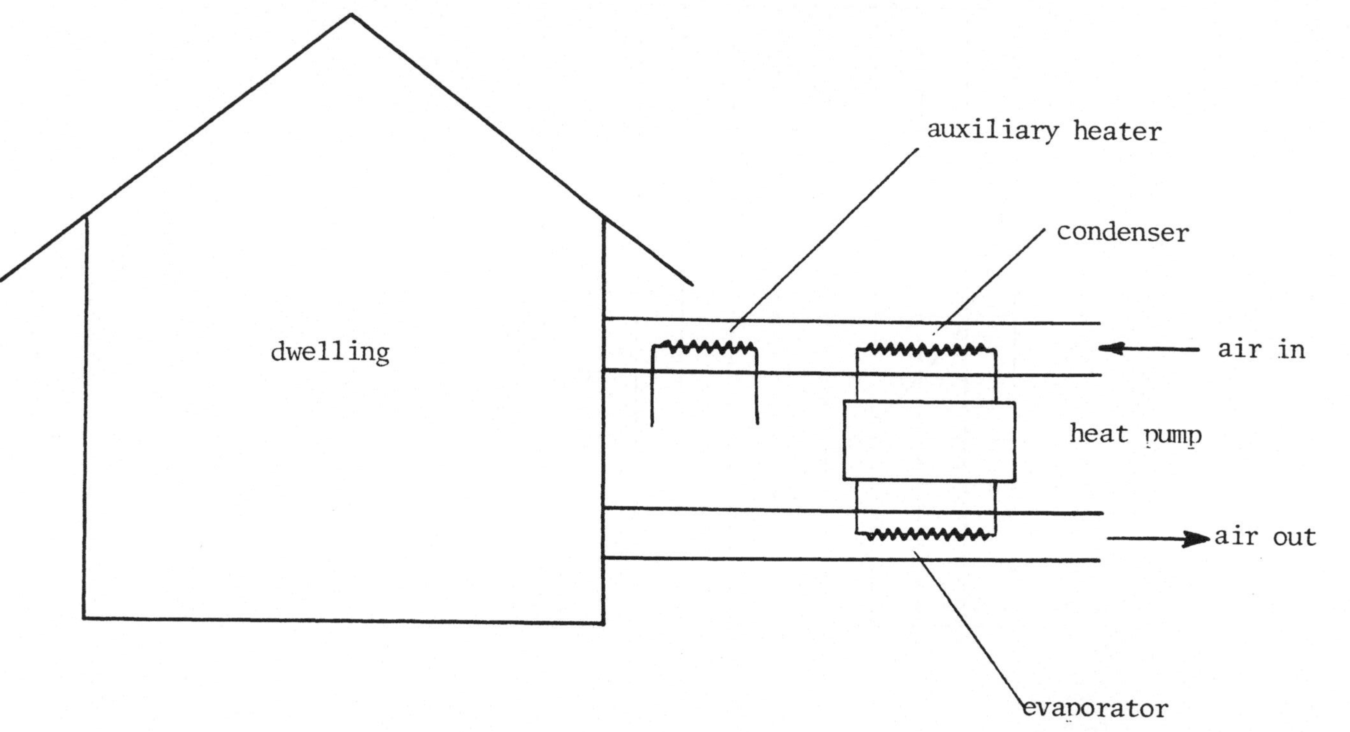
auxiliary heater
condenser
dwelling
air in
heat pump
air out
evaporator

Figure 9

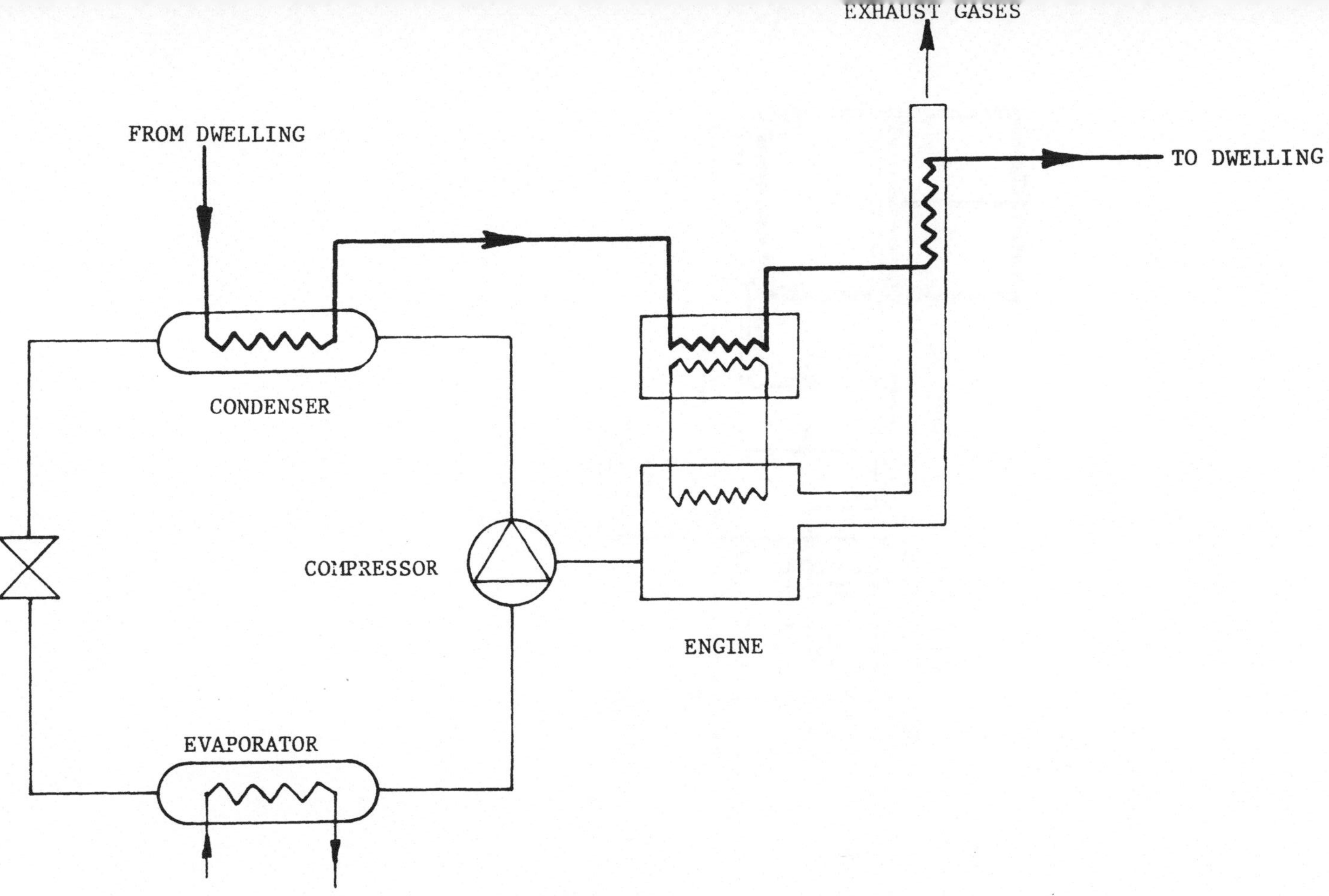
EXHAUST GASES
FROM DWELLING
TO DWELLING
CONDENSER
COMPRESSOR
ENGINE
EVAPORATOR

FIGURE 10

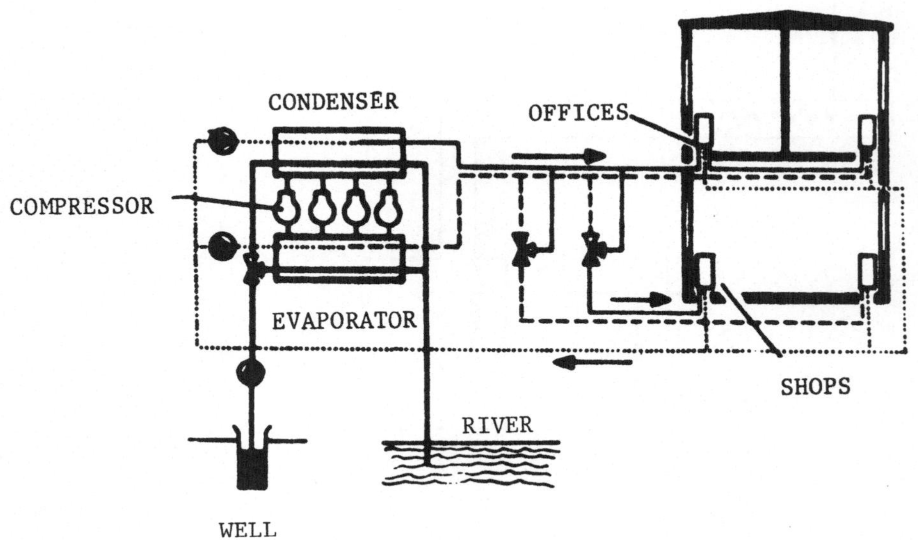
CONDENSER
OFFICES
COMPRESSOR
EVAPORATOR
SHOPS
RIVER
WELL

FIGURE 11

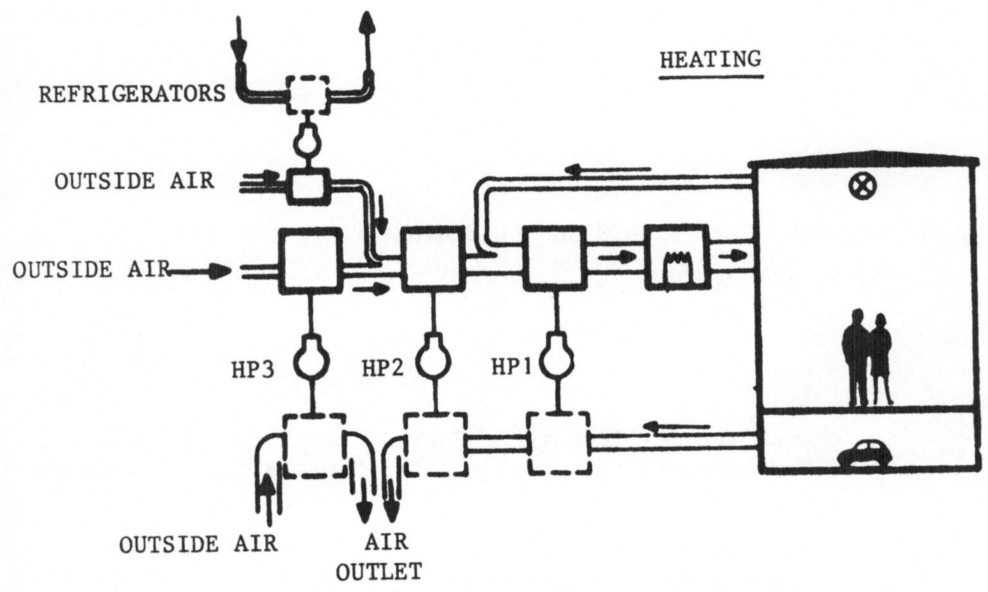
HEATING
REFRIGERATORS
OUTSIDE AIR
OUTSIDE AIR
HP3
HP2
HP1
OUTSIDE AIR
AIR
OUTLET

a

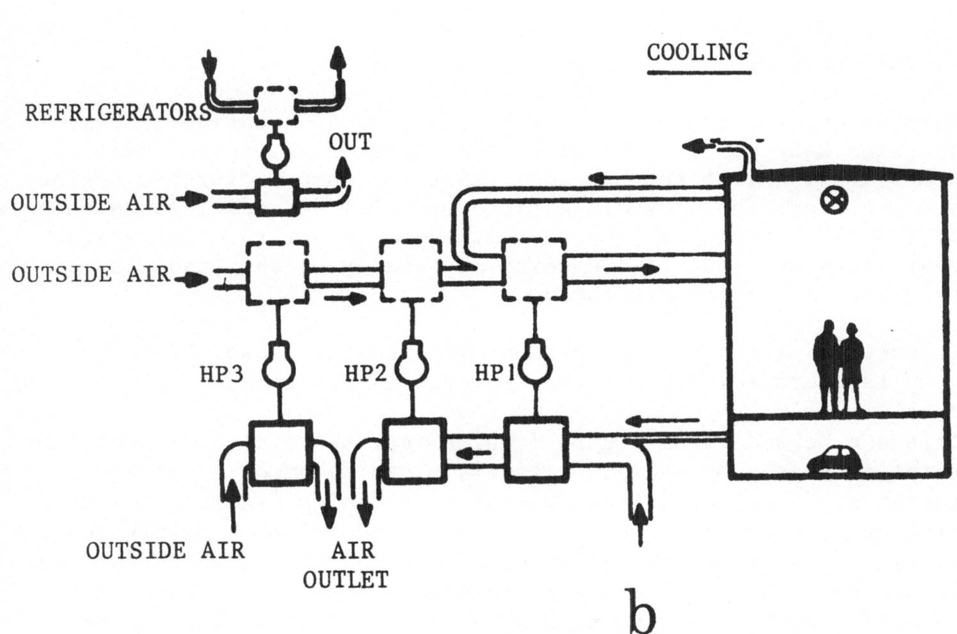
COOLING
REFRIGERATORS
OUT
OUTSIDE AIR
OUTSIDE AIR
HP3
HP2
HP1
OUTSIDE AIR
AIR
OUTLET

b

FIGURE 12

THERMODYNAMICS OF MIXTURES AND ELEMENTARY PROCESSES IN ABSORPTION HEAT PUMPS

H. Auracher, H. Glaser, K. Stephan

Institut für Technische Thermodynamik und Thermische Verfahrenstechnik, Universität Stuttgart, F.R. Germany

PART I : THERMODYNAMICS OF MIXTURES WITH SPECIAL APPLICATION TO ABSORPTION HEAT PUMPS

(K. Stephan)

1. INTRODUCTION

Absorption heat pumps are a very effective alternative to compression heat pumps. The main advantages of absorption heat pumps are : except for the small solution pump no further engines are required; the over-all efficiency (COP) depends less on changes of the ambient temperature than is the case for the compression heat pump and values above unity may easily be obtained also at very low ambient temperatures.

In contrast with the compression heat pump the absorption heat pump is operated with mixtures as working substances.

This can be seen from Figure 1 which represents an elementary absorption heat pump cycle. A liquid mixture, ammonia-water, for instance, is heated in the generator at a high pressure and at a high average temperature T_z. The vapor thus formed mainly consists of the lighter volatile component (ammonia), which is then condensed. The liquid, after expansion to a low pressure in a throttle valve, is evaporated at a low temperature T_o and the vapor then flows into the absorber where it is absorbed by the weak or poor solution with mass fraction ξ_p coming from the generator. The strong or rich solution with mass fraction ξ_r is pumped back to the generator where the process continues. On the whole, heat at a high temperature T_z in the generator and heat at ambient temperature T_o in the evaporator are flowing into the process and are transformed into heat used at a medium temperature T_h.

As can be seen from this simplified process, the absorber, the generator and also the condenser and evaporator use mixtures of different composition as working substances. It seems to be useful therefore to recapitulate the fundamental phenomena of thermodynamics of mixtures. This will facilitate understanding of the sorption heat pump processes discussed in the following chapters. We will also study the main elementary processes with mixtures which occur in absorption heat pump systems.

Special reference is made for this purpose to the books by E. Altenkirch [1], F. Bosnjakovic [2], and W. Niebergall [3].

2. EVAPORATION AND CONDENSATION OF MIXTURES

It is well known that the boiling temperature T of mixtures for a given pressure is not constant, but depends also on the composition $T = T\ (p, x_1, x_2 \ldots x_{k-1})$, where k stands for the number of components and x_i is the mole fraction of an arbitrary component i. For a binary mixture we have $T = T\ (p, x)$. Accordingly, for a given pressure, evaporation and also the processes in the generator of an absorption heat pump do not occur at constant temperature. They take place within a temperature range. The same holds for the processes of condensation, absorption and the so-called resorption. Figures 2 and 3 present evaporation and condensation for a typical binary mixture in p-x and T-x diagrams.

In general completely miscible substances are used in absorption heat pumps. Furthermore, up to now, no azeotropic mixtures have been used, although a ternary mixture consisting of a binary azeotropic mixture as working substance and a third component as solvent may exhibit the same advantages as azeotropic mixtures in refrigeration cycles. Mixtures of this type will be discussed in the chapter on working substances.

The boiling temperatures of mixtures used in absorption heat pump cycles therefore always lie between the boiling temperature of the pure components. The boiling temperature of an ammonia-water-mixture for instance is above the boiling temperature of pure ammonia ($T = -34°C$ at $p = 1$ bar) and below the boiling temperature of pure water ($T = 100°C$ at $p = 1$ bar). Also when using salt solutions as mixtures the change of phase occurs within a certain temperature range. Only in the condenser and evaporator, where pure substances undergo a phase-change, the temperature at a given pressure is constant.

3. HEAT OF SOLUTION AND SOLUBILITY DIAGRAM

3.1. Heat of Solution

Sorption processes mostly require a considerable heat of solution to be transmitted. This heat of solution is often called the enthalpy of mixing. From thermodynamics we know that the enthalpy of mixing is zero for ideal mixtures, whereas for the real mixtures as e.g. used in absorption heat pumps, the enthalpy of mixing cannot be neglected compared to the enthalpy of condensation. For ammonia-water mixtures the ratio between the enthalpy of mixing and the enthalpy of evaporation of pure ammonia is 0,537 for a mass fraction of ammonia of 0,25 and it decreases to 0,035 for a mass fraction of 0,75.

The specific enthalpy of mixing

$$\Delta h = h - \Sigma h_{ok}\, \xi_k \tag{1}$$

is defined as the difference between the enthalpy h of a mixture after isobaric-isothermal mixing and the enthalpy of the pure components $\Sigma h_{ok} \xi_k$ before mixing. Notice that the indices "ok" stand for pure "o" component "k". If we consider for instance an isobaric-isothermal mixing process of two components 1 and 2 we have to transfer a certain heat Δh per mass unit of the mixture in order to keep the temperature constant (Figure 4). The heat to be transferred is equivalent to the enthalpy of mixing. It can be negative (exothermal mixture) or positive (endothermal mixture) and it vanishes for ideal mixtures.
One may put :

$$h = \Sigma\, h_k\, \xi_k, \tag{2}$$

in which $h_k = \left(\frac{\partial H}{\partial m_k}\right)_{T,p,m_{j \neq i}}$

is the partial specific enthalpy, which says how much the enthalpy of a mixture changes at $T,p,m_{j \neq i}$ = const., when small amounts of one component i are added. Equation (1) can then be transformed into :

$$\Delta h = \Sigma\, (h_k - h_{ok})\xi_k \tag{3}$$

The difference

$$h_k - h_{ok} = \Delta h_k \tag{4}$$

in this equation is the so-called partial specific enthalpy of mixing. For a binary mixture, independent of the type of the mixture, e.g. liquid-liquid, gas-liquid or liquid-solid, equation (3) can be written as :

$$\Delta h = \xi_1 h_1 + \xi_2 h_2 - \xi_1 h_{o1} - \xi_2 h_{o2} \tag{5}$$

Let us assume that component 1 is the solvent and component 2 the solute. As a pure substance component 2 exists either in the gaseous or in the solid state at pressure p and temperature T. The enthalpy h_{o2} of component 2 therefore must be taken for this state and not for the liquid state of the solution. In order to avoid confusion with the enthalpy h_{o2} of the pure liquid component 2 we further-on denote the enthalpy of the pure component 2 in a gaseous or solid state by $h_{o2'}$. Also the specific enthalpy of mixing Δh is further-on written as $\Delta h'$. The prime reminds one that during the mixing process the solute undergoes a change of phase. The usual enthalpy of mixing Δh does not include a change of phase. Equation (5) therefore is written as

$$\Delta h' = \xi_1(h_1 - h_{o1}) + \xi_2(h_2 - h_{o2'}) \text{ or}$$

$$\Delta h' = \xi_1(h_1 - h_{o1}) + \xi_2(h_2 - h_{o2}) + \xi_2(h_{o2} - h_{o2'}). \tag{5a}$$

In this equation h_{o2} now stands for the enthalpy of pure component 2 in the same state as the mixture, i.e. in the liquid state, whereas $h_{o2'}$ stands for the enthalpy of component 2 in its original gaseous or solid state before mixing.

The term

$$h_{o2} - h_{o2'} = \Lambda_2 \tag{6}$$

is named the specific enthalpy of fusion when a solid is dissolved, and specific enthalpy of condensation, when a gaseous substance is dissolved in a liquid. We have therefore :

$$\Delta h' = \xi_1(h_1 - h_{o1}) + \xi_2(h_2 - h_{o2}) + \xi_2 \ \Lambda_2 \tag{7}$$

or

$$\Delta h' = \Delta h + \xi_2 \ \Lambda_2 \tag{7a}$$

The enthalpy difference $\Delta h'$ is the so-called integral enthalpy of mixing or the heat of solution. The value $h_1 - h_{o1}$ indicates how much the partial enthalpy of the solvent within the mixture differs from the specific enthalpy of the pure solvent for a given temperature T and a given pressure p.

As can be seen from equation (7) the evaluation of the heat of solution requires the specific partial enthalpies $h_1 - h_{o1}$, $h_2 - h_{o2}$ and also the enthalpy of fusion or condensation Λ_2 to be known. Whereas $h_1 - h_{o1}$ and $h_2 - h_{o2}$ mostly are known or may be evaluated from the enthalpy of mixing in liquid-liquid mixtures (laid down in many tables e.g. [4]), only few experimental data are known about Λ_2. It can, however, be determined approximately according to H. Mollier [5]. As indicated in Figure 5 the vapor pressure of the pure liquid component 2 at the temperature T is p_{2s}. Due to the isothermal decrease of vapor pressure in solutions, the vapor pressure p of the solution is smaller than p_{2s} and therefore $p_{2s} > p$. The enthalpy h_{o2} (T,p_{2s}) of the pure liquid is represented by point 1 in Figure 5. We construct now a cycle as shown in Figure 5. The pure liquid is cooled along an isobar 1-2, which practically coincides with the saturated liquid line, until it reaches the temperature T_s belonging to pressure p. The liquid is then evaporated along the saturation line 2-3 and the vapor is superheated along the isobar 3-4, where its specific enthalpy is $h_{o2}'(T,p)$. Introducing the enthalpy of evaporation Δh_V of component 2 at pressure p, the isobaric specific heat capacity c_L of the liquid and the corresponding isobaric specific heat capacity of the vapor c_V (both at pressure p) and taking into account that $h_{o2}(T,p_{2s}) \approx h_{o2}(T,p)$, we obtain from an energy balance for the cycle

$$h_{o2}' (T,p) = h_{o2} (T,p) - c_L (T-T_s) + \Delta h_V + c_V (T-T_s) \quad (8)$$

From which we obtain :

$$h_{o2} (T,p) - h_{o2}'(T,p) = \Lambda_2 = - [\Delta h_V - (c_L-c_V)(T-T_s)] \quad (9)$$

If pure component 2 is a solid, an analogous derivation leads to

$$\Lambda_2 = [\Lambda h_s - (c-c_L) (T-T_s)] \quad (9a)$$

where Δh_s is the enthalpy of melting and c the specific heat capacity of the solid, both to taken at pressurep. Inserting equation (9) into equation (7) yields for the enthalpy of solution :

$$\Delta h' = \xi_1(h_1-h_{o1}) + \xi_2(h_2-h_{o2}) - \xi_2[\Delta h_V - (c_L-c_V)(T-T_s)] \quad (10)$$

In cases of solid liquid-mixtures we obtain from equations (9a) and (7) :

$$\Delta h' = \xi_1(h_1-h_{o1}) + \xi_2(h_2-h_{o2}) + \xi_2[\Delta h_s - (c-c_L)(T-T_s)] \quad (10a)$$

By definition $\Delta h'$ is the enthalpy change of 1 kg of solution when the liquid solvent 1 and the gaseous solute 2 are mixed at constant temperature T and constant pressure p. It is identical to the heat (per mass unit of the solution) to be removed from the absorber. In the generator the heat $-\Delta h'$ has to be added in order to split up the liquid mixture into vapor and weak solution.

Equation (10) may still be simplified, because usually the temperature difference $T-T_s$ is small and therefore the term $(c_L-c_V)(T-T_s)$ is negligible in comparison to Δh_V. Also the enthalpy difference h_1-h_{o1} of the specific enthalpy h_1 of the solvent within the mixture and the specific enthalpy h_{o1} of the pure solvent is negligible provided that $\xi_2 \to 0$ or $\xi_1 \to 1$. Taking into account that in equation (10) we have $\Delta h' = \Delta H'/(m_1+m_2)$, $\xi_1 = m_1/(m_1+m_2)$ and $\xi_2 = m_2/(m_1+m_2)$ we get thus the approximative formula for the heat of solution of gas-liquid-mixtures :

$$\Delta H'/m_2 = (h_2 - h_{o2}) - \Delta h_V \tag{11}$$

This is the heat flowing to the absorber. The negative sign on the right hand side says that heat is transferred from the absorber. In some papers equation (11) is also written as :

$$-\frac{\Delta H'}{m_2} = q_G = (h_{o2}-h_2) + \Delta h_V = \lambda + \Delta h_V \tag{11a}$$

with the so-called "differential heat of solution" $\lambda = h_{o2} - h_2$ and the heat q_G transferred to a generator.

For solid-liquid mixtures we obtain from equation (10a)

$$\Delta H'/m_2 = (h_2-h_{o2}) + \Delta h_s, \tag{11b}$$

or

$$\Delta H'/m_2 = q_G = (h_{o2}-h_2) - \Delta h_s = \lambda - \Delta h_s \tag{11c}$$

Equations (11a) and (11c) are mostly used for calculation of the heat of solution. As shown, however, by the foregoing derivation they are only approximations valid for $T \to T_s$ and $\xi_1 \to 1$.

In fact for ammonia-water-mixtures h_1-h_{o1} is very small over a wide range of ammonia concentrations ($\xi_2 \lesssim 0.4$ or water concentration $\xi_1 \gtrsim 0.6$). Also in salt-solutions the term h_1-h_{o1} is mostly small because of the low salt concentrations $\xi_2 \to 0$, or $\xi_1 \to 1$.

In these cases the approximative formulas (11) and (11a) may be applied instead of the more accurate expressions (10) and (10a). It should be noted that for all working substances used in

refrigeration cycles and also for those proposed for absorption heat pumps the enthalpy difference $\lambda = h_{o2}-h_2$ turns out to be positive. This means that the heat q_G transferred to the generator consists of the enthalpy of evaporation Δh_V and the "differential heat of solution" λ.
For more accurate calculations also the term $m_1/m_2 \cdot (h_{o1}-h_1)$ according to equation (10) must be added. Again for the known working substances $h_{o1}-h_1$ is positive.

3.2. Solubility diagram

Whereas the vapor pressure of a pure substance is given in a p-T diagram by a single curve, we obtain for each given composition of a mixture a different vapor pressure curve. Thus for a binary mixture the vapor pressure curve is given by

$$p = p\ (T, \xi_L),$$

Where ξ_L is the mass fraction of one component in the liquid phase. The graphical representaion of this equation is also named the solubility diagram. As an example Figure 6 represents the solubility diagram of an ammonia-water mixture. According to the Clausius-Clapeyron equation the vapor pressure curve of the pure components at low pressure is given by

$$\log p = - \frac{\Delta h_V}{R\ T} + \text{const.}, \tag{12}$$

where h_V is the specific enthalpy of evaporation of the pure components. For a binary mixture the heat to be transferred for evaporating this mixture is given by $q_G = h_V + \lambda$, under the assumptions discussed before. The above equation (12) then becomes :

$$\log p = - \frac{\Delta h_V + \lambda}{R\ T} + \text{const.} \tag{12a}$$

As shown by thermodynamics this equation holds for moderate pressures and vanishing enthalpy of mixing in the gaseous phase. The slope of the different vapor pressure curves therefore depends on the differential heat of solution λ. For λ positive, as it occurs in practical applications, the lines for constant mass fraction in the log p-1/T plot become steeper. For λ negative they become flatter and for $\lambda = 0$ they remain parallel. Figure 7 demonstrates these different cases.

As a by-product it should also be mentioned already here that the heat ratio (heat transferred from the system divided by heat transferred to the system at an elevated temperature) of the reversible process can be read immediately from the diagram. For the heat pump process it is given by (see : chapter on absorption heat pump cycles) :

$$\zeta_{id} = \frac{1/_{T_o} - 1/_{T_z}}{1/_{T_o} - 1/_{T_h}} \quad \text{or} \quad \zeta_{id} = \frac{CB}{CA},$$

where CB and CA can be read from Figure 8. As we can see from Figure 8, we get :

$\zeta_{id} < 2$ for $\lambda > 0$, whereas

$\zeta_{id} > 2$ for $\lambda < 0$

and $\zeta_{id} = 2$ for $\lambda = 0$

For a heat transformer we find in a similar way :

$$\zeta_{id} = \frac{AC}{BC}$$

and therefore according to Figure 8

$\zeta_{id} > 0.5$ for $\lambda > 0$,

$\zeta_{id} < 0.5$ for $\lambda < 0$

and $\zeta_{id} = 0.5$ for $\lambda = 0$

For a heat pump it would be advantageous therefore to use working substances with $\lambda<0$, which at present are not available. For the heat transformation process the desirable substances with $\lambda > 0$ are available.

4. THE ENTHALPY-CONCENTRATION DIAGRAM

A very effective help for studying and evaluating absorption heat pump processes with binary mixtures is provided by the enthalpy-concentration diagram first introduced by Merkel [6] in 1928.

This diagram is based on experimental data of the mixture and allows a quick and accurate evaluation of the process. One does not have to refer to an approximative evaluation. Furthermore the influence of changes of parameters can easily be assessed. On the other hand h-ξ diagrams exist only for a few mixtures and in cases where comprehensive experimental data for drawing a diagram are missing approximative methods of thermodynamics must be applied. In some cases also a computer calculation of a process may easily be performed.

At any rate the h-ξ diagram is still a useful tool because it visualizes the process.

In order to construct a diagram for binary mixtures we start from the enthalpy

$$h = h_{o2}\ \xi + h_{o1}\ (1-\xi) + \Delta h\ , \qquad (13)$$

where h_{o1} and h_{o2} are the specific enthalpies of the pure components at pressure p and temperature T of the mixture. ξ is the mass fraction of component 2 (e.g. ammonia). Δh is the enthalpy of mixing, $\Delta h\ (T,p,\xi)$.

In the liquid state, far from the critical point, the enthalpies h_{o1} and h_{o2} of the pure components and also the enthalpy of mixing are practically independent of pressure p. In the gaseous state and for moderate pressures, the enthalpy of mixing is negligible. The enthalpy of the pure components is only a weak function of pressure and depends mainly on temperature. Isotherms in the vapor phase therefore are well approximated by straight lines (Figure 9). In the liquid phase they are straight lines for $\Delta h = 0$. The deviation from a straight line is determined by Δh, which for ammonia-water is always negative, as Figure 9 demonstrates. From the log p-1/T diagram we obtain the temperatures and concentrations of the boiling liquid at a given pressure (e.g. the boiling lines for constant pressure p_1, p_2, the thick lines in Figure 9). Above these boiling lines the isotherms for the indicated pressure do not exist any longer, because the solution is boiling already. For higher pressures however, boiling does not yet occur and therefore the isotherms represent existing states.

In the vapor region the isotherms are straight lines, as already mentioned. The vertical distance between vapor and liquid state points of the pure components at the same temperature corresponds to the enthalpy of evaporation. If furthermore the vapor composition $\xi''(p,T)$ is known, the dew line $\xi''(T)$ for a given pressure p can be plotted (see Figure 9).

For a given pressure the liquid region extends below the boiling line, the region of superheated vapor is situated above the dew line and between boiling and dew lines we find the region of vapor-liquid phase equilibrium. Isotherms in the vapor-liquid regions are straight lines connecting the boiling point and the dew point for given values of p and T. This follows from the mass and the energy balances :

total mass balance : $m = m_L + m_V$

mass balance for the lighter volatile component (e.g. ammonia) : $m\xi = m_V\xi_V + m_L\xi_L$

From both equations we obtain

$$\frac{m_L}{m} = \frac{\xi - \xi_V}{\xi_L - \xi_V} \quad \text{and} \quad \frac{m_V}{m} = \frac{\xi_L - \xi}{\xi_L - \xi_V}$$

The energy balance yields :

$$h = \xi_L h_L + \xi_V h_V = \frac{m_L}{m} h_L + \frac{m_V}{m} h_V$$

or using the above results from the energy balance we obtain :

$$h = \frac{\xi - \xi_V}{\xi_L - \xi_V} h_L + \frac{\xi_L - \xi}{\xi_L - \xi_V} h_V \qquad (14)$$

This is the equation of a straight line in the h-ξ diagram. Point M in Figure 9 represents the state of the vapor-liquid mixture. We obtain M by subdividing the distance AB according to

$$\frac{m_L}{m_V} = \frac{\xi - \xi_V}{\xi_L - \xi} = \frac{h - h_V}{h_L - h}$$

Figure 10 gives as example the enthalpy concentration diagram for ammonia water. The vapor-liquid isotherms are not plotted to achieve greater clarity. They may easily be plotted according to the construction ABC indicated in Figure 10.

PART II : ELEMENTARY PROCESSES IN ABSORPTION HEAT PUMPS.

(H. Glaser, H. Auracher)

1. VAPORIZATION AND CONDENSATION

1.1. Vaporization

An important elementary process in absorption heat pumps is the partial vaporization of a solution. Two limiting cases can be distinguished : the generated vapor and the solution may be either in parallel flow (Figure 11a) or in counter flow (Figure 11b). If equilibrium between vapor and solution is assumed, both limiting cases are different with respect to the concentrations of vapor and solution at the outlets of the generator. Both vaporization processes can be studied and evaluated very clearly and effectively by means of the enthalpy-concentration diagram due to Merkel [6].

Parallel flow (Figure 11a).

Assume that the solution enters the generator with concentration ξ_r and temperature t (state 1 in Figure 12) and a heat quantity Q_{pf} per kg of rich solution is supplied to the generator. The vapor thus generated flows parallel to the solution. At the end of this vaporization process the liquid-vapor mixture is assumed to be in state 2, just before leaving the generator. In the h-ξ diagram state 2 is situated in the region of vapor-liquid phase equilibrium. Note that the concentration ξ_r does not change from state 1 to state 2 whereas the temperature increases from t_s to t_{pf}. The isotherm t_{pf} through point 2 intersects the dew line in 3 and the boiling line in 4. Point 3 represents the state of the removed vapor (ξ_V, h_V) and point 4 that of the unevaporated poor solution (ξ_p, h_p).

According to Figure 11a, m_r kg of rich solution of concentration ξ_r are needed to generate m_V kg of vapor. The ratio $f = m_r/m_V$, often called the feed rate of the rich solution, follows from the mass balance for the lighter volatile component

$$m_r\xi_r = m_V\xi_V + (m_r - m_V)\xi_p \,, \qquad (15)$$

which gives :

$$f = \frac{m_r}{m_V} = \frac{\xi_V - \xi_p}{\xi_r - \xi_p} \qquad (16)$$

It is often useful to refer the heat input not to the mass of rich solution $m_r(Q_{pf})$ but to the mass of vapor m_V. Then we obtain (q_{pf}) :

$$q_{pf} = f \cdot Q_{pf} = \frac{\xi_V - \xi_p}{\xi_r - \xi_p} Q_{pf} \tag{17}$$

or

$$\frac{q_{pf}}{Q_{pf}} = \frac{\xi_V - \xi_p}{\xi_r - \xi_p} \tag{18}$$

If we extend line 4,1 in Figure 12 to the intersection with vertical line ξ_V = const. (point 5) and take into account the similarity of the triangles 3,4,5 and 2,4,1 we find with equation (18) that the heat q_{pf}, required to generate 1 kg of vapor, is represented by the interval 3,5. Thus, the required heat input in the limiting case of parallel flow and equilibrium between liquid and vapor can be immediately read from the h-ξ diagram.

Counter flow (Figure 11b).

In practice counter flow between liquid and vapor is more important than parallel flow. Again, the solution is assumed to enter the generator with concentration ξ_r, enthalpy h_r and temperature t (state 1 in Figure 13). After a heat input equal to $(h' - h_r)$ the liquid reaches the saturation state at a temperature t_s (point 2). The vapor in equilibrium with the saturated liquid is given by state 3, the concentration and enthalpy of which are ξ_V and h_V respectively. In a counter-flow vaporization process the vapor leaves the generator in state 3. A comparison between the states in Figures 12 and 13 indicates quite clearly that in counter flow the removed vapor has a higher fraction of the lighter volatile component than in parallel flow. This is an important conclusion with respect to the optimum design of the absorption heat pump process.

In parallel-flow vaporization the concentrations ξ_V of the vapor and ξ_p of the poor solution are determined by the inlet concentration ξ_r and the heat input Q_{pf} per kg of rich solution. In counter-flow vaporization the vapor concentration ξ_V depends only on ξ_r, whereas the concentration ξ_p of the poor solution is determined again by both the concentration ξ_r and the heat input Q_{cf} per kg of rich solution.

The mass balance for the lighter volatile component in the counter flow case (Figure 11b) is given by

$$m_r \xi_r = m_V \xi_V + (m_r - m_V) \xi_p \tag{19}$$

or

$$f = \frac{m_r}{m_V} = \frac{\xi_V - \xi_p}{\xi_r - \xi_p} \tag{20}$$

From an energy balance for the generator we obtain the heat q_{cf} (Figure 11b) required to generate 1 kg of vapor as follows :

$$m_r h_r + m_V q_{cf} = m_V h_V + (m_r - m_V) h_p \tag{21}$$

and with equation (20)

$$q_{cf} = \frac{\xi_V - \xi_p}{\xi_r - \xi_p} (h_p - h_r) + (h_V - h_p) \tag{22}$$

To present q_{cf} in the h-ξ diagram (Figure 13) we introduce the following abbreviation :

$$a = \frac{\xi_V - \xi_p}{\xi_r - \xi_p} (h_p - h_r) \tag{23}$$

or

$$\frac{a}{h_p - h_r} = \frac{\xi_V - \xi_p}{\xi_r - \xi_p} \tag{24}$$

Assuming the concentration ξ_p of the poor solution removed from the generator to be given, then its state is fixed in the h- ξ diagram by point 4 on the boiling line. A horizontal line through 4 intersects the vertical lines ξ_r = const. and ξ_V = const. in 5 and 6. If we extend further the line 4,1 to the intersection with ξ_V = const., we get point 7. From the similarity of the triangles 1,4,5 and 7,4,6 and from equation (24) it follows that the quantity a is represented by interval 6,7. The heat q_{cf}, required to generate 1 kg of vapor, can now be read immediately from the diagram as interval 3,7, because of

$$q_{cf} = a + (h_V - h_p) \ , \tag{25}$$

which follows from equations (22) and (23). The heat input to the generator per kg of rich solution is given by

$$Q_{cf} = q_{cf}/f \tag{26}$$

With equation (20) one finds :

$$\frac{Q_{cf}}{q_{cf}} = \frac{\xi_r - \xi_p}{\xi_V - \xi_p} \tag{27}$$

This relationship enables one to read also the heat Q_{cf} from the h-ξ diagram. Line 3,4 intersects with ξ_r = const. in point 8. The similar triangles 1,4,8 and 7,4,3 are thus obtained in which Q_{cf} is represented by the interval 8,1.

As a result we see that in the second limiting case of counter flow and equilibrium between liquid and vapor, the required heat input for generation of 1 kg of vapor can also be determined very easily from the h-ξ diagram. In practice the real heat quantities and vapor concentrations will lie somewhere between the two limiting cases (see e.g. [2]). This depends on the generator design and its operating conditions.

1.2. Condensation

The condensation process can also be studied very clearly by means of the h-ξ diagram. If, for example, saturated vapor of concentration ξ_V (point 1 in Figure 14) is to be condensed totally, the state of the resulting saturated solution is fixed by point 2 on the boiling line. The amount of heat q_c :

$$q_c = h_1 - h_2 \tag{28}$$

is to be extracted. During condensation, the temperature decreases continuously from t_1 at the beginning to t_2 at the end of the condensation process.

If only part of heat q_c is removed, e.g.

$$q_{cp} = h_1 - h_3, \tag{29}$$

the condensation process ends in the vapor - liquid region (point 3). Then the liquified fraction has a concentration ξ_3' whereas the remaining vapor is at concentration ξ_3''. The presentation in the h-ξ diagram indicates quite clearly that in this case the mass ratio between liquid and vapor is given by

$$\frac{m_L}{m_V} = \frac{\xi_3'' - \xi_V}{\xi_V - \xi_3'} = \frac{h_3'' - h_3}{h_3 - h_3'} \tag{30}$$

2. ABSORPTION OF VAPOR IN A LIQUID MIXTURE

The absorption of a vapor, having a certain concentration of the lighter volatile component, in a liquid mixture which is "poor" in the lighter volatile component, is the opposite of the partial vaporization of a liquid mixture. This absorption process may be modelled by two consecutive processes : the mixing of a vapor of concentration ξ_V with a liquid of concentration ξ_p resulting in a vapor - liquid mixture of concentration ξ_r and afterwards the removal of an amount of heat q_A such that a saturated liquid is obtained. This model indicates that, similar to condensation, the absorption process is accompanied by heat removal from the absorber (see also section 3.1, part I).

In the following considerations it is assumed that m_V kg superheated vapor of temperature t_1, concentration ξ_V and enthalpy h_V are fed to the absorber (Figure 15). Further it is assumed that a quantity $(m_r - m_V)$ of saturated poor solution of concentration ξ_p and enthalpy h_p is supplied to the absorber while m_r kg of rich solution of concentration ξ_r and enthalpy h_r are removed from the absorber. If this rich solution is to be saturated, a quantity of heat q_A per kg of vapor must be withdrawn, e.g. by cooling water.

Again the process can be presented in a very clear and simple way in the h-ξ diagram (Figure 16). The state of the superheated vapor is fixed by point 1, that of the poor solution by point 2. An adiabatic mixing of vapor and poor solution would result in a mixing state according to point 3 in Figure 16. In this case the following balance for the lighter volatile component holds :

$$m_V \xi_V + (m_r - m_V) \xi_p = m_r \xi_r \tag{31}$$

Therefrom it follows that :

$$f = \frac{m_r}{m_V} = \frac{\xi_V - \xi_p}{\xi_r - \xi_p} \tag{32}$$

Because usually all three concentrations in equation (32) are known, the feed rate f of the rich solution can be determined from this relationship.

In order to transfer the rich solution from the mixing state in point 3 to the saturation state in point 4 the heat q_A per kg of vapor has to be removed from the mixture. Thus the following energy balance holds true

$$m_V h_V + (m_r - m_V) h_p - m_V q_A = m_r h_r \tag{33}$$

Inserting equation (32) we obtain

$$q_A = (h_V - h_p) + \frac{\xi_V - \xi_p}{\xi_r - \xi_p} (h_p - h_r) \tag{34}$$

For brevity we define :

$$b = \frac{\xi_V - \xi_p}{\xi_r - \xi_p} (h_p - h_r) \tag{35}$$

so that

$$q_A = (h_V - h_p) + b \tag{36}$$

If in Figure 16 the line connecting points 2 and 4 is extended to the intersection with line ξ_V = const. (point 5) two similar triangles, 4,2,6 and 5,2,7 are obtained in which the quantity b appears as interval 5,7. According to equation (36) the heat q_A to be removed per kg of vapor is thus given by the interval 1,5.

If the heat is to be calculated per kg of rich solution,we have to divide q_A (equation 34) by the feed reate f (equation 32), so that :

$$Q_A = \frac{q_a}{f} = \frac{\xi_r - \xi_p}{\xi_V - \xi_p} (h_V - h_p) + (h_p - h_r) \tag{37}$$

The first term on the right hand side of equation (37) is equal to the interval 3,6 in Figure 16. This follows from the similarity of the triangles 2,6,3 and 2,7,1. Consequently the distance between point 3 and point 4 represents the heat Q_A to be extracted per kg of rich solution. It should be noted that Q_A is not identical to the heat of solution Δh' derived in part I section 3.1, because Δh' is the heat gain if saturated vapor is absorbed, whereas Q_A is the heat resulting from the absorption of a superheated vapor. The difference between them is equal to the sensible heat of the vapor between superheated and saturated state.

3. THROTTLING

An important process occurring in absorption heat pumps is the throttling process. It may be achieved in long capillary tubes or valves or other devices which have as a result a pressure reduction of the flow. If during throttling no heat is exchanged with the environment and inlet and outlet kinetic and potential energy differences can be neglected, the enthalpy of the fluid remains constant. We shall study this process in the h,ξ-diagram. In Figure 17 the dew and the boiling lines of two different pressures p and p_o are plotted. It is shown in the following that after throttling the state of the fluid may differ significantly.

In a first example we assume an initial liquid state represented by pressure p and temperature t_1. The mixture concentration is taken to be ξ, which is taken to be the same in this and the following examples. The initial state before throttling is represented by point 1 in the liquid region. The boiling line for pressure p is above point 1 (Figure 17). The concentration ξ and the enthalpy h_1 do not change during throttling. Consequently point 1 remains fixed in the h-ξ diagram. If the fluid reaches a pressure p_o after the throttling process, the mixture remains in the liquid state because point 1 lies, as before, below the boiling line for p_o. Thus only the pressure has changed whereas enthalpy and within reasonable approximation, also the temperature remains constant.

In the next example we consider a throttling process, the initial state of which is assumed to be given by point 2 in Figure 17 at pressure p and temperature t_2. This state lies again in the liquid region. However, after throttling to the pressure p_o, point 2 represents a state in the vapor-liquid region because the boiling line is in this case below the considered state. Contrary to the first example, the temperature has significantly changed now from t_2 to the lower value t_2' and part of the mixture is evaporated.

A similar situation exists when, already before throttling takes place, the state of the mixture lies in the vapor-liquid region (point 3). After throttling we have again a vapor liquid mixture but now with a lower temperature (t_3') and a higher vapor fraction.

As last example we consider a throttling process with an initial state according to point 4. In this case throttling starts in the vapor-liquid region and ends in the superheated vapor region because after throttling the state of the mixture, which does not move in the h-ξ diagram, lies above the dew line for pressure p_o.

4. RECTIFICATION

The vapor removed from a boiling solution in a generator has a much higher fraction of the lighter volatile component than the solution itself. However a higher purification of the vapor with respect to the lighter volatile component is often required. This can be achieved by a rectification process.

In a rectification column a liquid mixture (m_L) flowing downward is in close contact with the rising vapor (m_V) which is to be purified (Figure 18). Between liquid and vapor an intensive heat and mass exchange takes place so that the lighter volatile component is transferred from the liquid to the vapor and the less volatile component from the vapor to the liquid. To facilitate this transfer either a number of plates (bubble cap plates, sieve plates and the like) are installed in the column or the column is filled with packing material.

Assume a quantity of vapor m_V of temperatur t_V, concentration ξ_V and enthalpy h_V to enter the column from below (see Figure 18). At the top a fraction m_{RV} of the vapor, having the concentration ξ_{RV} and the enthalpy h_{RV}', leaves the column. The residual quantity is liquified in the reflux condenser by removal of the heat quantity $m_{RV} \cdot q_R$ and serves as reflux liquid. This leaves the bottom of the column with concentration ξ_L and enthalpy h_L. Of main interest is the reflux heat q_R removed from the column per unit mass of the outflowing vapor. It can easily be determined from a h-ξ diagram (Figure 19).

Assuming steady state operation the mass balance for the column shown in Figure 18 yields :

$$m_V = m_{RV} + m_L \tag{38}$$

Furthermore the balance for the lighter volatile component is given by

$$m_V \xi_V = m_{RV} \xi_{RV} + m_L \xi_L \tag{39}$$

For the energy balance we find :

$$m_V h_V = m_{RV} h_{RV} + m_L h_L + m_{RV} q_R \tag{40}$$

These last three equations yield

$$q_R = \frac{\xi_{RV} - \xi_L}{\xi_V - \xi_L}(h_V - h_L) - (h_{RV} - h_L) = a - (h_{RV} - h_L) \tag{41}$$

Point 1 in Figure 19 represents the state of the vapor entering the column from below. Point 2 marks the state of the removed vapor on the top and the state of the liquid flowing out on the bottom is represented by point 3. A horizontal line through point 3 intersects the vertical lines ξ_V = const. and ξ_{RV} = const. in 4 and 5. If one extends the line 3,1 to the intersection with the line ξ_{RV} = const. one obtains a point π. The triangles 1,3,4 and π,3,5 are similar and we can see immediately, that interval π,5 represents the first term on the right hand side of equation (41). Consequently the reflux heat q_R to be removed per kg of purified vapor m_{RV} is given by interval π,2. Point π is often called the pole of rectification.

The reflux ratio (f_r) of the column, i.e. the liquid per unit mass of leaving vapor is given by

$$f_r = \frac{m_L}{m_{RV}} = \frac{m_V - m_{RV}}{m_{RV}} \tag{42}$$

Substitution of eqûation (39) and rearranging yields

$$f_r = \frac{\xi_{RV} - \xi_V}{\xi_V - \xi_L} = \frac{b}{c} \tag{43}$$

The quantities b and c are plotted in Figure 19. The ratio of both intervals is according to equation (43) equal to the reflux ratio f_r which can thus be read very easily from the h-ξ diagram.

Normally the concentrations ξ_V and ξ_{RV} are fixed and the liquid concentration ξ_L can be varied. The following consideration shows that this can be done between the two limiting states 1' and 1" (see Figure 19). Assume the liquid concentration ξ_L to be equal to the concentration in point 1' which marks the equilibrium state belonging to the saturated vapor in 1 (temperature t_V) This limiting case would require the smallest possible reflux heat q_R. On the other hand, equilibrium must be established between the rising vapor and the falling liquid. This would require an infinitely large number of plates and therefore an infinitely long column. The second limiting case is given if point 3 coincides with point 1". Then the height of the column reaches a minimum. However its cross sectional area would have to be infinite if a finite mass m_{RV} is to be produced, because $\xi_V - \xi_L = 0$ and thus $f_r \to \infty$ (equations (42) and (43)). Likewise the reflux heat q_R would become infinite, because the pole π moves towards $h \to \infty$. Therefore the determination of the liquid concentration ξ_L in practice is a matter of economics. The situation is governed by the location of the pole π in the h-ξ diagram. If its distance from point 2 increases the reflux heat and the column diameter increase also but the required height of the column decreases. If the distance is reduced, q_R becomes smaller, the column height must be increased however whereas the diameter decreases.

REFERENCES

1. Altenkirch, E., *Absorptionskältemaschinen* (Berlin VEB Verlag Technik, 1954).

2. Bošnjaković, F. and Blackshear, P.L. *Technical Thermodynamics* (New York, Holt, Rinehart and Winston Inc., 1965).

3. Niebergall, W. Sorptionskältemaschinen *Handbuch der Kältetechnik* ed. R. Plank, VII (Berlin Springer-Verlag, 1959).

4. Timmermans, J. *The physico-chemical constants of binary systems in concentrated solutions* vol. 4 (New York, Interscience, 1960).

5. Mollier, H. *VDI-Heft* 63/64 (1909) 107.

6. Merkel, F. *Zeitschrift Verein Deutsch. Ing.* 72 (1928) 109.

7. Schulz, S. *Eine Fundamentalgleichung für das Gemisch aus Ammoniak und Wasser und die Berechnung von Absorptionskältemaschinen-Prozessen* (Bochum, Habil.-Schrift, Ruhr-Universität, 1971).

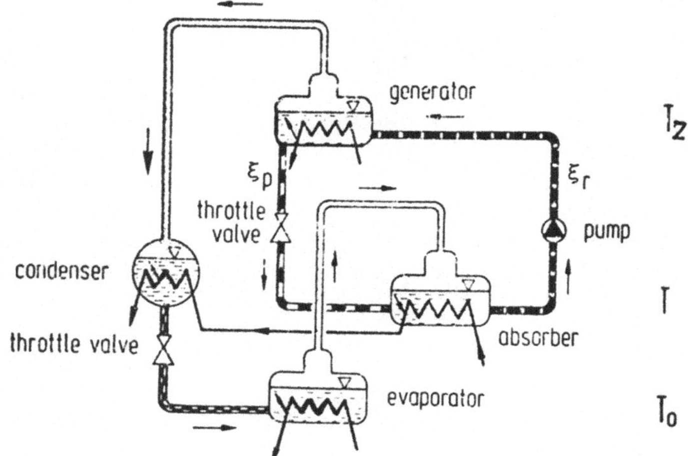

Figure 1 : Absorption heat pump

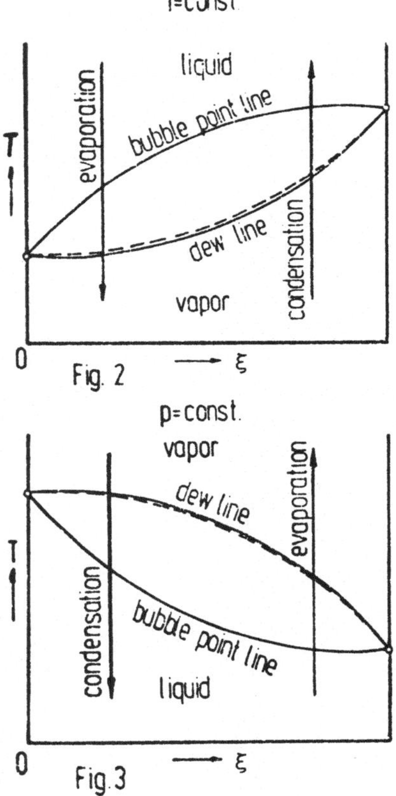

Figure 2 and 3 : Condensation and evaporation in the T-ξ diagram
Fig. 2 : T = const.; Fig. 3 : p = const.

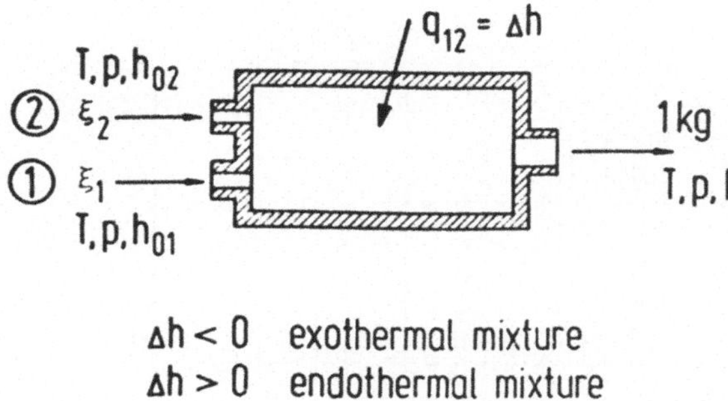

Figure 4 : Isobaric-isothermal mixing process to determine the enthalpy of mixing

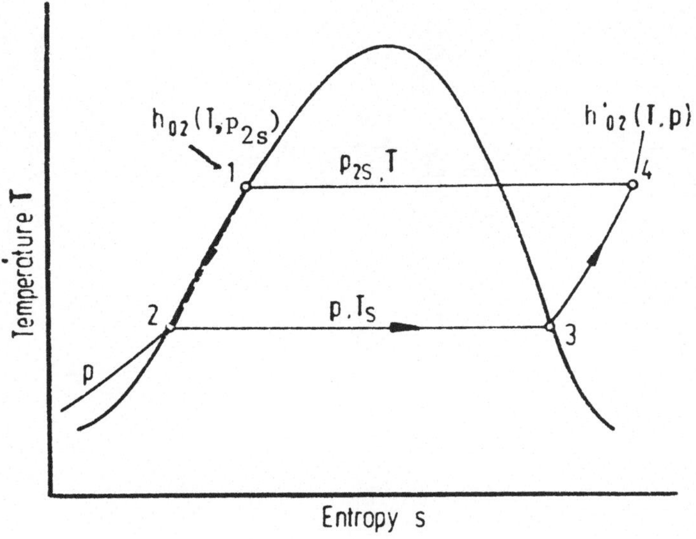

Figure 5 : Determination of the enthalpy of fusion according to H. Mollier [5]

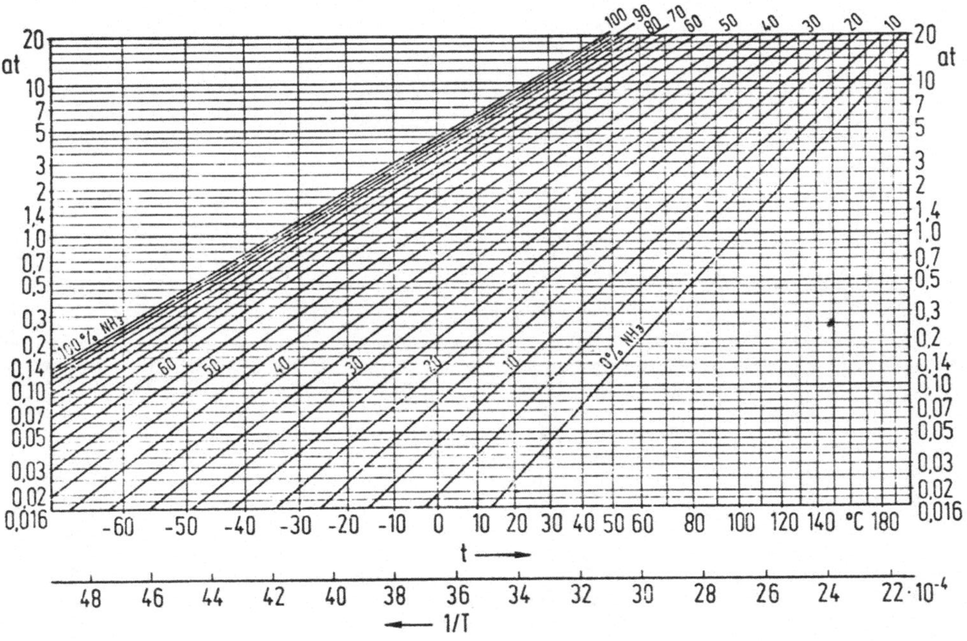

Fig. 6: log p - 1/T diagram for ammonia-water mixtures according to Bošnjaković and Wucherer [2].

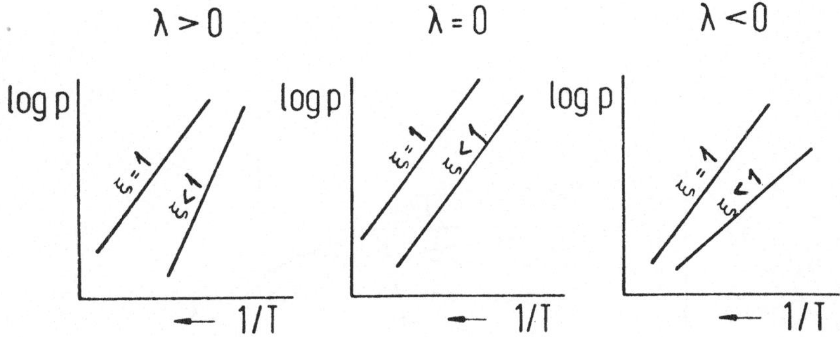

Figure 7 : Solubility diagrams for different values of the differential heat of solution

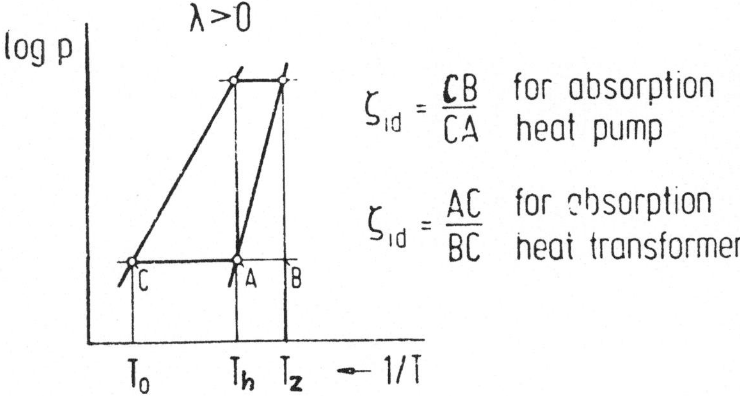

Figure 8 : Heat ratios from the log p - 1/T diagram

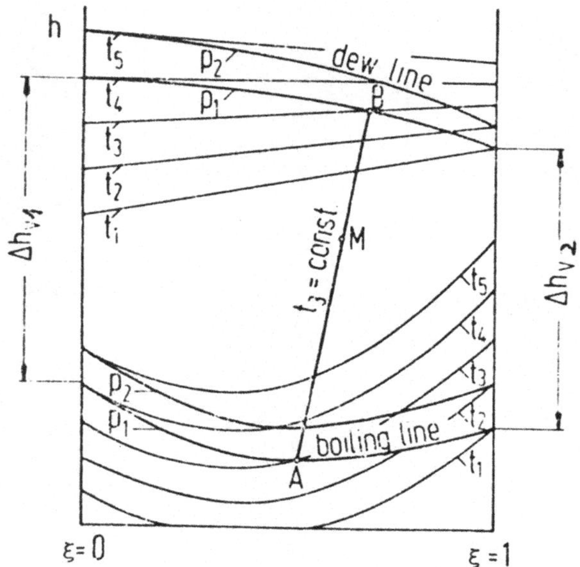

Figure 9 : h-ξ diagram

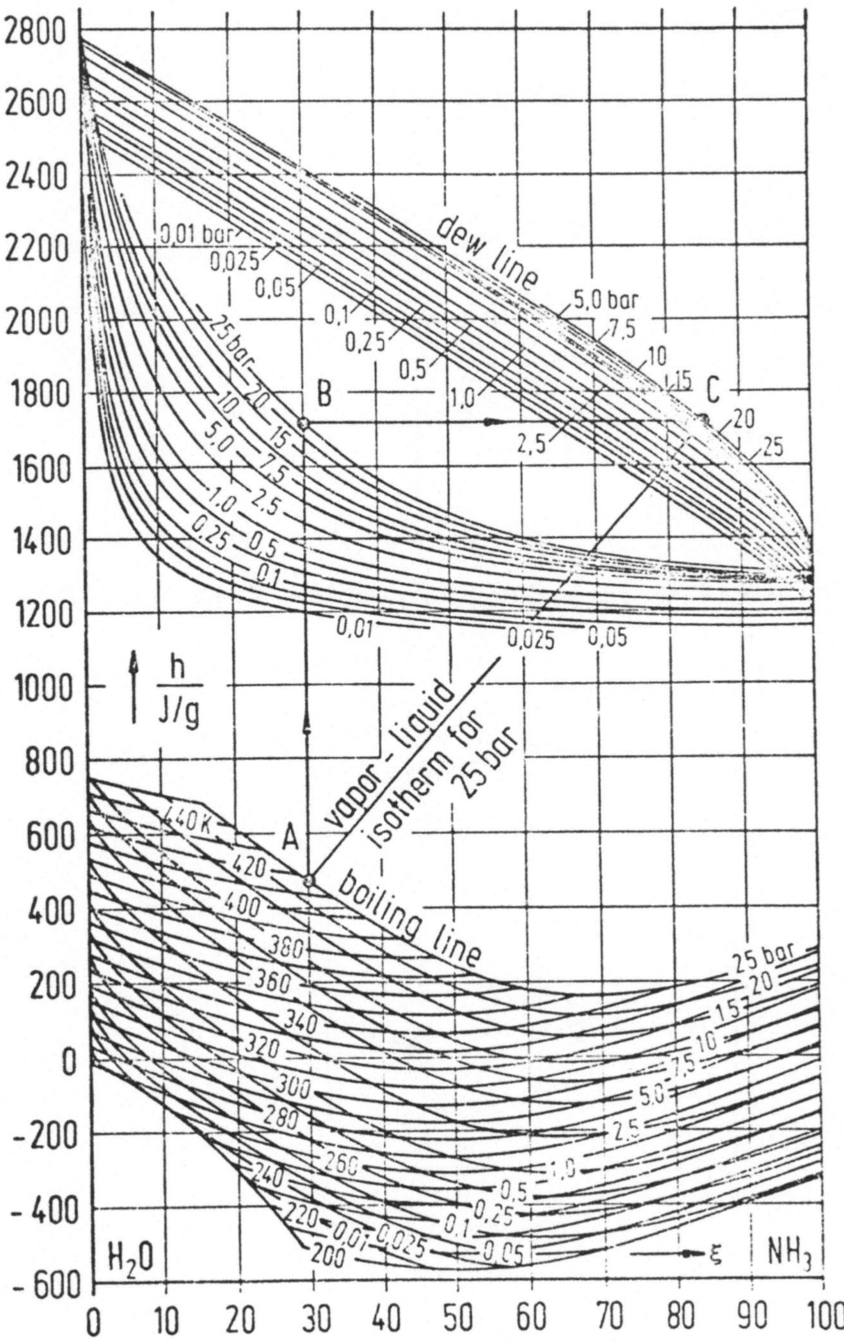

Figure 10 : h-ξ diagram for ammonia-water, according to Schultz [7]

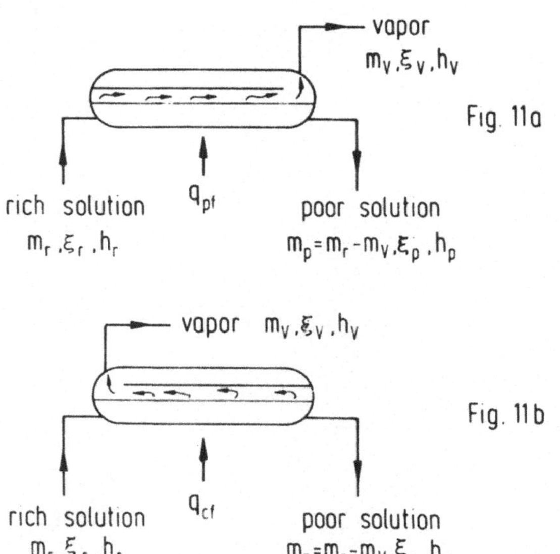

Fig. 11: Parallel flow (a) and counter flow (b) vaporization

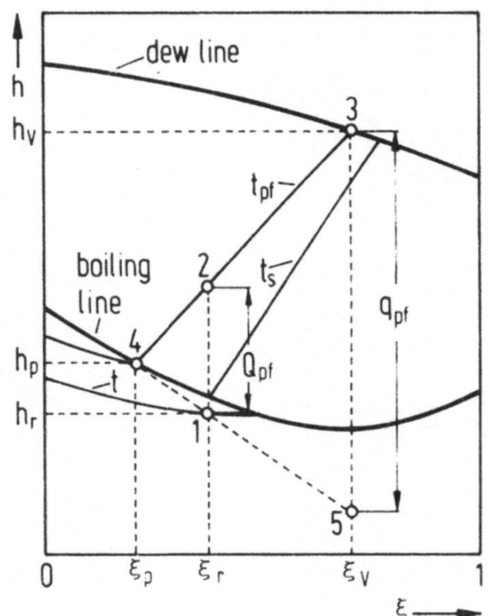

Fig. 12: Parallel flow vaporization in the h - ξ diagram.

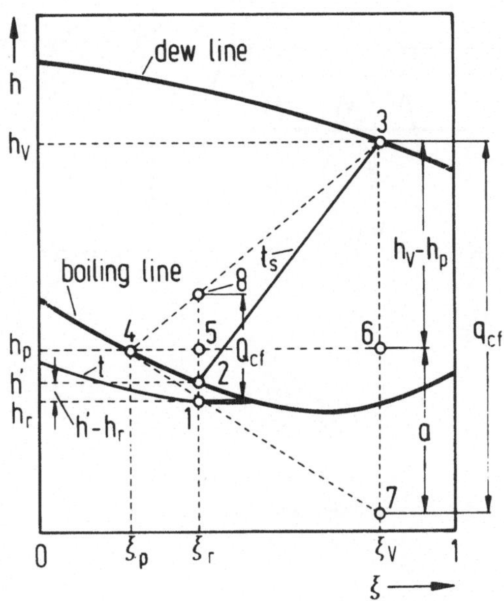

Fig. 13: Counter flow vaporization in the h-ξ diagram.

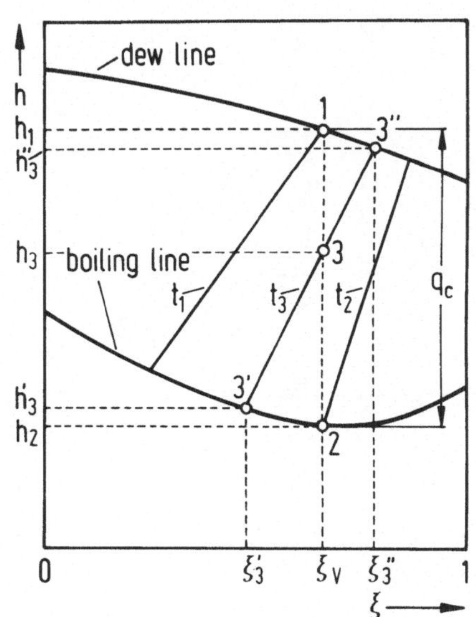

Fig. 14: Condensation in the h-ξ diagram.

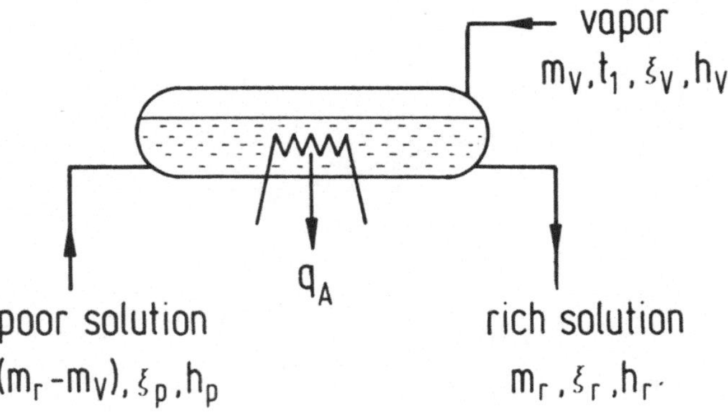

Fig. 15: Absorption.

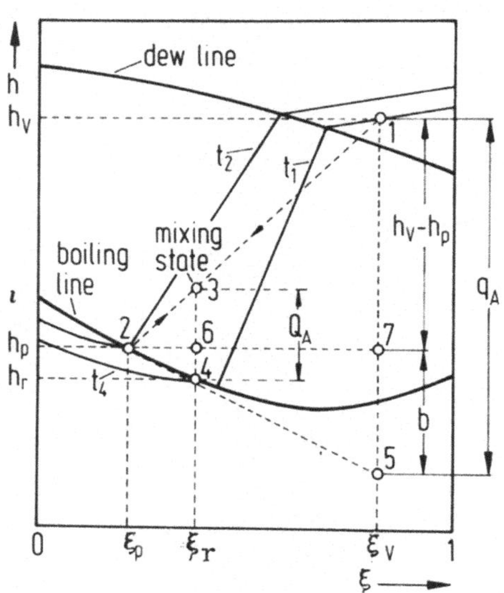

Fig. 16: Absorption process in the h-ξ diagram.

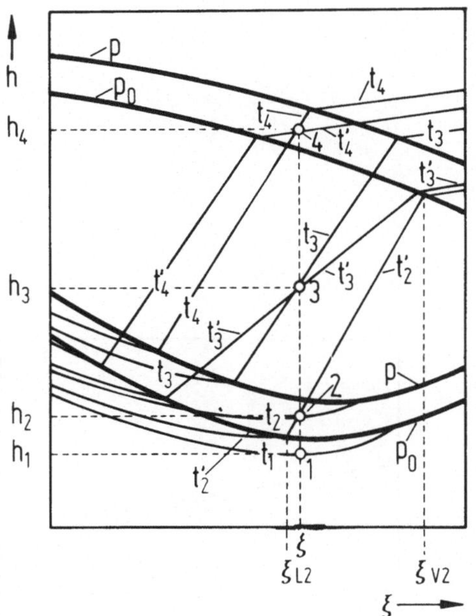

Fig. 17: Throttling process in the h-ξ diagram.

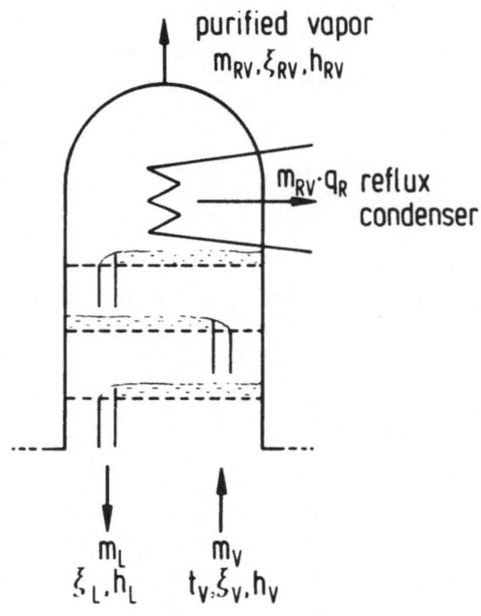

Fig. 18: Section of a rectification column.

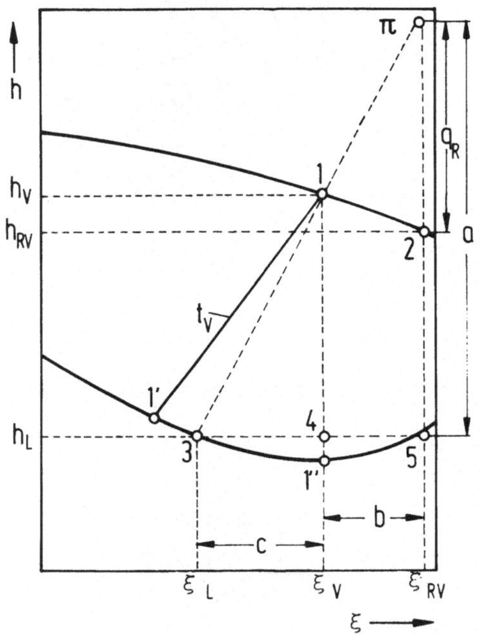

Fig. 19: Rectification in the h-ξ diagram.

THEORETICAL CYCLES OF ABSORPTION HEAT PUMPS

H. Glaser, H. Auracher

Institut für Technische Thermodynamik und Thermische Verfahrenstechnik, Universität Stuttgart, F.R. Germany

1. INTRODUCTION

A heat pump transforms heat from lower to higher temperature levels by means of available energy or, to say it in a modern way, by means of exergy. This exergy can be added in the form of mechanical or electrical energy as in a compression heat pump or in form of heat at temperatures above the ambient temperature. In the latter case cycles similar to the well known absorption refrigeration process must be used. The principe of absorption refrigeration was made available for technical application already more than hundred years ago by the works of Ferdinand Carré [5].

2. THE IDEAL PROCESS MODEL OF AN ABSORPTION HEAT PUMP

In an absorption heat pump an amount of heat Q_o at ambient temperature T_o is removed from the environment and a different amount of heat Q_h at a higher temperature level T_h is supplied for heating purposes. To operate the heat pump, a quantity of heat Q_z at temperature $T_z > T_h$ is needed. Its exergy E_z is given by :

$$E_z = (1 - \frac{T_o}{T_z})Q_z$$

In the case of an ideal reversible process the question of how many heat Q_z is required to produce a certain amount of heat Q_h can be answered easily by applying the 1st and the 2nd law of thermodynamics and without taking into account any technical requirements.

Suppose we have three heat sources at different temperatures T_o, T_h and T_z (Figure 1). These sources are assumed to be large enough such that any heat input or output does not change their temperatures significantly. All three sources are assumed to be in thermal contact with a system removing a quantity of heat Q_o from the source at T_o, Q_z from the source at T_z and supplying an amount of heat Q_h to the source at temperature T_h. No further assumptions concerning any technical details within the system are necessary for the following considerations.

Applying the 1st and 2nd law we can determine the amount of heat Q_z required to produce an amount of heat Q_h for heating purposes. For the system of Figure 1 the 1st law yields

$$Q_o + Q_z = Q_h \tag{1}$$

In this equation and the following equations all heat quantities - and also the other quantities - are taken as absolute values. According to the 2nd law, the total entropy of an adiabatic system including the heat sources can never decrease. In the limiting case of reversible processes within the adiabatic system the total entropy remains constant, i.e. the sum of all individual entropy changes in the system must be equal to zero

$$\Sigma \Delta S = 0 \tag{2}$$

The individual entropy changes are due to the heat removal from the environment :

$$\Delta S_o = - \frac{Q_o}{T_o} ; \tag{3}$$

due to the heat withdrawn from the high temperature source :

$$\Delta S_z = - \frac{Q_z}{T_z} \tag{4}$$

and due to the heat entering the lower temperature source :

$$\Delta S_h = \frac{Q_h}{T_h} \tag{5}$$

No entropy change takes place in the total reversible system.

Thus the entropy balance for the adiabatic system is given by

$$\Sigma\Delta S = \Delta S_o + \Delta S_z + \Delta S_h = 0 \qquad (6)$$

From equations (3), (4) and (5) we get

$$-\frac{Q_o}{T_o} - \frac{Q_z}{T_z} + \frac{Q_h}{T_h} = 0 \qquad (7)$$

Introducing equation (1) and rearranging, yields

$$Q_z \cdot \frac{T_z - T_h}{T_z} = Q_o \cdot \frac{T_h - T_o}{T_o} \qquad (8)$$

This relationship can be interpreted as follows : the left hand side is equal to the work W_z performed by a Carnot process operating between the temperatures T_h and T_z and receiving a quantity of heat Q_z at temperature T_z (Figure 2). The right hand side represents the work W_h which must be supplied to a Carnot process operating between the temperatures T_o and T_h and removing an amount of heat Q_o from the environment at temperature T_o. In a T-s diagram, the different quantities of heat and work are represented by the following areas : area 1, 2, 3, 4 = Q_z; area 3, 5, 6, 7 = Q_o; area 1, 2, 8, 9 = W_z and area 5, 6, 10, 8 = W_h. The heat Q_h, available for heating purposes, is, according to equation (1), equal to area 1, 2, 5, 6, 7, 4. On the other hand, we know from equation (8), that $W_z = W_h$, so that the heat Q_h is also represented by area 10, 9, 4, 7.

We thus found an ideal model of an absorption heat pump process, which transforms the heat Q_o from the environment to the higher temperature level T_h with the aid of high temperature heat Q_z. This model process consists of two Carnot-cycles in which the working fluid undergoes the following changes of state : 1, 2 : isothermal expansion with absorption of heat Q_z; 2,5 : isentropic expansion; 5, 6 : isothermal expansion with absorption of heat Q_o from the surroundings; 6, 10 : isentropic compression; 10, 9 : isothermal compression with removal of heat Q_h, the desired "product" of the process and finally to close the cycle an isentropic compression from 9 to 1. This ideal model of the heat pump process can serve as a basis to calculate the efficiency of real processes.

We come to an interesting conclusion when comparing this model to the ideal model of a compression heat pump process. The latter consists also of a Carnot-cycle (Figure 3) which must run along 10, 9, 11, 6 if the same amount of heat Q_h as in the ideal absorption process model is to be produced. The area 10, 9, 11, 6 in Figure 3, representing the required work to operate the Carnot cycle, is now by an amount 8, 9, 11, 5 greater than the area representing the required work W_h of the absorption process in Figure 2. Furthermore the compression heat pump cycle needs more heat from the surroundings than the absorption process. The difference is proportional to area 11, 5, 3, 4 in Figure 3. The explanation follows simply from thermodynamic considerations. The work required to operate the compression heat pump, due to the second law, can only be produced by transferring a certain amount of heat to the surroundings which very often takes place in a power plant far from the heat pump. In the absorption process the "waste heat" from the upper Carnot-cycle, represented by area 8, 9, 4, 3 is, in contrast to the compression cycle, a part of the desired product namely heat at temperature T_h for heating purposes. Consequently a compression heat pump providing the same amount of heat Q_h as an absorption heat pump, normally removes more heat from its direct environment than an absorption heat pump which in most cases is a disadvantage.

3. THE IDEAL HEAT RATIO

Insertion of equation (1) in equation (7) and eliminating Q_o yields

$$- \frac{Q_h}{T_o} + \frac{Q_z}{T_o} - \frac{Q_z}{T_z} + \frac{Q_h}{T_h} = 0 \qquad (9)$$

or

$$\zeta_{id} = \frac{Q_h}{Q_z} = \frac{T_z - T_o}{T_h - T_o} \cdot \frac{T_h}{T_z} \qquad (10)$$

The ratio ζ_{id}, i.e. the obtained heat Q_h of the process per unit of required heat Q_z, is called the ideal heat ratio of the absorption heat pump process. From equation (10) it follows that ζ_{id} is always greater than one if $T_z > T_h$, i.e. more heating energy Q_h is produced than "driving" energy Q_z is needed. By equation (10) it is thermodynamically proved that an amount of heat Q_z at temperature T_z can be transformed to a greater amount of heat Q_h at a lower temperature T_h without any further requirements except for the heat Q_o to be removed from the environment. The best transformation efficiency is given by ζ_{id} according to equation (10). The latter holds for reversible processes or in other words, for zero exergy losses, i.e. the exergy E_z of heat Q_z ($E_z = Q_z\,(1 - T_o/T_z)$) is equal to the exergy E_h of heat Q_h

($E_h = Q_h\ (1 - T_o/T_h)$). It should be noted that this result was obtained without any assumptions about technical details within the system.

Because reversible processes are ideal limiting cases, the real heat ratios ζ are always smaller than ζ_{id}. The reversibility ratio η [2] defined as

$$\eta = \zeta/\zeta_{id} \tag{11}$$

is a measure for the thermodynamic quality of a real transformation of Q_z to Q_h.

In Figure 4 the ideal heat ratio ζ_{id} is plotted versus temperature t_z with ambient temperature t_o as parameter. The heating temperature is fixed at $t_h = 70°C$. Figure 4 indicates that ζ_{id} increases with increasing temperature t_z . AT $t_z = 500\ °C$ we have, for instance, ζ_{id} values of about 3 to 4. The higher values are reached at higher ambient temperatures t_o. Thus even at moderate temperatures t_z considerably more than the "driving heat" Q_z can be provided for heating purposes at a lower temperature t_h. The dotted lines in Figure 4 mark the region of $t_z < t_h$. If this condition applies, we have to use a so-called heat transformer, which transforms lower temperature heat at t_z to heat which can be used at higher temperatures t_h for heating purposes.

Figure 5 shows the influence of the heating temperature t_h on the heat ratio ζ_{id} for a fixed ambient temperature $t_o = 0°C$. Particularly remarkable is the strong increase of ζ_{id} with decreasing heating temperature t_h. This is also clearly indicated by Figure 6, where the heating temperature t_h is plotted on the abscissa and t_z is the parameter at a fixed $t_o = 0°C$. In practice, the heating temperature t_h should be as low as possible. A floor heating system may therefore be advantageous. If in such a case the heating is required at about 35 °C and the supplied heat Q_z is available at 80 °C, which may be possible if solar systems are used, the ideal heat ratio would have the remarkably high value of 2.

The ideal model of the absorption heat pump process according to Figure 2 cannot be realized in practice. The real cycle is better approached when the isentropes in both Carnot cycles are substituted by isobars characterized by constant isobaric heat capacities. Assuming heat exchange between these isobaric processes characterized by zero temperature differences, we obtain a different cycle, which again is thermodynamically ideal and yield therefore the same results with respect to the ideal heat ratio. This new process model can however be realized approximately if a mixture is used as working fluid. We then come to the cycle of a real absorption heat pump, which is similar to that of an absorption refrigerator.

4. THE ABSORPTION HEAT PUMP PROCESS

Figure 7 shows a simple absorption heat pump process. The generator contains a liquid mixture in which one of the components is very volatile. Mixtures of ammonia-water fulfill this requirement for instance. Ammonia is the lighter volatile component. Its boiling temperature is - 33,4 °C at 1 bar. This mixture is often used in absorption refrigerators and we shall assume in the following that it is also applied in the absorption heat pump process.

In the generator (Figure 7) an amount of heat Q_G is supplied to the solution at pressure p and temperature T_z. As a result vapor is generated which is liquefied in the condenser by removing an amount of heat Q_C at temperature T_h. After throttling from p to the lower pressure p_o the condensate enters the evaporator where it evaporates by absorbing an amount of heat Q_E at the ambient temperature T_o. The low pressure vapor obtained then flows into the absorber where it is absorbed (at pressure p_o and temperature T_h) by the solution entering from the generator via a throttle valve. Due to the vaporization in the generator this solution is "poor" in the lighter volatile component (ammonia) and is thus able to absorb the ammonia vapor coming from the evaporator. The solution leaving the absorber is pumped to the higher pressure p and fed to the generator. This solution is "rich", i.e. it has a high ammonia concentration. During absorption an amount of heat Q_A is removed at the same temperature T_h as at which condensation occurs. Both the heat Q_A from the absorber and the heat Q_C from the condenser can for instance be utilized to heat water which then in turn is available for heating purposes. Following the notation introduced in the ideal process model, the heat Q_h produced by the absorption process at the heating temperature level T_h is given by

$$Q_h = Q_A + Q_C \tag{12}$$

correspondingly, the "driving heat" Q_z of the model is now the heat input Q_G to the generator and the heat Q_o from the environment is equal to the heat Q_E supplied to the evaporator.

For reasons of clearness the most simple version of an absorption heat pump process is shown in Figure 7. It would have a relatively small heat ratio because some irreversible processes with high exergy losses are involved. The rich solution coming from the absorber enters the generator at a temperature far below T_Z and must be warmed up before vaporization starts. On the other hand the poor solution leaving the generator at the high temperature level is only able to absorb the vapor coming from the evaporator if the solution is cooled to a lower temperature. The heating of the rich solution and the cooling of the poor solution are connected with high exergy losses and thus with a reduction of the heat ratio. This disadvantage can be eliminated if a heat exchanger I is installed between generator and absorber (Figure 8) to transfer heat from the poor to the rich solution. The necessary heat input Q_G to the generator is in this way reduced. It may seem to be a disadvantage that on the other hand the heat Q_A available for heating purposes is also reduced. A thermodynamic analysis shows however, that the efficiency of the whole cycle is improved by this internal heat exchange.

The same is true for heat exchanger II (Figure 8) in which heat is transferred from the warmer condensate to the colder vapor coming from the evaporator. Thus the vapor temperature at the inlet of the absorber is increased which results in the desired effect of a higher heat Q_A to be removed from the absorber by the heating water. This increase in the heating effect of the apparatus is connected with a higher input of heat from the surroundings to the evaporator because, due to cooling of the condensate, the heat quantity Q_E needed in the evaporator is increased. However heat from the surroundings contains no exergy so that the installation of heat exchanger II results in an increase of the heat supplied by the absorption cycle for heating purposes without an additional exergy input.

A further improvement is obtained, if the generator is coupled with a rectifier in which the vapor leaving the generator is purified almost completely to the lighter volatile component ammonia. Water accumulation in the evaporator where it would impair the operating conditions of the heat pump is thereby avoided. At the top of the rectification column a certain amount of so-called reflux heat Q_R must be removed. This heat Q_R can also be used for heating purposes.

5. THE PROCESS IN THE LOG p, 1/T-DIAGRAM

The basic operating conditions of an absorption heat pump can easily be studied in a log p-1/T plot of the working pair. Such a diagram enables only to determine the pressures and concentrations belonging to the temperatures T_O, T_h and T_Z in the different components.

We assume in the following that the temperature T_z in the generator as well as the ambient temperature T_o and the heating temperature T_h are known. Furthermore we assume that the vapor leaving the generator consists of pure ammonia, i.e. we assume an ideal rectifier. All heat transfer processes are ideal in the sense that the driving temperature difference vanishes at least in one cross section of the component. Finally we assume that the flow rate of the heating water is such that its temperature difference between inlet and outlet is negligible. Based on these assumptions we are now able to fix all significant equilibrium states of the absorption process in the log p-1/T diagram (Figure 9).

Pressure p in the generator and in the condenser is given by the intersection between the vertical line $1/T_h$ = const. and the saturation line for concentration ξ = 1,0 (pure ammonia - point 8). This is the pressure to be established if the pure ammonia vapor is to be liquified at temperature T_h in the condenser. Note that the numbering of the states in Figure 9 corresponds to that in Figure 8 (except for the states 1" and 2'). The states before (point 8) and after (point 9) condensation coincide in the log p-1/T diagram because temperature, pressure and concentration remain unchanged in the condenser. In the evaporator ammonia has to be evaporated at ambient temperature T_o. The corresponding pressure p_o is fixed by the intersection of the vertical line $1/T_o$ = const. with the equilibrium line for ξ = 1 (point 11). This pressure p_o prevails also in the absorber. As in condensation the states before (point 11) and after (point 12) evaporation coincide in the log p-1/T diagram. The maximum concentration to be reached in the absorber at pressure p_o is determined by the temperature T_h. This concentration is found by intersecting the line p_o = const. with the vertical line $1/T_h$ = const. (point 1). The equilibrium line passing through this intersection belongs to the concentration ξ_r of the rich solution leaving the absorber. The minimum possible concentration ξ_p of the poor solution coming from the generator is determined by temperature T_z. We find this concentration by intersecting the lines p = const. and $1/T_z$ = const. (point 2). The equilibrium line for ξ_p = const. passes through this intersection.

By means of the log p, 1/T-diagram a simple interpretation of the absorption heat pump process shown in Figure 8 is possible. The rich solution supplied to the generator with concentration ξ_r is split up into saturated vapor (point 8) and saturated poor solution of concentration ξ_p (point 2). The vapor is then condensed between state 8 and state 9 and the resulting liquified ammonia is cooled in the heat exchanger II to a state 10. After throttling the liquid reaches again the saturation state at the lower pressure p_o and temperature T_o on line ξ = 1 (point 11). The following evaporation produces saturated vapor of state 12, which is then, after warming up in heat exchanger II, supplied to the absorber. There it is absorbed by the poor solution coming from the generator so that rich solution of state 1 is obtained. It is, before entering the generator, pumped to the higher pressure p and preheated in heat exchanger I. The rich solution starts to vaporize after reaching the saturation state 1" and is split up again into saturated vapor of state 8 and poor solution of state 2. This poor solution, after leaving the generator, is first cooled in heat exchanger I to a state 3 in the subcooled liquid region, then throttled to pressure p_o and finally transferred into the absorber where it absorbs the vapor coming from the evaporator.

The log p-1/T diagram allows to clarify in a very simple way the effect of changes in the different temperatures on the process. If, for instance, the heating temperature T_h is to be increased point 1 would move on the p_o-isobar to a higher temperature which would result in a decrease of the ammonia concentration ξ_r of the rich solution. Furthermore state 8 would move along the saturation line for $\xi = 1$ to a higher temperature and thus a higher pressure p would be required. Consequently point 2 would also move upward along line $1/T_z$ = const. - provided temperature T_z remains unchanged -, resulting in an increase of concentration ξ_p of the poor solution. The influence of temperature changes in the generator (T_z) and in the evaporator (T_o) on the process can be discussed in a similar way. So the log p-1/T diagram allows a simple analysis of different process variations and their influence on temperatures, pressures and concentrations in the individual components of the heat pump.

6. THE PROCESS IN THE ENTHALPY - CONCENTRATION DIAGRAM

A clear survey on the heat inputs and outputs of the absorption process is obtained if it is studied in an h-ξ diagram [3]. The same assumptions regarding the operating conditions (vanishing temperature differences in certain cross sections of the components ideal rectifier etc.) as in section 5 are made in the following.

Figure 10 shows the h-ξ diagram of the modelled process. The dew lines and boiling lines for pressure p in the generator and the condenser as well as the corresponding lines for pressure p_o in the evaporator and absorber are plotted in this diagram. Furthermore the isotherms for the temperature t_z of "driving heat" Q_G' for the temperature t_h of "produced" heat Q_h and for the temperature t_o of environmental heat Q_E are plotted (see also Figure 8).

As already shown in section 5, the pressure p_o is fixed by the ambient temperature t_o. The boiling line for p_o intersects the ordinate on the right hand side ($\xi_V = 1$, i.e. pure ammonia) at the ambient temperature t_o. This intersection represents the state of the liquid in the evaporator. Because ammonia is liquified at temperature t_h in the condenser, the corresponding pressure p in the condenser and thus the state of the liquid is found by intersecting the t_h-isotherm with the right hand side ordinate (point 9). The intersection between the t_h-isotherm and the boiling line for p_o (point 1) marks the maximum concentration ξ_r which the solution can reach in the absorber. A higher concentration is impossible, if the lowest heating temperature is fixed at t_h. Point 1 in Figure 10 thus represents the state of the rich solution leaving the absorber. The solution is then pumped to the higher pressure p and flows via heat exchanger I into the generator. There the ammonia concentration can be reduced until the solution reaches the upper temperature limit t_z. This concentration ξ_p of the poor solution is thus given by the intersection between the boiling line for pressure p and the t_z-isotherm (point 2). In this state the poor solution enters heat exchanger I and is cooled by the rich solution. The lowest temperature to be reached is t_h, provided that ideal heat transfer conditions exist.

The generator is fed by a quantity m_r of rich solution (Figure 8) with concentration ξ_r, whereas a quantity m_V of vapor with concentration w_V and a quantity $(m_r - m_V)$ of poor solution with concentration ξ_p is flowing out. The mass balance for the lighter volatile component is thus given by

$$m_r \cdot \xi_r = m_V \cdot \xi_V + (m_r - m_V) \cdot \xi_p \tag{13}$$

The amount of rich solution needed per unit mass of generated vapor is often called the feed rate f of the rich solution. It is obtained by rearranging equation (13) :

$$f = \frac{m_r}{m_V} = \frac{\xi_V - \xi_p}{\xi_r - \xi_p} \tag{14}$$

This feed rate can be read from the h-ξ diagram (Figure 10) as ratio of two live segments, where according to our assumption of ideal rectification the vapor consists of pure ammonia, i.e. $\xi_V = 1$.

During compression from p_o to p in the pump the rich solution absorbs energy. The resulting enthalpy increase is, however, small enough that it can be neglected in the following. Therefore the rich solution enters the heat exchanger I in state 1 (Figure 10) and leaves it in a state 4. The energy balance of heat exchanger I reads :

$$(m_r - m_V)\,(h_2 - h_3) = m_r(h_4 - h_1), \tag{15}$$

where h denotes the enthalpy in the different states marked in Figures 8, 9 and 10. From equations (14) and (15) we obtain :

$$\frac{h_2 - h_3}{h_4 - h_1} = \frac{\xi_V - \xi_p}{\xi_v - \xi_r} \tag{16}$$

From this relationship state 4 in the h-ξ diagram can be found. We draw a line from point 3 via point 1 to the intersection with the right hand side ordinate (point a) and from there a line back to point 2. The line 2,a intersects the vertical line ξ_r = const. in point 4. This follows from equation (16) and the similarity of the triangles 1,a,4 and 3,a,2. Because state 4 lies in the vapor-liquid equilibrium region a small amount of rich solution evaporates already in heat exchanger I. At its outlet the temperature is t' (Figure 10) and the vapor concentration is ξ'_V (point 5).

If the generator is designed for counter-flow vaporization the vapor flowing into the rectifier has, in the theoretical limiting case, the temperature t' and the concentration ξ'_V, i.e. it is in the equilibrium state existing at the inlet cross section of the generator. Under this assumption the minimum reflux heat to be removed per unit mass of vapor in the rectifier

$$q_R = Q_R/m_V \tag{17}$$

can easily be determined from the h-ξ diagram. By extending the isotherm for t' to the intersection with vertical line $\xi_V = 1$ the so-called pole of rectification π is obtained. Its vertical distance to the dew line represents the reflux heat q_R. For generator and rectifier the following energy balance holds

$$m_r \cdot h_4 + Q_G = m_V \cdot h_V'' + (m_r - m_V) \cdot h_2 + Q_R \tag{18}$$

Introducing

$$q_G = Q_G/m_V \tag{19}$$

we obtain from equations (14), (17) and (18) :

$$\frac{q_G - q_R - (h_V'' - h_2)}{\xi_V - \xi_p} = \frac{h_2 - h_4}{\xi_r - \xi_p} \tag{20}$$

Taking into account the similarity of the triangles 4,2,b and a,2,c equation (20) indicates that the heat q_G to be supplied to the generator per unit mass of vapor (state 8) is proportional to the interval π,a in Figure 10.

The condenser is fed by saturated vapor of enthalpy h_V'' (point 8). After liquifaction the saturated liquid has an enthalpy h_V' (point 9). The heat to be removed from the condenser per unit mass of vapor is thus given by

$$q_C = h''_V - h'_V \tag{21}$$

In heat exchanger II the enthalpy of the liquid is reduced by an amount Δh so that it leaves the apparatus in state 10 (see Figures 8 and 10). The enthalpy remains constant during the adiabatic throttling. Therefore the states before (10) and after (11) the throttle valve coincide in the h-ξ diagram. Consequently point 11 lies in the two-phase region, because now the lower pressure p_o corresponds to this state. Thus a vapor-liquid mixture enters the evaporator where, through the heat input $q_E = Q_E/m_V$ from the surroundings, saturated vapor of state 12 is produced. Its temperature then increases from t_o (state 12) to t_h (state 13) in heat exchanger II, assuming ideal heat transfer conditions. Consequently the vapor enthalpy increases by an amount Δh. This enthalpy difference $\Delta h = h_{13} - h_{12}$ can be read directly from the h-ξ diagram, because state 13 is fixed by temperature t_h and state 12 by the saturation temperature t_o. The known enthalpy difference Δh now enables one to fix state 10 on the right hand side ordinate because point 10 lies by the same amount Δh below point 9. The heat q_E is now given either as interval 9,13 or as interval 11,12.

After generator, condenser and evaporator we shall finally analyze the situation in the absorber. It is fed by an amount $(m_r - m_V)$ of poor solution with enthalpy h_3 (provided adiabatic throttling) and by an amount m_V of vapor with enthalpy h_{13}. An amount m_r of rich solution with enthalpy h_1 leaves the absorber and, furthermore, the heat of absorption Q_A is removed. Therefrom we obtain the following energy balance

$$(m_r - m_V) \cdot h_3 + m_V \cdot h_{13} = m_3 \cdot h_1 + Q_A \tag{22}$$

Introducing the absorption heat per unit mass of vapor

$$q_A = \frac{Q_A}{m_V} \tag{23}$$

and taking into account equation (14) the following relation holds :

$$\frac{q_A - (h_{13} - h_3)}{\xi_V - \xi_p} = \frac{h_3 - h_1}{\xi_r - \xi_p} \qquad (24)$$

From this relationship and from the similarity of triangles 1,3,d and a,3,e we obtain q_A as interval 13,a in the h-ξ diagram.

An energy balance for the whole process must confirm the results derived for the individual components. This balance is given by (see Figure 8) :

$$Q_G + Q_E = Q_A + Q_C + Q_R \qquad (25)$$

Dividing by the amount m_V of vapor and rearranging yields :

$$q_G = q_C + q_R + (q_A - q_E) \qquad (26)$$

The different intervals plotted in the h-ξ diagram clearly indicate that this balance is fulfilled.

For heating purposes

$$q_h = q_A + q_C + q_R \qquad (27)$$

is available. Figure 10 and equation (27) prove that the heat ratio $\zeta = q_h/q_G$ is greater than 1, i.e. not only an ideal model but also a real absorption heat pump always produces more heat for heating purposes than it requires to heat the generator.

We have seen that all significant process quantities can be determined by a graphical procedure in the h-ξ diagram. A computer calculation of the whole process is, of course, also possible. There is no principle change in the procedure. Use of a computer is advantageous if parameter variations have to be carried out in order to optimize the system. The advantage of a graphical procedure mainly consists of the clear, quick and simple way to present the process and to estimate the result of process variations.

The aim of this chapter is to discuss the thermodynamic fundamentals of absorption heat pumps. The influence of finite temperature differences in the heat transfer processes, of pressure losses etc. have not been treated here. Furthermore some process variations are possible (two-stage operation for instance) which can improve the efficiency. Some of these subjects are treated in other chapters and in the literature [1], [4] . However a lot of research and development still has to be done.

7. CONCLUDING REMARKS

A large amount of primary energy-oil, gas etc. - today is used for direct heating. In all these cases the application of absorption heat pumps enables a drastic reduction of primary energy consumption. However, to make available the entire advantage of this heating system, new working substances for operation at high temperature levels must be found. In the near future the main effort in research must be directed towards this goal.

REFERENCES

1. Niebergall, W. Sorptions-Kältemaschinen *Handbuch der Kältetechnik* ed. by R. Plank, VII (Berlin Springer-Verlag, 1959).

2. Bošnjaković, F., Blackshear, P.L. *Technical Thermodynamics* (New York, Holt, Rinehart a. Winston, 1965).

3. Merkel, F. *Zeitschrift Verein Deutscher Ingenieure* 72 (1928) 109.

4. Eder, W., Moser, F. *Die Wärmepumpe in der Verfahrenstechnik* (Wien, Springer-Verlag, 1979).

5. Plank, R. Die geschichtliche Entwicklung und gegenwärtige Bedeutung der Kältetechnik *Handbuch der Kältetechnik* ed. by R. Plank, Vol. I (Berlin Springer-Verlag, 1954).

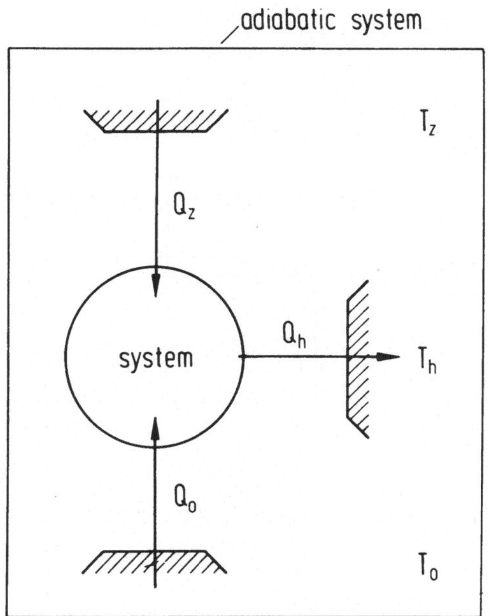

Fig. 1: For determination of the theoretical heat requirement

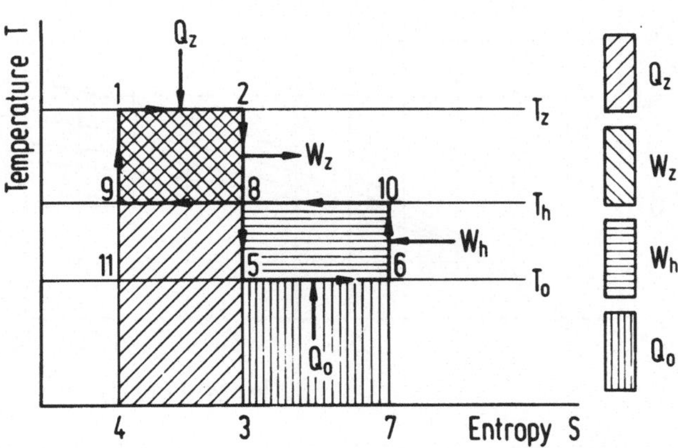

Fig. 2: Ideal model of an absorption heat pump process

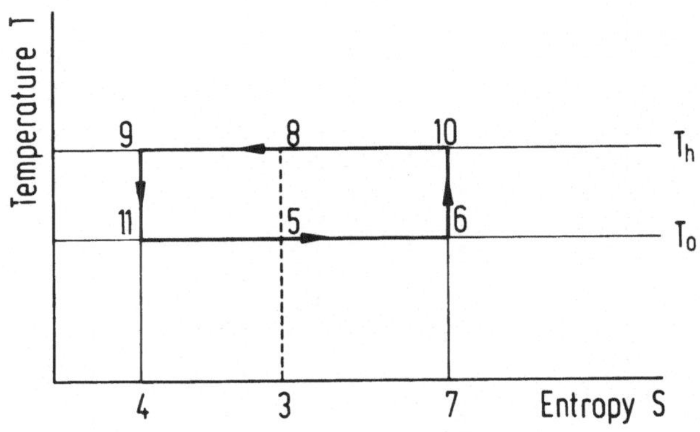

Fig. 3: Ideal model of a compression heat pump process

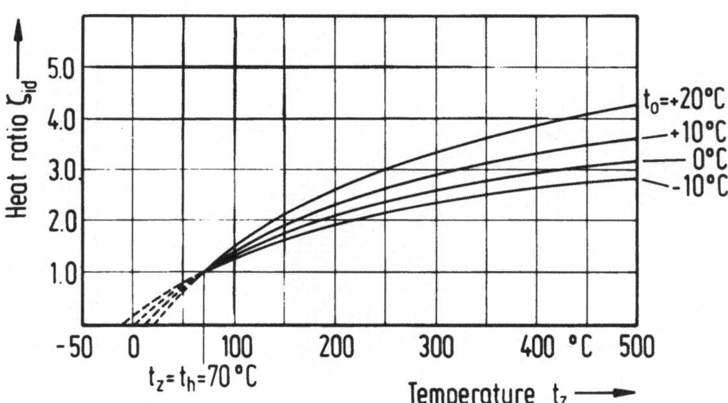

Fig. 4: Ideal heat ratio ζ_{id} vs. temperature t_z with ambient temperature t_0 as parameter

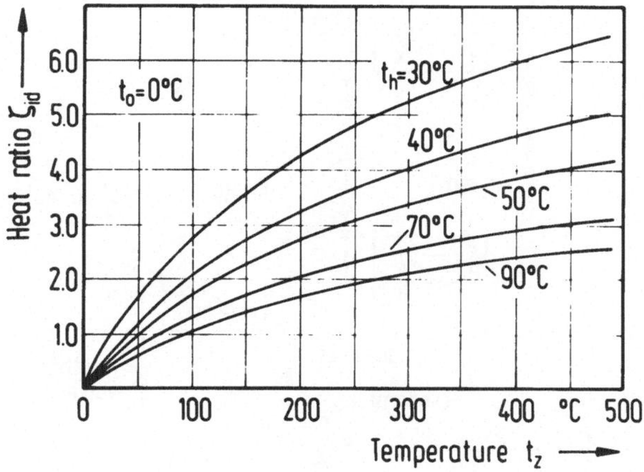

Fig. 5: Ideal heat ratio ζ_{id} vs. temperature t_z with heating temperature t_h as parameter

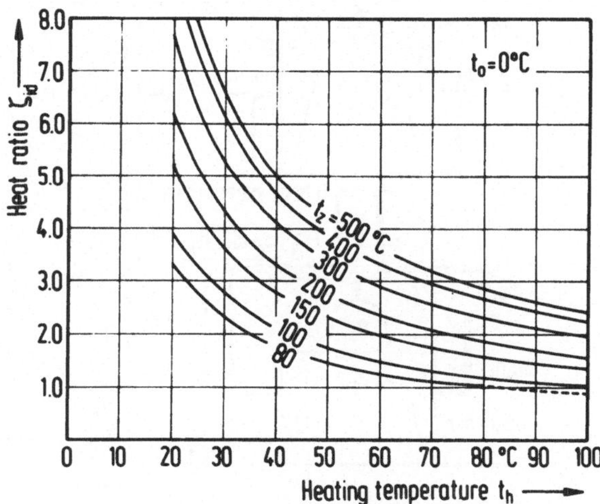

Fig. 6: Ideal heat ratio ζ_{id} vs. heating temperature t_h with temperature t_z as parameter

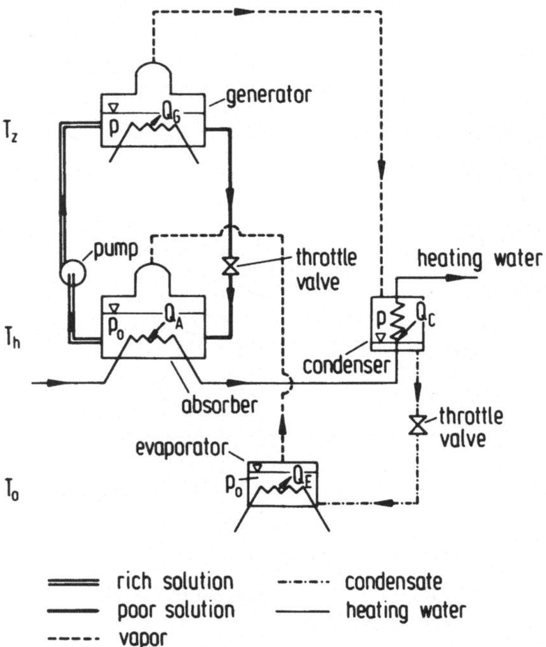

Fig. 7: Simple absorption heat pump process

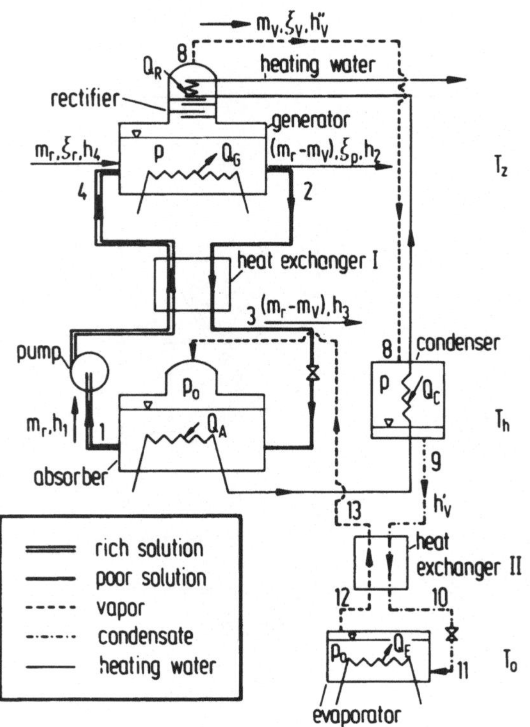

Fig. 8: Improved absorption heat pump process

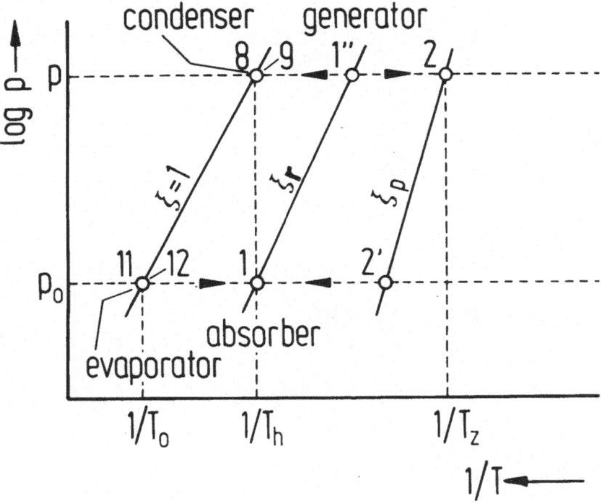

Fig. 9: Absorption heat pump process in the log p, 1/T - diagram

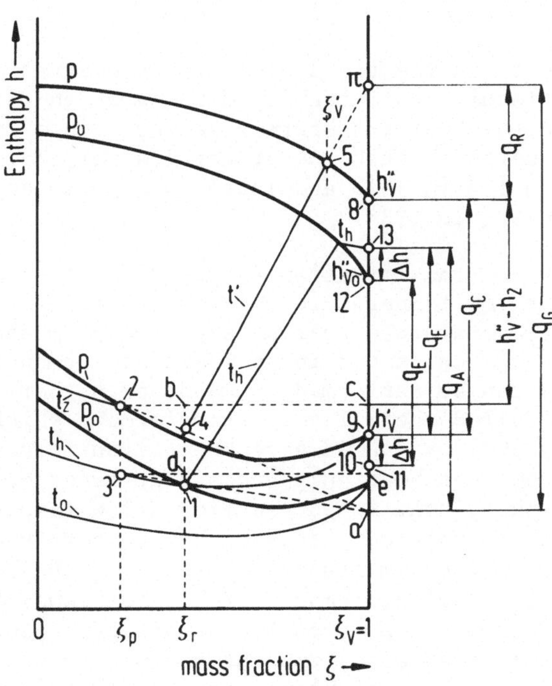

Fig. 10: Absorption heat pump cycle in the h, ξ-diagram

ABSORPTION HEAT TRANSFORMER CYCLES

K. Stephan

Institut für Technische Thermodynamik und Thermische Verfahrenstechnik, Universität Stuttgart, Stuttgart, F.R. Germany

1. INTRODUCTION

The extensive efforts in the field of energy conservation during recent years also require an intensified research and development activities in the heat transformer sector. This field has been neglected hitherto, though the principles were laid down first by Altenkirch [1] in 1918 and later in a more comprehensive form by Nesselmann [2] in 1933.

In general terms heat transformation is performed in a device receiving heat at a low temperature and transforming it into heat at a higher temperature either by addition of mechanical work or of available energy contained in an additional amount of heat. The latter process can be performed by means of a thermal heat pump or by the so called thermal heat transformer, which for reasons of brevity shall be named here "heat transformer". What is the difference between heat pump and a heat transformer?
In a thermal heat pump, heat is pumped from a low temperature source of ambient temperature T_o to a higher temperature T_h by means of heat from a high temperature source T_z. The temperature T_h is higher than the ambient temperature T_o but lower than the temperature T_z. This process is called a heat pump process.
In a heat transformer heat of a temperature T_z enters the process and is transformed into heat of higher temperature T_h, whereas the rest of the original heat is transferred to a low temperature sink, e.g. a sink at ambient temperature T_o. This process is called here "heat transformation". The principles of heat pumps and a heat transformer processes are laid down schematically in Figure 1.

In a heat pump, according to Figure 1a, heat at a high temperature T_z is required and transformed into heat at lower temperature T_h. Thus in terms of the second law a high quality heat is transformed into lower quality heat. A heat transformer works in the opposite direction : A lower quality heat of temperature T_z is transformed into higher quality heat of temperature T_h. Additional heat (or work) for performing this transformation process is not needed, which certainly is one of the main advantages of the process. Since the heat at temperature T_z is split up into two parts, one at temperature T_h and one at temperature T_o, it goes without saying that each of these parts is lower than the original heat at temperature T_z.

2. THE SINGLE STAGE PROCESS

The heat transformation process can be performed by means of a sorption process, which is a modification of the process of a sorption heat pump. In this chapter, for reasons of simplicity, a single stage process working with a binary mixture shall be described. In the generator, Figure 2, a binary mixture consisting of a light volatile component, for instance ammonia, and a heavier less volatile component, for instance water, is heated at a medium temperature T_z. The heat may for instance be waste heat at a temperature between 40 °C and 70 °C. Thus in the generator part of the light volatile component is evaporated and then condensed at lower temperature (e.g. the ambient temperature T_o) in a condenser. The condensate, in our case liquid ammonia, is then pumped into the evaporator where the pressure is higher than in the generator and also in the condenser. The evaporator is also heated at medium temperature T_z from a source.
The liquid ammonia is evaporated and flows into the absorber which has approximately the same pressure as the evaporator. Meanwhile the liquid ammonia-water mixture, which after part of the ammonia has been evaporated in the generator has become poor in ammonia, is pumped from the generator into the absorber. This weak solution absorbs the pure ammonia vapor coming from the evaporator. During the absorption process which occurs at high pressure heat is released at a temperature higher than T_z, for instance at an average temperature T_h. This heat of absorption may be used for heating a sink, thus producing hot water or steam for instance. The strong solution then flows from the absorber through an expansion valve into the generator, where the process starts again.

We note that during the process heat is transferred to the evaporator and the generator at a medium temperature. Part of this heat is rejected in the condenser to a cooling liquid, part is transferred to a sink at higher temperature in the absorber (heat to the load). Contrary to the heat pump which always requires an amount of additional available energy such as electrical or mechanical energy or as heating vapor, the heat transformer needs no additional available energy, except for the energy to operate the pumps. The latter however mostly is negligible compared to the energy transferred from the absorber to the load.

The temperature-entropy diagram (Figure 3) demonstrates why such a process does not violate the second law. In a first cycle a) a cycle 1234 is performed between the medium temperature T_z and the ambient temperature T_o and the work W is produced in the process. In a second cycle 1'2'3'4' (Figure 3b) this work W and heat from the source at ambient temperature T_o are transformed into heat at a higher temperature T_h.
Whereas in the first cycle the heat transferred to the source at ambient temperature T_o is represented by the rectangle 12ab, in the second cycle the heat from the source is represented by the rectangle 1'4' ac. Taking both processes together therefore the heat represented by the rectangle 4'2cb is transferred to the source. The other part of the total heat input at medium temperature is then being transferred to a source of higher temperature T_h. The heat ratio Q_A/Q_z for a reversible process is obtained considering that the amount of work W performed in Figure 3a and Figure 3b is the same :

$$W = (T_z - T_o)\,\Delta S_{ab} = (T_h - T_o)\,\Delta S_{ac}$$

or

$$\frac{\Delta S_{ac}}{\Delta S_{ab}} = \frac{T_z - T_o}{T_h - T_o}$$

Furthermore we have

$$Q_z = T_z\,\Delta S_{ab} \text{ and } Q_A = T_h\,\Delta S_{ac}$$

and therefore

$$\zeta_{id} = \frac{Q_A}{Q_z} = \frac{T_h}{T_z}\,\frac{T_z - T_o}{T_h - T_o} \qquad (1)$$

2.1. The process in the log p, 1/T-diagram and in the h,ξ-diagram

Pressures and temperatures for equilibrium may easily be determined from the log p, 1/T-diagram, Figure 4.

From this diagram however only temperatures and pressures in phase equilibrium states can be read. Processes after the pumps which are not in phase equilibrium cannot be plotted in this diagram. Assuming however phase equilibria at the exit cross section of the absorber, condenser, evaporator and generator, the dots in Figure 4 represent equilibrium states. Thus the temperature of the environment determines the lower pressure p_o, the medium temperature T_z determines the higher pressure p_1 and also the mass fraction ξ_p of the ammonia-weak solution. The temperature T_h determines the mass fraction ξ_r of the rich solution, or, if the spread $\xi_r - \xi_p$ is given instead of T_h, the temperature T_h may be read from the log p, 1/T-diagram.

The simple process in Figure 2 may be considerably improved. The concentrated solution leaving the absorber has a higher temperature than the weak solution from the generator. Also the pure vapor from the generator flowing into the condenser has a higher temperature than the liquid pumped from the condenser to the evaporator. The efficiency may be improved therefore by a heat exchanger preheating the weak solution pumped from the generator to the absorber and a second heat exchanger precooling the ammonia vapor before entering the condenser and simultaneously preheating the liquid pumped from the condenser to the evaporator. The arrangement of the apparatus is shown in Figure 5. The presentation of the process in the log p, 1/T-diagram remains unchanged.

Figure 6 represents the process in an entalpy-concentration diagram. In this diagram the boiling and dew lines for a pressure p_o in the condenser and generator are plotted and those for a pressure p_1 in the absorber and evaporator. Furthermore the isotherms T_o, T_z and T_h in the liquid state are plotted.

First the concentrations ξ_p and ξ_r of the weak and of the rich solution are determined. According to our assumption, the heat of absorption is transferred at the temperature T_h. The rich solution therefore leaves the absorber at T_h in a saturated condition, which is in Figure 6 given by the intersection of the isotherm T_h with the boiling line at the absorber pressure p (point 4). The rich solution flows through the heat exchanger HEI, where it is cooled to a temperature given by point 5 in Figure 6, then expands in the throttle valve where the enthalpy and mass fraction remain unchanged. Point 5 will be determined below. Thus point 6 after throttling and point 5 before throttling are identical, whereas the states 5 and 6 are different, because point 5 represents a subcooled state belonging to pressure p_1, and point 6 represents a state belonging to pressure p_o, where the mixture consists of saturated liquid of mass fraction ξ'_6 and saturated vapor of mass fraction ξ''_6. In the generator the lighter volatile component is evaporated. This process can take place until the temperature of the solution is equal to the temperature T_z of the heat source. The state of the weak solution leaving the generator is therefore given by point 1. It is compressed to pressure p_1. The new state is given by point 2, which coincides approximately with point 1, since the concentration ξ_p remains constant and the specific enthalpy is only slightly increased. The weak solution is then heated in the heat exchanger HEI to state 3 where its temperature is slightly below the temperature T_h of the rich solution leaving the absorber (point 4). The state of the poor solution entering the absorber is therefore on the line ξ_p below the temperature T_h.

Now also the state 5 of the rich solution leaving heat exchanger HEI can be determined from an energy balance for the heat exchanger.

Be $f = m_L/m_V$ the amount of liquid circulating per kg of vapor. When f kg of liquid enter the generator (state 6) 1 kg of vapor is produced and 1-f kg of liquid (state 1) leave it. Taking into account that f kg of liquid are flowing on one side of heat exchanger HEI and f-1 kg on the other side, the heat balance is given by

$$f(h_4 - h_5) = (f - 1)(h_3 - h_2) \tag{2}$$

On the other hand a mass balance of the generator yields

$$f\,\xi_r = \xi_V + (f - 1)\xi_p$$

or

$$f = (\xi_V - \xi_p)/(\xi_r - \xi_p) \tag{3}$$

Inserting this into equation (2), we obtain

$$\frac{\xi_V - \xi_p}{\xi_r - \xi_p}(h_4 - h_5) = \frac{\xi_V - \xi_p}{\xi_r - \xi_p}(h_3 - h_2) \tag{4}$$

According to equation (4) point 5 may be determined in the following way. A straight line through point 1 and 16 intersects the ξ_r-line in point A (Figure 7) and a straight line from 3 to A intersects the ordinate axis in a point B. From the similarity of the triangles 1, 3, A and A, 16, B in Figure 7, we obtain :

$$\frac{h_{16} - h_b}{\xi_V - \xi_r} = \frac{h_3 - h_2}{\xi_r - \xi_p}$$

or

$$h_{16} - h_B = \frac{\xi_V - \xi_r}{\xi_r - \xi_p} (h_3 - h_2)$$

Thus $h_{16} - h_B$ is equal to the right hand side of equation (4). On the other hand the straight line through points 16 and 4 intersects the ξ_p-line in a point D, and a straight line through point D and B intersects then the ξ_r-line in point 5. This follows from the similarity of the triangles D, 4, 5 and D, 16, B, from which it follows :

$$\frac{h_4 - h_5}{\xi_r - \xi_p} = \frac{h_{16} - h_B}{\xi_V - \xi_p}$$

or

$$\frac{\xi_V - \xi_p}{\xi_r - \xi_p} (h_4 - h_5) = h_{16} - h_B = \frac{\xi_V - \xi_r}{\xi_r - \xi_p} (h_3 - h_2) \quad ,$$

this is in agreement with equation (4).

f-1 kg of poor solution with mass fraction ξ_p (state 3) enters the absorber and also 1 kg of vapor with state 16. The heat of absorption q_A (in kJ per kg vapor) is transferred and f kg of rich solution with mass fraction ξ_r (state 4) is leaving the absorber. The energy balance therefore yields

$$(f - 1)h_3 + h_{16} = q_A + f\ h_4 \tag{5}$$

Together with equation (3) we obtain

$$\frac{\xi_V - \xi_p}{\xi_r - \xi_p} (h_3 - h_4) = q_A - (h_{16} - h_3) \tag{6}$$

Also the heat of absorption q_A can easily be determined by means of the enthalpy-concentration diagram, because the left hand side of equation (6) is given by Δh in Figure 6. This follows from the similarity of the triangles 3, 4, E and 3, F, G in Figure 6, according to which we have

$$\frac{\Delta h}{\xi_V - \xi_p} = \frac{h_3 - h_4}{\xi_r - \xi_p} \text{ or } \Delta h = \frac{\xi_V - \xi_p}{\xi_r - \xi_p} (h_3 - h_4)$$

Adding $h_{16} - h_3$ we obtain q_A, which is plotted in Figure 6.

Finally we have to consider the generator : the state of the rich solution entering the generator is characterized by point 6 (Figures 5 and 6). Part of the liquid is evaporated in point 6 and we have a saturated vapor-liquid mixture consisting of saturated liquid ξ_6' and saturated vapor ξ_6''. The generator is assumed to be coupled with the rectification column in order to get almost pure ammonia vapor, thus increasing the absorption heat and preventing water to be accumulated in the condenser, and also avoiding a reduction of heat transfer in the evaporator, because, as is well known, boiling heat transfer in mixtures is lower than that of the pure components.
A sketch of the generator with rectification column is given in Figure 8.

The overall energy balance yields

$$f\, h_6 + q_G = (f - 1)h_2 + h_{11} + |q_R| \;,$$

or taking into account equation (3) :

$$\Delta h^{*} = \frac{\xi_V - \xi_p}{\xi_r - \xi_p}(h_2 - h_6) + (h_{11} - h_2) = q_G - |q_R| \qquad (7)$$

The left hand side of equation (7) can easily be determined from Figure 6. A straight line through points 1, 6 intersects the ordinate in point H. A parallel to the abscissa through point 1 intersects the ξ_r-line in point L (which by chance coincides with point 4 in Figure 6) and the ordinate in point I. Similarity of the triangles H, I, 1 and L, 6, 1 yields

$$IH = \frac{\xi_V - \xi_p}{\xi_r - \xi_p}(h_1 - h_6)$$

The left hand side of equation (7) is therefore given by the line 11, H on the ordinate in Figure 6. The heat q_G transmitted in the generator depends according to equation (7) on the reflux heat $|q_R|$ per mass unit of vapor. The heat q_G should be kept as small as possible, therefore also $|q_R|$ must be small. From the theory of rectification columns, it is known that the so-called operating line determines also the reflux heat q_R according to Figure 9, where π' and π are the so-called poles of the operating line. As demonstrated in the theory of rectification (e.g. [3]) the reflux heat q_R and the heat q_G added in the generator are graphically determined according to Figure 9.
It is also shown in the theory of rectification that this operating line must be steeper than the saturation isotherm, in our case the isotherm T_6. It is evident from Figure 9 that the

smallest values of $|q_R|$ and q_G are obtained when the operating line and the isotherm T_6 coincide. This limiting case is also plotted in Figure 6.

The heat q_G of the generator and also $|q_R|$ therefore are obtained as constructed in Figure 9, and in Figure 6 respectively.

The vapor leaving the rectification column consists practically of the pure lighter volatile component (ammonia). It is saturated vapor of pressure p_o (state 11). Part of it is condensed in heat exchanger HEII (state 12) and finally it is completely condensed in the condenser.

The heat withdrawn from the condenser is given by

$$q_C = h_{12} - h_{13}, \tag{8}$$

whereas the heat inflow in the evaporator is given by

$$q_E = h_{16} - h_{15}. \tag{9}$$

Both q_C and q_E can easily be read from Figure 6.

The heat ratio ζ is given by

$$\zeta = \frac{q_A}{q_G + q_E} .$$

It is smaller than 1, because q_A is only part of the heat $q_G + q_E$ flowing into the process, the other part is transferred to the environment. As Figure 6 demonstrates qualitatively we have $\zeta < 1$. Thus from the enthalpy-concentration diagram of the transformer process all parameters can easily be determined. As a main advantage the diagram gives not only a quick overview of the process, it also shows in a very clear way how alterations in the decisive parameters, for instance in the temperatures or heat fluxes, change the whole process.

For a more detailed design, taking into account the finite temperature differences in the heat exchangers, a computer program seems to be useful. Such a program was developed for ammonia-water as working fluids by the author and his coworkers. Results are discussed in chapter 4 of this paper.

3. A TWO-STAGE PROCESS

As can easily be seen from the log p-1/T-diagram (Figure 4) there exists an upper limit for the temperature T_h in a single stage process. Assuming the temperature of the environment T_o and the medium temperature T_z to be given, the points 13, 16 and 1 are fixed in Figure 4. Then for a finite spread $\xi_r - \xi_p$, also point 4 is fixed. For the theoretically lowest spread (zero), the curves for ξ_r and ξ_p fall together and we obtain the upper limit for T_h.

A higher temperature T_h may only be achieved with a two-or multistage process. The heat ratio of multistage processes obviously must be lower than in a single stage process for thermodynamic reasons because, according to Figures 3a and 6 the heat from the environment becomes smaller (see Figure 3a) with increasing temperature T_h and thus the total heat transferred to the environment increases. The portion transformed to a higher temperature therefore becomes smaller. The increase of working temperature T_z goes at the expense of a lower heat ratio and also of a more complicated and expensive equipment. The thermal efficiencies in a two-stage process are still reasonable (see below). In a process with three or more stages however, apart from the expensive equipment, the losses of available energy endanger the whole transformation process. Only a two-stage process therefore shall be discussed here. Such a process was first proposed by Altenkirch and Niebergall [4] in 1956. Its basic idea is that the temperature T_h for a given spread $\Delta\xi = \xi_r - \xi_p$ may be increased for lower temperatures T_o. This effect can easily be understood, when considering a process where T_o, T_z and $\Delta\xi$ are given (see dashed line in Figure 10). For this process the attainable absorber temperature is T_h. If instead, we have a lower temperature $T_o' < T_o$, while T_z and $\Delta\xi$ remain unchanged, we obtain according to Figure 10 (full line), a temperature $T_h' > T_h$.

In order to decrease the temperature T_o, Altenkirch and Niebergall proposed to couple a compression heat pump, resp. a refrigeration device, with a heat transformer (Figure 11). The evaporator of the heat pump is combined with the condenser of the transformer. Thus the heat for condensing the working substance in the transformer is used to evaporate the working substance in the heat pump cycle. The heat released in condenser C_1 of the heat pump cycle is transferred to a sink of e.g. ambient temperature T_o. Assuming for instance T_o = 10 °C, T_z = 40 °C and $\Delta\xi$ = 0,05, one obtains for ammonia-water T_h = 70 °C. If instead T_o decreases to T_o' = 0 °C, while T_z and $\Delta\xi$ remain constant, a temperature T_h' = 78 °C may be attained. Altenkirch and Niebergall proposed the above process to be used as a heat pump, where T_z is the temperature of the soil or of the water of a well and T_o the temperature of the usually colder ambient air. As they pointed out the heat transmitted to the absorber increases with a decrease of the outside air temperature T_o, an effect which becomes apparent from Figure 1. This type of heat pump therefore delivers more heat when the outside temperature decreases. Its efficiency therefore increases, whereas a usual compression heat pump becomes inefficient at low temperatures due to the considerable amount of mechanical energy. A disadvantage of the device certainly is the fact that two heat sources of different temperatures T_z and T_o are necessary.

Additionaly some mechanical energy is required. This energy however is very small. For a reversible process running between $T_o' = 0$ °C, $T_o = 10$ °C, $T_z = 40$ °C, $T_h' = 78$ °C it is only $P = 0{,}0156\ (Q_G + Q_E)$, where Q_G is the heat flowing into the generator and Q_E that flowing into the evaporator. Under the assumption of a reversible process the heat ratio is

$$\zeta_{id} = \frac{Q_A}{Q_G + Q_E} = 0{,}57$$

A realistic value is estimated to be around 0,25 to 0,3.

Instead of the compression heat pump cycle coupled to the transformer, naturally also an absorption heat pump can be coupled.

Another two stage process, which is also obtained as a modification of refrigeration absorption processes, is given in Figure 12. The mixture of concentration ξ_r is heated in the generator G at temperature T_z, and the vapor ξ_r' (see also Figure 13) is condensed in the resorber R where it is absorbed by a weak solution ξ^*_p. The rich solution ξ^*_r from the resorber is then pumped into the heater (generator) G 1, where it is heated and separated into a pure vapor and the weak solution ξ^*_p flowing back over a throttle valve to the resorber. The pure vapor from the generator is then condensed in condenser C. The liquid is pumped to a higher pressure and evaporated in the evaporator E. From there the vapor flows to the absorber A where it is absorbed by the weak solution ξ'_p. Figure 13 gives the process in an log p-1/T diagram. The arrows indicate the direction of heat fluxes. In the generator G more vapor is produced than in the generator G_1. As a result a greater amount would be fed from the left side of the absorption cycle in Figure 12 into the right side than vapor is flowing back. In order to compensate this, the two-stages are connected through a pipe by which part of the weak solution ξ^*_p is fed back.

Beside the processes discussed in this chapter many other two-stage devices are conceivable. There exist no general rules on

which of the processes should be given preference. The kind of process to be chosen is largely determined by the conditions to be fulfilled, e.g. temperature range and whether there exist three sources and sinks at temperatures T_o', T_o and T_z or only two at temperatures T_o and T_z.
In the case of sink and source at two temperatures T_o and T_z, a modification of the well known two stage refrigeration cycle may be used, as it is indicated in the log p-1/T diagram (Figure 14). Another very effective means to increase the temperature T_h is the coupling of an absorption heat transformer to a compression heat pump. The heat released in the absorber at temperature T_h is flowing into the evaporator of a compression heat pump and then, by addition of mechanical energy through a compressor, transformed into heat at higher temperature T_h'.

4. RESULTS AND CONCLUSIONS

So far little research and development has been done on heat transformers, although the transformer is the most effective means to upgrade low temperature heat without the need of additional available energy and not withstanding the fact that the basic ideas were laid down by Altenkrich [1] and Nesselmann [2] more than half a century ago. The increasing demand for energy and at the same time the increasing energy shortage are now beginning to change this situation drastically and recently activities have been started in a few places of the world also in the field of the heat transformer technology. Hopefully we are not far from realizing the process for technical application, the more since most parts of the process are common and well explored technologically. Also reasonable efficiencies may be attained. Figure 15 gives the heat ratio ζ to be achieved in a process. It is evaluated according to

$$\zeta = \zeta_{id}\,\eta,$$

where ζ_{id} is the heat ratio of a reversible process (equation (1)) :

$$\zeta_{id} = \frac{T_h}{T_z}\,\frac{T_z - T_o}{T_h - T_o}$$

and η is the overall efficiency, which was assumed to be $\eta = 0.6$. Furthermore the curve ζ_{id} for the upper temperatures $T_h = 80\ °C$ and $T_h = 140\ °C$ are plotted as dashed lines. As the diagram demonstrates, the heat ratio increases for a given ambient temperature T_o with increasing medium temperature T_z. For a given temperature T_z and at constant value T_o the heat ratio becomes lower, when higher temperatures T_h are required. More realistic values for ammonia-water mixtures are given in Table I. They are calculated with a computer program using the equations of state of ammonia-water mixtures and taking into account finite temperature differences. Also some values for a two-stage process according to Figure 14 are given. The tendency of these results is the same as plotted in Figure 15. However the results also clearly indicate that the efficiencies strongly depend on the temperature differences in the heat exchanger, which therefore must be carefully designed. On the other hand the efficiencies and temperatures attainable in a single or in a two-stage process are very promising. It seems to be worthwile therefore to activate research and development on heat transformer processes.

REFERENCES

1. Altenkirch, E. *Zeitschrift f.d. gesamte Kälteindustrie* 25 (1918) no. 7, 49-53, no. 8 pp. 57-60.

2. Nesselmann, K. 12 *Wissensch. Veröffentl. Siemens Konzern* (1933), no. 2, 89-109.

3. Bosnjakovic *Technische Thermodynamik* Vol. II (Dresden Steinkopf-Verlag, 1971).

4. Altenkirch, E.; Niebergall, W., DB Pat. 95 3378.

TABLE I

Data for one and two-stage processes, according to Figure 14

ΔT = o K (T$_z$ = 313 K)

	T_o |K|	273	273	278	278	283	283	288	288
	$\Delta\xi$ |%|	5	8	5	8	5	8	5	8
stage 1	$T_{h1}=T_{z2}$ |K|	351	346	345	341	339	335	334	330
stage 1	ζ_1	0,478	0,487	0,478	0,486	0,478	0,485	0,477	0,485
stage 1	η_{ex1}	0,795	0,781	0,792	0,774	0,787	0,763	0,784	0,751
stage 2	$T_h=T_{h2}$ |K|	431	414	412	398	393	379	377	363
stage 2	ζ_2	0,440	0,460	0,454	0,469	0,463	0,476	0,467	0,479
stage 2	η_{ex2}	0,657	0,700	0,693	0,722	0,721	0,738	0,737	0,745
	$\zeta=\zeta_1\zeta_2$	0,210	0,224	0,217	0,228	0,221	0,231	0,223	0,232
	$\eta_{ex}=\eta_{ex1}\eta_{ex2}$	0,522	0,547	0,548	0,559	0,567	0,563	0,578	0,559

ΔT = 3 K (T$_z$ = 313 K)

	T_o |K|	273	273	278	278	283	283	288	288
	$\Delta\xi$ |%|	5	8	5	8	5	8	5	8
stage 1	$T_{h1}=T_{z2}$ |K|	338	334	332	328	327	324	-	-
stage 1	ζ_1	0,446	0,446	0,450	0,469	0,449	0,467	practically	
stage 1	η_{ex1}	0,646	0,650	0,631	0,625	0,608	0,608	no increase	
stage 2	$T_h=T_{h2}$ |K|	391	377	372	358	355	345	in temperature	
stage 2	ζ_2	0,417	0,450	0,428	0,459	0,440	0,461	-	-
stage 2	η_{ex2}	0,606	0,650	0,021	0,648	0,627	0,633	-	-
	$\zeta=\zeta_1\zeta_2$	0,186	0,210	0,193	0,215	0,198	0,215		
	$\eta_{ex}=\eta_{ex1}\eta_{ex2}$	0,392	0,423	0,392	0,405	0,381	0,385		

$\Delta T = 5$ K $(T_z = 313$ K$)$

T_o |K|	273	273	278	278	283	283	288	288
$\Delta\xi$ |%|	5	8	5	8	5	8	5	8
$T_{h1}=T_{z2}$ |K|	330	326	324	321	-	-	-	-
ζ_1	0,420	0,453	0,430	0,455	practically no			
η_{ex1}	0,548	0,564	0,530	0,534	increase in temperature			
$T_h=T_{h2}$ |K|	357	352	347	336	-	-	-	-
ζ_2	0,405	0,443	0,409	0,451	-	-	-	-
η_{ex2}	0,555	0,591	0,544	0,566	-	-	-	-
$\zeta=\zeta_1\zeta_2$	0,170	0,201	0,176	0,205	-	-	-	-
$\eta_{ex}=\eta_{ex1}\eta_{ex2}$	0,304	0,333	0,288	0,302	-	-	-	-

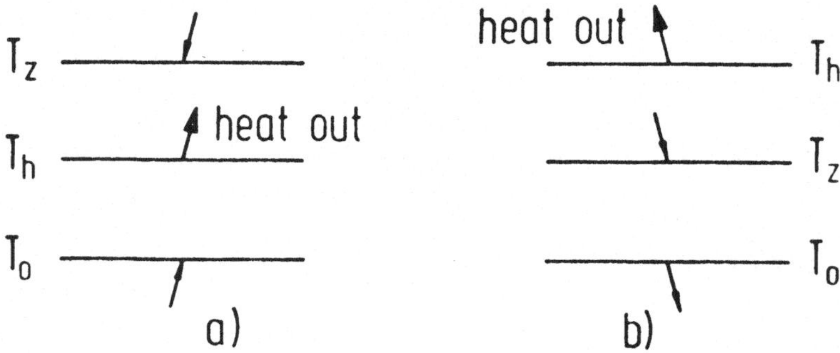

Fig. 1: Sketch for a thermal heat pump a) and a heat transformer b).

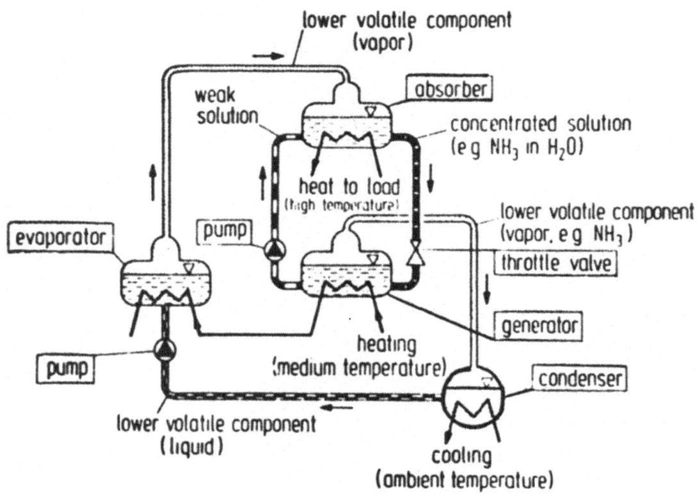

Fig. 2 : Absorption Heat Transformer

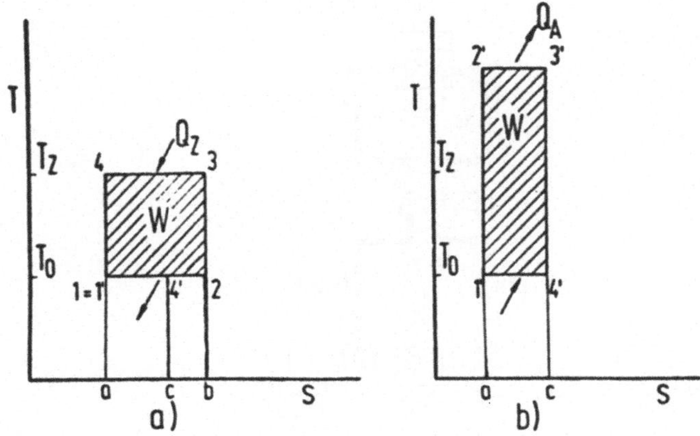

Figure 3: Absorption heat transformer process in T, s-diagram.

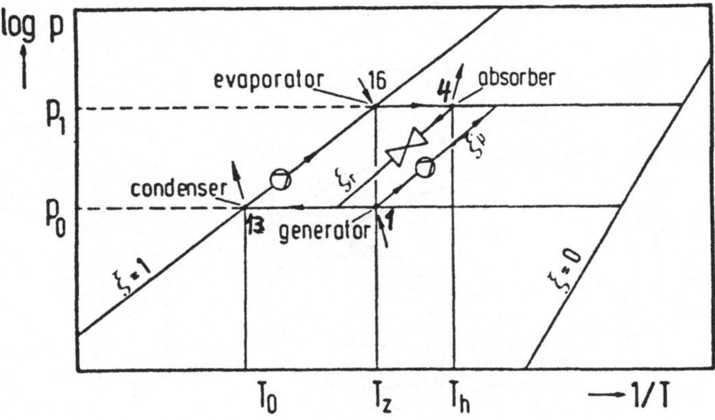

Fig. 4: Heat transformer in log p-1/T-diagram.

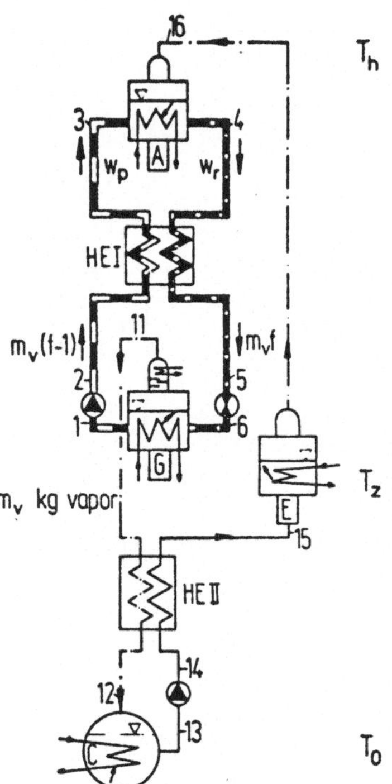

Fig. 5: Single stage process with internal heat exchange. A=Absorber, G=Generator, E=Evaporator, C=Condenser, HEI=Heat Exchanger I, HE II=Heat Exchanger II.

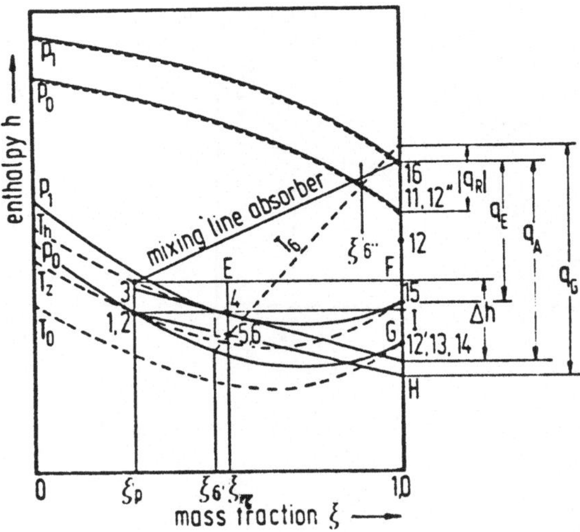

Fig. 6: Single stage process in the enthalpy-concentration diagram.

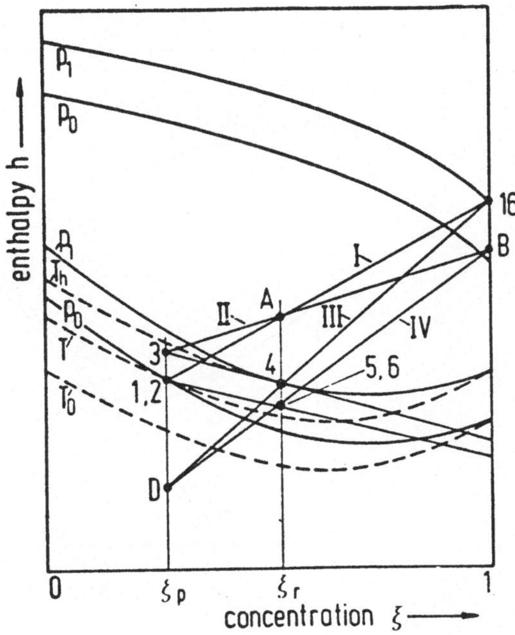

Fig. 7: State of rich solution (point 5 in Figs. 5 and 6) in the heat exchanger HE 1. Lines should be plotted in the order I, II, III, IV.

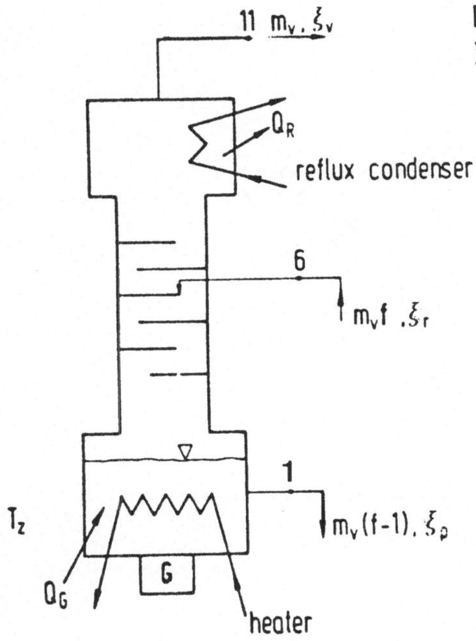

Fig. 8. Generator with rectification column

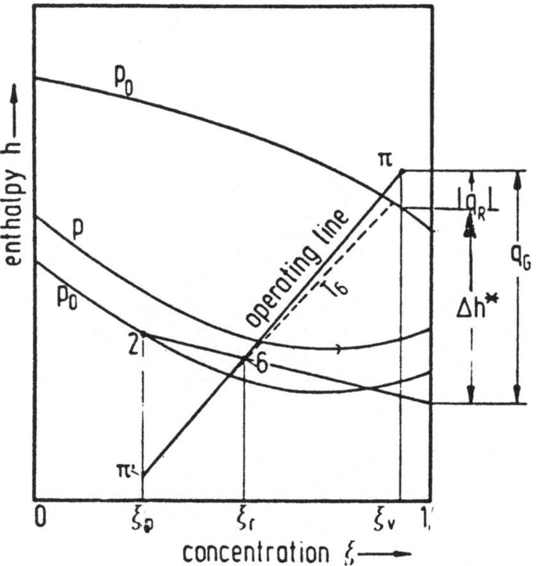

Fig. 9: Determination of $|q_R|$ and q_G

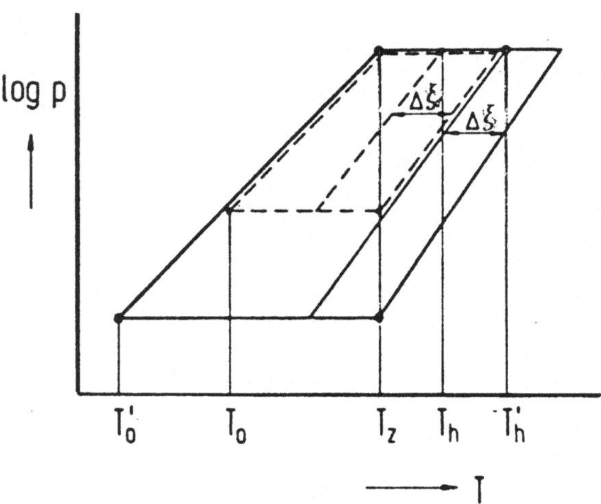

Fig.10 Increase of $T_h \rightarrow T_h'$ when temperature T_0 decreases $T_0 \rightarrow T_0'$.

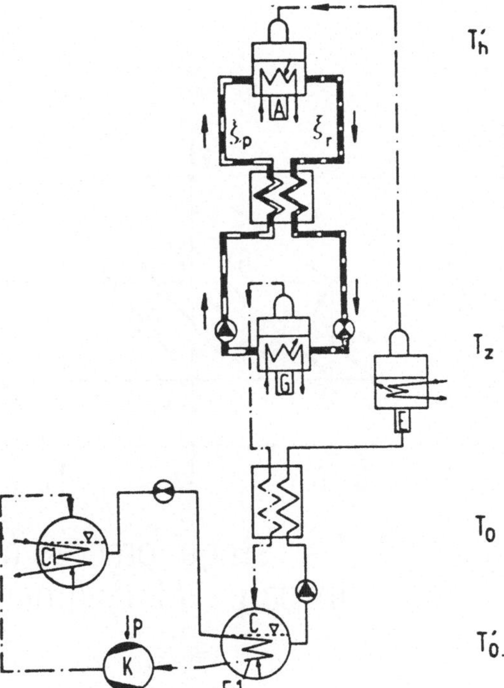

Fig. 11: Absorption heat transformer combined with heat pump.

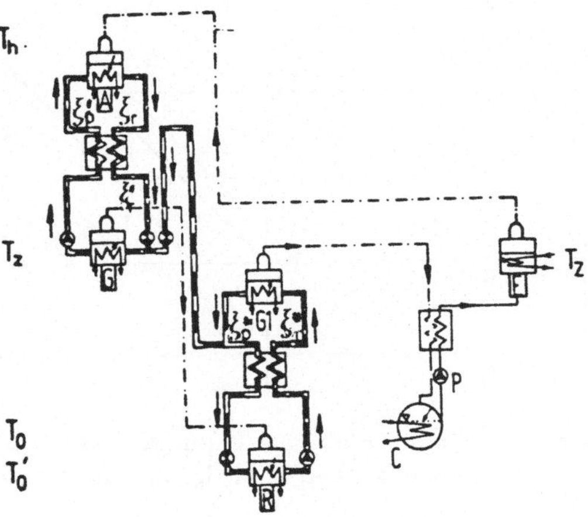

Fig. 12: Two-stage process with resorber.

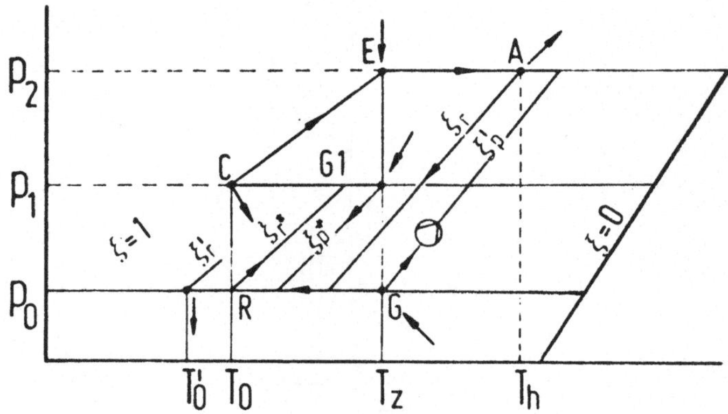

Figure 13: Two-stage process of Fig. 12 in log p, 1/T-diagram.

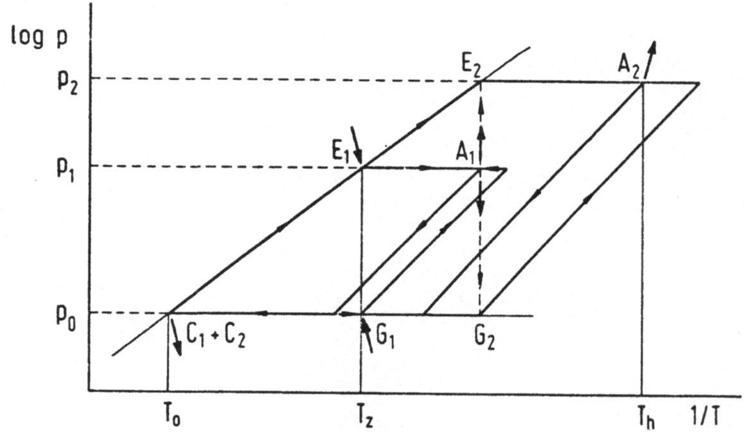

Figure 14: log p, 1/T - diagram for a two-stage process when only one source at T_h and one sink at T_0 exist.

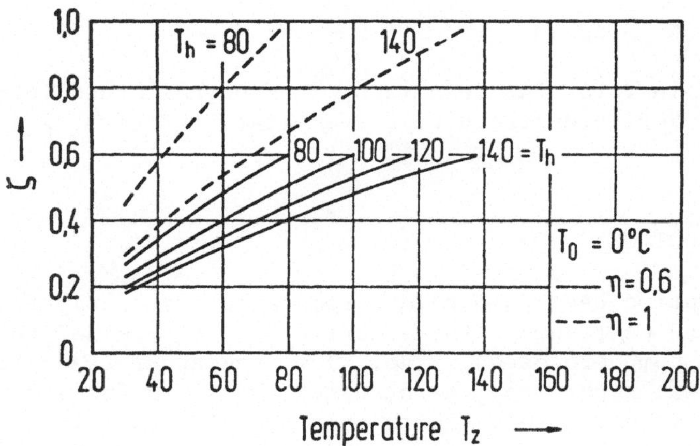

Fig. 15: Heat ratio $Q_A/(Q_E+Q_G)$ of a heat transformer.
---- ζ_{id}

WORKING SUBSTANCES FOR ABSORPTION HEAT PUMPS AND TRANSFORMERS

K. Stephan

Institut für Technische Thermodynamik und Thermische Verfahrenstechnik, Universität Stuttgart, F.R.Germany.

1. INTRODUCTION

The absorption heat pump offers some advantages compared with a compression heat pump : the electrical energy needed for absorption heat pump processes is very low, the apparatus relies upon conventional technology, maintenance costs are low and its expected life is very high. Furthermore its thermal efficiency depends less on the ambient temperature than that of a compression heat pump. This is demonstrated by Figure 1, where the ideal heat ratio :

$$Q_N/Q_Z = \zeta_{id} = \frac{1 - T_0/T_Z}{1 - T_0/T_h} \qquad (1)$$

of a sorption heat pump is compared with the efficiency

$$\zeta_{id} = \frac{\eta_K}{1 - T_0/T_h} \qquad (2)$$

of a compression heat pump for a variable ambient temperature. Herein Q_Z stands for the heat flux from a source into the system, Q_N is the heat flux to the load, T_0 the temperature of the environment, T_Z the temperature of the heat from source, T_h the temperature of the heat flux to the load. The efficiency (COP) of the Clausius-Rankine-cycle was assumed to be η_K = 0,3, which seems to be a realistic value([1], p. 308, Table I).
Figures 2 and 3 compare the processes of an absorption heat pump and a heat transformer in a log p, 1/T-diagram.

The processes according to Figures 2 and 3 operate between two different pressures. The lower one p_o is fixed by the ambient temperature (evaporator in Figure 2, condenser in Figure 3) and the saturation pressure of the working substance. The high pressure p is given by the state in the condenser in Figure 2, or evaporator in Figure 3. It is therefore determined by the temperature to the load in Figure 2 or the temperature from the source in Figure 3.

In the case of an absorption heat pump the concentration of the rich solution is fixed by the temperature and pressure in the absorber, i.e. by the load temperature and the lower pressure. The concentration of the weak solution is determined by the temperature and pressure in the generator, i.e. the source temperature and higher pressure.
In the case of the absorption heat transformer the concentration of the strong solution is also given by the temperature and pressure in the absorber, i.e. now the load temperature and the upper pressure, whereas the concentration of the weak solution is determinded by the temperature and pressure in the generator, i.e. now the source temperature and the lower pressure.
In both cases the vapour produced in the generator should be as pure as possible, in order to avoid an accumulation of the heavier volatile components in the evaporator.
There exists a comprehensive literature on working substances for absorption refrigeration processes, cf. reference [2]. Sorption heat pumps and transformers are operated, however at higher temperatures than refrigeration processes. Working substances used in refrigeration processes therefore are not readily appropriate for heat pumps or heat transformers.
The present paper therefore gives a review of the requirements to be fulfilled by working substances for absorption heat pumps and transformers and discusses the advantages and disadvantages of substances which seem to be appropriate.

2. REQUIREMENTS TO BE FULFILLED BY WORKING SUBSTANCES

In the following a catalogue of criteria and desired properties shall be established, first for the working substances, next for the mixture and finally for the process.

2.1. Thermal properties of the working substances

The melting point of a working substance must be at a low temperature in order to avoid freezing of the working substance or the mixture itself. Because the viscosity near the melting point of a liquid is very high, the lowest temperature in the process, usually the ambient temperature, should be sufficiently above the melting temperature of the working substance. Devices installed in houses may neither be operated at extreme low nor at excessive pressures. When working at too low pressures, air or water vapour can be sucked out of the atmosphere, thus causing a decrease of the efficiency and promoting corrosion. A too high pressure on the other hand requires thicker walls of the apparatus. The boiling point of the working substance therefore should coincide as well as possible with the working temperature of the process, whereas the boiling point of the solvent should be far above that of the working substance, otherwise a rectifier is needed for complete separation of the mixture in the generator. As experience shows a rectifier is unnecessary, when the difference of both boiling temperatures is above 200 K [3].

The enthalpy of evaporation should be as large as possible, because the larger it is, the more heat per unit of vapour mass produced in the generator, resp. per unit of liquid mass, can be transferred. A large enthalpy of evaporation is obtained for a sufficiently large distance of the process temperatures from the critical temperature. The process temperatures therefore should be far below the critical temperature.

Among the other physical properties, the viscosity, thermal conductivity and the thermal capacity are decisive for the heat transfer processes. Obviously thermal conductivity and thermal capacity should be sufficiently large, whereas a low viscosity is an advantage.

2.2. Thermal properties of the mixture

One of the conditions to be fulfilled by the mixture is the complete solubility of the working substance and the solvent over a large range of concentrations. This is especially required for operating the absorber. In cases where solids are used as solutes, for instance when salt solutions are used as mixtures, the chilling point should be beyond the working temperature.
The kinetics of solubility are of decisive importance to the size of the absorber. A long retardation in mixing solvent and solute and a low reaction rate prove to be very unfavorable, due to the long time required for the absorption process, which itself requires a large absorber. In order to accelerate the mixing and absorption process the sorption enthalpy must be large enough. Furthermore a large sorption enthalpy increases the heat transformed per unit mass of generated vapor.
The mixture can be easily separated into solute and solvent, when the fraction of the lighter volatile component in the vapor phase is high. Advantageous for this effect is an almost horizontal dew line in the temperature-concentration-diagram, until the mass

fraction approaches that of the pure vapor and then decreases as plotted in Figure 4. On the other hand a boiling line as straight as possible should be preferred, in order to obtain a large concentration spread and simultaneously a high solubility of the working substance in the solvent.

2.3. Process properties

An important factor for the choice of working substances are the temperatures to be achieved in the process and the required pressures. A single stage process operates between two different pressures. The upper one is given by the working temperature, the lower one by the temperature of the environment (Figure 5). The greater the pressure difference, the more pump energy is required for circulating the solution. The vapor pressure curves for constant mass fraction therefore should be as flat as possible in a vapor-pressure diagram. Due to a flat vapor pressure curve higher temperatures may be obtained. However according to the Clausius-Clapeyron equation mixtures with a flat vapor pressure curve exhibit a low enthalpy of evaporation and often also a low enthalpy of mixing. As a consequence a large absorber is needed because of the slow absorption process and also a large pump is required because of the large amount of solute to be circulated per mass unit of vapor.
Other important criteria for assessment of the sorption heat pump process are the heat ratio and the specific mass flow rate of the circulating liquid.
The heat ratio ζ is defined as the heat flux to the load divided by the heat flux from the source. For heat pumps its value is greater than unity. Its upper limit ζ_{id} is reached when all processes are reversible. It is given by :

$$\zeta_{id} = \frac{1 - T_0/T_Z}{1 - T_0/T_h}$$

A large value for ζ_{id} is obtained for high temperatures of the source, which may be achieved by using working substances with a flat vapor pressure curve. Furthermore it is desirable to realize a given heat ratio by circulating small amounts of liquid. The specific mass flow rate

$$f = \frac{\xi_v - \xi_p}{\xi_r - \xi_p} = \frac{m_L}{m_v}$$

therefore should be small. It indicates how much of the strong solution must be circulated per mass unit of vapor with mass fraction ξ_v. The difference $\Delta\xi = \xi_r - \xi_p$ between the mass fraction of strong and weak solution is called the spread. It should be sufficiently large in order to achieve a small specific mass flow rate.

2.4. Remarks on the structure of molecules

As already mentioned the enthalpy of evaporation of the working substance should be as large as possible. According to the Clausius-Clapeyron-equation

$$\Delta h_v = T\,(v'' - v')\,\frac{dp}{dT} \approx \frac{RT^2}{M\,p}\,\frac{dp}{dT}$$

this requires a small molar mass of the working substance. Unfortunately substances with a low molar mass exhibit a boiling point, which is too low for absorption heat pumps, a result which is clearly demonstrated for the homologeous series of alkanes in table I.

Butane for example with a boiling point of 272 K has the very low enthalpy of evaporation of 385,5 kJ/kg. In order to obtain a high boiling point for substances of low molar mass, polar molecules are required. Table 2 gives an example on how the boiling points of some substances of similar molar mass depends on the polarity.

Due to the polarity of molecules, clusters or so-called reversible complexes may be formed as a consequence of hydrogen bonds. With regard to the solvent, a high boiling point is required. Its enthalpy of evaporation is unimportant. The solvent therefore should have a high molar mass. Polar groups within the solvent molecules lead to an additional increase of the boiling point. Furthermore they enhance the interaction with polar molecules of the working substance and thus increase the solubility and also the mixing enthalpy.
Summarizing these results, one should prefer a working substance with small polar molecules and a solvent with large molecules containing polar groups.

2.5. Some further properties

Working substance and solvent are not allowed to decompose. Therefore they must contain chemically stable molecules. Flammable substances should not be used. Furthermore the molecules of the working substance and those of the solvent should not react chemically and form a new stable product, which cannot be separated in its original molecules. On the other hand the formation of compounds which may be separated by heat addition in the generator is desirable. Corrosive mixtures should be excluded. In some cases however corrosion may be considerably reduced by adding so-called inhibitors. Also the seals should not decompose. Furthermore poisonous substances should not be used, especially not in houses where usually the heat pump cannot be supervised. Hints and values concerning the toxicity of substances are given in different handbooks e.g. references [4] to [8].

3. A CATALOGUE OF SUITABLE MIXTURES

According to the hitherto discussed criteria, it seems reasonable to establish a catalogue of substances which are, at least in principle, appropriate for the absorption heat pump process. In a next step one should verify which of these substances fulfill the criteria best. An investigation of the existing literature revealed that there exist about 70 binary mixtures, which might be discussed. This catalogue still is by no means complete considering the large manifold of existing substances. Furthermore also some ternary mixtures are worthwile to be discussed.

Such a catalogue was recently established by Stephan and Seher [9]. It gives the data for boiling and melting points of working substance and solvent, enthalpy of evaporation of the working substance, general remarks concerning suitability for absorption heat pump processes and the references. This catalogue is subdivided in the following groups of substances :

- solutions of salts
- solutions of the refrigerant R22
- solutions of compounds with silicon
- solutions of ammonia and amines
- other mixtures

Only a few general remarks should be made here. Water as a working substance has a high enthalpy of evaporation, however, its melting point of 0°C is also high. Ammonia and the amines also have a high enthalpy of evaporation. They are however poisonous and flammable. Ammonia furthermore requires high pressures, which are not needed when using the mostly more poisonous amines. Freons are neither poisonous nor flammable, require however pressures of the same magnitude as ammonia. Their enthalpy of evaporation is fairly low, thus larger pumps are needed. The refrigerant R22 is superior to most other refrigerants with respect to its chemical stability and also with respect to corrosion [10]. Alcolhols are also suitable for absorption heat pumps. Methanol for example has a high enthalpy of evaporation and contrary to water a low melting point (- 97 °C). It is however poisonous. Also ketones may be discussed as working substances.

Due to its high polarity water may also be used as solvent, especially because of its relatively high boiling point of 100 °C at 1 bar. Mostly salts are taken as solutes.
The most important advantage of these mixtures is that the pure working substance leaves the generator and a rectification column is not needed. Instead of water also ammonia, methylamine and methanole can be used as solvents. Furthermore some other high boiling solvents consisting of organic compounds have been proposed in the literature. Examples are dimethylformamid [11] and tetraethylenglycoldimethylether (E 181) [12].

Summarizing the results, it is noteworthy that aqueous salt solutions are often very corrosive, an effect which can be reduced in some cases through inhibitors. Within the group of refrigerants in organic solvents, especially mixtures with E 181 are advantageous due to the high boiling point of E 181 (548 K), which makes a rectification column unnecessary. When using R 21 or R22 as working substances one has to take into account higher pressures because of the low boiling temperature (282 K resp. 232 K). Though the refrigerant R 21 allows lower pressures, it should not be considered as working substance, because it corrodes steel and copper at temperatures above 100 °C.

4. SOME SUITABLE MIXTURES

Among the 70 binary mixtures considered a few seemed to be favorable and were studied in detail in order to determine the extent to which the conditions described in chapter 2.
Unfortunately it turned out that the mixtures also had to be selected from the aspect of availability of measured thermal properties or whether these properties could be calculated. The investigation included a study of the properties of the working substance, thermal properties of the solvent, properties of the mixture and of the process. The following mixtures were studied :

1. ammonia-water
2. R 22 - E 181
3. methylamine-water
4. water-lithiumbromide
5. methanol-lithiumbromide
6. ammonia-lithiumnitrate
7. ammonia-sodiumthiocyanide
8. methanol-water
9. methanol-dimethylformamide
10. methanol - E 181
11. R 22 - dimethylformamide

The details are given in the paper by Stephan and Seher [9]. Only the essential results shall be discussed here.
When using the well known mixture ammonia-water, where the difference in atmospheric boiling points is only 133,6 K, a rectification column is required. On the other hand there exists an experience of long standing with ammonia-water mixtures in refrigeration. The thermal properties of ammonia and water are also well known. Only in the pressure range above 25 bar the thermal and caloric properties still have to be determined.

With the refrigerant R 22 similar pressures than with ammonia are obtained. The entalpy of evaporation of R 22 however is low, only 1/6 of ammonia. As a consequence large amounts of solvent must be circulated, which requires a relatively high power for the liquid pump. The critical temperature of R 22 (96°C) is very low. It is easily soluble in E 181 tetraethylenglycoldimethylether) as solvent. Working substance and solvent are not poisonous and chemically stable. A rectification column is unnecessary.

Methylamine also stands out for its high enthalpy of evaporation, which is about 2/3 of that of ammonia. Its vapor pressure curve is very flat. As a consequence it can be used at moderate pressures within a wide temperature range and insofar it is superior to ammonia. The difference in atmospheric boiling points, however (107 °C) is smaller than in ammonia-water mixtures and therefore a larger rectification column is required. Methylamine is chemically stable, only slightly corrosive, leakage can easily be detected due to its intensive smell. It is, however, poisonous and only 10 ppm are allowed.

As soon as the temperature of the environment approaches the freezing point of water, the viscosity of aquaous solutions increases rapidly together with the risk of frost formation in the condenser. Therefore the mixture water-lithiumbromide often used for refrigeration and air conditioning processes proves to be nonsuitable for absorption heat pumps. On the other hand at higher temperatures its use is limited due to the crystallisation of lithiumbromide.

In mixtures of methanol and lithiumbromide the methanol serves as the working substance. Its low freezing temperature (- 97 °C) evades the risk of solidification in the condenser. The enthalpy of evaporation is fairly high (1105 kJ/kg). However, methanol is poisonous, only 200 ppm are allowed, and it is easily ignited. Also the solubility decreases with temperature and at sufficient high temperatures crystals are formed.

In ammonia-salt solutions the solubility is higher than in methanol-lithiumbromide. Ammonia exhibits a high enthalpy of evaporation. On the other hand high pressures become unavoidable.

Furthermore ammonia is also poisonous, only 50 ppm are allowed, i.e. 1/4 of methanol and in certain concentration ratios with air it is flammable.

Mixtures consisting of methanol-dimethylformamide or of methanol-E 181 also seem to be favorable. Further experiments are still necessary for assessment of their solubility and their heat of mixing.

Also for methylamine-salt-solutions a rectification column is not required. For these solutions the solubility at higher temperatures, however, has not yet been explored.

With respect to heat pumps working at higher temperatures, substances with a sufficiently high boiling point, for instance water, are favorable. High boiling organic compounds may be used as solvents. Another possibility is the use of ternary mixtures. By addition of a further salt the solubility in a binary salt solution may be increased and thus crystallisation can be avoided. The effect is clearly indicated in a paper by Macriss [13].

Another group of ternary mixtures are solutions of a working substance in a binary mixture as solvent. By addition of a second solvent to a binary mixture consisting of working substance and solvent in some cases the solubility of the working substance can be increased. Thus higher temperatures can be achieved. Also the opposite cases, namely a binary mixture as working substance and a solvent consisting of a single component, or of two components, are conceivable. Thus we have a ternary or a quaternary mixture. When using these mixtures the temperature difference during the heat transfer in the evaporator and the absorber is shifting and the average temperature difference in both apparatus is lower than in the case of heat exchange between pure fluids. The losses of available energy therefore decrease and the efficiency may be remarkably increased. So far, however, little is known on the thermal properties of ternary and quaternary mixtures for their use in absorption heat pumps or absorption transformers and further investigations are necessary.

Nevertheless, the above results clearly indicate, that there exist a great variety of mixtures, worthwhile to be discussed. On the other hand a mixture, which is superior to all the others and therefore should be preferred, cannot be recommended. All mixtures which are at least in principle suitable for absorption heat pumps and heat transformers exhibit disadvantages and the choice of a mixture always requires a compromise, which may lead to different results according to the purpose of the absorption process.

REFERENCES

1. Braun, R. and Heß, R. *Brennstoff-Wärme-Kraft*, 29 (1977) No.8, 305.

2. Plank, R. *Handbuch der Kältetechnik, Band 7. Sorptions-Kältemaschinen* (Berlin, Springer Verlag 1957).

3. Renz, M. "Eignung von Arbeitsstoffpaaren für Absorptionswärmepumpenprozesse" *Wärmepumpen* (Essen, Vulkanverlag 1978).

4. Hommel *Handbuch der gefährlichen Güter* 2 (Berlin, Springer Verlag 1978).

5. Roth, Daunderer *Giftliste* (Verlag Moderne Industrie 1978).

6. *Hazardous Chemical Data 1976*, NFPA, no 49 (Boston, National Fire Protection Assoc. 1976).

7. *Manual of Hazardous Chemical Reactions*, NFPA, No. 491, Boston National Fire Protection Assoc. 1976).

8. Sax, N.I. *Dangerous properties of industrial materials*, (New York,Reinhold Publ. Corp. 3rd ed. 1968).

9. Stephan, K. and Seher, D. *Kälte-, Klima-Ing.* 1 (1980) 865.

10 Eisemann, B.J. "Why Refrigerant 22 should be favored for Absorption Refrigeration" *ASHRAE-J* (Dec. 1959) 45.

11 Borde, I., Jelinek, M. and Yaron, I. 84 *ASHRAE-Transact* (1978).

12 Kriebel, M., Löffler, H. *Kaltetechnik* 17 (1965) 266 ff.

13 Macriss, R.A. "Selecting Refrigerant-Absorbent Fluid Systems for Solar Energy Utilisation" 82 *ASHRAE-Transact*, 975.

TABLE I

Boiling point of some alkanes

Substance	molar mass kg/kmol	boiling point in K at 1 bar	enthalpy of evaporation at 1 bar in kJ/kg
Methane	16	112	510,2
Ethane	30	185	489,7
Propane	44	231	426,1
Butane	58	272	385,5

TABLE II

Boiling points of polar substances

Substance	molar mass kg/kmol	boiling point in K at 1 bar	enthalpy of evaporation in kJ/kg at 1 bar
CH_4	16	112	510,2 ↓ polarity increase
NH_3	17	240	1369
H_2O	18	373	2256

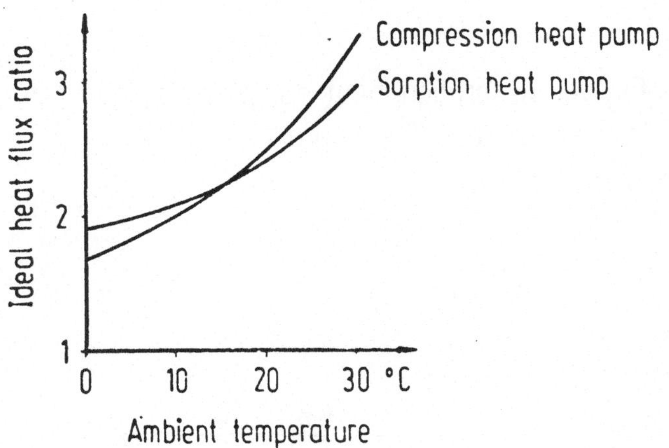

Fig 1 Ideal heat flux ratio as a function of ambient temperature for compression heat pump and for absorption heat pump

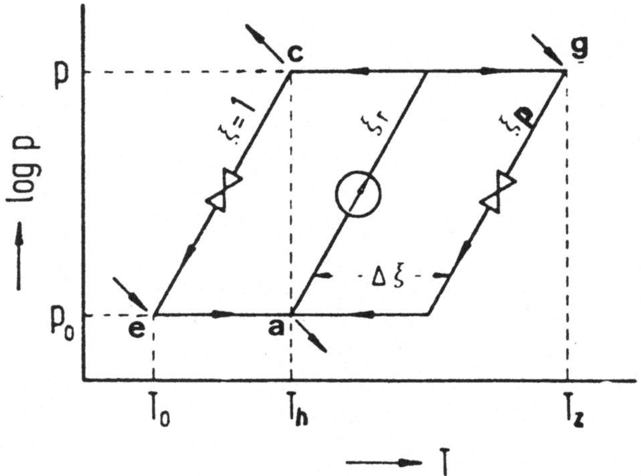

Fig 2: Absorption heat pump cycle in log p, 1/T - diagram.

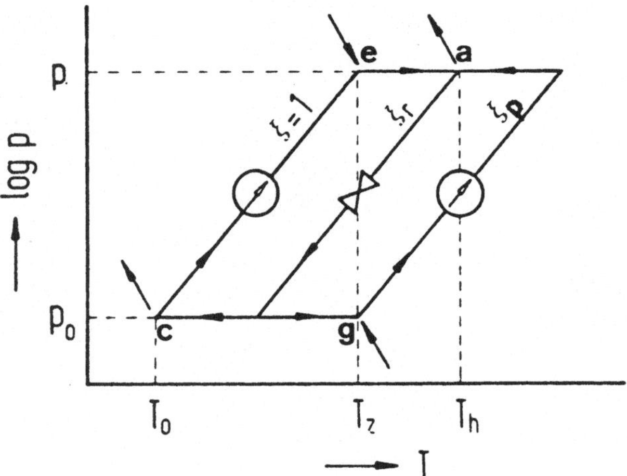

Fig. 3: Absorption heat transformer cycle in log p, 1/T - diagram.

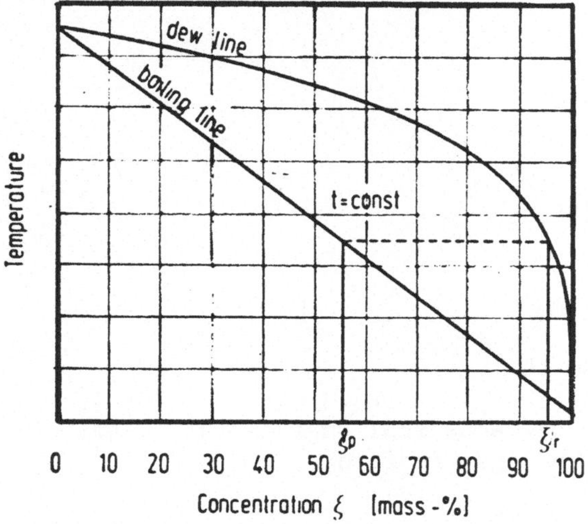

Fig 4 Ideal shape of boiling line and dew line in a T, ξ diagram

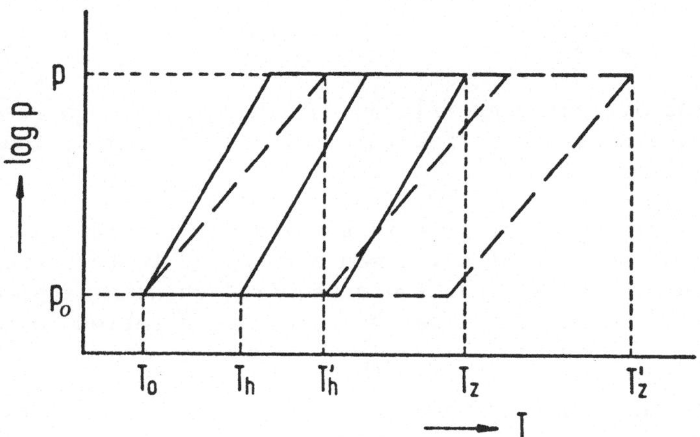

Fig. 5: Comparison of processes working between steep and flat vapor pressure curves.

HEAT PUMP ABSORBERS

I.E. Smith

School of Mechanical Engineering, Cranfield Institute of Technology, Bedford, U.K.

1. FUNDAMENTAL PRINCIPLES

The rate of absorption of a pure gas or vapour in a liquid is governed by the rate of diffusion of the absorbed gas within the liquid.

This diffusion can be expressed by :

$$R = - D. (\partial a/\partial x) \tag{1}$$

where D is the diffusion coefficient of the gas through the liquid (m^2/s) and a is the concentration ($mole/m^3$). R is thus expressed in $mole/m^2.s$.

For a gas undergoing absorption in a quiescent liquid, the surface concentration will at all times correspond to the saturation or equilibrium value, A^x, and the concentration at a large distance below the surface will be the bulk concentration in the liquid, A^o.

The concentration gradient immediately below the surface is given by :

$$da/dx = - (A^x - A^o)/\delta , \tag{2}$$

where δ is the thickness of the diffusion film, which clearly will be dependant on time.

A solution of these equations, with appropriate boundary conditions provides an expression of the absorption rate :

$$R = (A^x - A^o)\sqrt{D/\pi t} \qquad (3)$$

and an integration of this rate w.r.t. time provides a figure for the total quantity of gas absorbed after a given time interval.

$$M = \int_o^t R dt = 2\,(A^x - A^o)\sqrt{D.t/\pi} \qquad (4)$$

Thus for a gas having a diffusivity of 2 x 10^{-9} m^2/s (a typical value) diffusing into a gas-free liquid, after one hour the amount absorbed will only be equivalent to a saturated liquid layer 2.5 mm thick. The actual concentration profiles for short time intervals are shown in Figure 1 [1].

This small number amply illustrates the problem of the absorption of gases by liquids, and highlights the fact that only very thin, rapidly changing, liquid films are practicable if significant quantities of gas are to be absorbed in a given time.

In deriving equations (3) and (4) it has been assumed that the equilibrium concentration at the surface, A^x, is unaffected by the absorption process; in fact this is not so in the case of absorption heat pumps. As absorption occurs, heat is released and this leads to the establishment of a temperature gradient in the liquid as well as a concentration gradient. An increase in surface temperature leads, furthermore, to a lowering of A^x from the initial value, and this new A^x has to be calculated if equation (4) is to be employed meaningfully.

The rate of heat liberation at the surface will be given by :

$$H_s = RH \qquad (5)$$

where H is the heat of absorption of the gas (J/mole) and so

$$H = (A^x - A^o)\,H\sqrt{D/\pi t}$$

from equation (3).

The surface temperature rise, T^x, can be calculated for a given heat flux, H, knowing the thermal properties of the fluid. It is given by :

$$T^x = (A^x - A^o)\ H\ \sqrt{D/k\rho c} \qquad (6)$$

Thus for ammonia vapour being absorbed in water to 50 % concentration with $D = 2 \times 10^{-9}$, $k = 0.6$ w/m K etc, a 4 °C surface temperature rise will be encountered, which will significantly reduce A^x to less than the equilibrium value for the non-absorbing solution. This elevation of the surface temperature will not, of course, vary with time since the rate of absorption diminishes and the distance for heat conduction increases as a result of the reducing concentration gradient. So only a single set of iterations are necessary to establish the real value of A^x for a solution at a given bulk temperature.

Whilst it is apparant that a quiescent liquid layer presents few prospects in the context of a practical absorber, nevertheless quiescent film experiments can provide useful measurements of the absorption coefficient, k_L, and this provides a starting point for the design of practical absorbers.

For quasi-steady state systems containing quiescent elements that are periodically renewed or mixed with the bulk liquid, the absorption coefficient may be taken as

$$k_L = 2\sqrt{D/\pi t^x} \qquad (7)$$

where t^x is the lifetime of a liquid surface element.

Many steady and quasi-steady-state quiescent film absorbers have been described, but these will not be discussed here. However, surface element lifetimes have been evaluated for a number of configurations that are of practical interest in the design of absorbers, which enable the overall performance of the equipment to be calculated via equation (7).

For a falling laminar liquid jet of diameter d_j, height Z_j and volume flow rate G_L

$$t^x = \pi d_j^2\ Z_j/G_L \qquad (8)$$

whence

$$k_L = \frac{4}{\pi d_j}\sqrt{\frac{D.G_L}{Z_j}} \quad \text{(ref. [4])} \qquad (9)$$

An interesting aspect of this configuration is that since the total rate of absorption contains the product of k_L and the jet surface area, $\pi d_j Z_j$, the rate of absorption is independant of the jet diameter.

For the more practical case of a wetted wall column with laminar flow [4] :

$$t^x = \frac{2}{3} Z_c \left[\left(\frac{\pi d_c}{G_L}\right)^2 \frac{3\mu}{g\rho} \right]^{1/3} \tag{10}$$

where d_c is the column diameter, and for a wetted plate (one side) the plate length replaces the term πd_c.

It should be noted that for a wetted wall

$$k_L \propto G_L^{1/3}$$

whereas for a laminar jet

$$k_L \propto G_L^{1/2} .$$

In the case of absorbers in which the gas is bubbled through the solution, the element lifetime can be approximated as :

$$t^x = d / u_b \tag{11}$$

where d is the bubble diameter or length and u_b its velocity. Strictly speaking a factor of $\pi/2$ should preceed this expression if the bubble is spherical, but in forced or agitated bubbling systems bubble elongation frequently takes place.

For naturally rising bubbles the velocity is proportional to $d^{1/2}$ and therefore t^x varies as the square root of the diameter. Clearly bubbles having small diameter are desirable.

2. PRACTICAL ABSORBERS

A number of practical absorbers have been described in reference [2] and these have been reproduced.

The surface absorber (Figure 2) employs a semi-stagnant liquid in which is immersed a heat extracting coil. Although it operates isothermally, the surface area exposed for absorption is small in relation to the size of the equipment, and furthermore the average rate of absorption must be low for reasons already outlined previously. Only by employing a high ratio of liquid circulation/liquid content can a reasonable absorption rate be realized, but a low liquid content conflicts with the need for a reasonable area of the immersed cooling coil.

The packed bed (Figure 3) is an adiabatic system, and although packed towers are capable of providing both a large surface area per unit volume and a rapid renewal of the liquid film, they suffer the severe disadvantage that A^x diminishes as the absorbent temperature rises. They also require a separate heat exchanger and pump to return the cooled solution to the column. Whilst the temperature increase can be minimised by employing a high liquid circulation rate, the tower design provides a limitation in this respect in that liquid flooding occcurs. Data on permissable rates for a variety of packings are reviewed in reference [5]. It must not be forgotten that whilst packed beds or towers are simple in concept, there are problems in maintaining a uniform distribution of the liquid, and redistributors are required at regular intervals.

Although Figure 3 suggests that the gas and solution should be in counter flow, when dealing with a single vapour component there is no virtue in adopting such a flow arrangement.

In the immersion absorber (Figure 4) the gas is bubbled into the absorbing solution and the bubbles ascend through a finite depth providing a useful agitation of the liquid and enabling the removal of heat to be effected via the cooling coil so that isothermal conditions can be achieved.

For moderate bubbling rates the bubble diameter is related to the orifice diameter by the expression

$$d_b = (3\ do\ \sigma_L/4\ \rho_L\ g)^{1/3}$$

where σ_L and ρ_L are the liquid surface tension and density respectively. If the bubble formation rate exceeds the rise rate then coalescence of the bubbles may take place which reduces the surface area of exposed solution. This critical rate may easily be calculated from standard relationships for the bubble rise velocity.

If a bubbling rate greater than the critical is called for and mechanical devices can be indulged in them the Turbogas Absorber (Figure 5) can provide a solution. An open or semi-open impeller breaks up the gas stream which is directed against a target into fine bubbles, and if guide vanes are provided the bubbles may be directed downwards against the gravitational force, thus increasing their residence time. The function of the target is an important one, for if the vapour is directed at the centre of the impeller very large bubbles tend to be formed.

The gas handling capacity (volumetric) of turbo absorbers is given by an expression of the form [5] :

$$V = 7.7 \times 10^{-4} N^{2.5} d^{4.5}$$

where N is the rotational speed (s^{-1}) and d is the diameter (m).

In general bubble absorbers are restricted to systems where the pressure is moderate or high. At low pressures the sheer volume of vapour that must be handled leads to absorbers of impracticable size, and furthermore the hydrostatic head imposed on the vapour by the liquid severely attenuates the temperature difference that is possible between the evaporator and the absorber. The most commonly employed arrangement is the falling film or dribble absorber (Figures 6 and 7) which consists of a series of heat transfer surfaces over which the absorbent trickles.

The surface may be continuous, i.e., plate-like, or intermittent as shown in the illustrations. However, whatever form it takes it is vital that the liquid layer is frequently mixed in order to bring fresh solution to the absorbing surface, and maintain Z_c (equation(10)) as low as possible. Such absorbers operate isothermally with heat being transferred through the (normally) laminar layer of liquid into the heat absorbing surface.

The capacity of such absorbers may be limited by heat transfer consideration within the liquid film, for the liquid layer thickness (not to be confused with the absorbing film thickness) is given by

$$\delta = (3\, \mu_L\, G_L/\rho_L . \ell . g)^{1/3} \quad ,$$

and the Nusselt number, h δ/k, is numerically equal to about 4.

However, if the flow is not fully developed, as may frequently be the case if the height of the elements is not very great, considerably greater Nusselt numbers may be encountered [7]. Much ingenuity has been employed in the design of gas/liquid contacting systems with simultaneous heat removal and liquid mixing. Obtaining a uniform wetting of the elements is frequently a problem, and great lengths are taken to ensure that this occurs.

One proprietary unit, the Polyblock Absorber* is illustrated which

---

* Polyblock is the registered trade mark of absorbers manufactured by Robert Jenkins Systems Ltd., Rotherham, England.

utilises graphite blocks having vertical gas/liquid passages and radial coolant passages. Although relatively expensive as a material, graphite has the advantage of easy machinability and a complete resistance to virtually all substances that are likely to be encountered in heat pumps (see figure 8).

3. ADIABATIC ABSORBERS

Whilst most absorbers employed in the past have been designed for isothermal operation (heat transfer occurring simultaneously with mass transfer), there is no fundamental reason why an adiabatic absorbing system (e.g. the packed tower) should not be employed with the liquid being circulated through a separate heat exchanger.

Whilst the absorber performance must be degraded by the temperature rise that takes place in the solution (accompanied by the concentration change) if a sufficiently high liquid circulation rate can be tolerated this will be minimal.

Spray chambers offer one possibility, although there are problems of attaining a high spray concentration per unit volume, there are others. Spinning disc atomisers are attractive in that their power requirements are modest and the droplet size is independant of the liquid viscosity. However, they tend to produce a two dimensional sheet of droplets, making for large linear dimensions in the absorber.

Partially immersed rotating discs or gauzes offer another possibility of providing compact absorbers with the negligible motive power required coming from the solution flow.

There is much scope for ingenuity in the design of low cost, compact absorption units given that counterflow between the liquid and gas is not a requirement.

4. SOLUTION HEAT EXCHANGERS

It is seldom possible to carry out the designs of a component in isolation from the system as a whole and the absorber in a heat pump is no exception.

The generator/absorber solution flow rate is inversly related to the concentration change that takes place in the absorber. However, the greater the absorber concentration change, the larger the absorber must be since the average value of (A^x-A^o) is lower. On the other hand the size of the solution heat exchanger is directly related to the circulation rate. Thus there exists a trade-off situation between the absorber and the solution heat exchanger dimensions.

It must be stressed that the solution heat exchanger is a component of major importance, minimising, as it does, the irreversible transfer of heat from the generator to the absorber. For a typical heat pump design the capacity of this heat exchanger in relation to the total heat output of the system is illustrated in Figure 9 [6] . For absorber/generator concentration changes of the order of 1 % (implying a minimum absorber area) it can be seen that the heat exchanger must have a capacity 4 times that of the system output. If a 5 % concentration change is acceptable then the heat exchanger need only have a capacity similar to the system output, but the absorber area will have to be increased in area correspondingly.

Design calculations have suggested that absorber and solution heat exchangers may not have dissimilar areas, and the relationship that is the optimum will depend on economic considerations.

Not only do the economics involve the heat exchanger dimensions, but also the pumping power requirements. The solution viscosity exerts a dominant influence and for very viscous solutions laminar flow plate type heat exchangers may provide the best answer.

Since the fluids that deserve consideration for use in heat pumps vary widely in their physical properties as well as operating pressure levels it is impossible to generalise in respect of designs, and each must be examined on its merits.

REFERENCES

1. Dankwerts, P.V. *Gas-Liquid Reactions* (McGraw-Hill, 1967).

2. *Handbuch der Kältetechnik*, vol. 7 (1959)

3. Treybal,R.E. *Mass Transfer Operations*, 2nd Ed (McGraw-Hill, 1968).

4. Astarita *Mass Transfer with Chemical Reaction* (Elsevier, 1967).

5. Perry and Chilton *Chemical Engineers Handbook*, 5th Ed (McGraw-Hill/Kogakusha).

6. Smith and El-Shamarka "Absorption Heat Pumps for Space Heating" *I.E.E. Energy Conference* 1981.

7. Knudsden and Katz *Fluid Dynamics and Heat Transfer* (McGraw-Hill/Kogakusha, 1958).

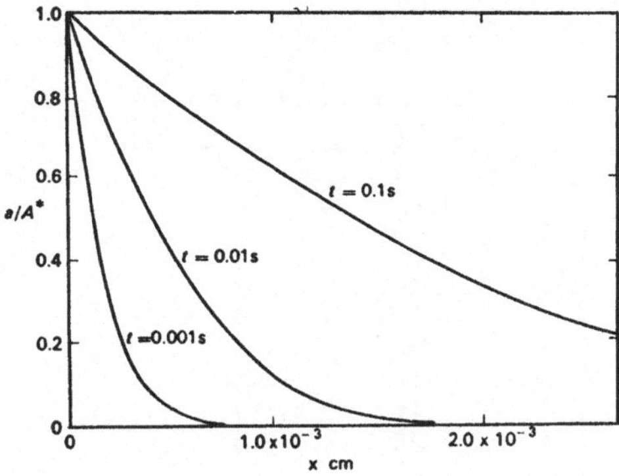

Figure 1 : Concentration-profiles for the absorption of a gas into water

c : cooler
rs : rich solution
ps : poor solution
v : vapour
w : cooling water

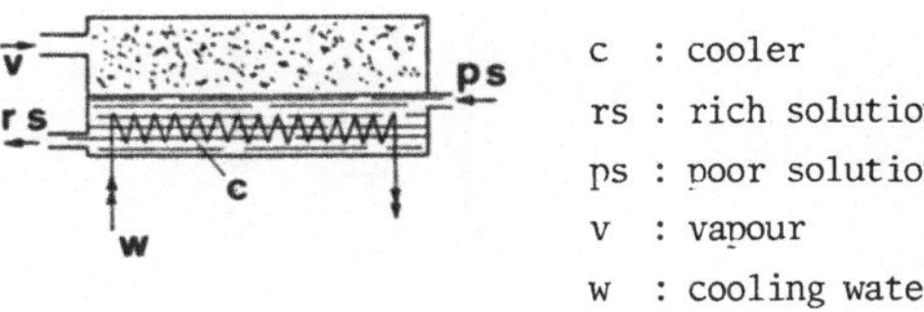

Figure 2 : Surface Absorber

Figure 3 : Packed Column

Figure 4 : Bubble Absorber

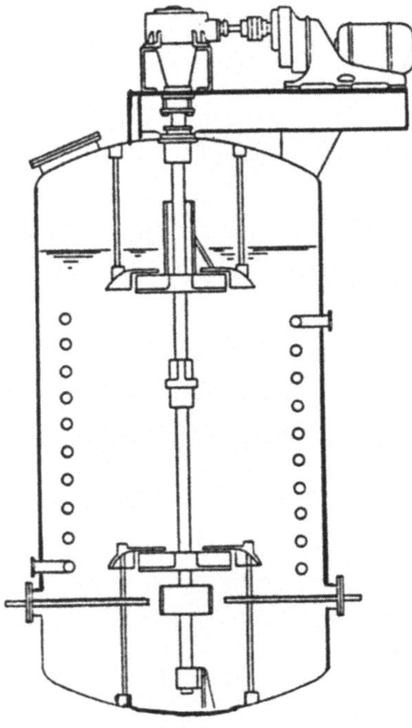

Figure 5

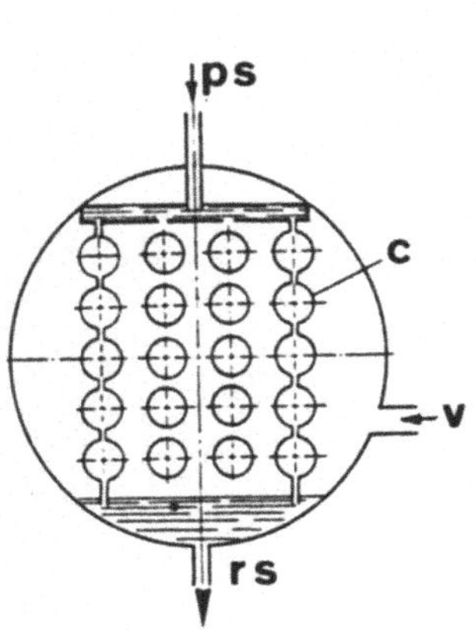

Figure 6 : Film Absorber

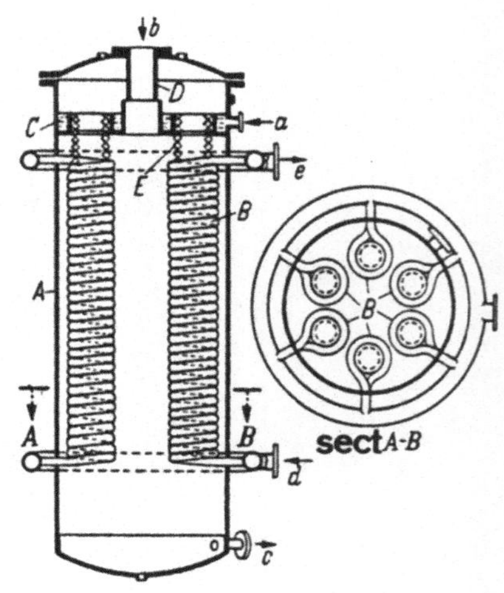

Figure 7

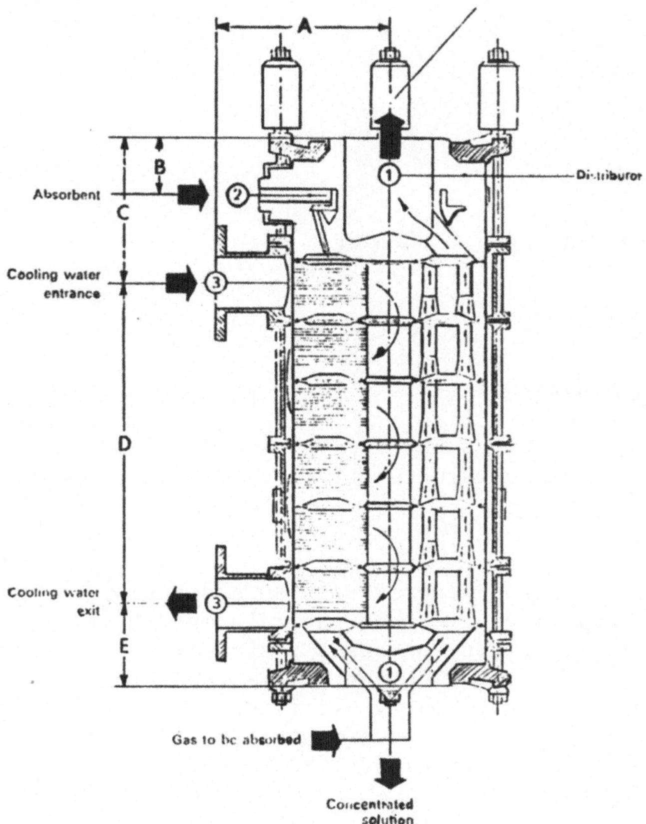

Figure 8

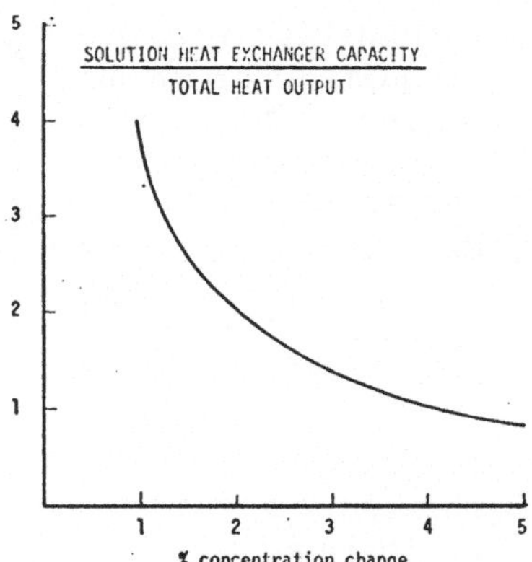

Figure 9

ABSORPTION HEAT PUMP GENERATORS

A.M.S. Qasrawi

Lucas Group Services Limited
Lucas Research Centre, West Midlands, United Kingdom

The function of the generator in an absorption cycle heat pump is described. The effect of the relative volatility of the working fluid mixture on the basic design of the generator is considered. Analysis and design of the two main types of rectifying columns used for binary distillation (plate or tray columns and packed columns) are then described with emphasis on the more important design parameters. The need for accurate thermodynamic data is first stressed. For plate columns, the methods of calculating the number of theoretical plates (using equilibrium - stage construction techniques) required for a specified separation, and the number of actual trays needed are outlined.

The analysis of the mass transfer process in packed columns, and the calculations of the required column height to achieve a specific separation are described. Finally the functions of the reboiler and the partial condenser are described and methods of heat transfer analysis for these two components are outlined with comments on the likely mass transfer performance.

1. INTRODUCTION

The generator, in an absorption cycle heat pump, is the component or group of components that serves to separate an absorbent solution, strong in the refrigerant, which is fed into the generator from the absorber, into a pure (or almost pure) refrigerant vapour stream and an absorbent solution stream weak in the refrigerant. The degree of separation is characterized by the difference between the weight concentrations, in the refrigerant, of the strong solution feed and refrigerant streams.

Binary mixtures used in absorption cycle heat pumps are usually separated by a distillation process which utilises high-grade heat directly. Distillation utilises vapour and liquid phases at essentially the same pressure and temperature as co-existing zones in which different environments are offered. Each molecular species in the mixture to be separated will establish, at equilibrium, a different concentration in each zone. Relative volatility of both components of the mixture is one of the important factors to be considered when selecting a working fluid pair. For a mixture with a high relative volatility (an essentially non-volatile absorbent; e.g. E181 or Lithium Bromide) distillation could be achieved simply by boiling a pool of solution in a heated still without a need for any further rectification of the ensuing vapour. For working fluid mixtures with moderate or low relative volatility (e.g. ammonia-water), it is necessary to further rectify the vapour that rises from the boiling pool of solution, since it contains a large proportion of the less volatile component (the solvent).

However in certain applications of water-ammonia absorption machines where the strong solution has a sufficiently high concentration, it is possible that the vapour in equilibrium with it need only to be passed in a partial condenser (rectifier) to produce refrigerant of adequate concentration for the desired cooling effect. In such machines, the generator consists of a liquid-filled tower with the strong solution entering at the top and the leaving vapour is caused to be in intimate contact, and hopefully therefore closely in equilibrium with it. The weak solution is extracted from the bottom of the tower, and the unit must be designed such that a liquid-concentration gradient exists from the top to the bottom. It is to be noted that the density gradient assists the establishment of a concentration gradient. The tower is heated at its lower end. The leaving hot solution is very often extracted via a vertical pipe passing up through the boiling pool to the top of the column, as shown in Figure 1.b. This achieves some degree of heat transfer between the leaving liquid and the boiling pool. This design of generator (known as analyser) will not be suitable at difficult running conditions when the feed may be unable to produce sufficiently-concentrated vapour, resulting in either an excessive rectifier load or an inability to achieve the required separation.

Various devices are used to bring the liquid and vapour phases into intimate contact; plates or trays are stacked one above the other and enclosed in a cylindrical shell to form a column. Distillation columns could alternatively be packed with a packing material that offers a large interfacial area for intimate contacting of the two phases.

Emphasis will be put in this lecture on the analysis and design of generators for mixtures of moderate relative volatility requiring substantial rectification.

A typical generator utilising a plate rectifying column is shown schematically in Figure 1a. The feed solution is introduced at one point along the column shell. The liquid runs down the column, cascading from plate to plate, while the vapour goes up the column contacting the liquid at each tray (plate). The liquid reaching the bottom of the column is partially vapourised in a heated still to provide reboil vapour which is sent back up the column. The remainder of the bottom liquid is withdrawn as the bottom product. The vapour reaching the top of the column is cooled and partially condensed to a liquid in the partial condenser (rectifier or dephlegmator). The condensate is returned to the column as reflux to provide liquid overflow. The remainder of the rising vapour steam is withdrawn to be further cooled and totally condensed in the heat pump total condenser. The function and performance of the partial condenser will be described later.

Countercurrent contacting of vapour and liquid streams occurs throughout the column, and vapour and liquid phases on a given plate approach thermal pressure and composition equilibrium to an extent dependent upon the efficiency of the contacting tray. The refrigerant (lower-boiling component) tends to concentrate in the vapour phase while the absorbent (higher-boiling component) tends towards the liquid phase. The result is a vapour phase which becomes richer in the refrigerant as it passes up the column and a liquid phase which becomes weaker in the refrigerant as it falls down.

The energy and mass-transfer processes in an actual distillation column (whether it is of the tray column or packed column types) are much too complicated to be readily modelled in any direct way. This difficulty is circumvented by the use of the equilibrium stage model for plate columns and mass transfer unit model for packed columns. In an equilibrium-stage, the leaving vapour and liquid streams are assumed to be in complete equilibrium with each other, and thermodynamic relationships can be used to relate the concentrations in the two streams.

The use of the equilibrium stage model or mass transfer unit concept separates the design of a distillation column into three main parts. The first part is concerned with the thermodynamic data methods needed to predict the equilibrium - phase compositions. The second part involves the calculation of a number of equilibrium stages or number of transfer units required to achieve a specified separation. The third part deals with the conversion of the number of equilibrium stages or transfer units to an equivalent number of actual contact trays or height of a packed column.

The rest of this lecture will be devoted to the description of the above mentioned parts.

2. THERMODYNAMIC DATA

Good thermodynamic data are essential for the accurate design and analysis of distillation columns. Failure of equipment to perform at specified levels is often attributable, at least in part, to the lack of such data.

Ideal-gas model for the behaviour of the individual components in the vapour phase is often assumed when predicting phase-equilibrium data. The adequacy of this model for design purposes depend on; the pressure levels, the complexity of molecules involved and the difficulty of separation to be achieved. For complex molecules,the deviation from ideal-gas behaviour can become non-negligible at relatively low pressures (7 to 10 bars). For ordinary cases of separation where volatilities are not close to each other, the deviation from the ideal-gas model is not important.

It is not yet possible to predict binary mixtures data from pure components data alone. However, it is possible to predict data over the entire binary composition range from only limited experimental data.

Accurate experimental vapour-liquid equilibrium data are difficult to obtain. Certain checks on the accuracy of data reported in the literature show that 25 to 90 % of it are of doubtful quality [1].

The best bibliography for vapour-liquid equilibrium data is given in Reference 2. Reference 3 lists the experimental x, y, p, t data for about five hundred selected binary and multi-component systems and tabulates the constants for five correlation equations.

For binary mixtures that could be used in absorption cycle heat pumps, ammonia-water is the combination that has been covered most extensively. Even then, the available data has to be extrapolated outside their limits for use at the operating pressures and temperatures encountered in heat pump cycles.

3. BINARY DISTILLATION METHODS

The design of multi-stage distillation columns can be accomplished by graphical techniques when the feed mixture contains only two components. Two diagrams are used; the x - y diagram (concentration of the liquid phase - concentration of the vapour phase in one of the two components) and the enthalpy-concentration diagram. The x-y diagram utilises only the equilibrium and mass-balance relationship and is accurate only when energy effects are negligible. The enthalpy concentration diagram utilises energy balance also and is accurate when enough calorimetric data are available to construct the diagram without assumptions. Both diagrams are useful for quick approximations and for demonstrating the effect of the various design variables. The x-y diagram is the more convenient and will be described in the following sections for these purposes.

3.1. Equilibrium Data

The x-y diagram is usually plotted for the more volatile component in either mole, weight or volume fractions. Figure 2 shows the diagram for the most usual case where component 1 remains more volatile over the entire compositions range.

The x-y diagram can be plotted at either constant temperature or constant pressure but not both, since the phase rule permits only two variables to be specified arbitrarily in a binary, two phase system at equilibrium.

The equilibrium curve can be approximated quickly by the assumption of constant relative volatility. The relative volatility of component 1 to component 2 is defined as

$$\alpha_{1-2} = \frac{K_1}{K_2} = \frac{y_1\, x_2}{x_1\, y_2} \tag{3.1}$$

where K_1 and K_2 are the distribution coefficients; which can be rewritten as

$$y_1 = \frac{x_1\, \alpha}{1 + (\alpha - 1)x_1} \tag{3.2}$$

3.2. Material Balance Equations

Material balance equations for a binary system could be represented as straight lines on the x-y diagram if the constant molal overflow assumption was made, which obviates the need for an energy balance. The liquid phase rate is assumed to be constant from plate to plate in each section of the column above or below the feed point. This results in the vapour rate also being constant in these sections of the column.

The assumption of constant molal overflow implies the following prior assumptions;equal molal heats of vapourisation for two components; adiabatic column operation and absence of heat of mixing or sensible heat effects.

The effect of the assumption on the calculation method can be illustrated with Figure 3 which shows a material balance envelope cutting through the top section above the feed.

Since liquid phase and vapour phase rates are assumed to be constant the component material balance for any such envelope can be represented by

$$y_n = \frac{L}{V} x_{n+1} + \frac{Vo}{V} y_o \tag{3.3}$$

where x and y have stage subscripts but L and V are identified only with the appropriate section of the column. Equation (3.3) represents a straight line with a slope of $\frac{L}{V}$ and y interest of $\frac{Vo\ y_o}{V}$ at $x_1 = 0$. This line on the x-y diagram is called an operating line and each point on it represents two passing streams at a certain point in that section of the column. The slope of the operating line is termed the internal reflux ratio. The ratio in the top section of the column is related to the external reflux ratio $R = \frac{L_{N+1}}{Vo}$ by :

$$\frac{L}{V} = \frac{L_{N+1}}{V_N} = \frac{RVo}{(1+R)Vo} = \frac{R}{R+1} \tag{3.4}$$

when the reflux stream L_{N+1} is a saturated liquid.

The slope of the operating line will change whenever a feed is passed. For a feed stream F a quantity q is introduced to define the thermal condition of the feed. It represents the moles of saturated liquid formed in the feed stage per mole of feed. q takes the following values for the various possible feed states.

Subcooled liquid $q > 1$
Saturated liquid $q = 1$
Partially flashed $1 > q > 0$
Saturated vapour $q = 0$
Superheated vapour $q < 0$

The liquid and vapour rates below the feed, L' and V', can be related to the corresponding rates above the feed, L and V, by :

$$L' = L + q\,F \tag{3.5}$$

$$V = V' + (1 - q)\,F \tag{3.6}$$

The quantity q can be used to derive the q-line equation for the feed, which is the locus of all points of intersection of the two operating lines which meet at the feed, as shown in Figure 6. It is easily shown that the q-line equation is given by

$$y = \frac{q}{q - 1}\,X - \frac{X_F}{q - 1} \tag{3.7}$$

and that the q-line intersects the diagonal at X_F.

Figure 4 shows a material balance envelope cutting through the bottom end of the column with the partial reboiler (still) representing the first equilibrium stage. For this envelope the material balance is represented by

$$y_m = \frac{L''}{V''}\,X_{m+1} - \frac{BX_B}{V''} \tag{3.8}$$

3.3. Plate or Tray Columns

3.3.1. Equilibrium - Stage Construction

The alternate use of the equilibrium curve and the operating line for construction of equilibrium stages is illustrated in Figure 5. Starting from the known compositions y_n, x_n at the stage n it is possible as shown to define the compositions at either stage n-1 or n+1.

The graphical construction of an entire column consisting of six equilibrium stages, a partial reboiler (still) and a partial condenser is shown in Figure 6. Each of the partial reboiler and partial condenser represents an equilibrium stage.

The pressure drops within the column are neglected and the equilibrium curve represents the data at the working pressure of the generator system. Heat losses are assumed to be zero.

The feed concentration, rate, location and state, and product rate B are known. The leaving vapour purity is specified. The external reflux ratio is set and the reflux is assumed to be saturated liquid in equilibrium with the leaving vapour V_o. The solution of the problem provides the bottom concentration and the top vapour rate V.

From the external reflux ratio, the internal reflux ratio L/D is calculated for the rectifying part of the column (above the feed). With X_D fixed, a material balance around the partial condenser provides the value of V and y_6. The top operating line can then be drawn. The value X_B which gives exactly six stages can only be found by trial and error. If X_B was known and the required number of equilibrium stages is to be found the bottom operating line could be drawn immediately since L'/V' could be calculated from L/V, the feed rate and state. If one arrives at a fractional number of equilibrium stages, this will be divided by a tray or plate efficiency and the resulting figure is rounded to an integral number of actual plates. An overall component balance gives the values of Vo and V.

Feed Location. The optimum feed location is that location which, given a set of operating conditions, will result in the widest separation between y_o and X_B with a given number of equilibrium stages or the one which requires the lowest number of stages to accomplish a specified separation. Either condition could be satisfied if the operating line farthest from the equilibrium curve is used for each stage.

A badly mislocated feed can be very wasteful in so far as the effectiveness of the stage is concerned.

Minimum Reflux Ratio. The minimum reflux ratio is defined as that ratio which if decreased by an infinitesimal amount would require an infinite number of stages to accomplish a specified separation.

Optimum Reflux Ratio. The general effect of the operating reflux ratio on the capital costs, running costs and total costs is shown in Figure 7 for a given feed and a specified separation.

3.3.2. Types and Characteristics of Plate Columns.

Plate columns utilised for liquid vapour contacting in distillation or absorption processes use cross-flow plates or counter-flow plates.

The cross-flow plate (shown schematically in Figure 8) utilises a liquid downcomer and is more generally used than the counter-flow plate (shown also schematically in Figure 9) because of transfer efficiency advantages and greater operating range. The liquid flow pattern on a cross-flow plate can be controlled by placement of downcomers in order to achieve the desired stability and transfer efficiency. Commonly used flow arrangements are shown in Figure 10.

The fraction of column cross-sectional area available for gas dispersers (perforations or bubble caps) decreases when more than one downcomer is used. Thus, optimum design of the plate involves a balance between liquid flow accomodation and effective use of cross-section for gas flow.

Most new designs of cross-flow plates employ perforations for dispersing gas into liquid on the plate. These perforations may be simple round orifices (resulting in sieve plates) or have movable valves that provide non-circular variable orifices (resulting in valve plates). For sieve plates, liquid is prevented from flowing through the perforations by the flowing action of gas. At low gas flow rates some or all of the liquid drain through the perforations and thus bypassing portions of the contacting zone. The valve plate minimises this drainage or "weeping" since the valve tends to close at low flow rates.

In counter-flow plates, liquid and gas utilise the same openings for flow. Openings are usually simple round perforations (dual-flow plate) in the 3-12 mm range or long slots (Turbogrid tray) 6 to 12 mm wide. In general gas and liquid flow in a pulsating fashion through the same opening.

Plate Column Capacity. The maximum allowable capacity of a plate for handling gas and liquid flow is of primary importance because it fixes the minimum possible diameter of the column. For a constant liquid rate, increasing the gas rate results in excessive entrainment and flooding. At the flood point it is difficult to obtain net downward flow of liquid, the column inventory of liquid increases, pressure drop across the column becomes quite large, and control becomes difficult. Design should call for operation at a safe margin below this maximum allowable condition, this condition is known as entrainment flooding or priming.

Flooding could also be brought about by increasing the liquid rate while keeping the gas rate constant, this condition is known as down flow flooding.

Minimum capacity of a column is determined by the need for effective dispersion and contacting of the phases. A cross-flow sieve plate can operate at reduced gas flow until weeping occurs. Valve plates can be operated at very low gas rates. All devices have a definite minimum gas rate below which good distribution is not possible. Counter-flow plate columns operate under the same constraints and a qualitative capacity diagram for all plates is shown in Figure 11; the shape and extent of the satisfactory operating zone depend on the type of plate used.

Entrainment Flooding. Based on a force balance on an average suspended droplet of liquid a capacity parameter C_{sb} was defined as [4]

$$C_{sb} = U_n \left(\frac{\rho_g}{\rho_l - \rho_g}\right)^{0.5} \left(\frac{20}{\sigma}\right)^{0.2} \qquad (3.9)$$

in which all quantities are expressed in British Units. The net area of the column (on which the linear gas velocity U_n is based) is the column cross-sectional area for counter-flow plates and that area less the area blocked by the downcomers (s) for cross-flow plates.

The maximum allowable values of the capacity parameter $C_{s,b,flood}$ have been correlated [5] against a flow parameter F_{lg} as shown in Figure 12. From this figure the flooding gas velocity could be determined,

$$U_{nf} = C_{sb,flood} \left(\frac{\sigma}{20}\right)^{0.2} \left(\frac{\rho_l - \rho_g}{\rho_g}\right)^{0.5} \qquad (3.10)$$

Figure 12 could be used for sieve plates, valve plates, bubble cap plates and counter-flow plates subject to certain restrictions [5] pertinent to different types of plates.

Information on weeping and down-flow flooding in sieve plates and cross-flow plates, and on pressure drop in all types of plates could be found in Reference [1].

Plate Efficiency. The efficiency of a plate for mass transfer depends upon three sets of design parameters :

1) Composition and properties of the mixture
2) Gas and liquid flow rates
3) Plate type and dimensions.

The aim of the design would be to maximise the overall column efficiency E_{oc} defined as

$$E_{oc} = \frac{Nt}{Na} \qquad (3.11)$$

or the ratio of theoretical plates to actual plates required to achieve a specific separation.

One of the empirical correlations for column efficiency that can be used for high-relative-volatility systems is O'Connell's modification [6] of Drickamer and Bradford correlation [7] as shown in Figure 13.

Relative volatility and feed viscosity are evaluated at the arithmetic average of top and bottom column temperatures.

A semi-theoretical method for predicting overall column efficiency is that of Bakowski [8]. It is based on the assumption that mass transfer rate for a component moving to the vapour phase is proportional to the concentration of the component in the liquid and to its vapour pressure. Interfacial area is assumed proportional to gas velocity.

The resulting equation for binary distillation is :

$$E_{oc} = \frac{10^5}{3.7 \dfrac{KM}{h' \rho'_1 T}} \tag{3.11}$$

where K = vapour-liquid equilibrium ratio ($= y^*/x$)
M = molecular weight
ρ'_1 = liquid density
h' = effective liquid depth
T = temperature

for sieve or valve plates, h' = outlet weir height.

3.4. Packed Columns Analysis and Design

3.4.1. Introduction

Packed columns are used extensively for absorption operations and, to a limited extent for distillation, by providing gas-liquid contacting through counter-flow of the phases. The column is usually filled with randomly orientated packing material supported by a plate at the bottom.

A liquid distributing device is used to provide effective irrigation of the packing. Devices may be added to provide redistribution of the liquid that might channel down the wall.

For heat pump generators, the conditions that favour the use of packed columns are :

1) Since diameters are small (for small and medium sized units), packings would be cheaper than plates.
2) Packings often have desirable efficiency - pressure drop characteristics.

3) Holp-up of liquid can be quite low, an advantage when the fluid is thermally sensitive.

Many packings are commercially available, each having specific characteristics regarding cost, surface availability, interface regeneration, pressure drop, weight and corrosion resistance. Typical packings are shown in Figure 14.

3.4.2. Mass Transfer Between Phases

Because vapour and liquid compositions change differentially in a packed column, rather than in a stepwise fashion as in a plate column, the difficulty of the separation to be accomplished is characterised in terms of transfer units rather than theoretical plates. The following is a brief description of the significance and derivation of the transfer unit concept.

When material is transferred from one phase to another across an interface, the resistance to mass transfer in each phase causes a concentration gradient in each as shown in Figure 15. Rates of transfer are assumed to be proportional to the difference between the bulk concentration and the concentration at the interface. Thus,

$$N_A = k_L\,(x - x_i) = k_G\,(y_i - y) \qquad (3.4.1)$$

This equation may be used to find the interfacial concentrations corresponding to any set of values of x and y provided that the ratio of the individual coefficients is known. Thus,

$$\frac{y_i - y}{x - x_i} = \frac{k_L}{k_G} = \frac{L_M\,H_G}{G_M\,H_L} \qquad (3.4.2)$$

Equation 3.4.2 could be solved graphically if the equilibrium liquid and vapour compositions are plotted and a point is also located representing the bulk concentrations x and y as shown in Figure 16.

The rate of mass transfer must be estimated from known or predicted values of the transfer coefficients and the bulk concentrations. This may be done by solving equation 3.4.2 simultaneously with the equilibrium relation $y_i = K_i\,(x_i)$ to obtain y_i and x_i.

If the equilibrium curve is simple e.g. a straight line, the rate of transfer is proportional to the difference in the bulk concentration in one phase and the concentration in that same phase which would be in equilibrium with the bulk concentration in the other phase, i.e. $y^* - y$ or $x - x^*$. In this case there is no need to solve for x_i and y_i since

$$N_A = k_G (y_i - y) = k_L (x - x_i) = K_G (y^* - y) \qquad (3.4.3)$$

where K_G is the overall gas-phase mass transfer coefficient expressed as

$$\frac{1}{K_G} = \frac{1}{k_G}\frac{(y^* - y)}{(y_i - y)} = \frac{1}{k_G} + \frac{1}{k_L}\frac{(y^* - y_i)}{(x - x_i)} \qquad (3.4.4)$$

This could be written as

$$\frac{1}{K_G} = \frac{1}{k_G} + \frac{m}{k_L} \qquad (3.4.5)$$

where m is the slope of the equilibrium line.

Similarly the overall liquid-phase mass transfer coefficient K_L is related to the individual coefficients by

$$\frac{1}{K_L} = \frac{1}{k_L} + \frac{1}{mk_G} \qquad (3.4.6)$$

Experimentally-observed rates of mass transfer are often expressed in terms of overall transfer coefficients, even when the equilibrium lines are curved. Since the available interfacial area in packed or tray columns cannot be determined, rates of transfer are usually reported in terms of transfer coefficients based on a unit volume of the column rather on a unit of interfacial area. Such volumetric coefficients are proportional to $k_G a$, $k_L a$ where (a) represents the interfacial area per unit volume of the column. Variations in the values of volumetric coefficients due to variations in size, type of packing etc., may be due as much to changes in the value of (a) as to changes in (k).

The overall coefficients are then calculated from the individual coefficients by the following equations :

$$\frac{1}{K_G a} = \frac{1}{k_G a} + \frac{m}{k_L a} \qquad (3.4.7)$$

$$\frac{1}{K_L a} = \frac{1}{k_L a} + \frac{1}{mk_G a} \qquad (3.4.8)$$

3.4.3. Height Equivalent to a Transfer Unit (H.T.U.)

Frequently the values of individual coefficients of mass transfer vary so rapidly with flow rates that the quantity obtained by the division of the flow rate by the coefficient is more nearly constant than the coefficient itself. This quantity is called the height of one transfer unit, since it expresses the height of column required to achieve a separation of standard difficulty.

The following relations between the transfer coefficients and the values of H.T.U. apply,

$$H_{OG} = \frac{G_M}{K_G a} \tag{3.4.12}$$

$$H_G = \frac{G_M}{k_G a} \tag{3.4.13}$$

$$H_{OL} = \frac{L_M}{K_L a} \tag{3.4.14}$$

$$H_L = \frac{L_M}{k_L a} \tag{3.4.15}$$

In terms of H.T.U.'s the equations that express the addition of resistance become

$$H_{OG} = H_G + H_L \frac{m\ G_M}{L_M} \tag{3.4.10}$$

and

$$H_{OL} = H_L + H_G \frac{L_M}{m\ G_M} \tag{3.4.11}$$

The number of overall gas-phase transfer units N_{OG} required for changing the concentration of the vapour stream from y_1 to y_2 is

$$N_{OG} = \int_{y_2}^{y_1} \frac{dy}{y^* - y} \tag{3.4.9}$$

The number of transfer units required for a given separation is closely related to the number of theoretical plates or stages required to achieve the same separation in a plate column.

Height equivalent to one theoretical plate (H.E.T.P.) is another quantity that is used occasionally to express the efficiency of a packing material for achieving a separation. If equilibrium and operating lines are parallel i.e.

$m\,G_M/L_M = 1$, H.E.T.P.'s and H.T.U.'s are equal. If the equilibrium and operating lines are straight but not parallel,

$$\frac{H_{OG}}{\text{H.E.T.P.}} = \frac{(mG_M/L_M) - 1}{\ln \dfrac{(m\,G_M)}{L_M}} \tag{3.4.16}$$

In distillation work, the number of transfer units is usually calculated as N_{OG}, i.e. it is based upon gas-composition changes, even though considerable transfer resistance lies in the liquid phase [1] . Required height of packing for a given design is obtained by multiplying the number of transfer units or theoretical plates by the height of packing equivalent to a transfer unit or theoretical plate discussed above.

Because experimental data often have to be extended it is helpful to list the effects of the controlling variables [1]. Increasing liquid rates cause values of H.T.U. to decrease, but the effect of gas rate is minor. Increased temperature causes H.T.U. values to decrease, but the effect of pressure is small or negligible, except for its effect on the boiling temperature. The smaller the size of packing the greater the surface area and one would expect lower values of H.T.U. In general this is true, although the effect is not direct.

There is considerable uncertainty regarding the effect of column diameter. It is generally believed that values of H.T.U. become slightly less favourable for a given packing the larger the diameter of the column, possibly owing to poorer liquid distribution.

In general, values of H.T.U. are found to become slightly less favourable for greater heights of packing, possibly because of progressive maldistribution effects.

Correlations of liquid-phase mass transfer experimental data have been widely reported. The generalised equation of Cornell et al [9] and Onda et al [10] are based on experimental data for water, and water and organic solvents respectively.

Figure 17 shows correlation curves for various sizes of Raschig rings while Figure 18 shows correlation curves for various sizes of Berl saddles.

Generalisation of mass-transfer data for the gas phase has been much less successful than it has for the liquid phase. The main difficulties preventing the development of a general relationship have been (1) the variation of effective area as a function of flow rates and surface tension and (2) the lack of a gas phase controlling test system unaffected by heat development or chemical reaction.

Figure 19 shows correleation of data on gas-film resistance to mass transfer in packed columns, with the straight line representing early correlation equation of Sherwood and Holloway [11]. Other correlation equations include those of Cornell et al [9] for Raschig rings and Berl saddles and that of Onda et al [10] for small and large ring and saddle packing in absorption and vapourization columns.

3.4.4. Packed Column Hydraulics

Pressure drop of a gas flowing upward through a packing countercurrent to liquid flow, is characterized graphically in Figure 20. At very low liquid rates, the effective open cross-section of the packing is not appreciably different from that of dry packing and pressure drop is due to flow through a randomly sized and located opening in the bed. Thus, pressure drop is proportional approximately to the square of the gas velocity as indicated in the region AB.

At higher liquid rates the effective open cross-section is smaller because of liquid and a portion of the gas stream energy is used to support the liquid in the column (Region A'B'). For all liquid rates, a zone is reached where pressure drop is proportional to a gas-flow rate power distinctly larger than 2; this is called the loading zone. The pressure drop increases due to the rapid accumulation of liquid in the packing-void volume.

As the liquid hold-up increases, one of two changes may occur. If the packing is composed essentially of extended surfaces, the effective orifice diameter becomes so small that the liquid surface becomes continuous across the section of the column, generally at the top of the packing. The change in pressure drop is quite great with only a slight change in gas rate (condition C or C'). The phenomenon is called flooding and is analogous to entrainment flooding in a plate column.

If the packing surface is discontinuous in nature, a phase inversion occurs, and gas bubbles through the liquid. The column is not unstable and can be brought back to gas-phase continuous operation by merely reducing the gas rate. The pressure drop also rises rapidly as phase inversion occurs.

Flooding and Loading.

Since flooding or phase inversion normally represents the maximum capacity condition for a packed column, it is desirable to predict its value for new designs. A modification by Eckert [12] of the first generalised correlation of packed-column flood points by Sherwood et al [13] is shown in Figure 21.

The packing factor F_p is a characteristic of the packing material. Factors for common packing materials are shown in Table 1. (General packing characteristics are shown in Table 2). All factors must be used with caution and the design capacity for a packed column should allow for ± 30 % error in the predicted flood points.

The loading point is a more nebulous quantity than the flood point. Since transition from preloading to loading conditions may be very gradual, the loading point can only be approximated.

Figure 22 shows loading velocities of some commercial packings based on the air-water system data of Tillson [14]. ϕ in this figure is given by :

$$\phi = \left(\frac{\rho_g}{0.075}\right)^{0,5}$$

In this last equation as well as in Figure 22 all quantities are to be expressed in British units.

Pressure Drop.

Generalised pressure drop correlations of Leva et al [15] are also shown in Figure 21. Packing factors for pressure drop estimation are different from those for flood-point estimation, and vary somewhat with flow conditions. Pressure drop packing factors are given in Table 3. Pressure drop data for some packings are shown in Figure 23 taken from Tillson [14] data for an air-water system.

Support Plates.

The main function of a packing support plate is to retain a bed of packing without excessive restriction to gas and liquid flow. The support plate also serves to distribute both gas and liquid streams. The design of the support plate affects the column pressure drop and stable operating range. Unless carefully designed, the plate can cause premature flooding.

There are two basic types of support plates :

(1) Countercurrent
(2) Separate flow passages for liquid and gas

The two types are shown in Figure 24.

The degree of open area on a support plate is the fraction of void inherent in the design of the plate minus that portion of the open area occluded by the packing. To avoid premature flooding, the net open area of the plate must be greater than that of the packing itself. With the countercurrent type support plate the free area for gas flow can be up to 90 % of the column cross-sectional area. However, such a plate could be easily occluded by the packing material resting on it.

The separate flow passage plate can have free areas up to 90 % of the column cross-sectional area and is not easily occluded by packing.

Liquid Hold-Up.

There are three modes of liquid hold-up in a packed column, these are : (1) Static, h_s (2) Total, h_t (3) Operating, h_o, where h values are in volumes of liquid per total volume of bed. Static hold-up is the amount of liquid remaining in a packing that has been fully wetted and then drained. Total hold-up is the amount of liquid in the packing under dynamic conditions. Operating hold-up is attributed to operation and is measured experimentally as the difference between total and static hold-up.

Thus,

$$h_t = h_o + h_s \qquad (3.4.17)$$

Typical total hold-up data for packings (taken from Reference 16) are shown in Figure 25. Operating hold-up contributes effectively to mass transfer rate since it is a reflection of the residence time for phase-contact and surface regeneration.

Liquid Distribution.

It is essential to have uniform distribution of the liquid at the top of the packed bed and at the feed point for efficient operation. Several types of liquid distributors are used (shown in Figure 26). The perforated pipe distributors offer minimum restriction to gas-flow and can be used for high liquid flows. Operating pressure drop for liquid ranges from 0.3 to 1.5 bar. Trough-type distributors are used for large installations (column diameter of 1.0 m and larger). Those distributors are not subject to clogging and have a wide operating range but accurate levelling is required at low liquid flow rates.

The orifice type distributor offers some resistance to gas-flow because of the relatively low gas riser area.

Maldistribution of the liquid could result from initial uneven distribution; improper installation of packing; column being out of vertical alignment; liquid migration towards the wall; and packing geometry inhibiting lateral distribution.

Liquid migration towards the wall (channelling) tends to occur more at small column diameter/packing size ratios [17] (less than 10) and could be rectified by the use of side wipers or redistributors as shown in Figure 27. Geometry of certain rigid packing materials can cause inhibition of lateral distribution especially at column diameter/packing size ratios greater than 30 [18].

With careful consideration of all possible causes of maldistribution, it is possible to design commercial packed columns of heights up to 9.0 m between redistributors.

End Effects.

Mass transfer can take place outside the bed in a packed column, i.e. at the ends of packed sections. Inlet gas may contact exit liquid below the bottom support plate and exit gas can contact falling liquid from some types of distributors (e.g. spray nozzles). Silvey and Keller [19] found that the reboiler plus the bottom end effect could give up to two or more theoretical plates.

Interfacial Area.

The effective area of contact between gas and liquid, which participates in the mass exchange process, may be less than the actual interfacial area because of stagnant pools of liquid. The effective area is difficult to measure directly since it includes contributions from film flow of liquid across packing surface, liquid rivulets, drippings and gas bubbles.

An example of the effective interfacial areas as obtained by Shulman et al [16] for vapourisation and based on the calculated ammonia absorption data of Fellinger [20] are shown in Figure 28.

Effective interfacial areas could be different between absorption and vapourisation and for systems with different surface tensions, but when corrections for surface-tension differences are made, the areas are effectively the same.

3.5. The Still (Reboiler)

The still or partial reboiler is the part of the generator where a pool of the working fluid solution is boiled to drive out the refrigerant vapour (in case of non-volatile solvents) or relatively refrigerant-rich vapour for rectification in the column. The heat-exchanger arrangement to provide the necessary heat to the still may take several forms. For small generators a jacketed still (Figure 29a) may be adequate. The heat transfer surface area is relatively small and could possibly be increased by adding fins to the still outer surface. A tube-in-shell heat exchanger could be built in the still to provide a much larger surface area (Figure 29b), but cleaning of the heat exchanger would require shut down of the generator's operation and draining of the working fluid.

In the types of reboiler described overleaf, the vapour entering the solumn is essentially in equilibrium with the bottom product, so that the reboiler represents one equilibrium stage (21) (a theoretical plate).

External reboilers of several varieties are commonly used for large installations. The most frequently used type is the kettle reboiler shown in Figure 29c which also produces vapour in equilibrium with the bottom product. The vertical thermosiphon reboiler (Figure 29d) vapourises all the liquid entering and hence produces a vapour of the same composition as that of the bottom product. The reboiler of Figure 29e receives liquid from the trap-out of the bottom tray, which it partially vapourises to provide a vapour and a bottom product which are essentially in equilibrium.

When external reboilers are used, the reservoir at the bottom of the column usually holds 5 to 10 minutes flow of liquid to achieve reasonably steady operation of the reboiler. In small units, the amount of liquid in the reboiler might have to be limited to the minimum needed to provide the heat transfer area. This is true (e.g.) in case of ammonia-water system intended for indoor domestic use when the ammonia charge should not exceed a recommended or obligatory maximum amount.

Calculation of Heat Transfer Surface Area of the Still

The heat transfer coefficient α for the still is calculated with the assumption of pool boiling. The effect of the binary mixture is included by approximating liquid and gas phase properties for the composition predicted for the bottom solution and the rising vapour.

For pool boiling [22] ,

$$\alpha = 0.00122 \frac{k_L^{0.79}\, c_L^{0.45}\, \rho_L^{0.49}}{\sigma^{0.5}\, \mu_L^{0.29}\, (\lambda \rho_G)^{0.24}} (T_W - T_4)^{0.24} (P_W - P_4)^{0.75} \quad (3.5.1)$$

where k_L is the liquid thermal conductivity P_W and P_4 are the saturation pressures corresponding to T_W and T_4.

The value of T_W should be reasonably selected and the corresponding value of the heat transfer area A could be found from

$$\dot{Q} = \alpha\, A\, (T_W - T_4) \quad (3.5.2)$$

Wall Heat Transfer.

The still wall thickness t is estimated from design stress considerations. The temperature rise through the wall ΔT_W could then be calculated from

$$\Delta T_W = \frac{\dot{Q}}{A \frac{kw}{t}} \quad (3.5.3)$$

Burnout Point Estimation.

It is important that the actual heat flux per unit area of the still remains below the critical value q_{crit}.

From reference [22] :

$$q_{crit} = 0.131\, \lambda\, [\sigma\, g\, (\rho_L - \rho_G)\, \rho_G^2]^{0.25} \quad (3.5.4)$$

this value of q_{crit} should be much larger than $\frac{\dot{Q}}{A}$

3.6. The Partial Condenser (Rectifier or Dephlegmator)

The partial condenser serves to further enrich (in refrigerant) the vapour rising from the rectifying column to achieve the required composition of vapour entering the total condenser of the heat pump.

A partial condenser may produce any of several effects [21].

(1) If the time of contact between the vapour product and the liquid reflux is sufficient, the two will be in equilibrium and the partial condenser provides an equilibrium condensation (one equilibrium stage or theoretical plate).

(2) If the condensate is removed as rapidly as it forms, a differential condensation may occur.

(3) If cooling is very rapid, little mass transfer between vapour and condensate results and the two will have essentially the same composition.

Figure 30 shows a schematic of a water-cooled rectifier; from equilibrium condensation assumption.

$$T_7 = T_{15} \tag{3.6.1}$$

All flow parameters shown are calculated from a heat pump cycle analysis. The cooling water passes through a circular horizontal tube on the outside surface of which vapour condenses.

The overall heat transfer coefficient U is calculated from [22],

$$\frac{1}{U} = \frac{1}{\alpha_o} + \frac{1}{\alpha_i}\frac{Do}{Di} + \frac{y_w}{kw} \cdot \frac{Do}{Dw} \tag{3.6.2}$$

where

$$D_w = \frac{Do - Di}{\ln \frac{Do}{Di}}, \quad y_w = \frac{Do - Di}{2}$$

$$LMTD = \frac{(T_{17} - T_{out}) - (T_7 - T_{in})}{\ln \frac{T_{17} - T_{out}}{T_7 - T_{in}}} \tag{3.6.3}$$

The condensation coefficient α_o on outside of a horizontal tube is calculated from [22] :

$$\alpha_o = 0.951\ k_L \left\{\frac{\rho_L (\rho_L - \rho_g)}{\mu_L\ \Gamma}\right\}^{1/3} \tag{3.6.4}$$

where Γ is the condensate rate per unit tube length.

Some of the vapour and liquid properties at the temperatures and pressures in the rectifier might have to be approximated from the available data on the binary mixture utilised.

Water-side Surface Heat Transfer Coefficient.

Cooling water mass flow rate is calculated from the rectifier load

$\dot{Q}_D$ and the temperature rise $T_{out} - T_{in}$.

Selecting a reasonable water speed in the tube (this could be calculated by assuming a sensible internal diameter of the tube to be used) would enable the designer to calculate the Reynolds number Re for the water flow.

Prandtl number Pr could also be calculated.

The Nusselt number Nu is calculated from [22], assuming fully turbulent flow :

$$Nu = 0.023\ R_e^{0.8}\ P_r^{0.4} \tag{3.6.5}$$

with

$$\alpha_i = \frac{Nuk}{l'} \tag{3.6.6}$$

where l' is the characteristic length (internal diameter of the tube).

Furthermore :

$$\dot{Q}_D = U.A_o\ LMTD \tag{3.6.7}$$

From equations (3.6.2) through (3.6.7), l' could be calculated.

3.7. Dual-Effect (Two-Stage) Generators

Some large absorption machines (utilising water-lithium bromide) intended for water chilling have been built with dual-effect (two-stage) generators. Figure 31 shows a schematic of such a machine (26). The first-effect generator receives the external heat which boils refrigerant from the strong solution. This hot refrigerant vapour is piped to a second generator, supplying heat for further refrigerant vapourisation from the absorbent of intermediate concentration (in the refrigerant) which has flowed from the first generator. This intermediate concentration solution is usually cooled by passing through a first stage heat recuperator as shown in Figure 31.

The complexity of dual-effect generators make them unattractive for heat pump applications especially for small and medium sized units.

ACKNOWLEDGEMENT

Much of the material in this lecture is based on some of the material presented in Sections 13, 14 and 18 of the "Chemical Engineers Handbook" 5th Edition, edited by R. Perry (Reference 1).

REFERENCES

1. Perry, R. and Chilton, C. *Chemical Engineers Handbook* 5th Edition.

2. Hala, Pick, Fried and Vilim, *Vapour-Liquid Equilibrium* 2nd Edition (New York, Pergamon, 1967).

3. Hala, Polak, Wichterle and Boublik, *Vapour-Liquid Equilibrium Data at Normal Pressures* (New York, Pergamon, 1968).

4. Sanders and Brown, *Ind. Eng. Chem.* 26, 98 (1934).

5. Fair, *Petrol/Chem. Engr.* 33 (10), 45 (1961).

6. O'Connel, *Trans. Am. Inst. Chem. Engrs* 42, 741 (1946)

7. Drickamer and Bradford, *Trans. Am. Inst. Chem. Engrs.* 38, 319 (1943).

8. Bakowski, *Brit. Chem. Eng.* 8, 384, 472 (1963); ibid, 14, 945 (1969).

9. Cornell et al, *Chem. Eng. Progr.* 56 (8) 68, (1960).

10. Onda et al, *J. Chem. Eng. Japan* 1, 56 (1968).

11. Sherwood and holloway, *Trans. Am. Inst. Chem. Engrs.* 36, 21 (1940).

12. Eckert, *Chem. Eng. Progr.* 66 (3), 39 (1970).

13. Sherwood, Shipley and Holloway, *Ind. Eng. Chem.* 30, 768 (1938).

14. Tillson, *S M Thesis* (M.I.T., 1939).

15. Lea et al, *Chem. Eng. Progr. Symp. Ser.* 50 (10), 51 (1954).

16. Shulman et al, *Am. Inst. Chem. Engrs.* J. 1, 247 (1955).

17. Baker, Chilton and Vernon, *Trans. Am. Inst. Chem. Engrs.* 31, 296 (1935).

18. Huber and Hilt Brunner, *Chem. Eng. Sci* 21, 819 (1966).

19. Silvey and Keller, *Proc. Intern. Symp. Dist.* (Brighton,

England, 1970).

20. Fellinger, *Sc. D. Thesis* (M.I.T., 1941).

21. Treybal, *Mass Transfer Operations, Chemical Engineering Series* (New York, McGraw-Hill, 1955).

22. Butterworth, "Introduction to Heat Transfer" *Engineering Design Guides* (Oxford University Press, 1977).

23. Smith, *Design of Equilibrium Stage Processes* (McGraw-Hill, 1963).

24. Sherwood and Pigford, *Absorption and Extraction* (McGraw-Hill, 1952).

25. *Chem. Eng. Progr.* 62 (1), 59 (1966).

26. *ASHRAE Equipment Handbook* (1975).

TABLE I

Packing Factors F_p (Wet and Dump Packed) [1]

Type of Packing	Material	Nominal packing size, in										
		$\frac{1}{4}$	$\frac{3}{8}$	$\frac{1}{2}$	$\frac{5}{8}$	$\frac{3}{4}$	1	$1\frac{1}{4}$	$1\frac{1}{2}$	2	3	$3\frac{1}{2}$
Intalox saddles	Ceramic	725	330	200		145	98		52	40	22	
Intalox saddles	Plastic						33			21	16	
Raschig rings	Ceramic	1600	1000	580	380	255	155	125	95	65	37	
Berl saddles	Ceramic	900		240		170	110		65	45		
Pall rings	Plastic				97		52		32	25		16
Pall rings	Metal				70		48		28	20		
Raschig rings 1/32 in wall	Metal	700	390	300	170	155	115					
Raschig rings 1/16 in wall	Metal			410	290	220	137	110	83	57	52	

TABLE II

General Characteristics of Dumped Packings [1]

Packing Type	Nominal size in.	Wall thickness in.	o.d. and length in.	Approx. number per cu.ft.	Approx. weight per cu.ft. (lb)	Approx. surface area sq.ft/cu.ft.
Ceramic Raschig Rings	1/4	1/16	1/4	85,600	60	217
	1/2	3/32	1/2	10,700	55	112
	3/4	3/32	3/4	3,090	50	74
	1	1/8	1	1,350	42	58
	1 1/2	3/16	1 1/2	387	43	37
	1 1/2	1/4	1 1/2	381	46	36
	2	1/4	2	164	41	28
	3	3/8	3	48	37	19
	4	3/8	4	20	36	14
Carbon Raschig Rings	1/4	1/16	1/4	85,000	46	212
	1/2	1/16	1/2	10,600	27	114
	3/4	1/8	3/4	3,140	34	75
	1	1/8	1	1,325	27	57
	1 1/2	1/4	1 1/2	392	34	38
	2	1/4	2	166	27	28
	3	5/16	3	49	23	19
Steel Raschig Rings	1/2	1/32	1/2	11,400	75	122
	1/2	1/16	1/2	10,900	132	111
	5/8	1/32	5/8	6,130	62	103
	3/4	1/32	3/4	3,340	52	81
	3/4	1/16	3/4	3,140	94	75
	1	1/16	1	1,310	71	56
	1 1/2	1/16	1 1/2	400	49	39
	2	1/16	2	168	37	29
	3	1/16	3	51	25	20
Ceramic Pall Rings	2	1/4	2	164	38	29
	3	3/8	3	49	40	20
Steel Pall Rings	5/8	26 gage	5/8	5,930	37	104
	1	24 gage	1	1,405	30	63
	1 1/2	22 gage	1 1/2	377	26	39
	2	20 gage	2	171	24	31
Polypropylene Pall Rings	5/8		5/8	6,050	7 1/4	104
	1		1	1,420	5 1/2	63
	1 1/2		1 1/2	385	4 3/4	39
	2		2	180	4 1/2	31
	3 1/2		3 1/2	33	4 1/2	26

TABLE II (Second Part)

Ceramic	1/4			107,000	56	274
Berl	1/2			16,700	54	142
Saddles	3/4			5,000	49	87
	1			2,180	45	76
	1 1/2			645	40	46
	2			250	39	32
	1/4			117,500	54	300
	1/2			20,700	45	190
	3/4			6,500	44	102
	1			2,385	44	78
	1 1/2			709	42	59
	2			265	42	36
	3			53	37	28
Low Density Polyethylene Tellerettes	1			1,125	10	76

TABEL III

Packing Factors for Pressure Drop Estimation [25]

$L/G \sqrt{\rho_g/\rho_1}$	> 3.75			0.5 - 3.75			< 0.5		
Δp = in.H_2O/ft.depth	0.5	1.0	1.5	0.5	1.0	1.5	0.5	1.0	1.5
Raschig rings, steel :									
1/16 in wall, 1 in	89	77	67	139	122	101	158	144	142
1/16 in wall, 1 1/2 in	40	43	42	75	65	65	94	85	80
1/16 in wall, 2 in	38	38	37	60	54	50	72	71	66
Pall ring, steel :									
0.024 wall, 24 gage 1 in	45	42	42	54	52	53	52	48	47
0.030 wall, 22 gage, 1 1/2 in	34	31	27	39	36	34	30	28	29
0.036 wall, 20 gage, 2 in	21	20	20	26	24	22	25	28	23
Intalox, Ceramic :									
1 in		84	75		91	93		97	98
1 1/2 in	40	39	35	53	49	48	50	49	52
2 in	31	31	30	39	34	33	38	39	40
Raschig ring, ceramic :									
1/8 in wall, 1 in	135	120	110	175	153	144	165	155	156
3/16 in wall, 1 1/2 in	75	65	60	90	82	80	92	88	90
1/4 in wall, 2 in	52	50	47	69	64	62	73	73	72

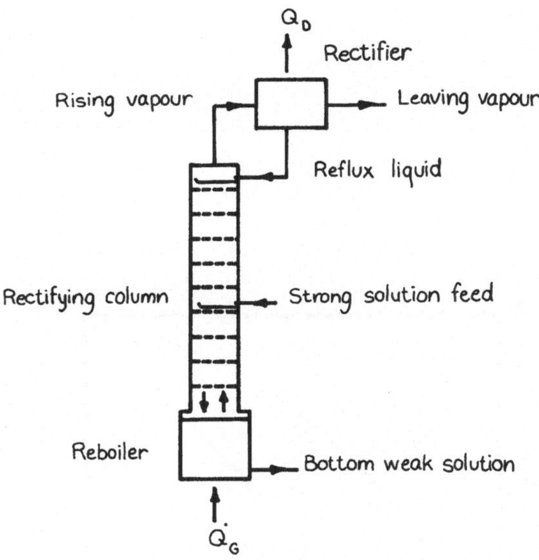

Figure 1a : Schematic diagram of a generator system

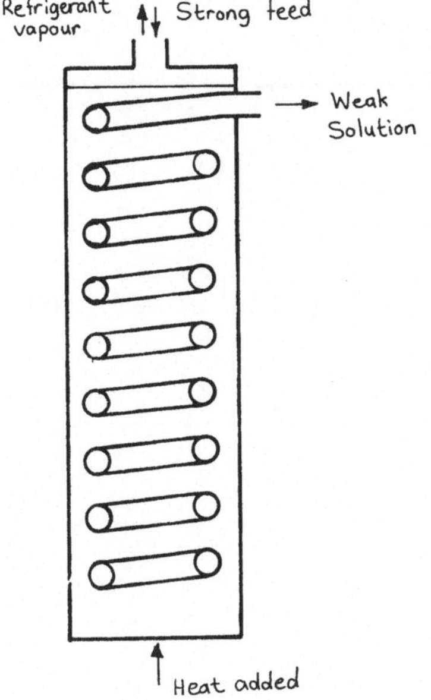

Figure 1b : Schematic of an analyser generator

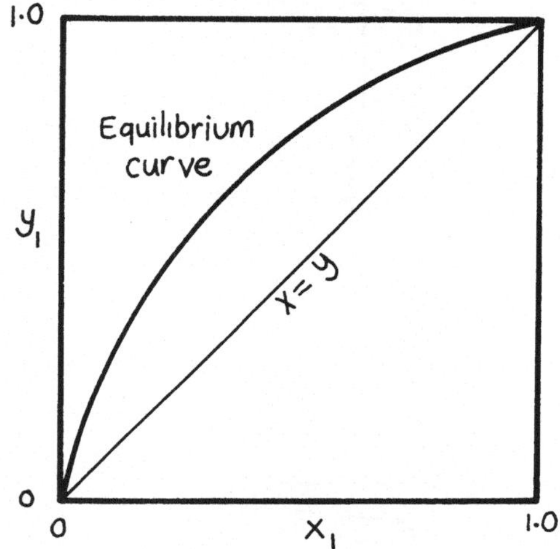

Figure 2 : Typical binary equilibrium curve

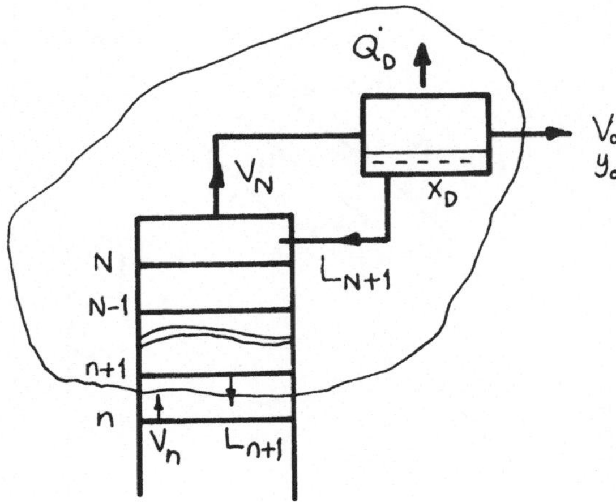

Figure 3 : Material-balance envelope around the top section of a column

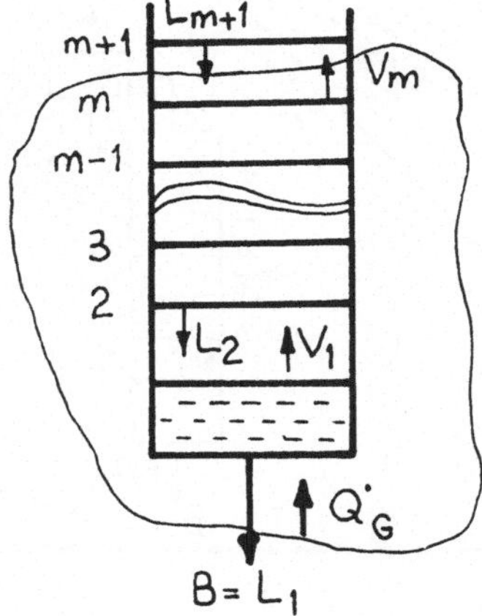

Figure 4 : Material-balance envelope around the bottom of a column

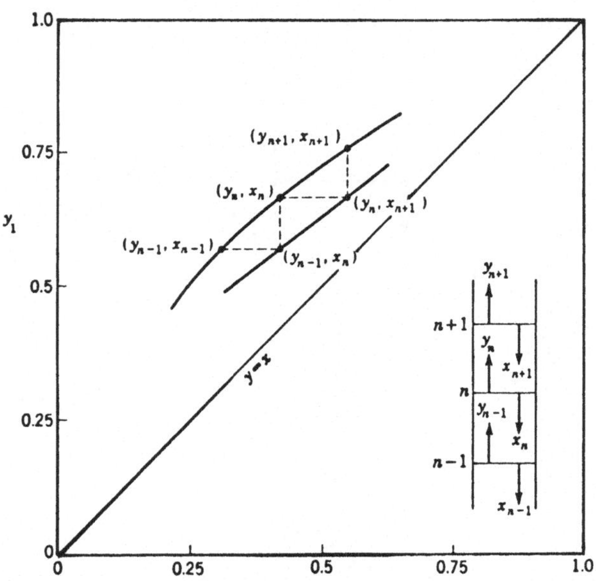

Figure 5 : Construction of equilibrium stages [1]

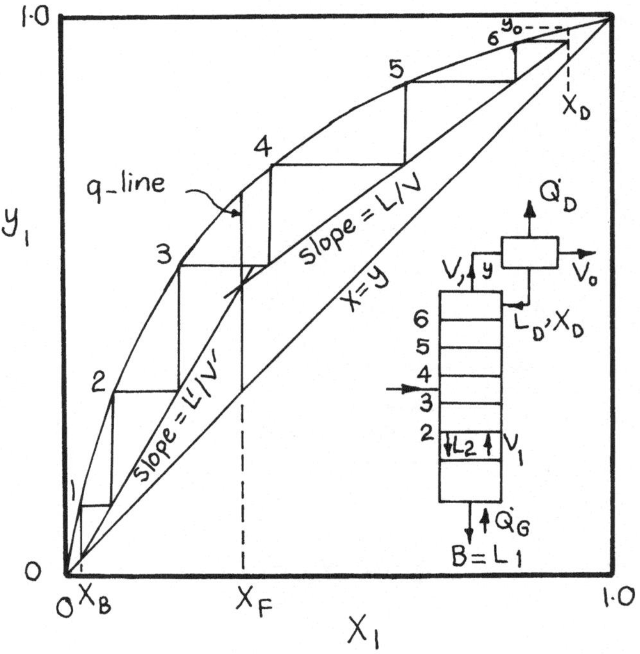

Figure 6 : Construction of equilibrium stages for a complete generator

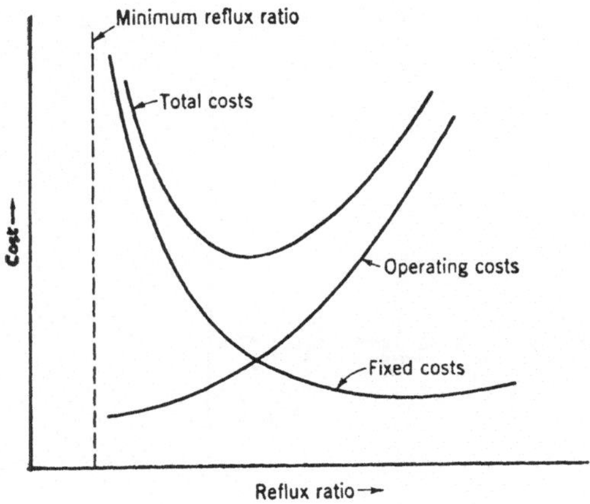

Figure 7 : Location of the optimum reflux for given conditions [1]

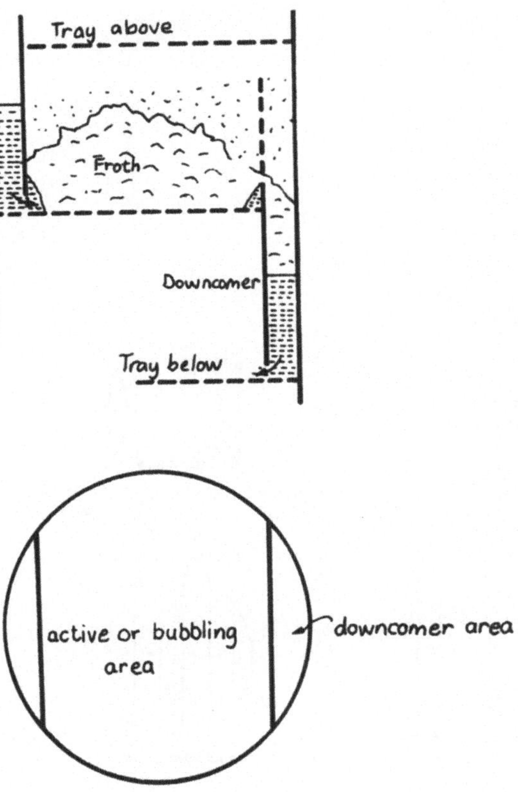

Figure 8 : Crossflow sieve-plate [23]

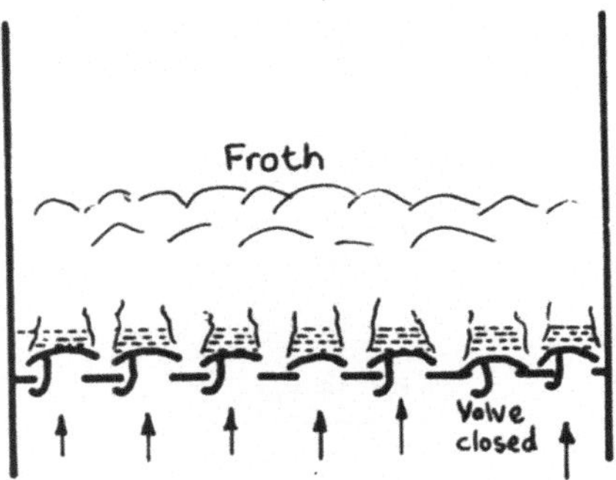

Figure 9 : Countercurrent valve plate [23]

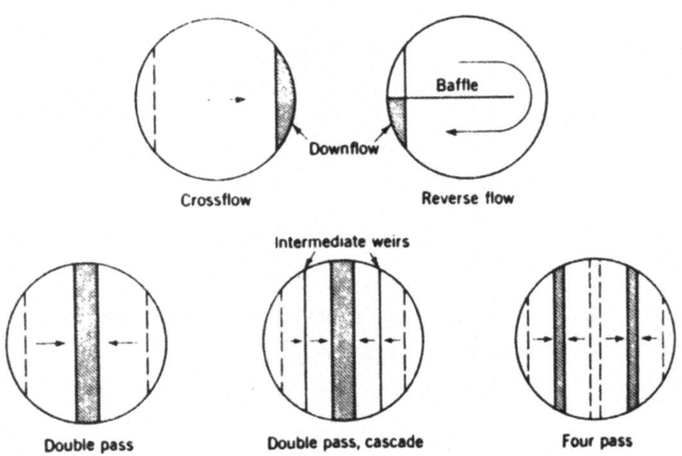

Figure 10 : Flow arrangements in plate columns [23]

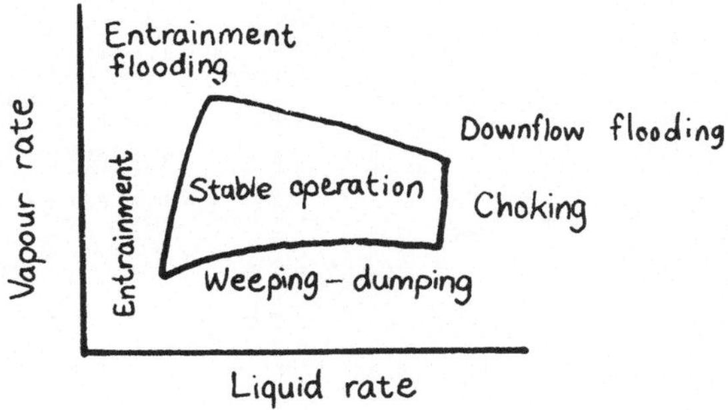

Figure 11 : Stable operating region, plates [23]

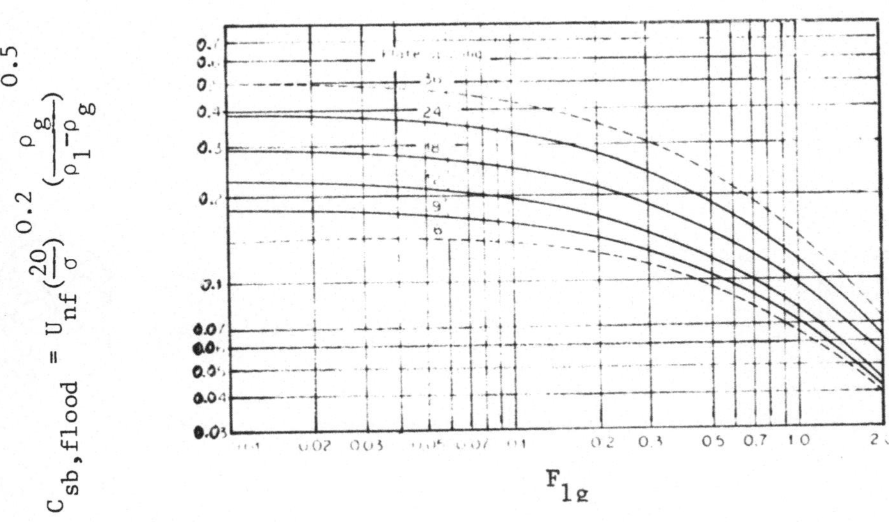

Figure 12 : Flooding limits for bubble-cap and perforated plates [5]

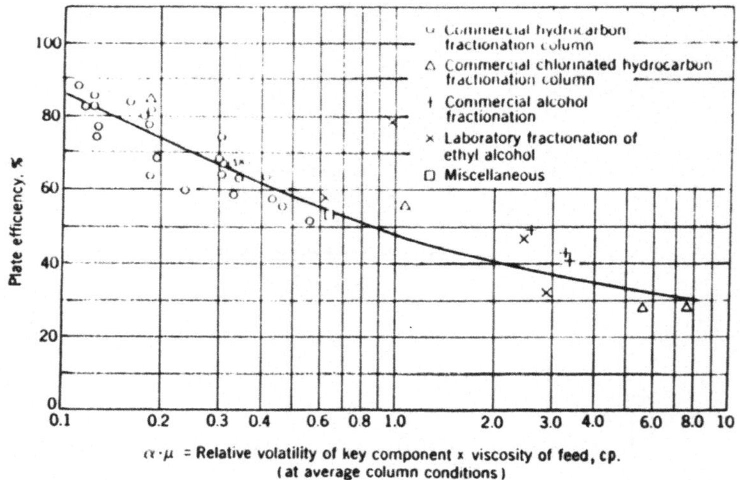

Figure 13 : Overall column efficiency vs. relative volatility and viscosity [6]

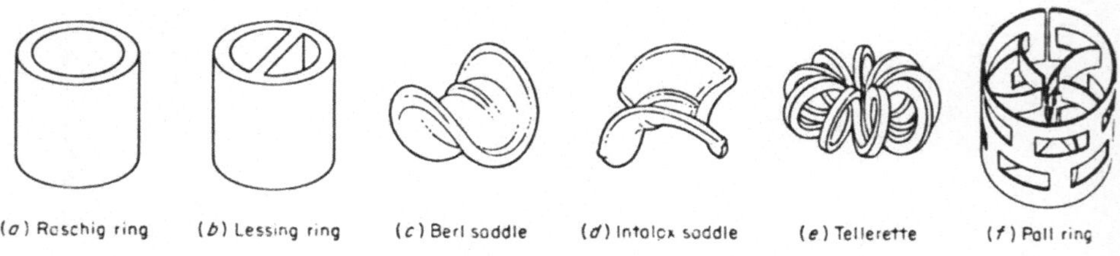

Figure 14 : Typical commercial packings [1]

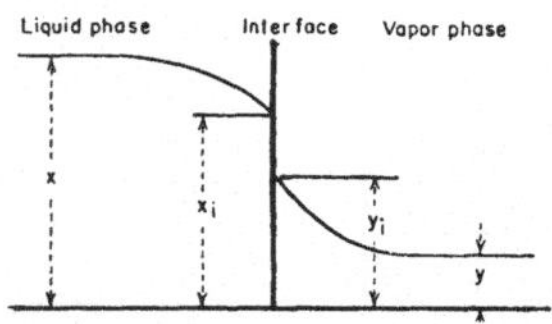

Figure 15 : concentration distribution near an interface [1]

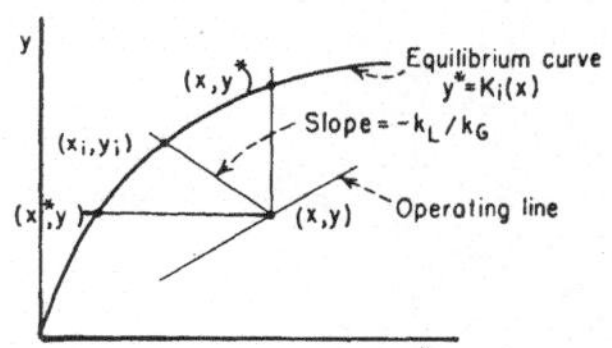

Figure 16 : Location of equilibrium concentrations for a point in countercurrent flow [1]

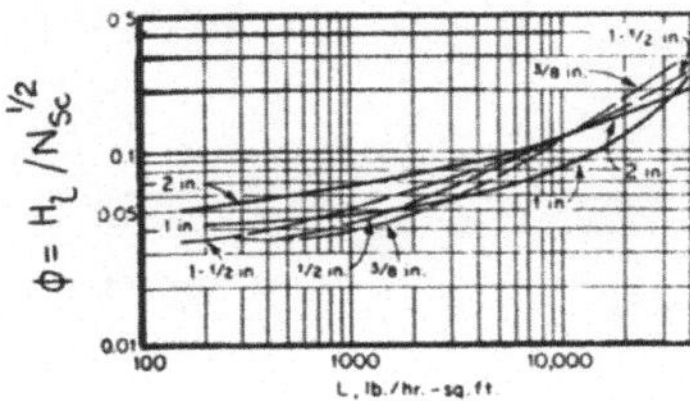

Figure 17 : H_l correlation for Raschig rings [9]

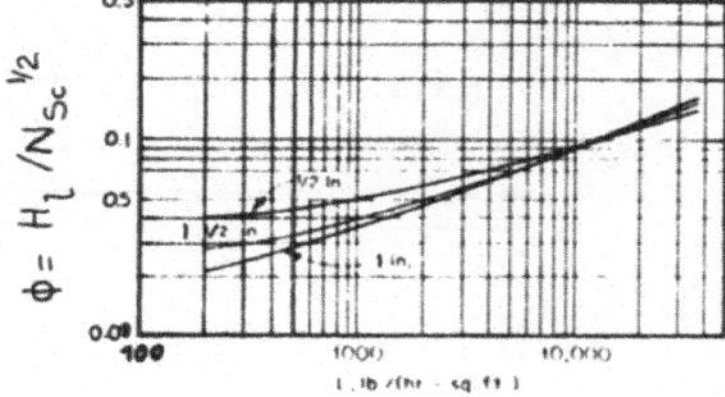

Figure 18 : H_l correlation for Berl saddles [9]

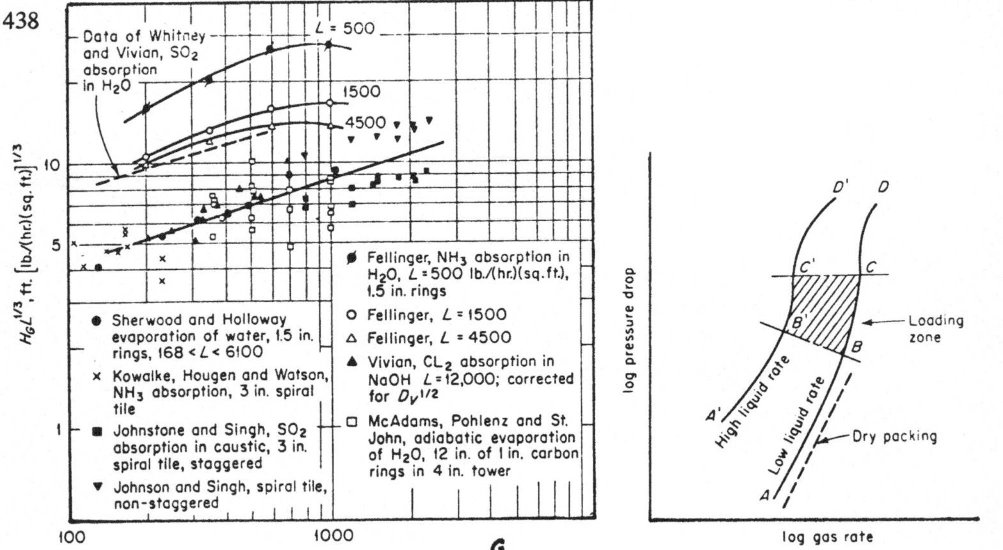

Figure 19 : H_G correlation in packed columns [24]

D_v = Gas diffusivity (D_g)

Figure 20 : Pressure-drop characteristics of packed columns [1]

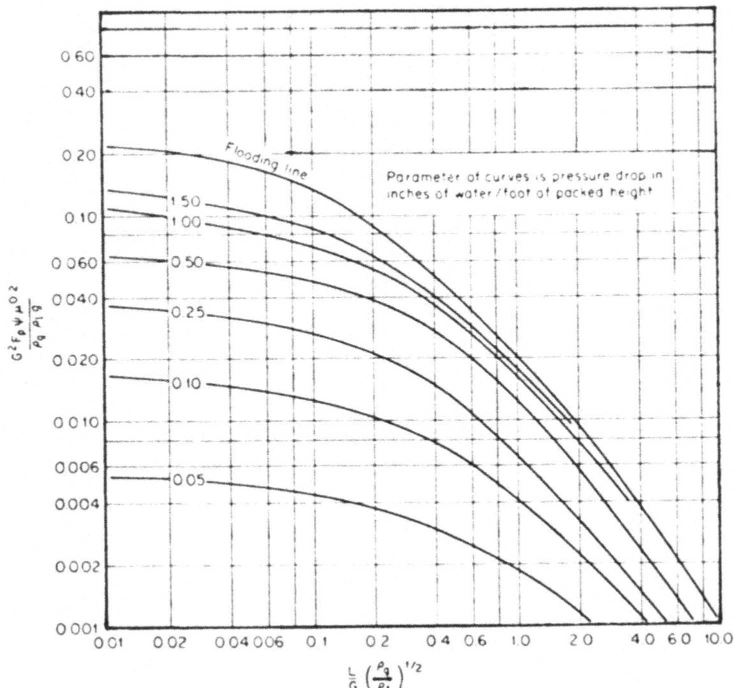

Figure 21 : Generalized flooding and pressure drop correlation for packings [12]

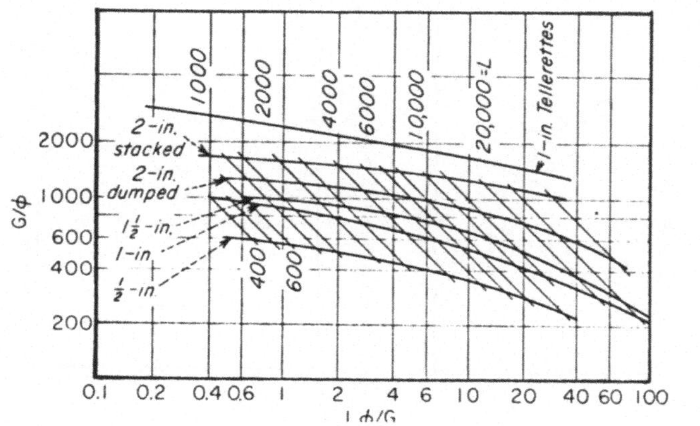

Figure 22 : Loading velocities for Raschig rings and Telleretts [14]

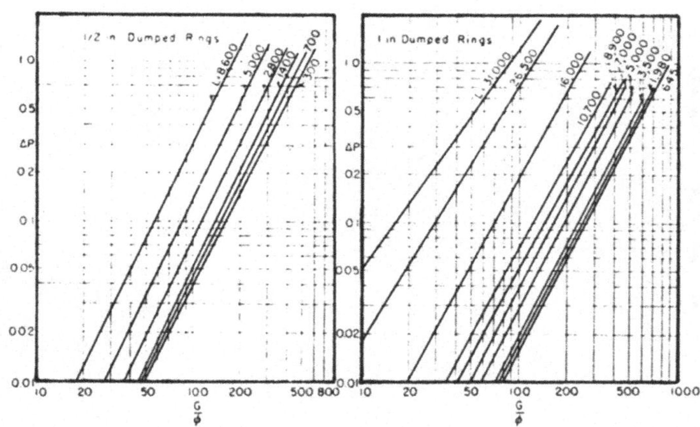

Figure 23 : Pressure-drop data for ceramic Rashig rings [14]

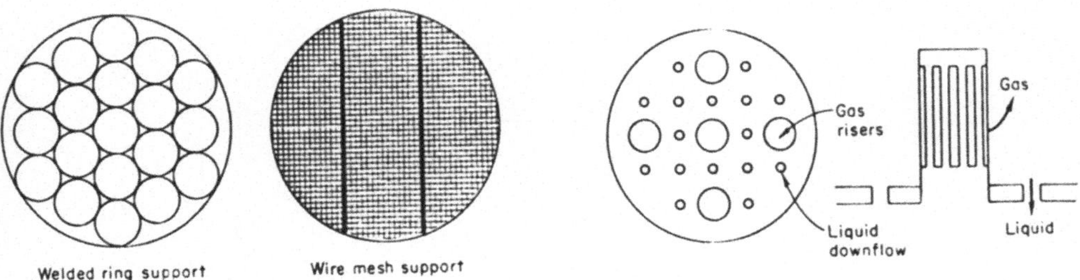

Figure 24 : Packing support plates [1]

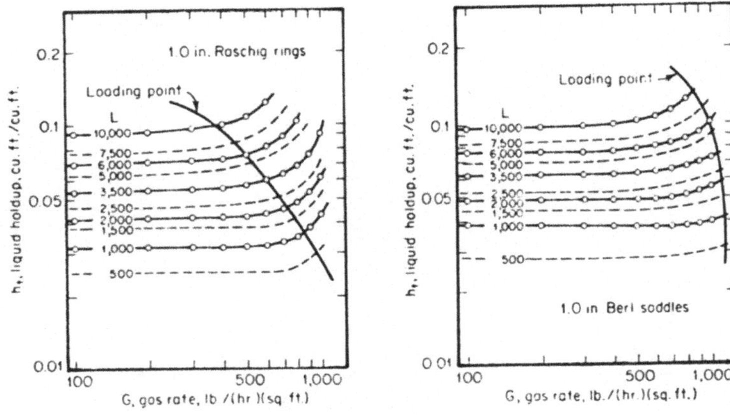

Figure 25 : Typical total hold-up data for packings [16]

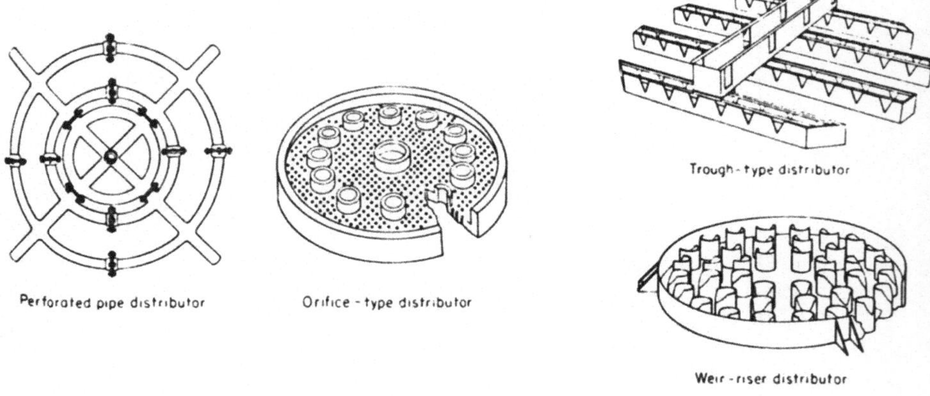

Figure 26 : Typical liquid distributors [1]

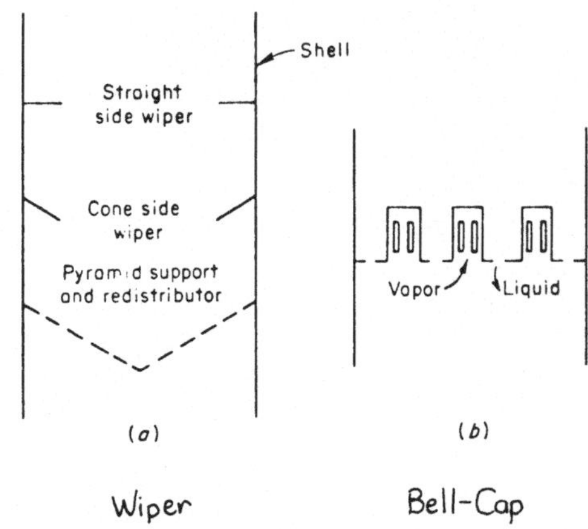

Figure 27 : Redistributors [1]

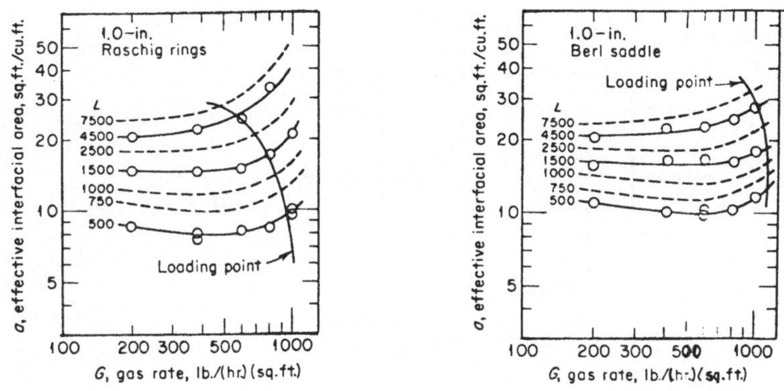

Figure 28 : Effective interfacial area of packings [16]

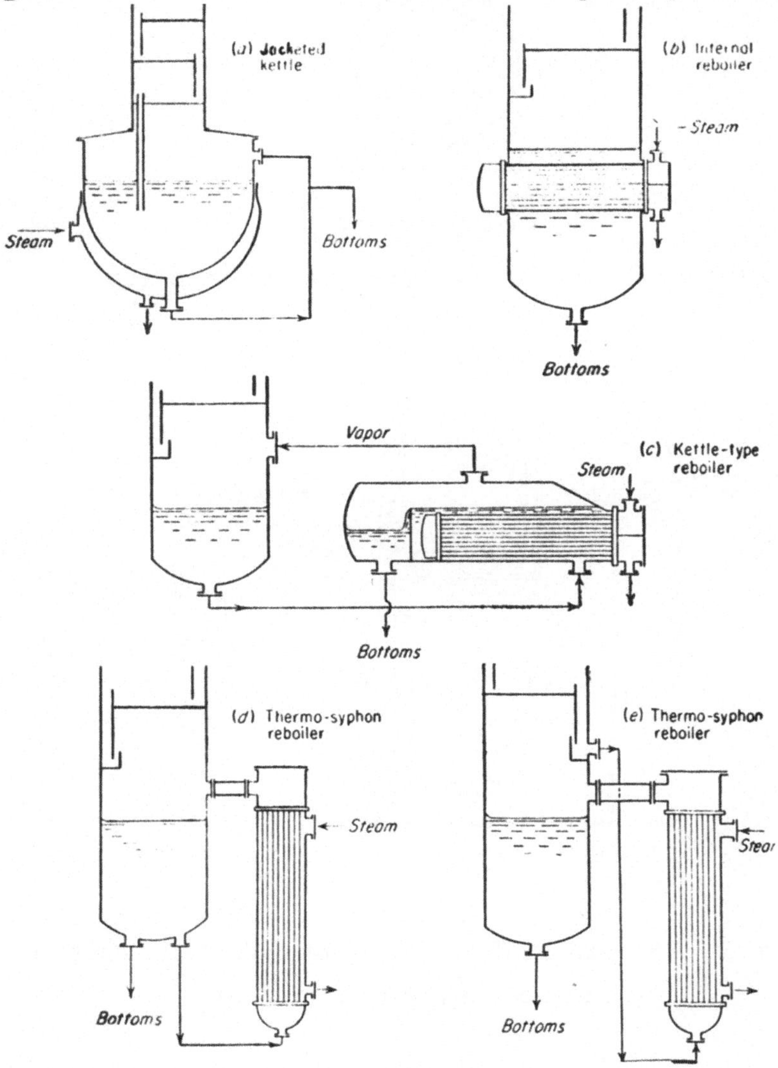

Figure 29 : Reboiler arrangements [21]

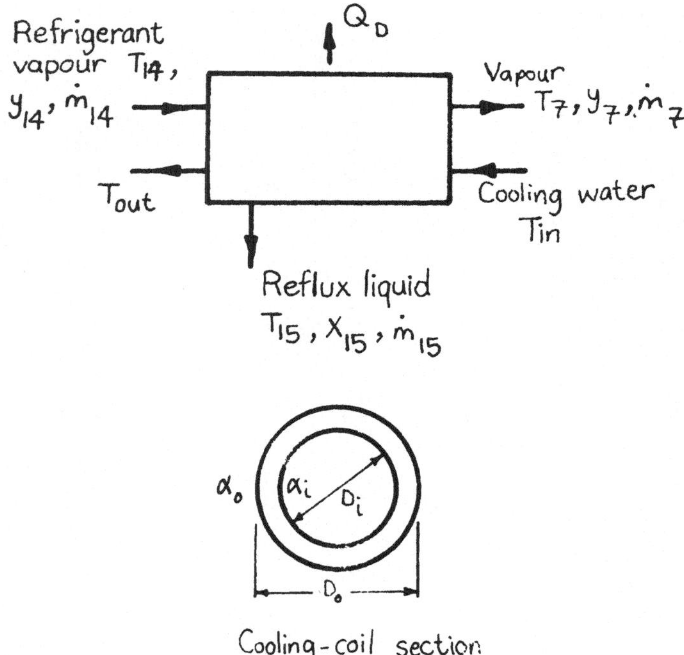

Figure 30 : Schematic of a water-cooled rectifier

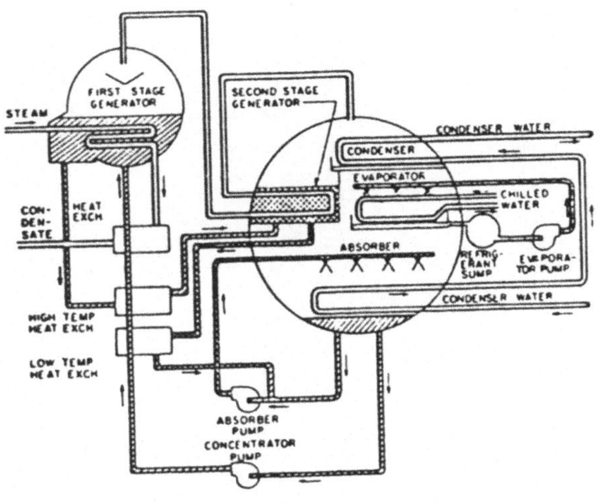

Figure 31 : Schematic of a lithium bromide cycle chiller with a two-stage generator [26]

ABSORPTION HEAT PUMP DESIGN

A.M.S. Qasrawi

Lucas Group Services Limited
Lucas Reserach Centre, West Midlands, United Kingdom

This lecture will describe the important factors that would affect the design of an absorption cycle heat pump. The influence of these factors on the design and selection of the main components of such a heat pump will be briefly outlined. A design example will be given to illustrate the procedures and methods used to select and size the main components of the heat pump for one particular application, namely, a heat pump for water and space heating only of a unitary domestic dwelling.

1. FACTORS AFFECTING THE DESIGN OF ABSORPTION HEAT PUMPS

There are four main factors that affect the design of an absorption heat pump, these are as follows :

(1) Temperature range of low-grade heat source
(2) Temperature range and medium of heat sink
(3) Working fluid pair
(4) Heating output of the pump (capacity)

1.1. Temperature Range of Low-Grade Heat Source

The low-grade heat source temperature range affects the selection of flow variables at various points in the cycle and the possible resultant performance of the heat pump. A wide range of source temperature will tend to complicate the process and equipment used to control the operation of the heat pump to achieve optimum or near optimum performance throughout the temperature range. When ambient air is used as a low-grade heat source, performance is reduced due to the relatively significant amount of mechanical energy needed to drive the fan for air circulation, and the amount of heat needed for defrosting the evaporator. This need for evaporator defrosting further complicates the heat pump control system.

1.2. Temperature Range of Heat-Sink Medium

The temperature range of the heat-sink medium strongly affects the design of an absorption heat pump. This is not only because the condensing temperature and pressure are related to the sink delivery temperature, but also because heat is extracted from the absorber, refrigerant condenser and the partial condenser (when one is needed in the generator). When these components are water-cooled e.g. there exist many options as to the order of cooling water passage through the above mentioned components.

1.3. Working Fluid Pair

The choice of a working fluid pair for an absorption heat pump will have a considerable affect on its design, the effect could fall in one or more of the following areas :

1.3.1. The choice of construction materials

Corrosive properties of the fluid pair and the appropriate cycle pressures and temperatures would limit the choice of materials that could be used in constructing an absorption heat pump. This would reflect on the cost of materials used and the manufacturing techniques adopted. Ammonia/water for example, would preclude the use of copper and its alloys.

1.3.2. The selection and design of a solution circulating pump

The choice of the working fluid pair will affect the choice of solution pump as related to the maximum pressure, safety requirements and the pump capacity aspects. These aspects will be expanded on in 2.2.

1.3.3. Heat pump operation safety requirements

Safety requirements pertinent to the operation of heating, chilling or air conditioning equipments vary from one country to another. Ammonia for example, cannot be used indoors in the USA. In Germany, an indoor system cannot contain more than 2.5 kg of ammonia, whilst in the UK this is only a code of practice rather than a legal requirement. This requirement would effectively limit the capacity of indoor units.

1.4. Heating Capacity of the Pump

The capacity of an absorption heat pump will significantly affect the design of a system with view to maximising the overall coefficient of performance. With large systems for example, more complex cycles could be employed. These also can afford the use of motorised refrigerant and solution expansion devices, metering solution pumps, modulated gas or oil burners and the necessary flow sensing devices, all of which would permit capacity modulation and operation at optimum performance throughout the ranges of source and sink temperatures.

2. PRINCIPLES OF DESIGN OF ABSORPTION HEAT PUMP COMPONENTS

This section will briefly describe design principles pertinent to the main components of an absorption heat pump and as affected by the factors outlines in 1.

2.1. The Generator

As was mentioned in the previous lecture the relative volatility of the working fluid pair will largely influence the general design of an absorption heat pump generator. When fluid pairs have moderate or low relative volatility (e.g. ammonia-water combination) a generator would consist of a still, a rectifying column and a partial condenser (rectifier).

2.1.1. The still (reboiler)

The function and types of generator stills were described in the previous lecture. The main design criterion for the still is that it should have an adequate surface area for transfer of the generator heat requirement at a heat flux per unit area much below the critical value. Different reboiler arrangements for the heat exchanger were described in the previous lecture.

Two other design points pertinent to the still have to be considered; firstly the need to maintain a relatively constant level of solution in the still (and more importantly preventing burnout of still and dangerous pressure build-up in case of failure of solution feed to the still), and secondly the need to control the heat addition to the still (for example, to maintain a preset bottom solution temperature at a particular outflow rate).

2.1.2. The rectifying column

The size of the heat pump (its heating capacity) and the fluid pair used are likely to dictate the choice of type of rectifying column to be used. Several factors favour the use of packed columns against tray columns especially for small and medium sized units, related to performance and cost. Some of these factors were described in the previous lecture. The methods of analysis and design of both types of rectifying columns were also briefly outlined.

Having decided on the type of column to be used, the designer has to consider the following parameters : diameter of the column and its height, type of plates or packing, liquid distribution, feed position(s) and heat addition to the column.

Column Diameter and Height

From a structural point of view, especially if the operating pressures are high, a small column diameter is desirable. The minimum column diameter will be dictated by the plate capacity for a plate column and by the flooding or loading point for a packed column. Moreover, the column diameter will have to be decided with regards to its likely effect on the Height Equivalent to a Transfer Unit (H.T.U.). Smaller diameters result in more favourable H.T.U.'s as a result of better liquid distribution and increased liquid rate (per unit area).

The height of the column will be defined by the number of actual trays or the height of packing needed to achieve the most severe separation (usually at the lowest source temperature and highest delivery temperature of the sink medium). It is worth remembering that (for a packed column) the H.T.U. becomes slightly less favourable [1] for greater heights of column possibly because of progressive maldistribution effects.

Type of Plates or Packing

When plate columns are chosen, the type of plates to be used is likely to be chosen with regards to the operating range and heating capacity of the unit. Cross-flow plates would be more appropriate for the larger units because of higher transfer efficiency and wider operating range. A wider operating range also favours valves or bubble caps for gas dispersers. For smaller units or those with narrower operating range counter-flow plates with simple perforations for gas dispersion would be more cost-effective.

Type of packing is chosen for its mechanical strength, resistance to corrosion, cost, capacity and efficiency. Most economical packings are 2,5 to 5 cm ceramic or carbon rings (1 cm size for columns under 10 cm in diameter); 2,5 cm saddles and 8 cm spiral or partition rings.

Liquid Distribution

The devices that are usually employed to distribute the solution at the feed point and the reflux liquid at the top of the column were described in the previous lecture. The size of the unit, its operating range and the rate of liquid flows will influence the selection of the type of liquid distributor. For small and medium systems the perforated pipe distributor offers a simple and cost-effective distributor.
For large installations that require a wide operating range, trough-type distributors would be more suitable.

If adequate care was taken in designing the column (of the packed type), it is likely that, for column heights encountered in heat pump application, liquid redistributors or side wipers would be necessary.

Feed Position(s)

For small systems and those with a narrow operating range, a single feed point (accurately placed for a selected design condition) would be adequate. For large systems with a wide operating range, the complexity of incorporating more than a single feed point with the necessary controls would have to be justified on the grounds of increased efficiency of the rectifying column.

Heat Addition to the Column

There is little data on the effect of adding heat to the rectifying column on its efficiency. For heat pump applications the direct use of the residue heat in the fuel exhaust gases to raise the sink's temperature is likely to be more beneficial to the overall performance than the addition of some of that heat to the rectifying column.

2.1.3. The partial condenser (rectifier)

The design of the rectifier should aim for the achievement of equilibrium condensation by ensuring adequate time of contact between the vapour product and the liquid reflux. Vapour or (and) liquid baffles could be used to ensure this time of contact. The rectifier should also have an adequate heat transfer surface area for cooling the vapour at the maximum rectifier load to be encountered throughout the operating range.

2.2. The Solution Pump

The main design considerations related to the solution circulating pump are as follows :

(1) Maximum cycle pressure : when the cycle operating pressures are very small (as in the case of water-lithium bromide cycle), the need for a solution circulating pump may be obviated by careful arrangement of the heat pump components so that gravitational forces could be utilised for solution circulation. When cycle pressures are moderate (up to 7 bars) gear or vane type pumps could be used, and for corrosive, inflammable or toxic fluids, magnetic coupling of the drive to the pumping component would provide a leak-proof system.

(2) Hazards of working fluids : as was mentioned above, hazardous working fluids should be circulated using a leak-proof pump. When high cycle pressures are also involved, hydraulically-driven diaphragm pumps are the only reliable solution pumps that are commercially available at the present time. These are, however, costly and complex.

(3) Pump capacity and capacity modulation : the maximum capacity of the solution circulating pump is dependent on the working fluid pair and the heat pump's operating range. For a particular type of solution pump, the circulation pump's power requirement would be consequently defined. For small and medium size heat pumps, especially those with a narrow operating range, a single-speed or a double-speed solution pump would satisfy the requirement. For large units, especially those with a wide operating range a variable output could be necessary for efficient operation.

2.3. The Absorber

There are three main design considerations related to the absorber. These are :

(1) Mass transfer consideration
(2) Heat transfer requirement
(3) Solution storage requirements

2.3.1. Mass transfer considerations

Systems which use fluid pairs that exhibit strong affinity for each other (like water-ammonia) could have a relatively simple design of the absorber. The different possible designs of the absorber that could be utilized are described in the lecture on absorbers. Systems which utilise working fluid pairs which exhibit low affinity for each other (methanol-lithium bromide/zinc bromide for example) might require the provision of more intimate contact of the phases through the use of plate or packed towers. The methods described in the lecture on absorbers should be used to arrive at a satisfactory design of the absorber for a particular application.

2.3.2. Heat transfer requirements

Since on average 50 to 60 % of the heating output of an absorption heat pump will be extracted from the absorber, it is important to transfer the heat to the sink medium at useful temperatures. The absorber vessel would be sized to accommodate the water cooling coil(s) (if internal cooling is adopted) needed for this heat transfer. More than one cooling coil with separate supplies of the cooling medium at different temperatures could be utilised in medium and large systems to make use of any temperature gradients within the absorber [2].

2.3.3. Solution storage requirements

There is a need for a solution collection chamber to act as a header tank for the solution circulating pump and to accomodate any changes in the solution concentration at extreme operating conditions (see 2.9). This warm solution could also be used to defrost an air heated evaporator by internal heating of the evaporator coil. The solution collection chamber would be sized to satisfy both requirements for normal operation and defrost cycles. If an internally-cooled absorber is selected it is sensible to use the bottom part of the vessel as the solution collecting chamber.

2.4. The Evaporator and Precooler

There are common design considerations, related to the evaporator, to both vapour compression and absorption heat pumps. These are : minimising fan power requirement for an air-heated evaporator, and minimising the energy used for defrosting the evaporator. As for the fan power requirement experience shows that large diameter slow running fans consume smaller amounts of power than smaller faster ones for the same air flow rate [3] and are also quieter. For absorption heat pumps, the evaporator defrosting could be effected by either internal of external heating. Hot exhaust gases from the burner could be utilised for external defrosting, however, this could result in corrosion problems with particular types of fuels. Alternatively, the warm strong solution could be diverted from the collection chamber to the evaporator for rapid and efficient partial melting of the ice layer on the coil.

It is in the aria of selection of the evaporator operating variables that the evaporator design in an absorption heat pump could markedly differ from that of a vapour compression cycle evaporator. This is true for systems which use working fluids with low volatility ratio. The presence of a binary mixture in the evaporator gives the designer some degree of freedom in selecting an evaporator pressure corresponding to a particular source temperature. The relevance of this to the performance of an absorption heat pump could be explained as follows.

When pure refrigerant is evolved from the generator (e.g. with water-lithium bromide and methanol-lithium bromide systems) the evaporator pressure will have to be reduced considerably for evaporation to take place at the lowest source temperature in the design range. When the absorber is effectively at the same pressure as the evaporator, this results in a reduction of the concentration of the strong solution in the absorber and consequently an increase in the required solution circulation rate and a reduction in the cycle coefficient of performance. This drawback could be partially overcome in such systems by raising the pressure of the vapour refrigerant leaving the evaporator to an intermediate pressure at the absorber. Such a solution was adopted by Battelle-Institut Frankfurt for their R22/E181 absorption heat pump [4] where a jet pump, operated by some of the refigerant liquid diverted from the condenser is used to raise the pressure of the vapour refrigerant leaving the evaporator. This however entails complexity and some loss of performance.

When working fluid pairs with moderate or low relative volatility (ammonia-water for example) are used, the evaporation process in the evaporator does not take place at a constant temperature because of the presence of some small amounts of solvent in the refrigerant stream entering it. The liquid entering the evaporator (and similarly that entering the throttling device) flashes into a vapour richer in the refrigerant and a liquid weaker in the refrigerant than the entering liquid. This is accompanied by a rise in temperature; a drawback for cooling systems but of less significance for systems intended for heating only. With such systems the evaporator pressures (and hence the absorber pressures) needed to evaporate most of the mixture at moderate source temperatures (5 < T < 15 °C) are adequate to achieve high concentrations of the strong solution in the absorber. At lower source temperatures it is more desirable to allow some liquid to leave the evaporator at moderate evaporator pressures than attempt to fully evaporate the mixture at much reduced evaporator (and hence absorber) pressure. The liquid carried over from the evaporator could then be vapourised in the refrigerant precooler at elevated temperatures (not much below condensing temperatures in fact). The precooler would have to be sized to complete the vapourisation of the maximum amount of liquid carryover that is allowed from the evaporator at the minimum source temperature. Baffles might have to be incorporated in the precooler to provide lengthy passage time of the liquid thus ensuring vapourisation before entry into the absorber.

2.5. The Solution Pre-Heater (Recuperator)

The size of the recuperator is determined by the desired state of the feed to the generator and the temperature range of the bottom solution leaving the generator still. Increasing the feed enthalpy reduces the still heat requirement but also could make the separation of relatively volatile solvents from refrigerants more difficult which tends to counteract to a degree the performance improvement from the first effect. There is no evidence to suggest that there is much advantage in increasing the enthalpy of the feed beyond that of a saturated liquid.

Economic factors will have to be taken into consideration when deciding on the recuperator capacity. A trade-off between the cost of the heat exchanger and the performance improvement (resulting from raising the feed enthalpy) has to be made.

2.6. The Condenser

Economic considerations are also likely to play an important role when selecting the type of condenser to be used. For small water-cooled condensers e.g., a single helical coil in a shell (in which the cooling water flows) might be the most cost-effective type. With larger units the more costly tubes-in-shell arrangement would produce a more compact condenser. Bundles of finned tubes are appropriate to air-cooled condensers.

2.7. The Solution Throttle

The solution throttle is simply a pressure-reducing device through which the weak solution could partially flash before entry into the absorber. For small systems with an on-off mode of operation, a fixed constriction (chosen for the required solution flow rate at the design operating condition) could be adequate. If the heat pump is to have a capacity-modulation capability or when large systems have a wide operating range, then a motorised variable constriction valve would be needed to control the flow to match the different operating conditions and the requirements during start up and shut down.

2.8. Refrigerant Control Devices

The choice of a refrigerant control device for a particular type of heat pump will depend on the size of the unit and the working fluid pair through its effect on the design approach adopted to select the evaporator operating conditions (as was described in 2.4).

For small systems with a narrow operating range, a capillary tube selected to operate at the specified design condition is likely to be adequate.

When a wider operating range is required an expansion valve that could be manually adjusted would be necessary. For working fluid pairs where a pure refrigerant is evolved, a thermostatically-controlled expansion valve which maintains a preset degree of vapour superheat at the exit of the evaporator could be used. With an ammonia/water system, an automatic expansion valve (a pressure regulator) which maintains a preset pressure in the evaporator (and hence in the absorber) would be more appropriate.

For large systems with a wide operating range motorised expansion valves (whether thermostatically-controlled or pressure-regulating) might be cost-effective but they add to the complexity of the system which could have an adverse effect on its reliability.

2.9. Liquid Collection Chambers

When an absorption heat pump has to operate through a wide operating range, the concentrations of the refrigerant, weak solution and strong solution streams could vary considerably between the most severe and least demanding operating conditions. These variations lead to a need to provide in the system two liquid collection chambers; one for the solution and another for the refrigerant. The solution collection chamber could be placed on the suction side of the solution pump to act also as a header tank for the pump. The regrigerant collection chamber could be placed downstream of the condenser and be integrated with the condenser itself if a tubes-in-shell arrangement was used. One method of sizing the collection chambers will be described in section 3 (A Design Example).

2.10. Capacity Modulation

Full capacity modulation of an absorption heat pump necessitates the control of the flow rates of refrigerant and weak and strong solution streams and the rate of heat input to the generator still. As was mentioned earlier, when the design principles of the refrigerant control device, solution throttling device and solution pump were discussed, these devices need to be automatically controlled to vary their throughput to suit the different load conditions. The cost and complexity of providing two motorized expansion devices, a variable discharge solution pump and a variable output burner (gas or oil fired) makes full capacity modulation of absorption cycle heat pumps possible only in medium and large units (output > 100 kW), when the improved performance resulting from capacity modulation could justify the increased capital cost. It is also to be remembered, in this context, that the absorption cycle heat pump has inherently less transient losses than a vapour compression cycle heat pump, because of the substantial volume of high pressure vapour in the generator system. This could mean that the penalty of adopting an on-off mode of operation for small systems would be relatively small.

3. A DESIGN EXAMPLE

The following design example will not attempt to answer all the questions that arise when an absorption cycle heat pump is to be designed for a particular application. It will attempt to provide only a general outline of the design procedures that could be utilised. These design procedures could vary according to the application. It is also to be remembered that the data on the processes involved and working fluids to be used could be scant or non-existent in some cases.

3.1. Definition of the System under Consideration

The system to be designed has the following characteristics :

- an absorption heat pump for domestic water and space heating;
- nominal heating capacity 10 kW
- low grade heat source : ambient air at temperatures from - 5 °C to + 15 °C; design source temperature is 0 °C;

- heat sink : water, delivery temperature 60 °C and return temperature 40 °C.

These temperatures are compatible with space heating radiators which are used with present heating systems utilising gas-fired or oil-fired boilers.

- High-grade heat is supplied by a gas burner.
- On-off mode of operation at the rated heating capacity.

3.2. Selection of a Working Fluid Pair and a Cycle Design

Taking the requirements of the system into consideration and studying all the factors that influence the selection of a working fluid pair, the designer arrives at a compromise decision. A cycle design follows to a great extent from this selection.

For the purpose of this design example, it is assumed that ammonia/water combination was selected. The cycle design chosen is the classical single-stage generator cycle supplemented by two heat exchangers (a refrigerant precooler, and a solution preheater or recuperator). This cycle design is shown schematically in Figure 1, the numbers shown refer to the different state points within the cycle and for the cooling water circuit. These numbers will be used throughout the design example when the values of flow variables at these points are calculated.

Because of the relatively high value of relative volatility ratio α for ammonia-water combination, the separations required for this application necessitate the use of a partial condenser (rectifier) and a rectifying column. The performance of the heat pump will largely depend on the reflux ratio required at the different operating conditions and the design procedure should aim to minimise this ratio (i.e. minimise the rectifier load).

For the example considered it was decided to cool the rectifier using the water circuit rather than the strong solution since it could be reasoned that the latter option provides no relative thermodynamic advantage. This is true because the amount of heat taken out of the rectifier, using the solution, will effectively be added to the absorber at point 6. This means that the heat exchanger surface in the absorber will have to be increased.

It is worth noting that cooling the rectifier by water (as was adopted in this example) would reduce the mass flow rates around the cycle since the output of the heat pump is the sum of the heat quantities removed from the condenser, the absorber and the rectifier.

3.3. Cycle Analysis

The cycle analysis starts by ignoring the rectifier heat load and assuming that the heat pump's output is the sum of heat quantities removed from the condenser and the absorber only. It is also assumed that the refrigerant flow rate through the condenser is equal to unity (1 kg/s). The mass flow rates are later corrected for the output required (taking also into account the rectifier heat output).

Many assumptions concerning the cycle are taken, these are as follows :

(1) Heat exchange in all vessels and heat exchangers occurs adiabatically, i.e. there are no heat losses to the environment.

(2) There are no pressure changes in the vessels and heat exchangers. Pressure drops occur only across the two throttles and the system is characterised by two pressure regions.

(3) Saturated liquid conditions are assumed at the following state points : 1, 3, 4, 8 and 15.

(4) Saturated vapour conditions are assumed at the following state points : 7, 12 and 14.

(5) Two assumptions are made about the generator system :

- the liquid reflux stream is in equilibrium with the vapour leaving the partial condenser. This means that the rectifier accomplishes mass transfer equivalent to one theoretical plate (equilibrium condensation occurs).
- the vapour rising from the boiling pool in the still is in equilibrium with the weak solution extracted from the bottom of the still. This means that the still is also equivalent to a single theoretical plate.

(6) Cooling water is arranged in parallel through the absorber, condenser and rectifier.

(7) Minimum limits for the temperature differences across the heat exchangers are assumed. In the present example, the temperature differences for the condenser inlet, evaporator, absorber, rectifier, precooler HX2 and recuperator HX1 are set to 5 °C minimum. The temperature difference at the condenser outlet is set to 15 °C minimum.

(8) The temperature of the vapour entering the condenser (T_7) and that of the weak solution leaving the bottom of the still (T_4) are taken as boundary conditions. T_7 should be selected with

regards to the water delivery temperature required. For the present example it is set to 80 °C. T_4 is set to a value that is likely to produce the highest coefficient of performance (everything else being constant) taking into consideration likely limits imposed by the thermal stability of a working fluid (an organic solvent for example when used with a fluorcarbon refrigerant) or the formation of solids with salts' solutions (lithium bromide-water system for example). In the system under consideration such limitations do not apply and T_4 is chosen as to its possible effect on performance. For the present example T_4 was taken to be 175 °C.

(9) Phases are assumed to be in equilibrium at the exit from both throttles.

(10) The evaporator pressure could be considered as an independent variable that could be calculated if the refrigerant is to be fully evaporated in the evaporator (then 11 is assumed to be a saturated vapour state) or could be treated as a boundary condition to be pre-selected. For reasons explained in 2.4, the evaporator pressure, for the purpose of this example, will be considered as a boundary condition to be preset. A value of 2.5 bar was chosen for the absolute value of the evaporator pressure.

The cycle analysis is carried out in two areas; cycle parameters calculations, and components design and sizing. The first area utilises the cycle boundary conditions, thermodynamic assumptions, and fluids' properties to establish all flow parameters within the cycle (including the different components of the generator system) and in the cooling water circuit. The second area involves the design and sizing of all vessels, heat exchangers and throttles to satisfy the requirements of the cycle at the most demanding condition.

3.3.1. Cycle parameter calculations

- Calculations start at the condenser : minimum temperature differences are assumed for both ends of the condenser as shown in Figure 2. This defines the water exit temperature and the refrigerant exit temperature. The condensing pressure and vapour concentration are then calculated iteratively from the mixture properties. The condenser heat output and its cooling water flow rates are then found.

- From the high-side pressure and T_4 (with the assumption of saturated liquid), the weak solution concentration is calculated.

- The absorber exit state is analysed next, the minimum temperature difference is assumed at exit as shown in Figure 2. From the lowside pressure and the exit temperature, the strong solution concentration is calculated.

- Mass flow rates of the weak and strong solutions per unit mass flow rate of the refrigerant are calculated from mass conservation of ammonia and water. The solution pump work is then calculated assuming an efficiency of 75 %. State point 2 is then completely defined.

- The solution recuperator is then analysed; state point 3 is defined by the high side pressure, the concentration and the assumption of saturated liquid. Heat exchange from the strong solution stream is then calculated. Since state point 4 is also defined (from the pressure, concentration and the assumption of saturated liquid) T_5 is then calculated from the heat exchange across the recuperator.

- At this point the generator system is analysed. Having defined the different concentrations within the cycle and approximate values of the flow rates (neglecting the rectifier load), the methods described in the previous lecture could be used to arrive at the required number of theoretical plates or transfer units using a preliminary value of the external reflux ratio (a value less than 0.5 is desirable) and column diameter. The resulting column height should be reasonable for the environment in which the unit is to operate. This number of theoretical plates or mass transfer units should now be used for a complete analysis of the generator system which would produce the values of the reflux ratio, the temperature and concentration of the reflux liquid, the temperature and concentration of the vapour leaving the rectifying column, the position of the feed and the generator and rectifier heat loads per unit refrigerant flow rate. For the present example, a packed column was selected and a non-constant molal overflow mass transfer analysis was carried out. The cooling water flow through the rectifier is defined by the maximum water temperature allowed (95 °C in this example) or by the water exit temperature as related to T_{14}.

- The refrigerant precooler is next analysed; T_{12} is related to T_8, hence state 12 is completely defined (it is to be noted that when the evaporator pressure is preset, state 12 is not necessarily a saturated vapour state). T_{11} is related to the source temperature ($T_{amb.}$); this together with the pressure and concentration, defines state 11 (liquid carryover is then determined). Heat exchange across HX2 is then known and this defines state point 9.

- The evaporator is then analysed; the throttling process through the refrigerant control device defines state 10 (flash and temperature). The evaporator heat exchange is also defined.

- The throttling process through the solution throttling valve defines state 6. The absorber heat exchange could now be calculated and the water exit temperature for the absorber cooler is found.

A computer program that performs a comprehensive cycle analysis along the preceeding lines was used to carry out the cycle calculations for the present example. Table 1 shows some of the input data to the program and Table 2 shows the output. These tables are applicable to the design ambient temperature of 0 °C and assuming 2 mass transfer units. This number of mass transfer units was chosen after a preliminary calculation (1) (5) has shown that for a reflux ratio of about 0.5 a column height of about 1.0 m (diameter of 0.1 m) would be adequate to produce the required separations.

3.3.2. Components' design and sizing

This section will describe the design and sizing of the major components of the heat pump to provide the required heat, mass and flow transfers as calculated in the previous section. Emphasis would be put on those components that are unique to an absorption cycle heat pump (the generator and the absorber) and that would not be commercially available, especially when an experimental or pilot plant is to be designed and constructed. All vessels and heat exchangers are to be constructed from good commercial quality carbon steel. For experimental rigs stainless steel construction is recommended to minimise the possibility of corrosion during manufacture or modifications. It is to be stressed here that the calculations of heat exchange areas should be treated as first order estimations : since fouling resistances were ignored and heat transfer coefficients were occasionally assumed. The designs of many of the components should also be treated as schematic and not as engineering designs.

The Generator.

The Still (reboiler).

The still in this example transfers a maximum of about 7.12 kW of heat to the solution at a temperature of 175 °C. Using the method described in the lecture on generators, the heat transfer coefficient [6] (assuming free pool boiling) is calculated to be approximately 2.35×10^4 W/m^2 °K.

Assuming an internal wall temperature of 180 °C (5 °C higher than the boiling solution temperature), the necessary heat transfer area is calculated to be 0.061 m^2. The heat flux for this surface area is about 1.2×10^5 W/m^2.

A jacketed cylindrical shell (having 5 mm wall thickness and 0.1 m diameter) to be heated by the burner's flue gases is used as a reboiler as shown schematically in Figure 3. The wall heat transfer enables one to calculate the temperature difference between the outer wall surface and the inner surface. This difference for the present case is about 12.5 °C.

The critical value for the heat flux per unit area is calculated to be 3.4×10^6 W/m^2, a figure much larger than the actual value of 1.2×10^5 W/m^2; indicating a safe operation much below burnout point.

No attempt will be made in this example to indicate the possible designs of the gas burner to supply the necessary heat, but it will suffice to say that an advanced burner will have to be used; with a provision for recuperation of some of the residual heat in the flue gases.

A Chromel-Alumel thermocouple sensing the bulk temperature of the liquid in the still as shown in Figure 3 could be used to control the gas burner used to heat the still such that T_4 is maintained at a preset level. Two insulated-electrode type level sensors could be used together with an electronic level controller to maintain the solution level in the still between the desired minimum and maximum levels.

The Rectifying Column

A packed column with a single feed point is selected for the present example. The column will consist of a shell (0.108 m outside diameter and 3.25 mm wall thickness) filled with Raschig rings having a 9.5 mm diameter.
Using the methods described in the previous lecture the HETU's above and below the feed are estimated (based on overall gas phase resistance) to be 0.75 m and 0.32 m respectively. Hence the overall column height will be 0.90 m and the feed position is at 0.45 m from the bottom of the column. A wire mesh support plate and a perforated pipe for feed liquid distribution will be used. A perforated drip tray at the top of the packed column distributes the reflux liquid as shown in Figure 4.

The Rectifier (Partial Condenser)

The rectifier is arranged as a separate vessel as shown in Figure 5 to effect control of the reflux liquid back to the top of the rectifying column. To simplify the construction of the rectifier, the pipe that connects the rectifying column to the rectifier carries both the vapour rising from the column and the reflux liquid flowing back to it.

The rising vapour has the following properties :
temperature : 134.4 °C; concentration : 0.885; mass flow rate : 0.0035 kg/s.

The reflux liquid has the following properties :
temperature : 80 °C; concentration : 0.626; mass flow rate : 0.001 kg/s.

A tube (0.00635 m outside diameter and 0.00452 m inside diameter) wound, as shown in Figure 5, is used for the cooling water circuit. Using the method described in the lecture on generators [6], the tube length needed to transfer the 2.19 kW of heat from the rectifier is estimated to be 0.6 m. A needle flow control valve placed downstream of the cooling water tube is used to adjust the rectifier load since this will have a strong influence on the operation and performance of the heat pump.

The Absorber.

The absorber for the present example is of the wetted-wall dripple type. It will take the form of a vessel in which the work solution is distributed (after being throttled outside the absorber) by a drip tray on a helical cooling coil as shown in Figure 6. The refrigerant vapour is admitted through a perforated pipe to provide cross and counter flow of the phases. With this arrangement heat and mass transfers occur simultaneously. The bottom part of the absorber is utilised as a header tank for the solution pump.

The heat exchange from the absorber is about 5.0 kW. It is here assumed (for the purpose of calculating the dimensions of the cooling coil required) that absorption takes place prior to heat removal.
State point 13 represents the state of the mixture prior to cooling. The solution and cooling water temperatures are shown in Figure 2.

For the purpose of heat transfer calculations, the method described in reference [7] was used to estimate the coil dimensions based on a falling-film heat transfer correlation. These are shown together with other dimensions of the absorber in Figure 6. The header tank was sized to provide a maximum pump flow duration of three minutes.

As for the mass transfer capability of the absorber, the appreciable affinity of water for ammonia makes the arrangement described above adequate for this purpose, hence no check on the design was made from this aspect. For other working fluids that do not possess such affinity, the methods described in the lecture on absorbers should be used to ensure that the mass transfer requirements are met.

The Condenser.

Two phase flow occurs in the condenser. A vertical tubes-in-shell arrangement was chosen for this example, with downward flow of vapour in the tubes and upward flow of the cooling water in the single pass shell, as shown in Figure 2. The methods described in reference [6] were used to estimate the condensing coefficient for the heat exchanger. The required number of tubes and all dimensions are shown in Figure 7 as arrived at using the method mentioned above.

The Recuperator (HX1).

The solution temperatures at both ends of the recuperator are shown in Figure 2. A tube-in-tube arrangement was chosen for this heat exchanger with the weak hot solution flowing in the inner tube. Choosing the following dimensions for the tubes : inner, internal diameter : 0.01384 m, outside diameter : 0.01714 m; outer tube, internal diameter : 0.02093 m, outside diameter : 0.02667 m, the required heat exchanger length for a heat exchange rate of 4.58 kW is found to be 8,7 m, using the method described in reference [6]. The heat exchanger could be arranged as seven lengths of tubes standing vertically along the generator as shown in Figure 13. The total tube length could alternately be coiled around the rectifying column, in two segments (each having ten turns with a mean diameter of about 0.14 m) above and below the feed pipe.

The evaporator.

The refrigerant and air streams' temperatures for the evaporator are shown in Figure 2. A finned-tubes bundle in cross air flow was chosen for the evaporator as shown in Figure 8. Forced draught of the air is generated by an electrically driven fan.

An approach air velocity of 5 m/s is assumed. The methods described in reference [6] were used to determine the required length of finned tubes in the bundle. The dimensions of the bundle and the fins are shown in Figure 8. The power needed to drive the fan is estimated to be 125 Watts. Defrosting of the evaporator is achieved by periodic internal heating of the evaporator using the warm solution in the bottom of the absorber. This solution is diverted to the evaporator through a three-port two-way solenoid valve as shown in Figure 1. The frequency of defrosting the evaporator could be changed as a function of the air temperature in the range $-5 < T < 7$ °C for example. This function could be tailored to the prevailing humidity levels in the locality in which the heat pump is to be used.

The Precooler (HX2)

The precooler in this example has the capacity to evaporate the maximum liquid carry over that is allowed, at the minimum design temperature of - 5 °C, of 0.069 and raise its temperature to 50 °C (5 °C below the temperature of the condensed refrigerant leaving the condenser). A single pass tubes-in-shell arrangement is used for this heat exchanger with the hot liquid flowing vertically downwards in the tubes and the wet cold vapour entering the shell and flowing upward around the baffles as shown in Figure 2. The methods described in reference [6] were again used to estimate the dimensions of the heat exchanger. These are shown in Figure 9.

The Solution Pump.

There are very few designs of a solution pump that could serve the duty required in this design example. Hydraulically driven diaphragm type pumps are the only commercially available pumps that could be utilised at the maximum discharge pressures needed. The solution pump selected for this example is a scaled-down version of the pump used on the Arcla ACB air-cooled water chillers. The hydraulic pump is a 'Saginaw" rotor vane pump that delivers 104 pulsations per minute, and the pressure goes from 0 to a possible 28 bar and back to 0.

This pressure is transmitted, through a connecting line, to one side of a Teflon diaphragm in the solution pump.

The hydraulic pump is belt driven by an AC electric motor rated at 0.095 kW. For long life and reliability the pump is sized to satisfy the maximum solution circulation requirement of about 0.8 l/min. at about 60 % of its maximum capacity. Figure 10 is a schematic of the solution and hydraulic pumps.

The Refrigerant Control Valve.

A pressure regulating valve is used for refrigerant control. This valve (shown schematically in Figure 11) maintains a preset pressure in the evaporator; 2.5 bar in this example. It is a proprietary item, supplied by the Refrigerating Specialities Company of the USA.

The valve could be readjusted manually for optimum performance at different values of the ambient temperature or could be set at the required pressure of 2.5 bar for the design ambient temperature of 0 °C.

The Solution Throttling Device.

A fixed restrictor acts as a solution throttling device as shown in Figure 12. The orifice is designed to allow the required flow rate of weak solution of about 0.57 l/min. at the rated high side and low side pressure.

Refrigerant and Solution Collection Chambers.

If the heat pump is to operate at different running conditions (manifested in variations in the strong and weak solutions'mass concentrations), variations in the liquid volume requirements will exist. These variations could be accommodated by the provision of two collection chambers; one for the solution and another for the refrigerant liquid.

Equations may be written representing the total charge of water and ammonia contained within the heat pump. These expressions are products of volume and density of the liquid for water and ammonia at the extreme operating conditions (giving maximum and minimum solution concentrations). Solving the resulting four simultaneous equations gives the sizes of the solution and refrigerant collection chambers and the total masses of ammonia and water in the unit.

The assumptions made when writing the equations mentioned overleaf are as follows :

1. Masses of vapour are neglected, refrigerant vapour and liquid are assumed to be pure ammonia.

2. Total liquid hold up in the packed column is estimated for the packing used and assumed to be weak solution.

3. The evaporator is assumed to be one quarter full of liquid.

4. The solution collection chamber is assumed to be at its minimum level for the operating condition producing minimum values of concentrations of strong and weak solutions.

5. The refrigerant collection chamber is assumed to be at its minimum level for the operating condition producing the maximum values of concentrations of strong and weak solutions.

For the present example, one operating condition is maintained throughout the ambient temperature range, and hence there is strictly no need for having collection chambers in the unit. However the bottom of the absorber was used as a solution collection chamber to act as a header tank for the solution pump and to be used for evaporator defrosting purposes. It was on the basis of these functions that this chamber was sized.

The sizes of all vessels and heat exchangers' tubes and connecting piping was estimated. From these and the properties of the refrigerant and weak and strong solutions, it was calculated that the required masses of ammonia and water in the unit were approximately 1.8 kg and 3.25 kg respecitively.

Filters.

Two filters are used in the heat pump to protect the solution pump and the refrigerant control valve. These are in-line strainers of stainless steel gauze having 0.1 mm diameter wire. The solution filter is fitted on the suction side of the solution pump and the refrigerant filter is fitted immediately upstream of the refrigerant control valve. The strainers are proprietary items (Type FA15) supplied by Danfoss of Denmark.

The Heat Pump General Arrangement

Figure 13 shows a proposed general arrangement of the heat pump components. This arrangement aims at utilising gravitational forces to assist the proper flow of the working fluid and at optimising the space requirements of the unit. A split arrangement was considered preferable to a packaged one to minimise the structural modifications needed when installing the heat pump indoors of an existing dwelling; a factor that is likely to influence customers' acceptance of such a unit.

Figure 13 does not show details of the water-cooling circuit or those of the monitoring and control systems. There are also many unanswered questions regarding the operation and control of the heat pump. The purpose of this design example was, as was mentioned before, only to outline possible design procedures for one particular application of the absorption cycle heat pump. A considerable amount of research and operating experience is still to be accumulated by manufacturers of such systems before the optimum operation and control procedures are arrived at.

ACKNOWLEDGEMENTS

The author would like to express his gratitude to the Management of Lucas Research Centre for providing time and resources for preparation of these lectures, and to Mr. R.J. Treece of Newcastle upon Tyne Polytechnic for his many helpful comments.

REFERENCES

1. Perry, R. and Chilton, C. *Chemical Engineers Handbook* 3rd and 5th Editions.

2. Stoecker, W.F. and Reed, L. *Effect of Operating Temperatures on the Coefficient of Performance of Aqua-Ammonia Refrigerating Systems* Ashrae Paper N° 2183 (1971).

3. Gordian Associates *Heat Pump Technology* Inc. for US Department of Energy (1978).

4. Jansen H.A. and Oelert G. *Development of a Primary Driven Absorption Heat Pump for Domestic Heating* EEC Contractors Meeting Brussels (1978).

5. Fellinger *Sc. D. Thesis* MIT (1941).

6. Butterworth, D. *Introduction to Heat Transfer* Engineering Design Guide N° 18 Oxford University Press (1977).

7. Mc Adams,W.H.; Drew, T.B. and Bays, G.S. *Trans. ASME*, 62, 627-631 (1940).

TABLE I

Input Data for Computer Cycle Analysis Program

Cooling Water Circuit (Parallel Arrangement)

Delivery Temperature 60 °C		Return Temperature 40 °C
Heating Output	:	10 kW
Generator Bottom Temperature	:	175 °C
Condensing Temperature (Inlet)	:	80 °C
Solution Pump Efficiency	:	0.75
Ambient Air Temperature	:	0.0 °C
Number of Mass Transfer Units	:	2.0

Minimum Heat Exchangers Temperature Differentials :

Absorber : 5 °C, Condenser Inlet : 5 °C, Condenser Outlet : 15 °C, Recuperator : 5 °C, Precooler : 5 °C, Rectifier : 5 °C, Evaporator : 5 °C

Evaporator Pressure	:	2.5 bar

Non-constant molal overflow model of the rectifying column.

TABLE 2

Output of the Cycle Analysis Program

Number of Mass Transfer Units :

Above the Feed : 0.6   Below the Feed : 1.4

Reflux Ratio : 0.41

Liquid Carryover : 0.032

Evaporator Throttle Flash : 0.89

Absorber Throttle Flash : 1.00

Cycle Point	Pressure bar	Temperature °C	Concentration by weight	Mass Flow Rate kg/s	Enthalpy kj/kg
1	2.5	45.0	0.342	0.0112	- 21.3
2	23.0	47.0	0.342	0.0112	- 18.3
3	23.0	130.0	0.342	0.0112	389.5
4	23.0	175.0	0.155	0.0087	665.2
5	23.0	63.7	0.155	0.0087	139.8
6	2.5	60.0	0.155	0.0087	139.8
7	23.0	80.0	0.991	0.0025	1392.3
8	23.0	55.0	0.991	0.0025	259.1
9	23.0	17.6	0.991	0.0025	24.3
10	2.5	-12.6	0.991	0.0025	24.3
11	2.5	- 5.0	0.991	0.0025	1154.2
12	2.5	50.0	0.991	0.0025	1389.0
13	2.5	71.1	0.342	0.0	419.5
14	23.0	134.4	0.885	0.0035	1650.3
15	23.0	80.0	0.626	0.0010	150.4

Heat Exchange Rates (kW) :

Evaporator : 2.84, Generator : 7.12, Condenser : 2.85, Rectifier : 2.19, Absorber : 4.86, Recuperator : 4.58, Precooler : 0.59 (at - 5 °C).

Pump Power : 0.034 kW

Coefficient of Performance : 1.40

Cooling water flows :

	Mass Flow Rate (kg/s)	Inlet Temp.(°C)	Outlet Temp.(°C)
Condenser	0.0195	40	75
Rectifier	0.0095	40	95
Absorber	0.0904	40	53.1

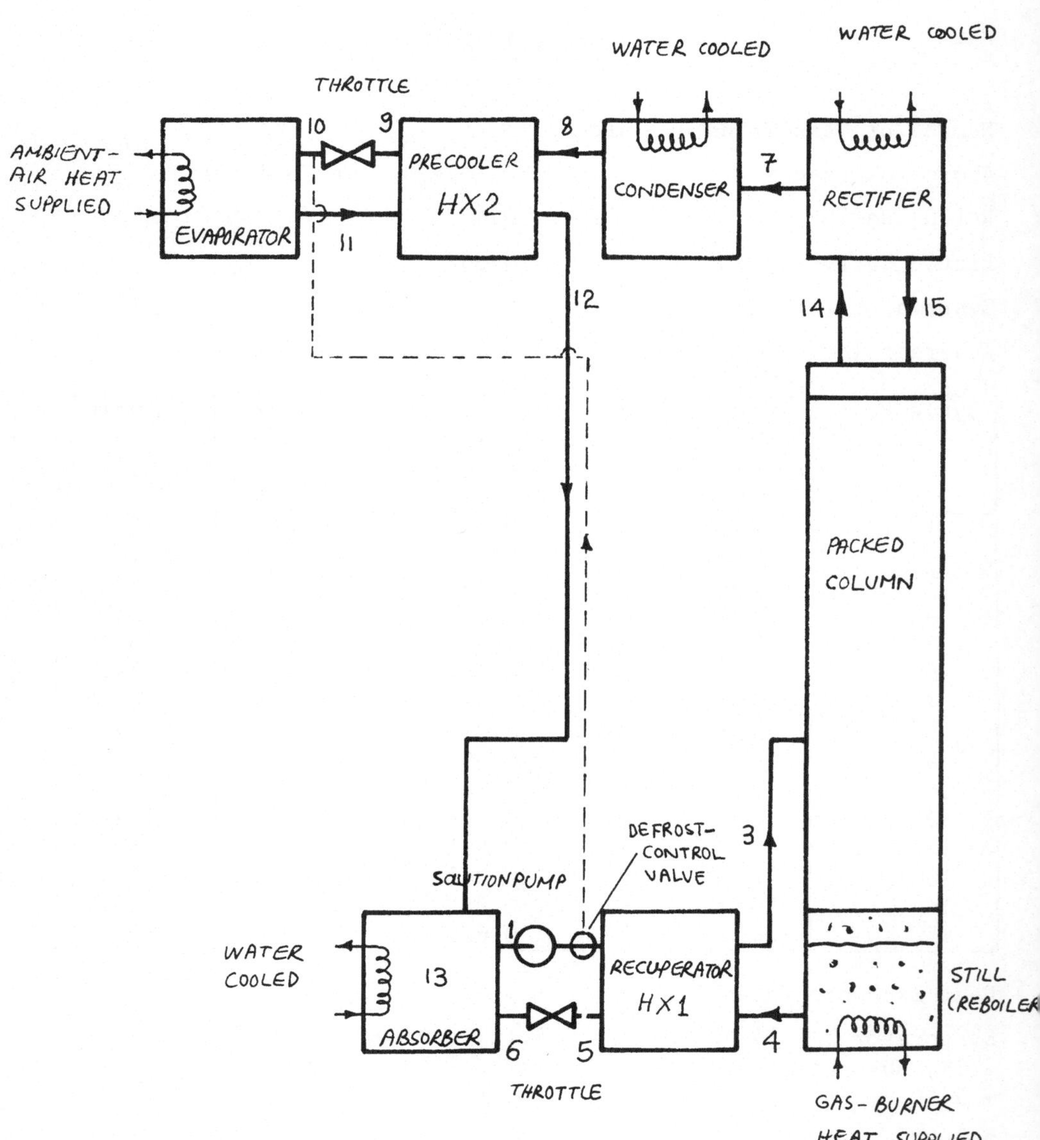

Figure 1 : Schematic diagram of heat pump cycle

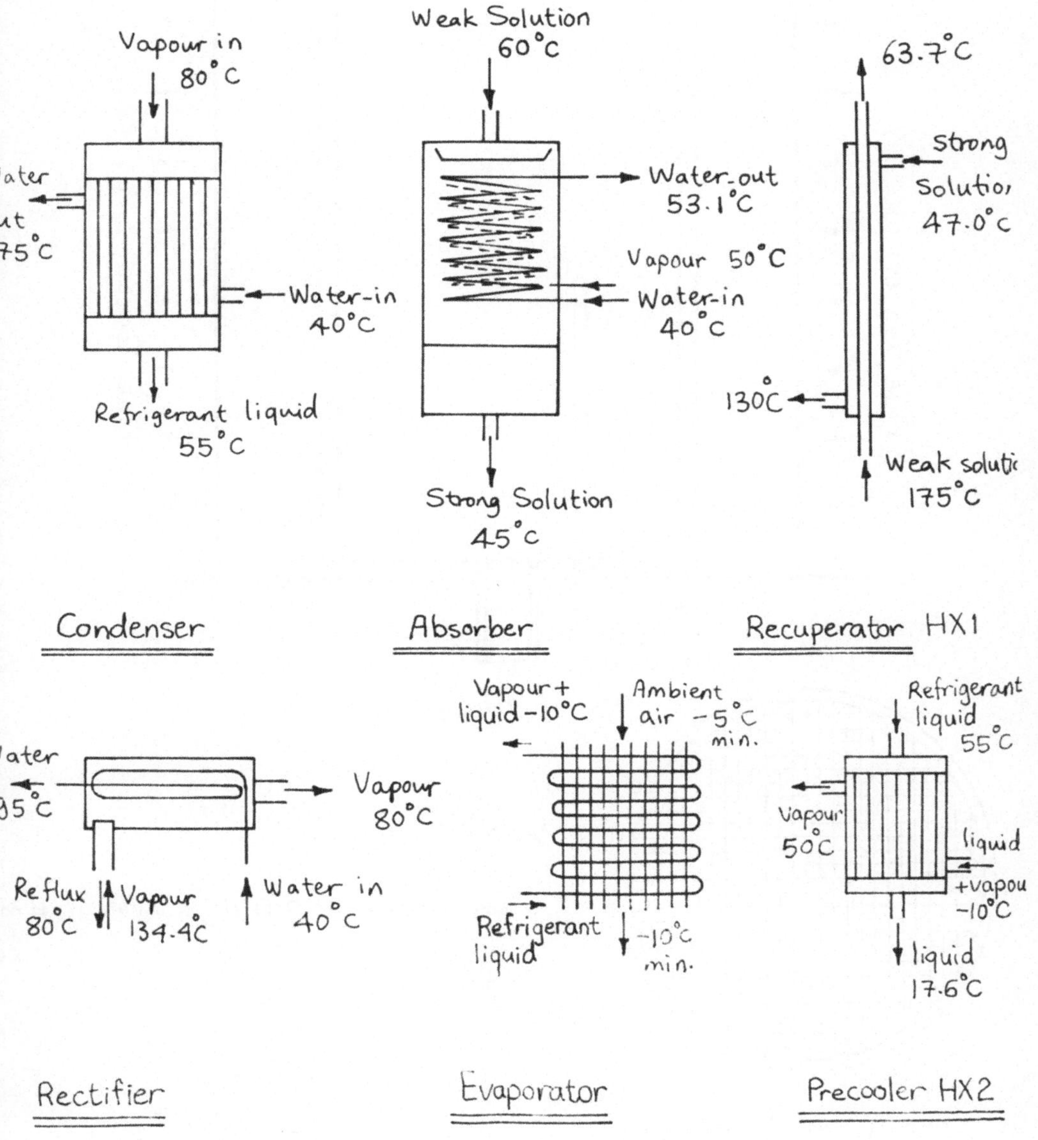

Figure 2 : Schematic diagrams and flow temperatures of heat exchangers

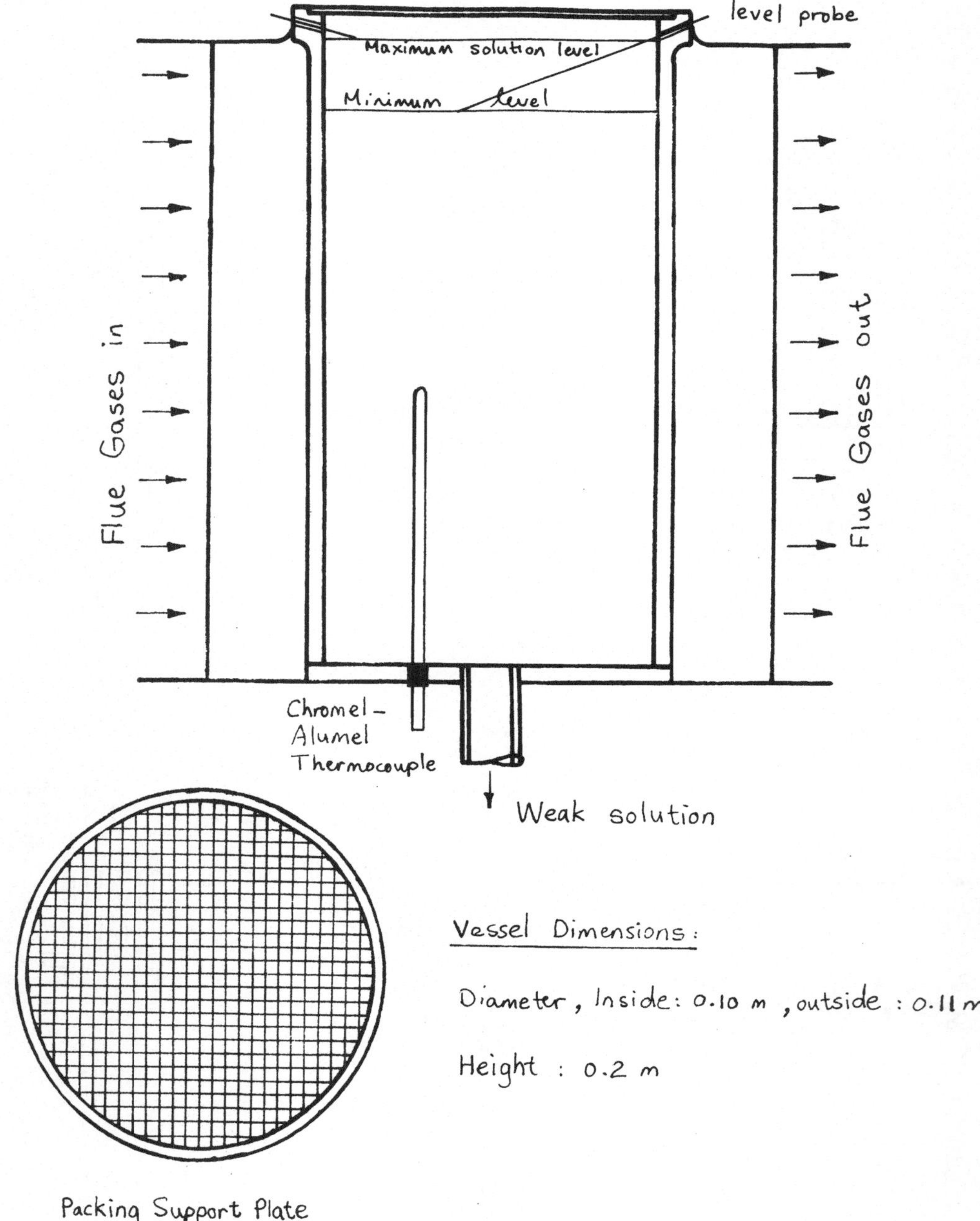

Figure 3 : Design of the generator still (reboiler)

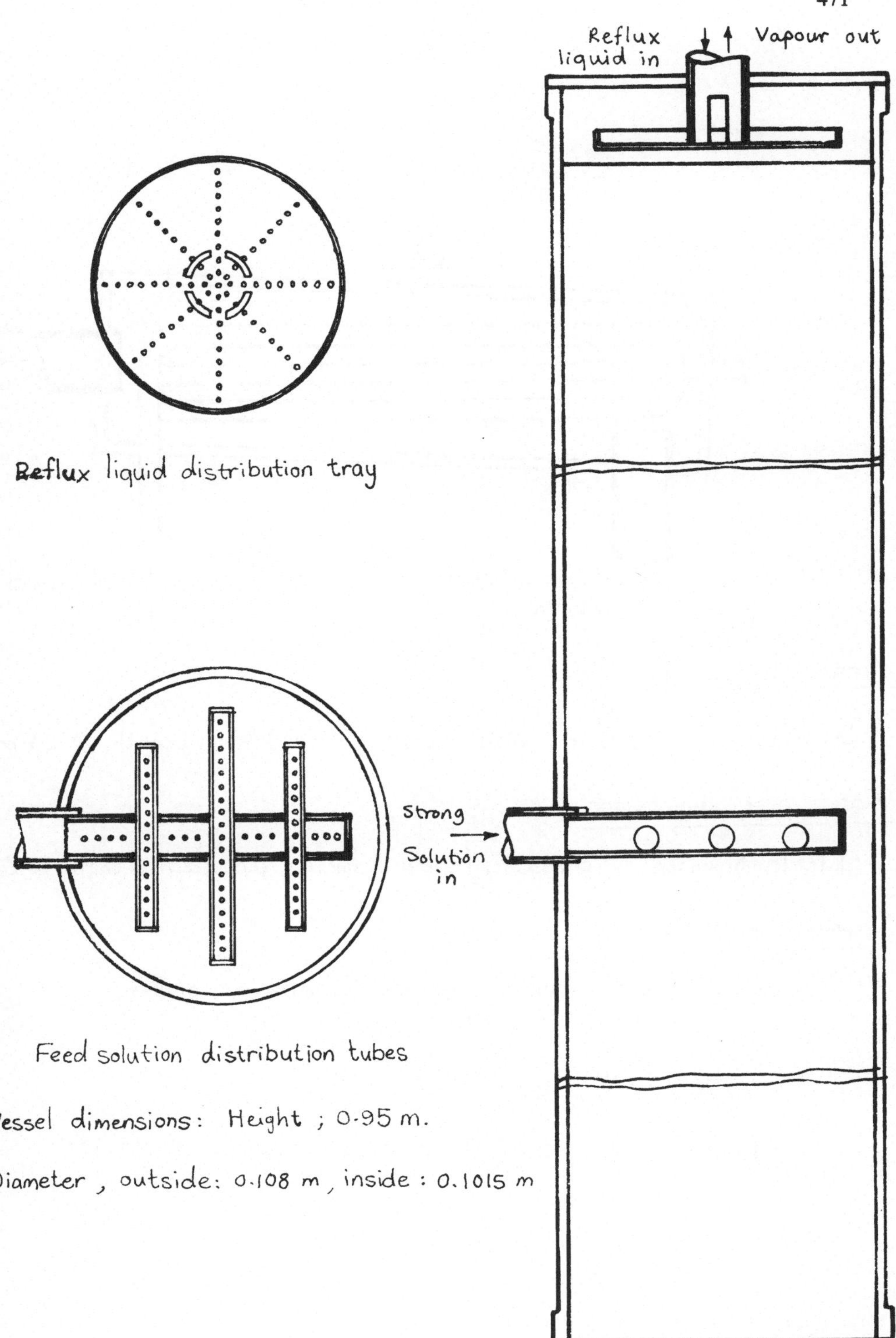

Figure 4 : Design of the generator packed column

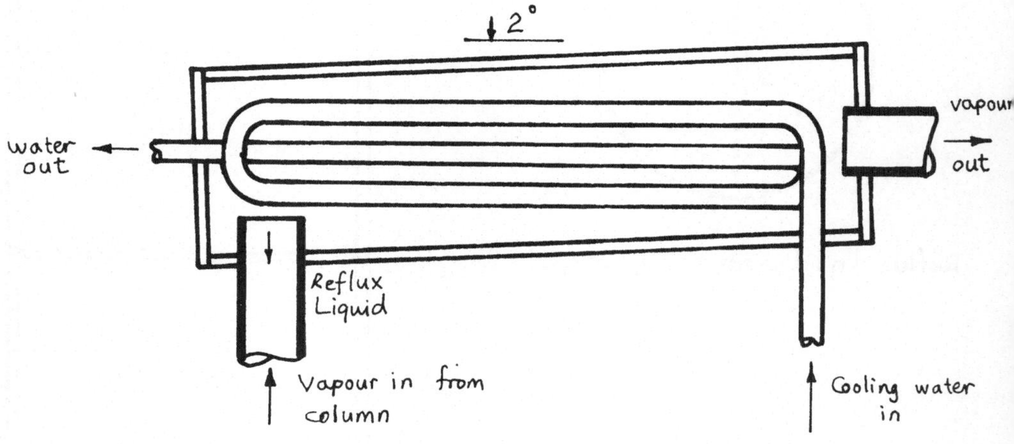

Vessel dimensions:

Diameter : 0.05715 m, Wall thickness : 0.00325 m, Length : 0.23 m

Water pipe dimensions:

Outside diameter : 0.00635 m, Inside diameter : 0.00452 m, Length : 0.6 m

Vapour pipe dimensions:

Outside diameter : 0.0195 m, Inside diameter : 0.0158 m.

Figure 5 : Design of the partial condenser (Rectifier)

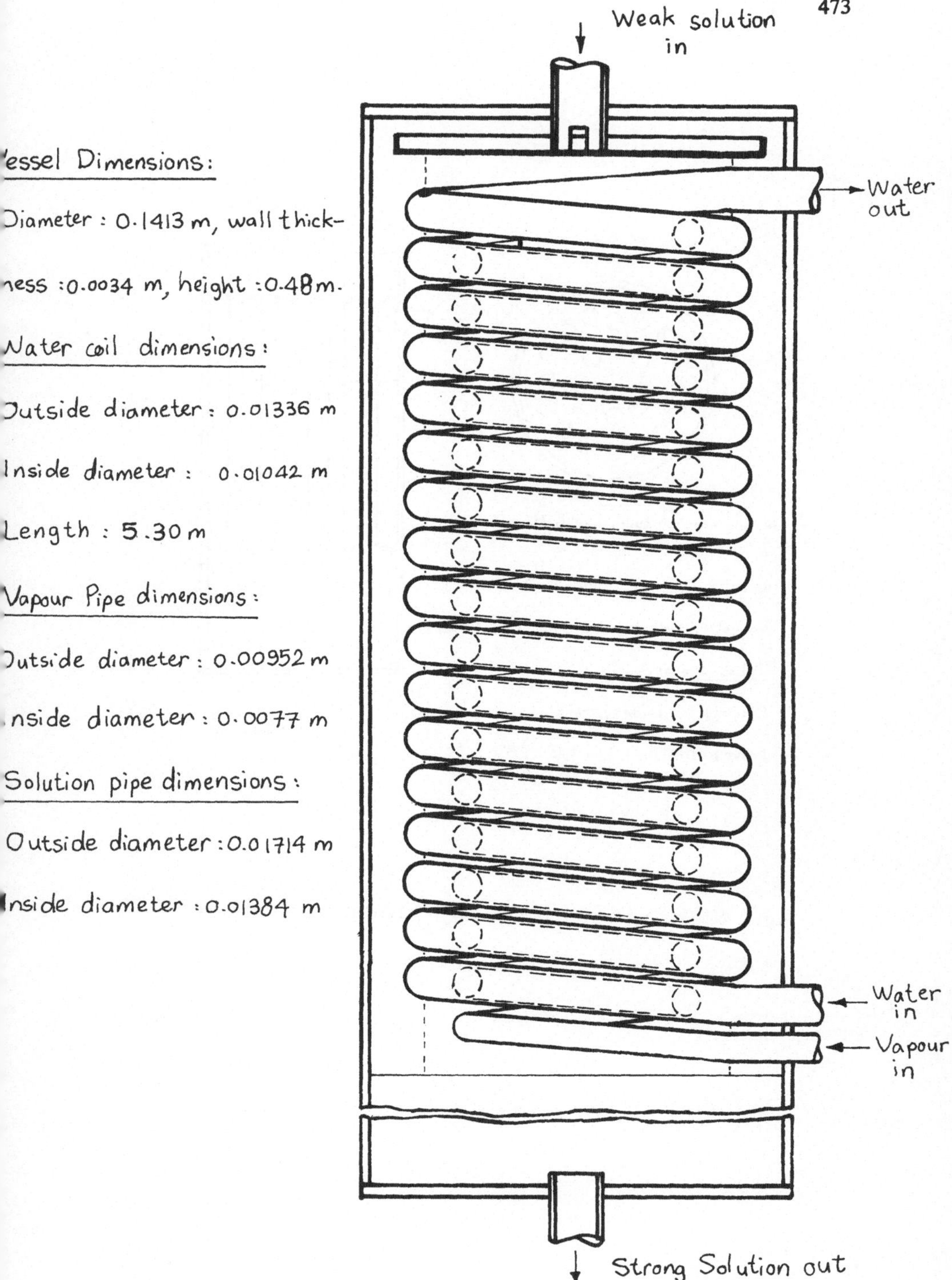

Weak solution in
essel Dimensions:
Diameter : 0.1413 m, wall thick-
ness :0.0034 m, height :0.48m.
Nater coil dimensions:
Outside diameter: 0.01336 m
Inside diameter: 0.01042 m
Length : 5.30 m
Vapour Pipe dimensions:
Outside diameter: 0.00952 m
Inside diameter: 0.0077 m
Solution pipe dimensions:
Outside diameter: 0.01714 m
Inside diameter: 0.01384 m
Water out
Water in
Vapour in
Strong Solution out

Figure 6 : Design of the absorber

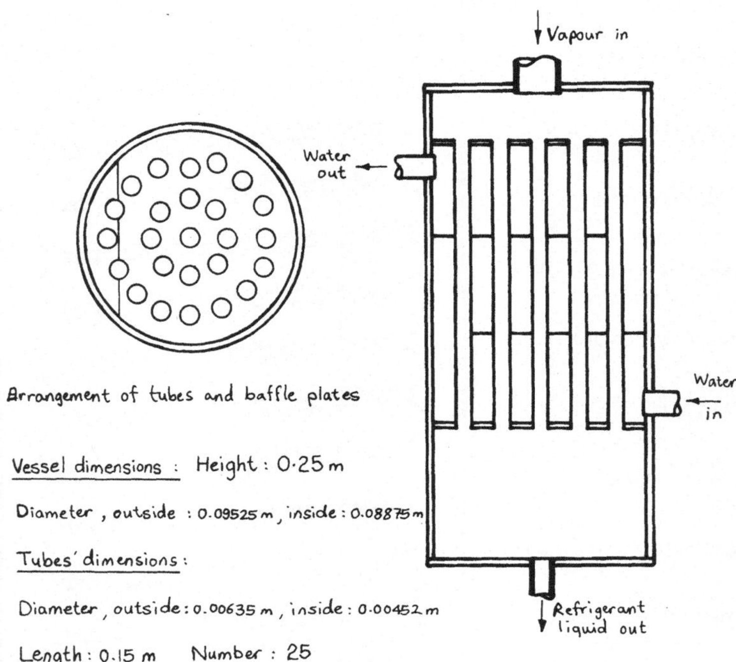

Figure 7 : Design of the condenser and refrigerant collection chamber

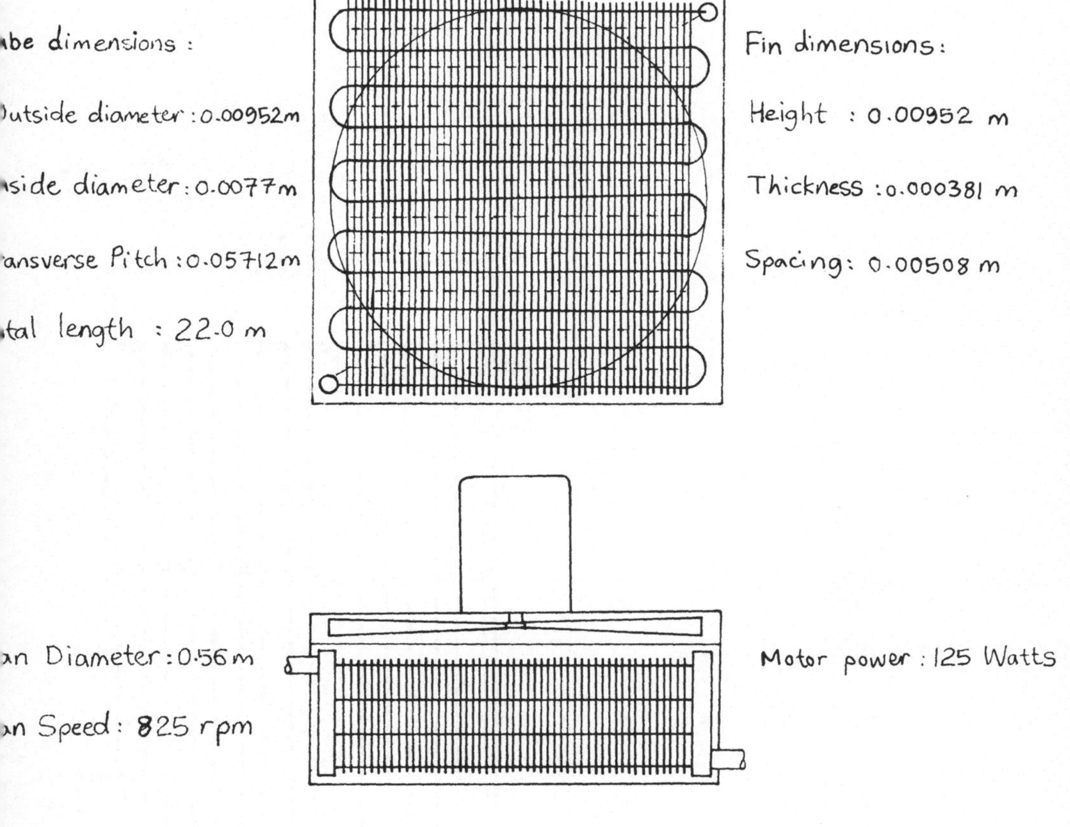

Figure 8 : Schematic of the evaporator design

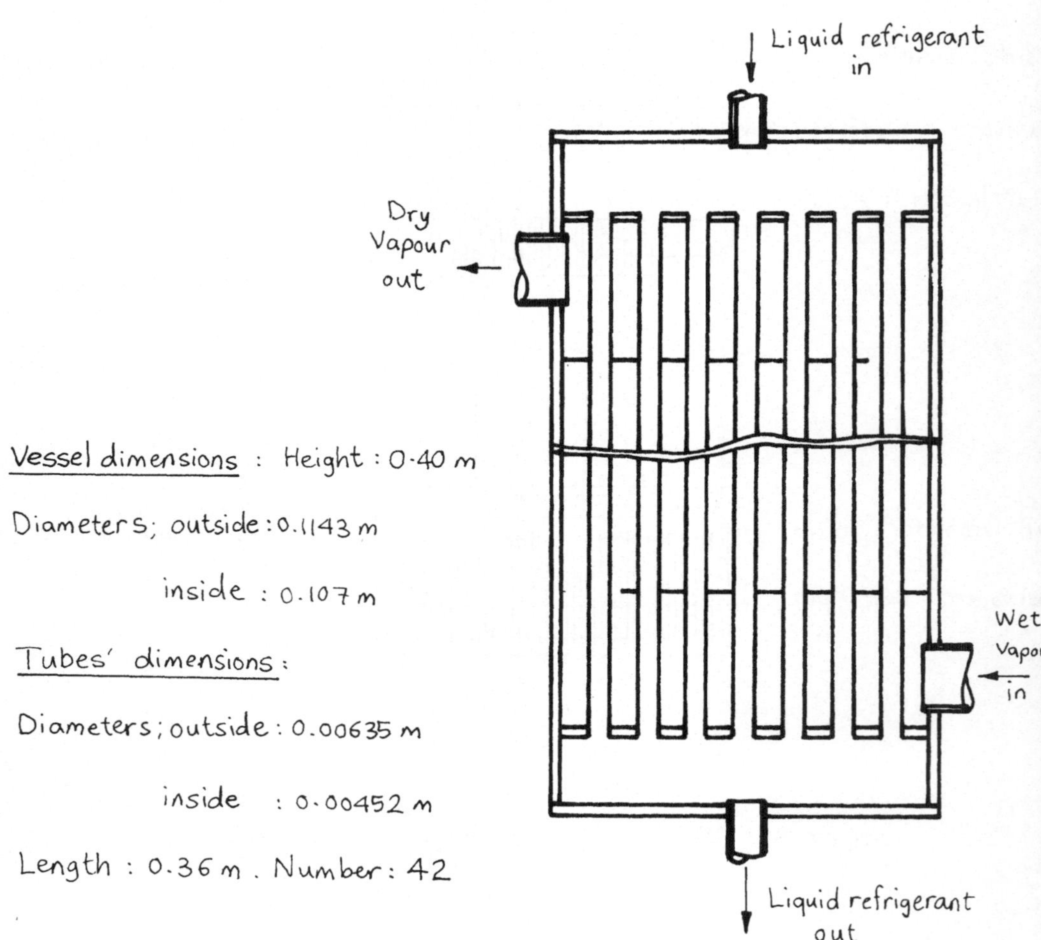

Figure 9 : Schematic of the refrigerant precooler (H x 2)

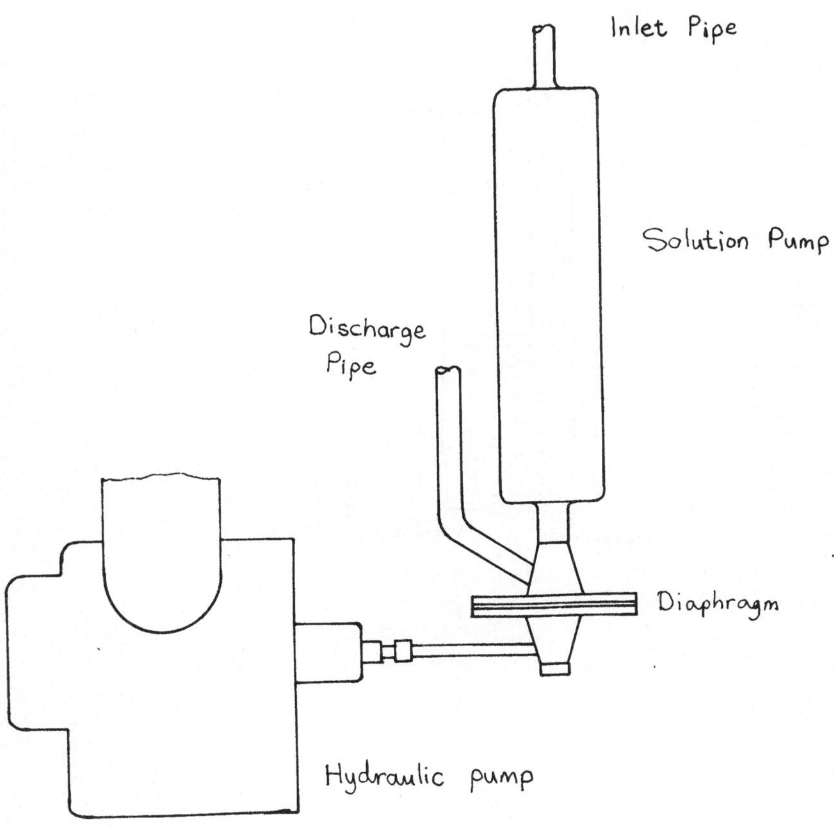

Figure 10 : Schematic of Solution and hydraulic Pumps

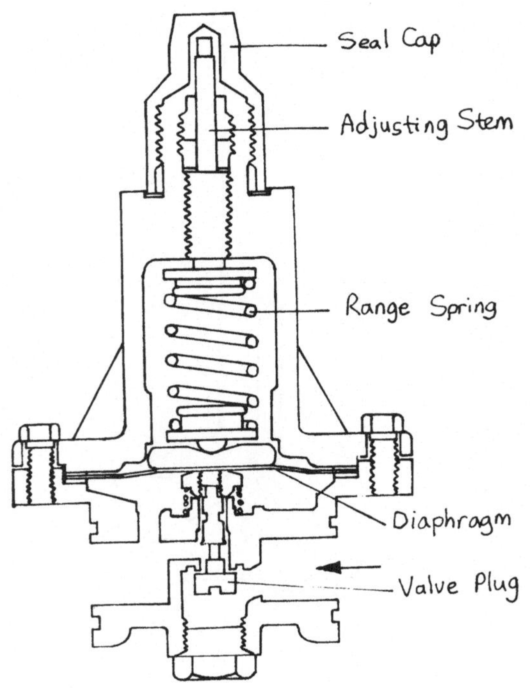

Figure 11 : Refrigerant control valve (A2B04 Refrigerating Specialties Company, U.S.A.)

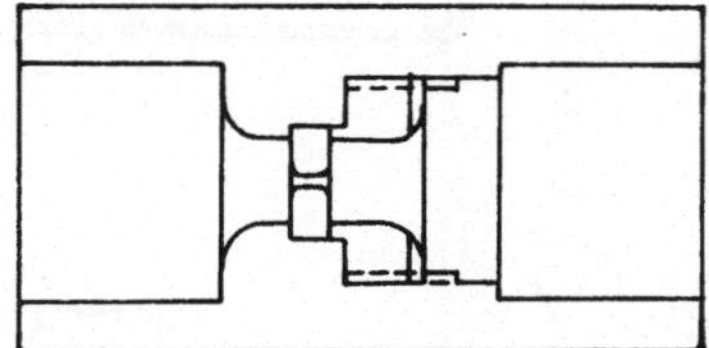

Figure 12 : Schematic of the solution throttle

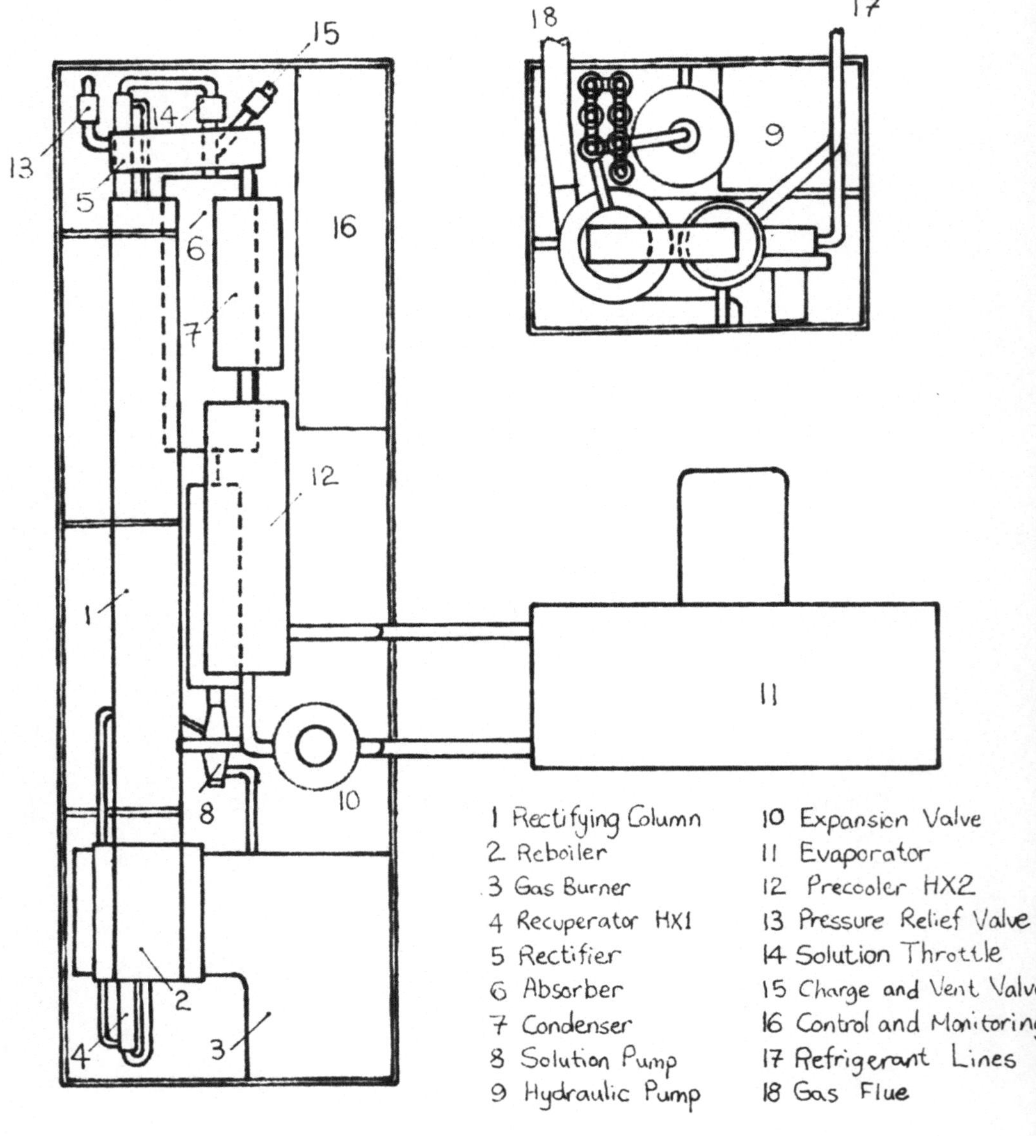

Figure 13 : General arrangement of the heat pump

RESORPTION HEAT PUMPS

H. van der Ree, P.A. Oostendorp

TNO, Apeldoorn, Netherlands

SUMMARY. The condensor and the evaporator are heat pump parts operating at constant temperatures of the internal fluids. When the external fluids are not available in infinite mass flows, their temperatures will change during the heat exchange within these devices. The discrepancy between the constant internal temperatures and the gliding external temperatures causes thermodynamical second law losses. In this contribution the use in heat pumps of devices with gliding temperatures, having the potential of avoiding these losses partly, is analysed.
Starting with idealized systems like the Carnot process and the Lorenz process, and assuming infinite heat exchanging surfaces, it turns out to be very favourable to apply gliding temperatures for the internal media in heat pumps, especially when the external media have to be cooled or heated over large temperature trajectories.
Bringing the processes with and without gliding internal temperatures closer to reality by introducing finite surfaces for heat transfer, the processes with gliding temperatures appear to be by far the most sensitive for this intervention and the advantage of applying them is drastically reduced.
Due to the smoothing effect of the heat produced by the power (positive) cycle, the heat-actuated heat pump has less benefit from the Lorenz process in the negative cycle than the mechanical heat pump.

When absorption and resorption heat pumps, being technical approaches of the Carnot process and the Lorenz process respectively, are compared, no energetic advantage is generally obtained. Using the binary mixture NH_3/H_2O the heat ratio of resorption heat pumps for heating only is lower than the heat ratio of the corresponding

absorption heat pumps. When sorption heat pumps are designed for heating and cooling external fluids over relatively large temperature trajectories, as occurs for instance when a simultaneous heating and cooling demand has to be covered, the heat ratio of the resorption heat pump can be slightly better compared with the corresponding absorption heat pump.
A rough analysis of sorption heat pumps using R22 - E181, being a binary mixture in which the absorbent has a negligible vapour pressure,yields about the same heat ratio for resorption and absorption heat pumps.
The only perspectives for the resorption principle appear to lie in its application in the low pressure stage of the two-stage absorption heat pump.

1. INTRODUCTION

In the last few years an increasing interest in sorption heat pumps has emerged. Attention is being paid almost exclusively to the absorption heat pump in this respect. One modification which has hardly come forward so far is the resorption heat pump. This device bears a large resemblance to the absorption machine, but differs from the latter by the fact that sorption processes have taken the place of condensation and evaporation of the refrigerant (Fig. 1a).

In refrigerating technology the resorption machine has already been known for a long time. In his famous publication "Reversible Absorptionmaschinen" Altenkirch, as early as 1913, gives a detailed description of this modification of the absorption machine, and recommends a systematic set of measures to decrease the thermodynamic losses of sorption machines [1]. Altenkirch points out that resorption need not be exclusively linked to the absorption machine, but that this process can be realized in the compression cycle as well (Fig. 1b). Tens of years later he made a thorough investigation of this latter combination for an application in the field of refrigerating technology [2].

At the moment there is a renewed interest in "resorption" with compression, this time with the specific aim of arriving at more energy efficient heat pumps [3]. In spite of the fact that they have a few advantages over other types, resorption machines have hardly been built for actual application. One of the scarce applications concerns a two-stage sorption refrigeration machine, where resorption occurs in the low-temperature stage, which has proved to give a very favourable heat ratio [4].

An essential characteristic of the resorption machine is that inside all components gliding process temperatures are realized. As will be explained later, this causes the resorption machine to be better adapted to the temperatures of the boundary fluids than the absorption machine. In principle this can result in a saving of energy which mainly determines the perspectives of resorption heat pumps. In this contribution the aspect of gliding process temperatures will therefore receive due consideration.

This will be achieved by considering the energetical merits of idealized cycles incorporating both fixed and gliding process temperatures, for mechanically-driven as well as for heat-actuated heat pumps. As mentioned before, the principle of replacing the condenser and evaporator by sorption devices can be applied in just the same way in these two heat pump categories, thus making it possible to compare them. Firstly idealized systems having infinite heat exchanging surfaces will be analysed. Secondly the processes are brought closer to reality by introducing finite surfaces for heat transfer.
Finally, the heat ratio (COP) of actual processes in resorption and absorption heat pumps will be compared.

2. CARNOT VERSUS LORENZ PROCESS

2.1. Introduction

The Carnot process is often considered as the best process to obtain reversible energy conversions between systems. If the Carnot process can interact with external fluids by reversible heat transfer processes, this is certainly true. The condition then has to be fulfilled that these fluids have constant temperatures throughout the entire heat exchange process, which means that they must be available with infinite mass flow rates (finite specific heats being assumed).
In practice, however, the boundary conditions mostly imply limited heat capacity flow rates of external fluids which are to be heated or cooled over a certain temperature range. In this situation the application of the Carnot process gives no optimal solution, as the discrepancies between the gliding external temperatures and the internal isotherms incorporate thermodynamic losses. A process by which these losses are avoided is the Lorenz process. This is defined as a reversible process which is performed in such a way that the interactions with external fluids are reversible as well.
When gliding external fluid temperatures occur, the Lorenz process is characterized by corresponding gliding internal fluid temperatures, eliminating all temperature differences between outside and inside where heat is transferred.

Regarding heat pumps it might be worthwile, in the light of the above, to consider Lorenz-like cycles aiming at obtaining gliding temperatures of the internal working fluid in the various heat exchangers. In this respect, resorption heat pumps are appealing, as gliding internal temperatures are realized in all components, as mentioned previously. Consequently, energy savings might be expected with respect to the absorption cycle, in which the refrigerant is bound to constant temperatures in condenser and evaporator.

2.2. The mechanical heat pump

Figure 2 illustrates the mechanical heat pump boundary conditions. The heat pump exchanges heat with two external fluid flows, which are available with limited flow rates. Therefore the outlet temperatures of these fluids differ by a value of ΔT_f from the inlet temperatures. When the heat pump operates on the Lorenz principle all conversions are by definition reversible, and no entropy is produced in the system.

From this condition and an energy balance, taking the specific heats of the external fluids to be constant, the following expression can easily be derived for the heating coefficient of performance (COP) of the Lorenz cycle :

$$COP_{L,H} = \frac{\overline{T}_2}{\overline{T}_2 - \overline{T}_1} \quad (1)$$

in which

$$\overline{T}_x = \frac{T_{xu} - T_{xi}}{\ln \frac{T_{xu}}{T_{xi}}} \qquad x = 1 \; ; \; 2 \quad (2)$$

$\varepsilon_{L,H.}$ apparently corresponds to the COP for a Carnot process, occurring between the logarithmic mean values of the inlet- and outlet temperature of the two fluid flows.

The advantage obtained by using the Lorenz process can be judged by comparing it with the Carnot process (Fig. 3a). Assuming in the first instance the heat exchanging surfaces to be infinitely large, the process temperatures in het Carnot process will have to be at the levels T_{2u} and T_{1u}, respectively. The heating $COP_{C,H}$ therefore depends on the extreme temperatures of the external fluids instead of on the average temperatures, as in the previous case :

$$COP_{C,H} = \frac{T_{2u}}{T_{2u} - T_{1u}} \qquad (3)$$

The Lorenz process therefore gives a saving in driving energy compared with the Carnot process. In this analysis this energy savings will be related to the specific energy consumption of the Carnot cycle to obtain a quantity which can be used later to compare the mechanical heat pump results with those of the heat-actuated heat pump :

$$s_{C/L} = \frac{1/COP_{C,H} - 1/COP_{L,H}}{1/COP_{C,H}} = 1 - \frac{COP_{C,H}}{COP_{L,H}} \qquad (4)$$

The relative specific energy saving $s_{C/L}$ is illustrated in figure 4 (curve log ΔT_m = 0 K). Use is made of a sample calculation, which has been partly derived from Lotz [3].

The conditions used are representative for heat pumps for house heating. It is evident that the Lorenz process applied to the mechanical heat pump, even for small ΔT's, gives an appreciable saving in energy. The advantages offered by this process increase with increasing temperature, changes to which the external media are subjected.
However, if finite heat transfer surfaces are introduced to approach the actual situation, the above advantage becomes far less important.
To elucidate this, the two processes are compared again, but this time a logarithmic mean temperature difference ΔT_m between the process fluids and the external fluids is introduced. This gives rise to a Lorenz-like cycle and a Carnot cycle in which the temperatures are further apart (Fig. 3b).
The first process still has the characteristic of gliding temperatures, and is also internally reversible, but essentially is no longer a Lorenz process, because, per definition, this process excludes the temperature difference that is introduced now. In order to make a fair comparison between the two processes equal logarithmic temperature differences for heat supply (ΔT_{2m}) respectively heat extraction (ΔT_{1m}) have been assumed. The temperatures T^* that now determine the heating COP's can be calculated by means of the following equations :

$$\text{Lorenz-like process : } \overline{T}^*_x = \frac{T_{xz} - T_{x1}}{\ln \frac{T_{xz}}{T_{x1}}} \qquad x = 1\ ;\ 2 \qquad (5)$$

$$\Delta T_{x1} = \Delta T_{xz} = \Delta T_{xm} \qquad x = 1 \; ; \; 2 \tag{6}$$

Carnot process :

$$T_1^* = T_{1i} - \Delta T_{1c} \tag{7}$$

$$T_2^* = T_{2i} + \Delta T_{2c} \tag{8}$$

$$\Delta T_{xc} = \frac{\Delta T_{xf}}{1 - 1/e^{\Delta T_{xm}}} \qquad x = 1 \; ; \; 2 \tag{9}$$

How, for the sample calculation introduced previously, the relative energy savings will turn out, is depicted again in figure 4. The by no means unrealistic values (5 K and 10 K respectively) chosen for the logarithmic mean temperatures appear to account for an appreciable decrease in energy saving.
This confirms what is pointed out by Altenkirch in [2] that the process with gliding internal temperatures is more sensitive to temperature differences for heat transfer than the simple compression process with constant temperatures during condensation and evaporation.

2.3. The heat-actuated heat pump

In heat-actuated heat pumps there are often two temperature levels at which the useful heat is produced (Fig. 5). For example, in the case of absorption heat pumps, the heat set free in the absorber will, in a number of cases, be produced at a temperature level different from that in the condenser.
As contrasted with the situation prevailing with the mechanical heat pump, the 1st end 2nd laws of thermodynamics, on the assumption that the system in its entirety functions without any losses, do not completely suffice to explicitly determine the heat ratio. This requires an other condition to be satisfied, like, for instance, specification of the ratio between the heat productions $\dot{Q}_2$ and $\dot{Q}_3$ (Fig. 5) or of the conversions in the process. The usual approach is to select a model which comprises a combination of a positive and a negative cycle in which all the work created by means of the first cycle is transferred to the second one (Fig. 6a, left-hand diagram). It can then be derived that the heat ratio for reversible (Lorenz) processes is given by :

$$\zeta_{LL,H} = \frac{\dot{Q}_2 + \dot{Q}_3}{\dot{Q}_4} = \frac{\overline{T}_4 - \overline{T}_3}{\overline{T}_4} \cdot \frac{\overline{T}_2}{\overline{T}_2 - \overline{T}_1} + \frac{\overline{T}_3}{\overline{T}_4}$$

$$= \frac{\overline{T}_4 - \overline{T}_3}{\overline{T}_4} \frac{\overline{T}_1}{\overline{T}_2 - \overline{T}_1} + 1 \qquad (10a,b)$$

with again :

$$\overline{T}_x = \frac{T_{xu} - T_{xi}}{\ln \frac{T_{xu}}{T_{xi}}} \qquad x = 1;\ 2;\ 3;\ 4 \qquad (11)$$

In (10a) the first term is the "specific heat" produced by the negative cycle, and the second term is the "specific waste heat" of the positive cycle. The combination shown in the left hand diagram of Fig. 6a is representative of the resorption machine, for in this machine gliding temperatures occur in all components. The generator/absorber part includes the possible cycle, the resorber/desorber part the negative cycle.
In a similar way the absorption machine can be modelled. Since in condenser and evaporator of the absorption machine the temperatures are constant, the negative cycle now has to comprise the Carnot cycle however. The remainder of the absorption machine essentially resembles the resorption machine, such that, in principle, the positive cycles may be assumed to be identical (Fig. 6a , right hand diagram). It can be derived that the heat ratio of this Lorenz/Carnot process combination is given by :

$$\zeta_{LC,H} = \frac{\overline{T}_4 - \overline{T}_3}{\overline{T}_4} \frac{T_{2u}}{T_{2u} - T_{1u}} + \frac{\overline{T}_3}{\overline{T}_4} = \frac{\overline{T}_4 - \overline{T}_3}{\overline{T}_4} \frac{T_{1u}}{T_{2u} - T_{1u}} + 1 \qquad (12a,b)$$

Proceeding as in the previous part, the energetical advantage of the Lorenz/Lorenz (resorption) model over the Lorenz/Carnot (absorption) model will be expressed as a specific energy saving, related to the specific heat consumption of the Lorenz/Carnot process :

$$s_{LL/LC} = \frac{\frac{1}{\zeta_{LC,H}} - \frac{1}{\zeta_{LL,H}}}{\frac{1}{\zeta_{LC,H}}} = 1 - \frac{\zeta_{LC,H}}{\zeta_{LL,H}} \qquad (13)$$

Figure 7 shows values of $s_{LL/CL}$ (curve log ΔT_m = 0 K) for a numerical example, which, as far as the negative cycle is concerned, is similar to the previous example for the mechanical heat pump. This figure demonstrates that the heat-actuated heat pump derives less benefit from the Lorenz process in the negative cycle than the mechanical heat pump (dashed lines). As is also evident from relationship (12), this is caused by the influence of the waste heat from the positive cycle contributing to the heat production of the system. This influence, in principle, applies to all heat-driven systems.

In Figure 7 also values of $s_{LL/LC}$ are depicted for the more realistic situation of distinct temperature differences across the heat transfer surfaces (log. ΔT_m = 5K and 10K) in the negative cycle (fig. 6b). Like in the previous example it can be seen that the specific energy saving decreases considerably when finite instead of infinite heat exchanging surfaces are applied.

When the positive cycle is not free from losses, as considered above, but operates at an efficiency η_L with respect to the Lorenz cycle, equations 10b and 12b respectively have to be changed to :

$$\zeta_{LL,H} = \eta_L \frac{\overline{T}_4 - \overline{T}_3}{\overline{T}_4} \frac{\overline{T}_1}{\overline{T}_2 - \overline{T}_1} + 1 \qquad (14)$$

and

$$\zeta_{LC,H} = \eta_L \frac{\overline{T}_4 - \overline{T}_3}{\overline{T}_4} \frac{T_{1u}}{T_{2u} - T_{1u}} + 1 \qquad (15)$$

Figure 8 indicates the effect on the relative energy saving of a 30 % loss in the positive cycle and also shows the influence of 10 K logarithmic temperature difference for heat transfer to and from the negative cycle.
This latter influence is apparently the most dramatic one. For the slight remaining energy saving, the generating efficiency is of minor importance.

From the previous it can be concluded that, at least with heat pumps for house heating, only little may be left in practice from the theoretical energy benefit of a process with gliding internal temperatures. Moreover application of gliding instead of constant temperatures in condenser and evaporator has less advantage with the heat-actuated heat pump than with the mechanically-driven heat pump. On the basis of the comparative models chosen, resorption therefore proves to be hardly better than absorption.

3. THE RESORPTION HEAT PUMP

3.1. Description

In Figure 9a the components that could possibly form part of a simple resorption heat pump (rhp) have been arranged in a p, T-diagram, as firstly suggested by Altenkirch [1]. The part of the rhp consisting of the generator (with rectifying columm if present), the solution heat exchanger and the absorber, which will be referred to as the absorption loop or positive cycle of the rhp, will be easily recognized as identical to that of the absorption heat pump (ahp). As mentioned before, the difference consists in the fact that in the negative cycle the condenser and the evaporator are replaced by a second solution loop called the resorption loop, with a second absorber called resorber and a second generator called desorber. Instead of condensing in a condenser at a constant temperature, the vapour from the generator in a rhp is absorbed by the solution in the resorber at a decreasing temperature. Analogously, the enriched solution from the resorber desorbs a part of the refrigerant in the desorber at lower pressure and increasing temperature. In the triple solution-vapour heat exchanger, the vapour and poor solution from the desorber are heated, resulting in higher values of the heat produced at the absorber and resorber. The subsequent cooling of the rich solution leads to an increase of the amount of heat extracted from the heat source in the desorber.

As will be known, the two pressure levels in an ahp are fixed by the choice of the temperatures at which the condenser and the evaporator operate. In a rhp on the contrary one is free, within certain limits, to choose one of the pressures. The pressure levels in the resorber and the desorber can be varied by changing the concentrations of the solution at fixed temperature ranges in the resorber and the desorber. As a consequence, where only one ahp-cycle can be designed when the temperature levels of heat source and heat sink are given, a range of rhp-cycles can exist. In Figure 10 this is visualized using a ln p, - 1/T - diagram. This diagram is known to be very suitable to present all kinds of sorption processes, because many of the temperatures, concentrations and pressures occurring can roughly be read from it. Moreover, when the sorption heat pump process is assumed to occur between the three temperature levels T_o, T_W and T_H, its Carnot-efficiency expressed by

$$\zeta_c = \frac{\frac{1}{T_o} - \frac{1}{T_H}}{\frac{1}{T_o} - \frac{1}{T_W}} \qquad (16)$$

can be visualised as the ratio a/b. Expression (16) can be derived from (10a,b) by putting $\overline{T}_1 = T_o$; $\overline{T}_2 = \overline{T}_3 = T_W$ and $\overline{T}_4 = T_H$.

3.2. Rectification

When in an ahp an absorbent is used with a substantial vapour pressure at the conditions occurring, rectification to a certain degree is necessary to obtain a reasonable heat ratio, especially when high temperatures have to be produced at low source temperatures. In the, not abundant, literature on resorption machines it is often stated that in these machines rectification can be omitted. With respect to this there are two relevant effects. On the one hand, the heat Q_R to be extracted from the top of the rectifying column causes a loss of efficiency. On the other hand, without rectification the vapour from the generator will generally contain much more absorbent than the vapour from the desorber does, because the desorber operates at higher solution concentrations and lower pressures compared with the generator. Consequently, a nett flow of absorbent from the absorption loop to the resorption loop will cause the heat pump to drift away from its thermodynamical point of operation. This drift can be prevented by bleeding as shown in Figure 9a (dashed line), carrying a portion b of the poor solution from the resorption loop (concentration $\xi_{p,res}$) to the absorption loop. Considering mass balances over a control surface crossing the two vapour conduits and the bleeding, and relating all quantities to 1 kg of vapour produced at the generator, the following expression for b can be derived :

$$b = \frac{\xi_{v,des} - \xi_{v,gen}}{\xi_{v,des} - \xi_{p,res}} \qquad (17)$$

where $\xi_{v,gen}$, $\xi_{v,des}$ and $\xi_{p,res}$ represent the concentrations of the vapour from the generator and desorber respectively and the poor solution of the resorption loop. Not very extreme values of these concentrations (e.g. $\xi_{v,des} = 1$, $\xi_{v,gen} = 0.8$ and $\xi_{p,res} = 0.5$) yield b = 40 %. This amount of liquid, not being used for heat extraction from the heat source, represents a considerable loss of process efficiency. When a rectifying column, cooled by the external medium to be heated is added to the generator the higher vapour concentration will lead to a value of b which can be orders of magnitude smaller. However, as already mentioned, also the rectifying process causes a loss of efficiency. To quantify these counteracting effects, calculations will be performed in the next chapter assuming different degrees of rectification.

3.3. The rhp in the h, ξ-diagram

The fundamentals and the application of the h, ξ-diagram for analysing sorption processes are thoroughly treated in [5] and [6]. In the following the emphasis will be put on the resorption loop.

In Figure 9b the rhp process according to Figure 9a is depicted in an entalpy-concentration diagram. The corresponding thermodynamic states are indicated with encircled numbers. The construction starts at (9), representing the end of the desorption process, where the desorber temperature has reached its maximum value. The position of (9) follows from the pressure p_o (freely chosen) and the temperature $t_9 = t_{14}$, which has to be lower than the temperature of the source medium delivering the heat Q_o for the desorption process. The position of (7), representing the end of the resorption process at which the resorber temperature has reached its minimum value, is influenced by a number of variables such as the high pressure p_{gen}, the subcooling in the resorber, the temperature of the entering external fluid to which the heat of resorption Q_{res} has to be transferred, and the desirable width of desorption $\xi_{r,res} - \xi_{p,res}$.
As already mentioned, some care is necessary in dealing with the concentrations of the vapours from the generator and the desorber. In an ahp it is always possible to reach a vapour concentration at the end of the rectifier very close to 100 %. This is due to the fact that all the heat of rectification Q_R involved can be transferred to the external fluid to be heated because in an ahp the temperature trajectory at which Q_R comes out always lies above the condensation temperature. In an rhp, however, due to the lower pressures, the temperature level of Q_R can be rather low compared with the temperatures occurring in the absorber and the resorber. As a result, the heat Q_R to be carried off in order to obtain a high vapour concentration cannot be transferred completely to the fluid to be heated, and consequently only a limited vapour concentration will be attainable at the top of the rectifying column. In Figure 9b it is shown that for a lowest temperature at the top of the rectifying column of t_R the vapour concentration has a value $\xi_{v,T}$. If the vapour leaving the desorber is in equilibrium with the poor solution at concentration $\xi_{p,res}$ (9), its state can be represented by (14), situated at the two-phase isotherm t_{14}. The difference between $\xi_{v,R}$ and ξ_{14} illustrates the problem concerning the net flow of absorbent from the absorption loop to the resorption loop.
The balance can be restored by adding a controlled quantity b of the poor solution (9) to the vapour (14) through the bleeding, leading to a liquid-vapour mixture of concentration $\xi_{10} = \xi_{v,R}$. The mixture (10) is heated in the triple heat exchanger to state (11). The amount of heat $Q_{3,v}$ involved is constrained by two factors. To heat the poor solution from state (9) to state (8) an amount of heat $h_7 - h_{12}$ rich solution is used. The maximum amount

of heat available for heating the liquid-vapour mixture is then $h_{12} - h_{13}$ rich solution, corresponding to $h_M - h_N$ kJ/kg vapour. The second constraint is formed by the condition $t_{11} < t_7$, in order to have a driving force for the heat flow. In Figure 9b the first constraint is determining the process, as is mostly the case when the absorbent content of the mixture (10) is rather high.

Considering heat and mass balances over suitable control surfaces, the amounts of heat transferred to and from the external fluids can be determined easily. A control surface wrapped around the resorber and crossed by the poor solution (8), the rich solution (7) and the vapour (4) yields the resorption heat Q_{res} (kJ/kg vapour) represented by (4) M. A control surface around the desorber and the triple heat exchanger, crossed by the liquid-vapour mixture (11), the poor solution (8) and the rich solution (7) yields M (11) as the desorber heat Q_o (kJ/kg vapour). Analogously Q_{gen} and Q_{abs} can be evaluated.

Finally the heat ratio follows from

$$\zeta = \frac{Q_R + Q_{res} + Q_{abs}}{Q_{gen}} = 1 + \frac{Q_o}{Q_{gen}} \tag{18}$$

3.4. The influence of some parameters on the heat ratio of rhp's and ahp's

In order to compare ahp's and rhp's calculations will be presented for heat pumps of comparable complexity. For the rhp's the pressure levels as well as the degree of rectification will be varied. Unless stated otherwise, the refrigerant-absorbent pair in the examples is ammonia-water. To examine the effect of using an absorbent having a negligible vapour pressure, also a rough calculation will be presented using the binary mixture R22 - E181 (difluoromonochloromethane - tetraethyleneglycoldimethylether). The calculations have been carried out by graphical constructions in the enthalpy-concentration diagram, as described above.

Unless stated otherwise, the following assumptions form the general basis for the calculations :

- The outside air is the low temperature source of the heat pumps
- Evaporator temperature : 6 °C below air temperature
- Temperature at the end of the desorber : 4 °C below air temperature
- Logarithmic mean temperature difference : 5 °C and 10 °C for liquid-liquid, respectively liquid-vapour heat exchangers, but : $t_6 - t_5 = 5$ °C; $t_7 - t_8 = 5$ °C (Fig. 9b)

- Subcooling at the end of absorber and resorber : 3 °C
- The fluid to be heated is water
- $\xi_{r,abs} - \xi_{p,abs} \approx 0{,}1$; $\xi_{r,res} - \xi_{p,res} = 0{,}06$

3.4.1. Ahp versus rhp at different pressure levels. The heat pumps operate under the following conditions :

- Outside air temperature : $t_a = -3$ °C
- Water to be heated from $t_r = 31.5$ °C to $t_s = 38.5$ °C
- Maximum rectification is applied both in the ahp and in the rhp's.

In Figure 11 it is shown by Q, t-diagrams to what extent and in what order the several parts of the heat pumps contribute to the heating of the water and at what temperature trajectories they operate. In the ahp the return water is firstly fed to the condenser in order to keep its pressure, being the maximum pressure in the ahp, as low as possible. In the rhp the return water is firstly fed to the rectifier in order to obtain a vapour as pure as possible, as explained earlier.
Strictly speaking the ratio at which the different components of the ahp and the rhp heat the water has to be established by an interative method . In the examples investigated this ratio appeared to be nearly constant within the limits of accuracy required per type of sorption heat pump.

In table I the calculated values of the heat ratio and a number of other important variables are given for one ahp and for four rhp's. Some secondary variables are specified to facilitate the reconstruction of the results. The independent parameter for the different rhp's is the desorber pressure p_o.

It can be seen that the heat ratio of all rhp's is lower than the heat ratio of the ahp, and moreover decreases with the pressure level in the rhp's. In Figure 12 the heat ratio of the heat pumps is presented as a function of p_o. This graph suggests that the ahp can be considered as a special (border line) case of the more general rhp, in which $\xi_{p,res} = \xi_{r,res} = 1$. The fact that the heat ratio of the heat pumps decreases with the pressure p_o can be easily understood from Figures 13, 14 and 15, representing the cases ahp 1, rhp 1 and rhp 4. Firstly, maintaining the temperature level in the absorber at lower pressures leads to lower concentrations in the absorption loop, which moves to the left. As a result, the primary vapour from the generator (in thermodynamic equilibrium with the rich solution $\xi_{r,abs}$) will get less pure (see $\xi_{v,gen}$ in Table I). This causes an increasing effort for rectification (Q_R in Table I). Secondly, lowering the pressure p_o causes the point L in Figures 13 to 15 to move downwards. From L the generator heat Q_{gen} is measured. This fact coheres with the increasing differential isothermal mixing heat at lower concentrations.

Summarizing, it can be stated that although applying the Lorenz principle in a negative cycle can yield an improvement of the efficiency in that cycle (see above), the net result for the sorption heat pumps described in this paragraph is negative. This is due to the fact that the change-over to a resorption loop in the negative cycle causes dominant negative effects in the positive cycle. The degree to which these effects play a role depends on the thermodynamic properties of the binary mixture used.

3.4.2. Variation in level of rectification. In order to examine the effect of the level of rectification, the heat ratios of the cases ahp 1, rhp 1 and rhp 4 have been recalculated assuming no rectification. Moreover, case rhp 1 has been investigated for a "medium" level of rectification. The results are listed in Table II, while in Figures 16 and 17 the plots are presented of rhp 1 without and with "medium" rectification, respectively. Moreover in Figure 12 the heat ratio of the rhp's without rectification has been plotted (dashed line).
It turns out that the rhp is far less sensitive to the level of rectification than the ahp. When no rectification is applied, rhp 1 even has a higher heat ratio than ahp 1. As can be expected, the heat ratio of the ahp decreases rather sharply by lowering the level of rectification. Obviously in the rhp's the positive effect of the decreasing of the bleed b obtained by rectification, is for the greater part balanced by the simultaneously occurring increase of Q_R.

It may be concluded that rectification is not very appropriate for the rhp's described in this paragraph.

3.4.3. Heating and cooling the external fluids over greater temperature trajectories. As shown in the previous chapter, there is a greater advantage of switching over from a Carnot-process to a Lorenz-process when the external fluids have to be cooled or heated over a longer temperature trajectory ΔT_f. To examine this further, an ahp and a rhp have been depicted in the h,ξ-diagrams of Figures 18 and 19, both of them having been designed for covering a simultaneous heating and cooling demand. The desorber cools the external fluid from + 4 °C to - 16 °C, while in the rectifier, resorber and absorber the external fluid is heated from 20 °C to 50 °C. The values of some characteristic variables are presented in Table III. Taking into account the accuracy of the graphical calculation method, the choice of a larger ΔT_f now results in about the same heat ratios for both the rhp and the ahp.

It must be stressed that the rhp results have been obtained only because in this paragraph a need to heat and cool external fluids over a wide temperature range has been assumed. However, in applications of heat pumps for heating only it is not advantageous at all to cool the heat source down over a large temperature trajectory ΔT_{1f}.

This can be clarified by considering the source medium as an infinite reservoir with temperature T_{1i}, from which a mass flow $\dot{M}_f$ is first cooled to a temperature $T_{1i} - \Delta T_{1f}$ by the heat pump and thereupon heated again to its original temperature T_{1i} by flowing back into the infinite reservoir. The reheating of the source medium is an irreversible process in which the entropy increases as follows :

$$\Delta S = \left[\dot{M}_f \cdot c_p \cdot \int_{T_{1i} - \Delta T_{1f}}^{T_{1i}} \frac{dT}{T} \right] - \frac{\dot{Q}_o}{T_{1i}} \tag{19}$$

The first term on the righthand side of (19) represents the change of entropy of the source medium. The second term gives the change of entropy of the infinite reservoir at T_{1i}. Because $\dot{Q}_o = \dot{M}_f . c_p . \Delta T_{1f}$, (19) can be transformed into

$$\Delta S = \dot{Q}_o \left[\frac{1}{\Delta T_{1f}} \ln \frac{T_{1i}}{T_{1i} - \Delta T_{1f}} - \frac{1}{T_{1i}} \right] \qquad \text{or}$$

$$\Delta S = \dot{Q}_o \left[\frac{1}{T_{1i}} + \frac{1}{2} \frac{\Delta T_{1f}}{T_{1i}^2} + \frac{1}{3} \frac{\Delta T_{1f}^2}{T_{1i}^3} + \ldots - \frac{1}{T_{1i}} \right] \tag{20}$$

Consequently, when an amount of heat $\dot{Q}_o$ has to be extracted from the heat source, the accompanying entropy production goes to zero with ΔT_{1f}. The situation thermodynamically most favourable is therefore characterized by

$$\left. \begin{array}{l} T_{1f} \rightarrow 0 \quad \text{and consequently} \\ \\ \dot{M}_f = \dfrac{\dot{Q}_o}{c_p \cdot \Delta T_{1f}} \rightarrow \infty \end{array} \right\} \tag{21}$$

3.4.4. The binary mixture R22 - E181. As shown in the previous paragraphs, the dissappointingly low heat ratio of rhp's is largely due to the absorbent (water) often having a rather high vapour pressure under the conditions in question, which leads to either a high value of the bleeding b or a high value of the heat of rectification Q_R. For this reason it seems interesting to compare an ahp and a rhp when a binary mixture is used in which the absorbent has a negligible vapour pressure. For the present example the binary mixture R22 -E181 (difluoromonochloromethane - tetraethylene glycodimethylether) has been chosen, for which an enthalpy-concentration diagram, is available from [8].
In Table IV the mean parameters of the heat pumps are given, whereas in the Figures 20 and 21 the corresponding plots in the h,ξ-diagram are shown. Comparing the results of Tables I and IV leads to the conclusion that the use of the mixture R22 - E181 brings the values of the heat ratios of the ahp and rhp closer together. However, changing over from an ahp to a rhp, there is still no question of improvement of the heat ratio.
A point upon which the attention might be focussed is the fact that when a binary mixture is used in which the absorbent has a negligible vapour pressure, the generated vapour is superheated. The amount of heat to be extracted from the vapour to reach saturation (Q_S) has to a certain degree the same effect as the heat of rectification Q_R : it causes the heat ratio to decrease.

3.5. The two-stage ahp

A special application of the resorption principle, not resulting from the wish to switch over from a Carnot to a Lorenz process, is its use in a two-stage ahp. Under certain conditions (e.g. when the temperature difference between the evaporator and the condenser/absorber is not too large) it is possible to supplement the ahp with a low-pressure resorption stage, as is shown in Figure 22. The vapour from the evaporator is absorbed in a resorption loop, resulting in a second extraction of heat from the heat source in the desorber. The vapour generated in the desorber is fed to the absorber. The net result, under favourable circumstances, is a doubling of the heat extracted from the heat source requiring a slightly higher heat input to the generator.
Referring to the binary mixture NH_3-H_2O with its rather high pressures in heat pump applications, the main advantage of this set-up is the fact that a two-stage ahp is obtained by adding a low-pressure stage to the normal ahp which does not lead, like other kinds of two-stage operation, to an increase of the maximum pressure in the system.

The limitations to application of two-stage ahp of this type emanate from the following considerations. When a resorption stage is added to a normal ahp, the pressure of the absorber will decrease. Assuming that the temperature trajectory of the absorber will be the same, the concentrations in the absorption loop will decrease and the temperature in the generator will rise. Obviously the concentrations cannot be lower than 0 and the highest admissible generator temperature is limited by the chemical stability of the binary mixture.

When a binary mixture with a need for rectification is used, the lower concentrations in the absorption loop may lead to a strong increase of the rectification losses, especially under winter conditions when the heat pump has to generate high temperature heat at low temperatures of the heat source (outside air). These losses can diminish the heat ratio to a level even lower than the heat ratio of the corresponding single-stage ahp. From that point on it is preferable to change over to single-stage operation, which can be realized simply by short-circuiting the resorption stage (dashed conduit in Figure 22).

4. CONCLUSIONS

Assuming infinitely large surfaces for the heat exchange with the external fluids, the application of the Lorenz process results in substantial improvements of the efficiency compared with the Carnot process. This is especially so when the temperature changes of the external fluids are large.

Assuming finite surfaces available for the exchange of heat with the external fluids the improvements are reduced drastically, although they are still maximal for large temperature changes of the external fluids.

When absorption and resorption heat pumps using the binary mixture NH_3-H_2O are investigated as technical approaches of the Carnot- and Lorenz-process in the negative cycle, the heat ratio of the resorption heat pumps is generally lower than the one of the absorption heat pumps.

When a resorption heat pump is designed for simultaneous heating and cooling of external fluids over a sufficiently large temperature trajectory, the application of gliding temperatures may be advantageous : in such a case the (NH_3-H_2O) resorption heat pump need not be inferior to the corresponding absorption heat pump.

A real advantage of the resorption heat pump in comparison with the absorption heat pump is the fact that, within certain limits, the pressures can be freely chosen. However, in the case of

NH_3-H_2O the heat ratio decreases with the pressure.

A rough analyses of sorption heat pumps using the mixture R22 - E181, which unlike NH_3-H_2O does not require rectification, results in about the same heat ratio for both the absorption and the resorption heat pump.

On the whole resorption heat pumps don't have higher heat ratios than absorption heat pumps.

A useful application of the resorption principle, although not arising from the wish to have gliding temperatures, is its use in the low-pressure stage of a two-stage absorption heat pump.

REFERENCES

1 Altenkirch, E.,"Reversible Absorptionsmaschinen", *Zeitschrift für die gesamte Kälte-Industrie* 20 (1913) nrs. 1, 6, 8 and 21 (1914) nrs. 1, 2.

2 Altenkirch, E.,"Die Kompressionskältemaschine mit Lösungskreislauf", *Kältetechnik* (1950) nrs. 10, 11, 12.

3 Lotz, H., "Heat pumps, energy saving by use of refrigerant mixture compressor system", *IIR Commission B2*, 1978-3, pp. 175-187.

4 Gompertz, M. and Niebergall, W., "Untersuchung an einer Zweistufigen Ammoniak-Absorptions-Kältemaschine System Dr. Altenkirch", *Zeitschrift für die gesamte Kälte-Industrie*, 39 (1932) nrs. 5, 7, 8, 10, 11, 12.

5 Bošnjaković, F., *Technische Thermodynamik II*, (Dresden 1971, Verlag Theodor Steinkopff).

6 Niebergall, W., *Handbuch der Kältetechnik, Band VII : Sorptions-Kältemaschinen*, (Berlin 1959, Springer-Verlag).

7 Weise, A., "Das Verhalten der Absorptions-Kältemaschine", *Zeitschrift für die gesamte Kälte-Industrie*, 36 (1929) 169-173.

8 Latishew, W.P., Wnichi Nr. 3243, Moscow (1968).

TABLE I

Ahp versus rhp at different pressure levels (maximum rectification)

		ahp 1	rhp 1	rhp 2	rhp 3	rhp 4
p_o	(bar)	3	2	1.4	1	0.6
ζ	(-)	1.53	1.47	1.42	1.41	1.36
p_{gen}	(bar)	15	11	9	7.5	5.5
Q_R	(kJ/kg vapour)	482	502	628	783	1248
Q_c or Q_{res}	"	1109	1038	1038	1109	1160
Q_{abs}	"	1800	1959	2001	2076	2123
Q_o	"	1176	1121	1084	1147	1194
Q_{gen}	"	2215	2378	2583	2821	3337
$\xi_{p,abs}$	(kg NH_3/kg)	0.280	0.220	0.171	0.120	0.060
$\xi_{r,abs}$	"	0.388	0.320	0.270	0.220	0.155
$\xi_{p,res}$	"	-	0.668	0.570	0.510	0.438
$\xi_{r,res}$	"	-	0.728	0.628	0.570	0.498
$\xi_{v,gen}$	"	0.945	0.920	0.885	0.840	0.730
$\xi_{v,R}$	"	1	≃ 1	0.999	0.998	0.997
b	(kg/kg vapour)	-	≃ 0	0.002	0.004	0.005

TABLE II

Ahp's and rhp's at different levels of rectification
(* optimized evaporator pressure for ahp without rectification according to [7].)

rectification level		ahp 1		rhp 1			rhp 4	
		max.	no rect.	max.	medium	no rect.	max.	no rect.
p_o	(bar)	3	2*	2			0.6	
ζ	(-)	1.53	1.41	1.47	1.46	1.45	1.36	1.31
p_{gen}	(bar)	15	15	11			5.5	
Q_R	(kJ/kg vapour)	482	-	502	222	-	1248	-
Q_c or Q_{res}	"	1109	1494	1038	1265	1411	1160	1804
Q_{abs}	"	1800	1478	1959	1801	1608	2123	1306
Q_o	"	1176	866	1121	1039	938	1194	741
Q_{gen}	"	2215	2106	2378	2249	2081	3337	2369
$\xi_{p,abs}$	(kg NH_3/kg)	0.280	0.200	0.220			0.060	
$\xi_{r,abs}$	"	0.388	0.320	0.320			0.155	
$\xi_{p,res}$	"	-	-	0.668			0.438	
$\xi_{r,res}$	"	-	-	0.728			0.498	
$\xi_{v,gen}$	"	0.945	0.907	0.920	0.921	0.922	0.730	0.740
$\xi_{v,R}$	"	1	-	≃ 1	0.960	-	0.997	-
b	(kg/kg vapour)	-	-	≃ 0	0.121	0.236	0.005	0.461

TABLE III

Ahp and rhp for simultaneous heating and cooling over a relatively large temperature trajectory

		ahp 2	rhp 5
p_o	(bar)	1.8	1
ζ	(-)	1.47	1.49
p_{gen}	(bar)	14	6
Q_R	(kJ/kg vapour)	632	720
Q_c or Q_{res}	"	1122	1319
Q_{abs}	"	1926	2005
Q_o	"	1181	1331
Q_{gen}	"	2499	2713
$\xi_{p,abs}$	(kg NH_3/kg)	0.168	0.100
$\xi_{r,abs}$	"	0.309	0.235
$\xi_{p,res}$	"	-	0.460
$\xi_{r,res}$	"	-	0.625
$\xi_{v,gen}$	"	0.900	0.860
$\xi_{v,R}$	"	1	≃ 1
b	(kg/kg vapour)	-	≃ 0

TABLE IV

Ahp and rhp, using R22 - E181. Source temperature ≃ 5°C (air)
Water heated from ≃ 30°C to ≃35°C

		ahp 3	rhp 6
p_o	(bar)	5	1.5
ζ	(-)	1.53	1.49
p_{gen}	(bar)	15.4	6.5
Q_s	(kJ/kg vapour)	45	78
Q_c or Q_{res}	"	163	167
Q_{abs}	"	309	340
Q_o	"	178	193
Q_{gen}	"	339	392
$\xi_{p,abs}$	(kg NH_3/kg)	0.25	0.04
$\xi_{r,abs}$	"	0.39	0.14
$\xi_{p,res}$	"	-	0.41
$\xi_{r,res}$	"	-	0.52
$\xi_{v,gen}$	"	1	1
$\xi_{v,R}$	"	-	-
b	(kg/kg vapour)	-	-

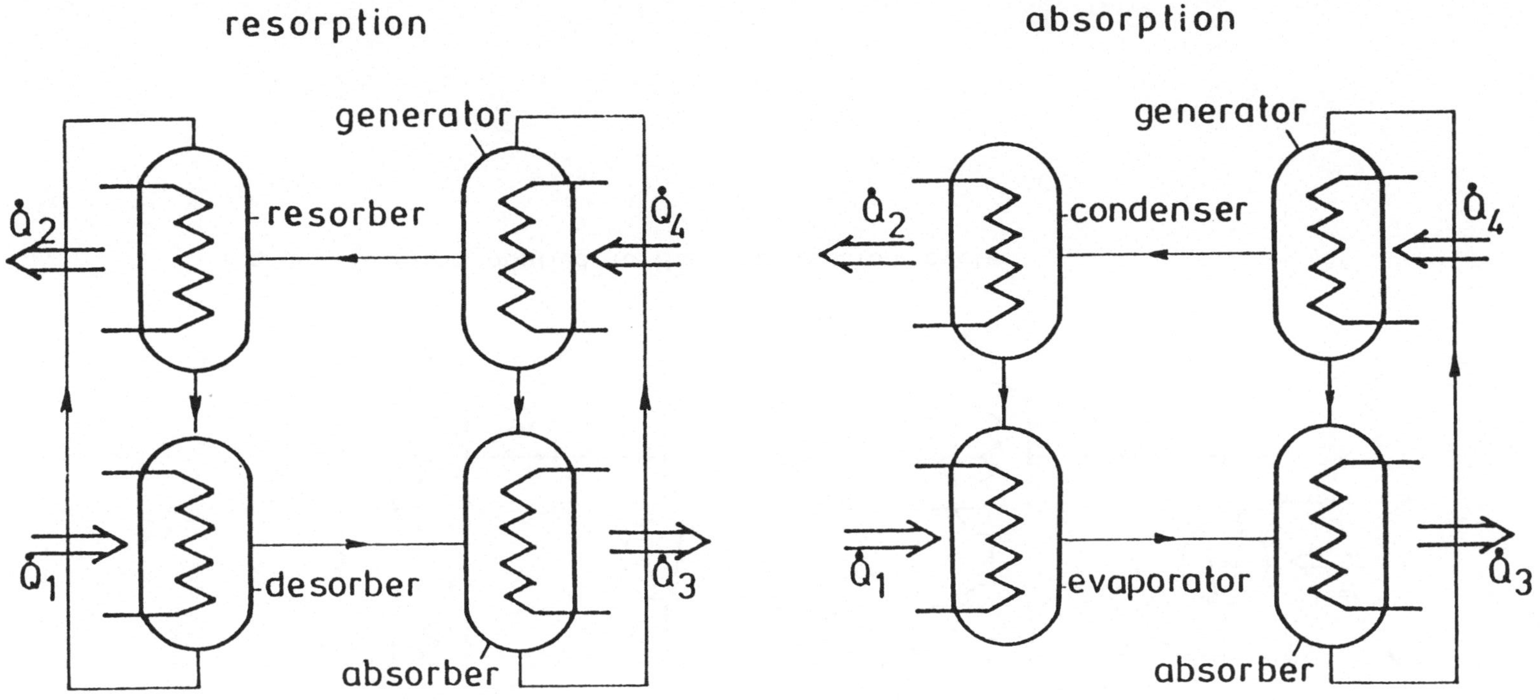

Fig. 1a Resorption heat pump versus absorption heat pump

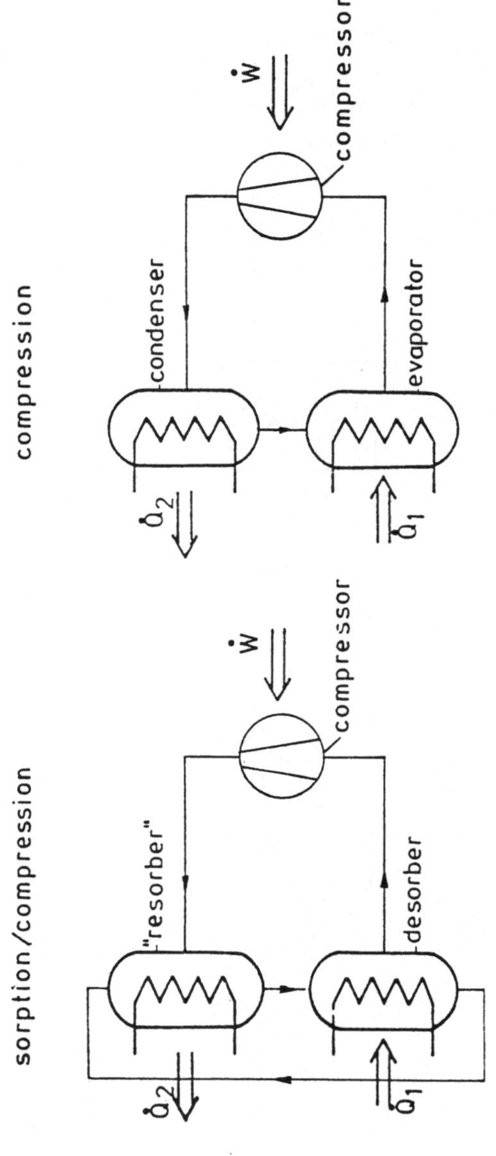

Fig.1b Sorption applied to the compression cycle.

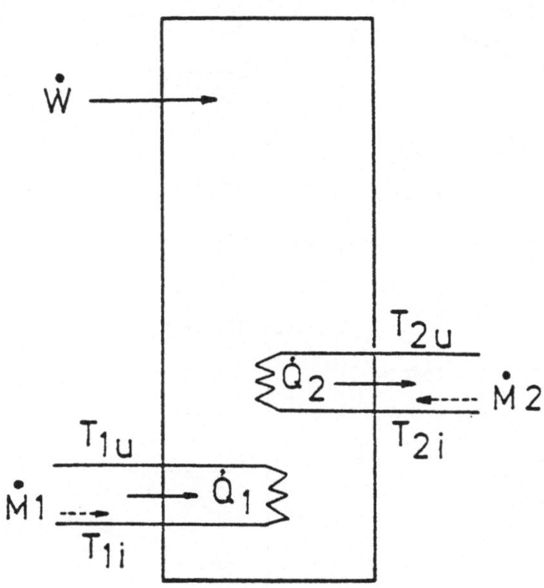

Fig. 2 Mechanical heat pump boundary conditions

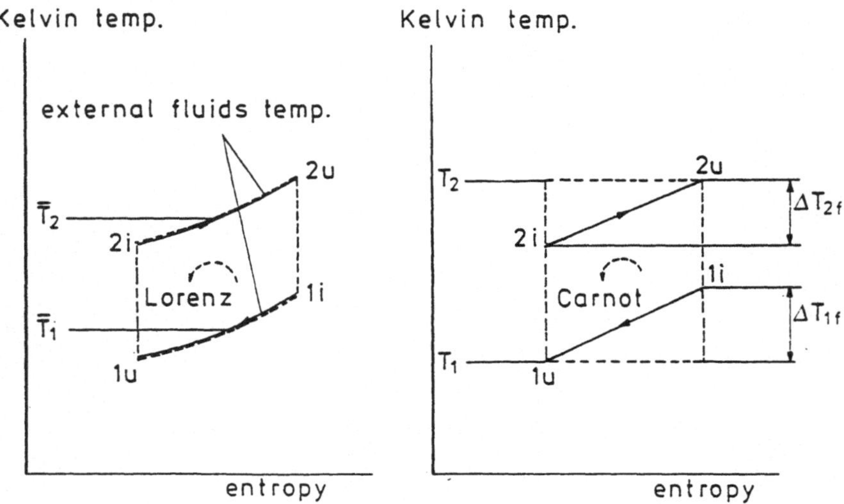

Fig.3a Lorenz and Carnot cycles (infinite heat exchanging surfaces)

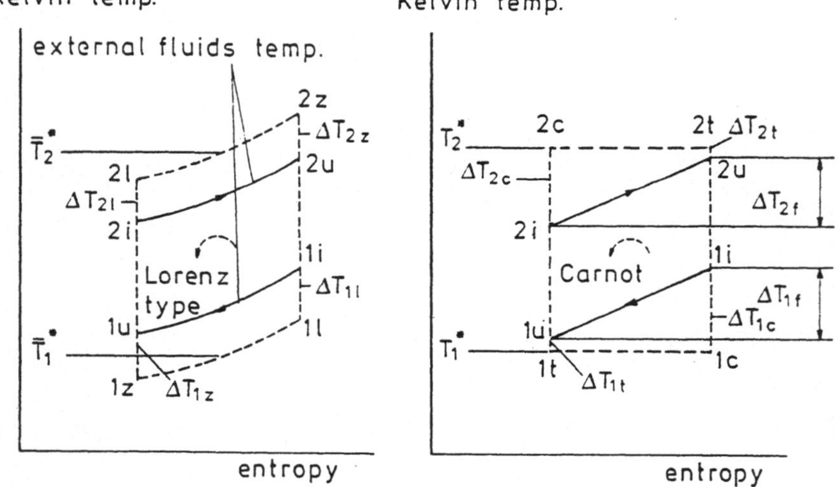

Fig.3b Lorenz-type and Carnot cycles (finite heat exchanging surfaces)

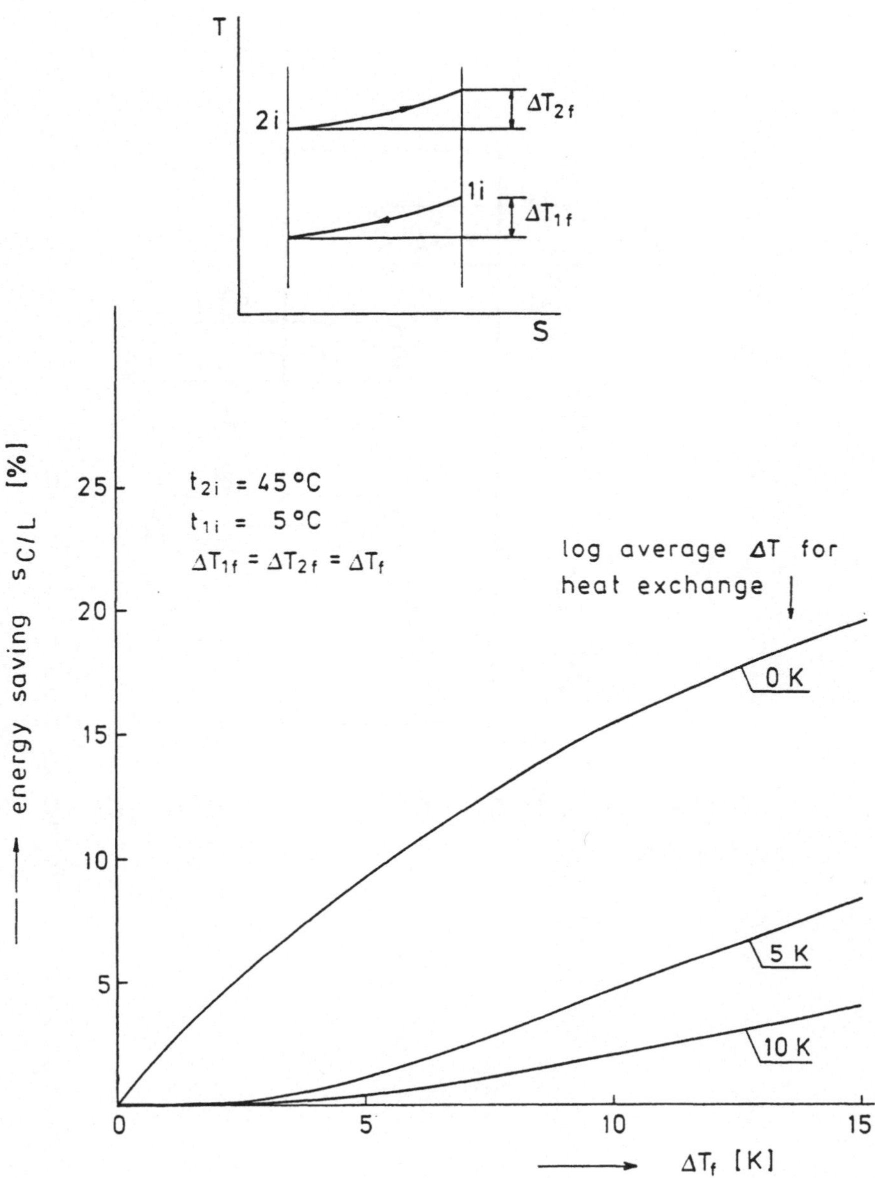

'ig. 4 Energy saving by the Lorenz cycle, as related to the Carnot cycle

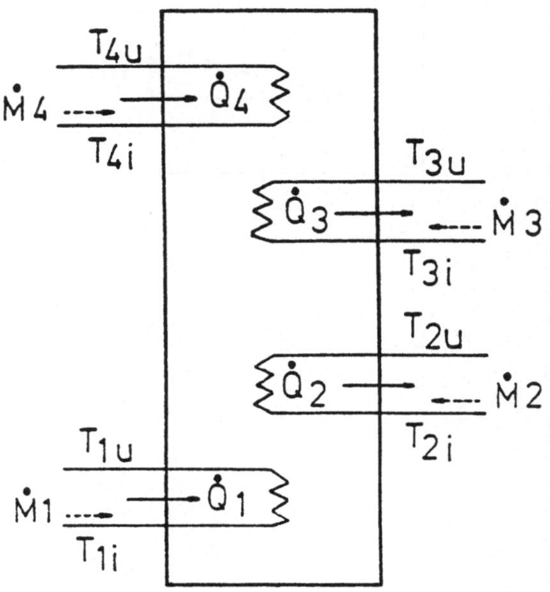

Fig. 5 Heat-actuated heat pump boundary conditions

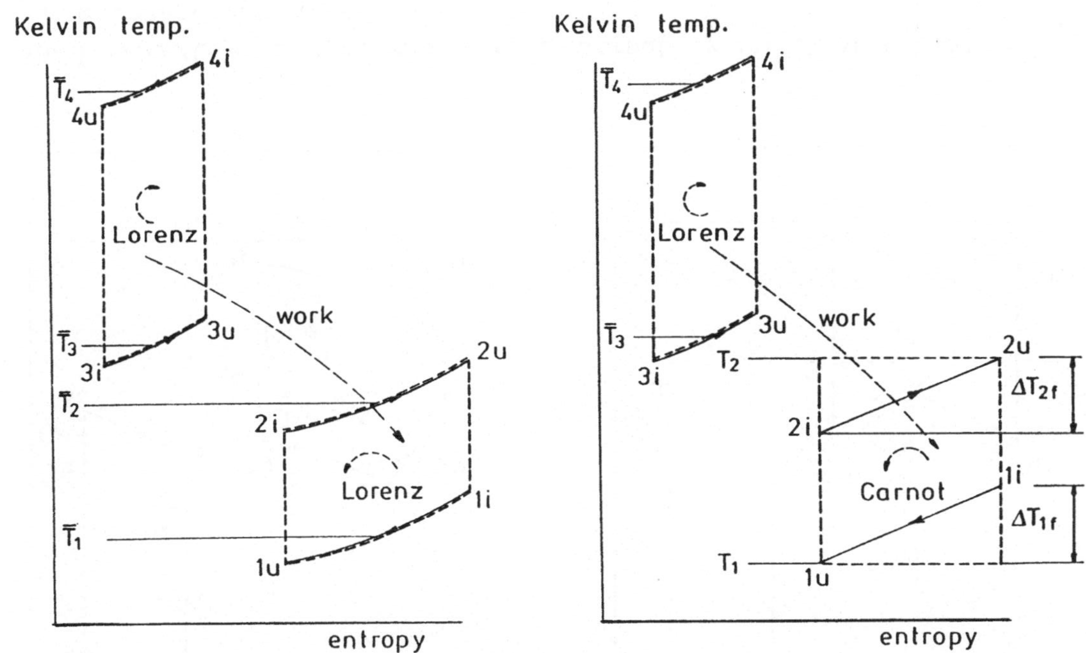

Fig. 6a Lorenz / Lorenz and Lorenz / Carnot cycles (infinite heat exchange surfaces)

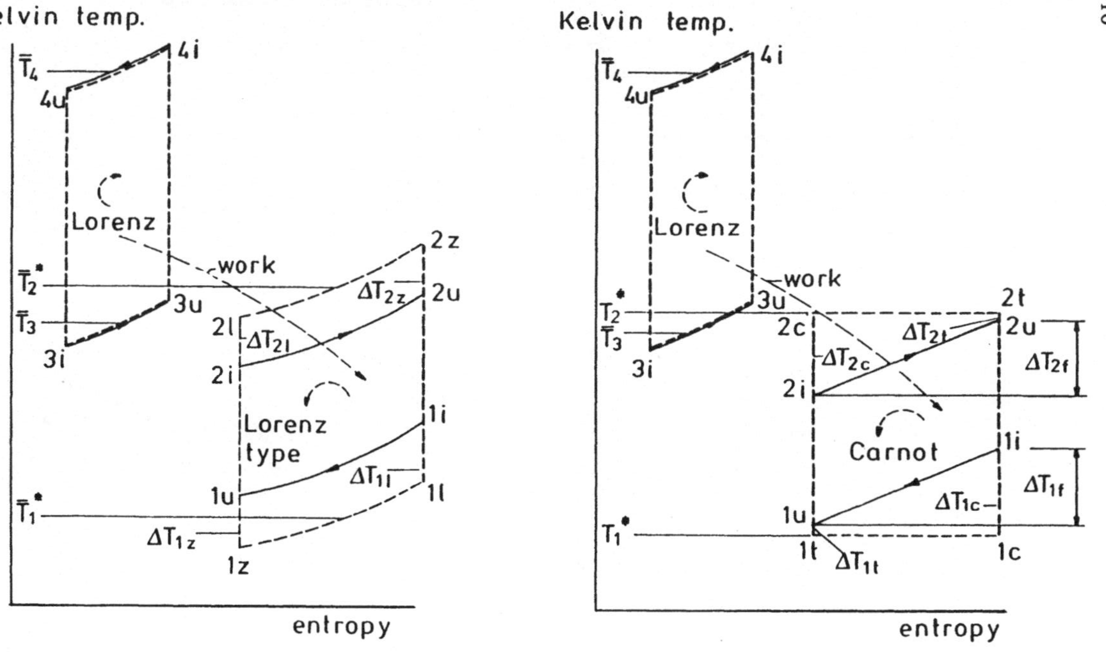

Fig.6b Lorenz/Lorenz type and Lorenz/Carnot cycles (finite heat exchange surfaces)

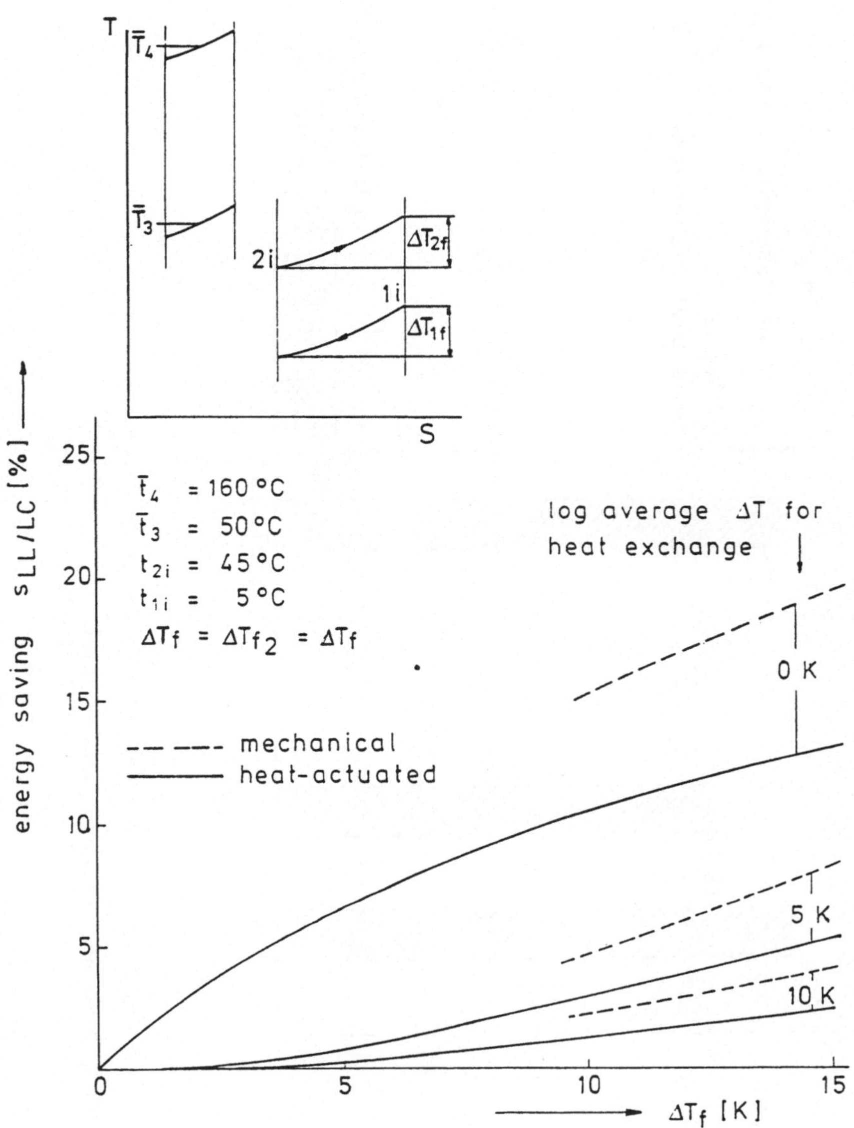

Fig. 7 Energy saving by Lorenz/Lorenz cycle as related to Lorenz/Carnot cycle

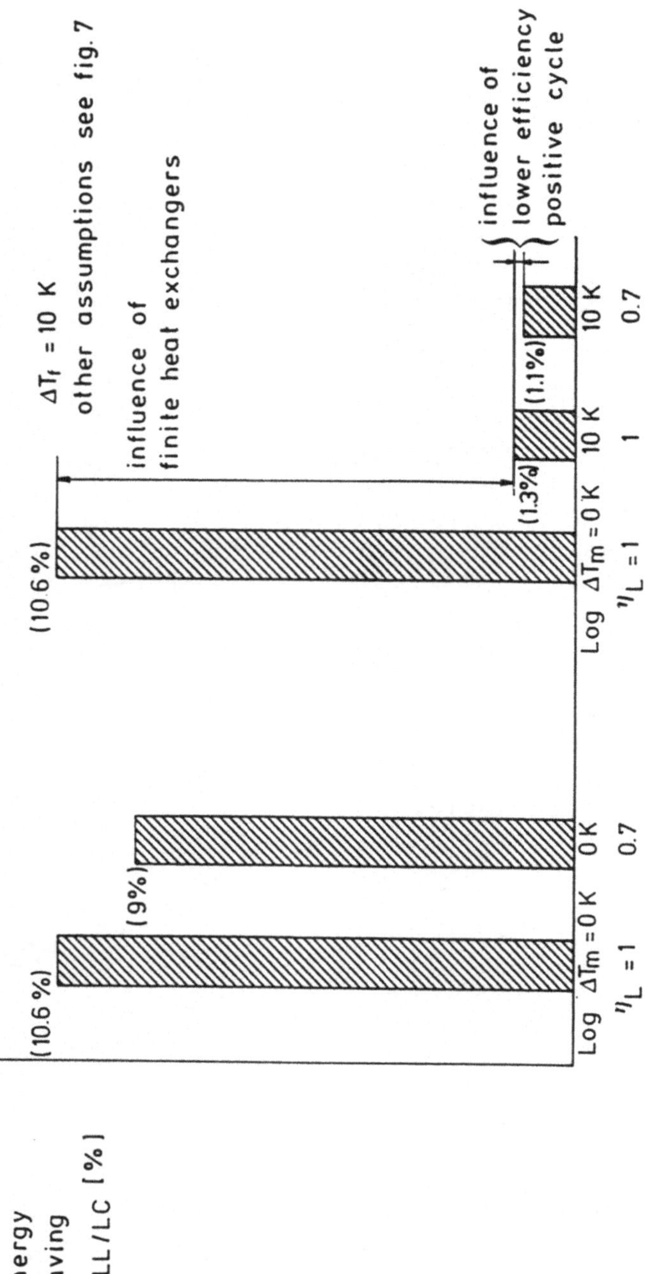

Fig. 8 Influence of a less efficient positive cycle

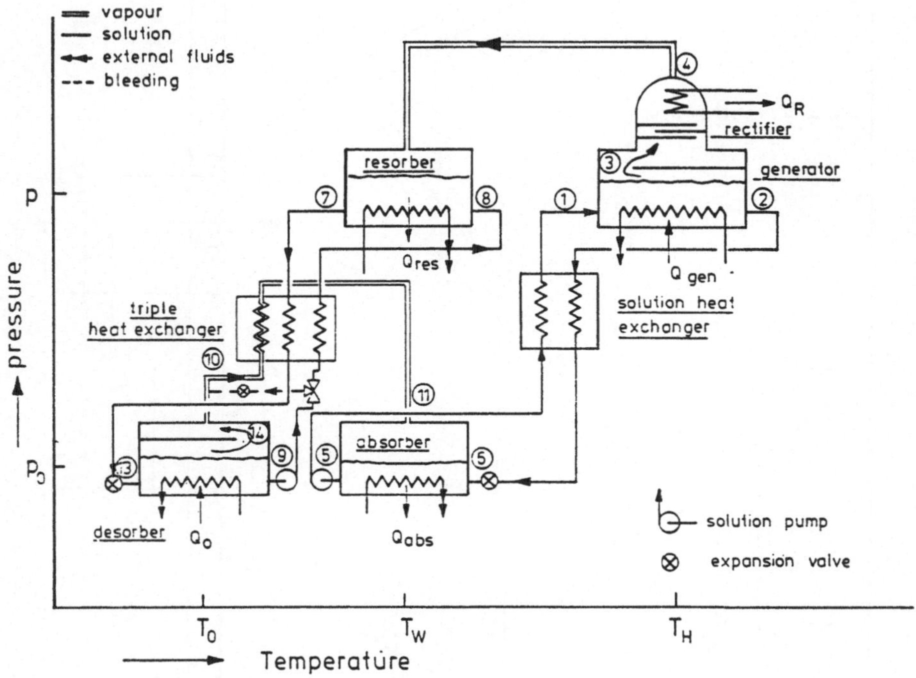

Fig. 9a A simple resorption heat pump arranged in a p,T-diagram

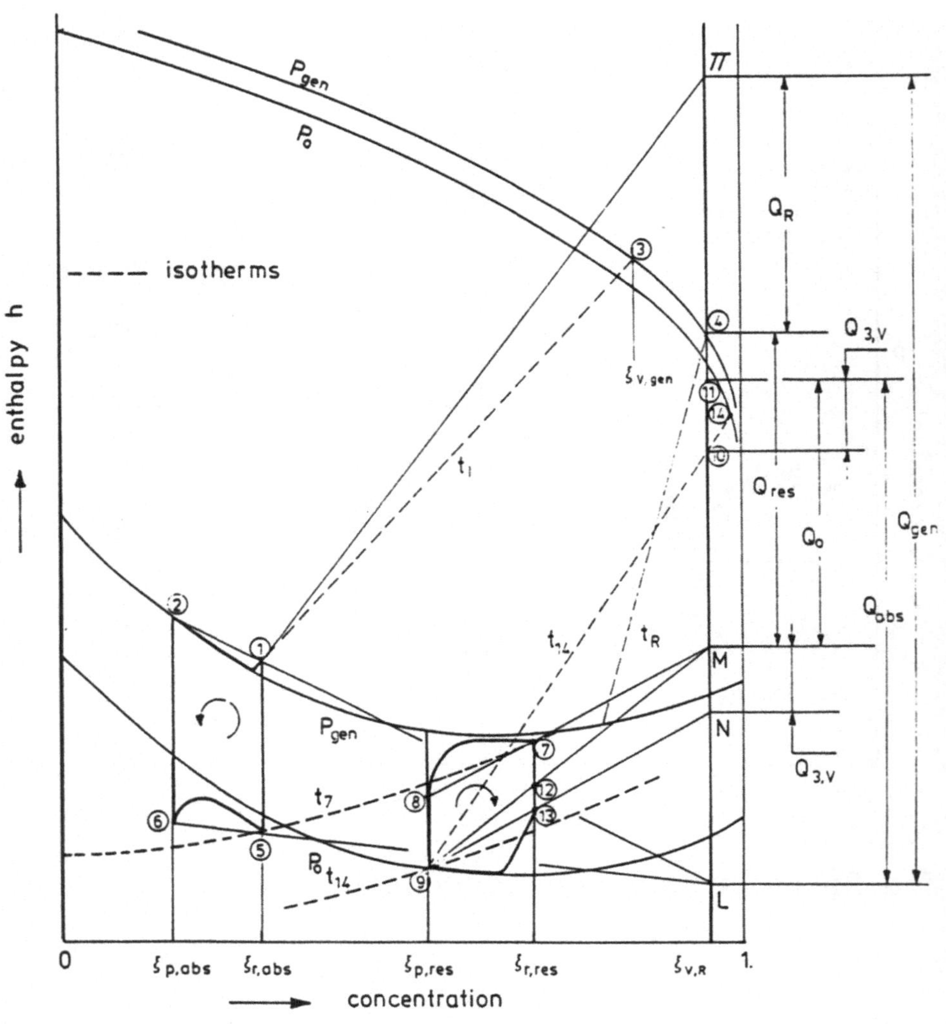

Fig. 9 b The same heat pump depicted in an h, ξ-diagram

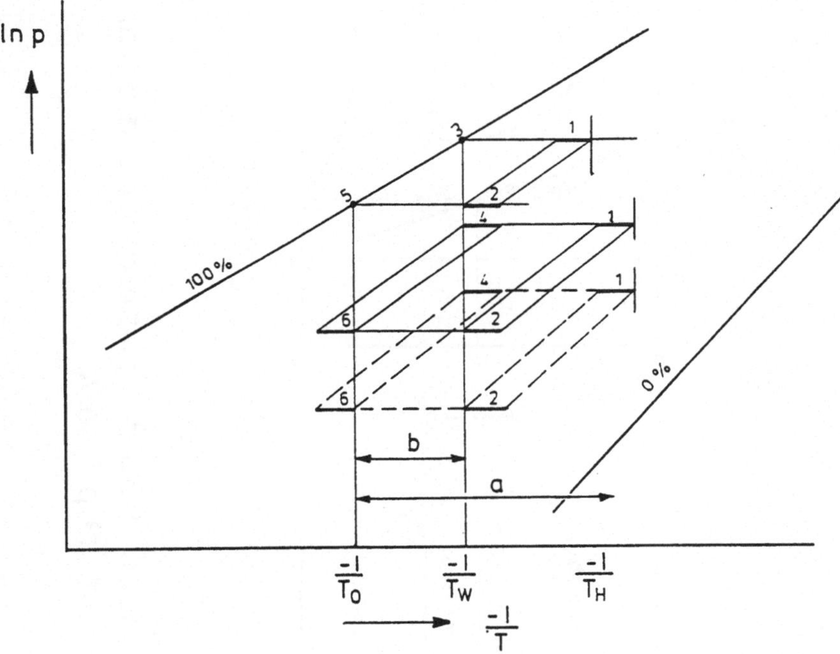

Fig.10 One ahp and two rhp's, all operating at the same temperature levels of heat source (T_0) and heat sink (T_W)
1 generator ; 2 absorber ; 3 condenser ; 4 resorber ; 5 evaporator ; 6 desorber

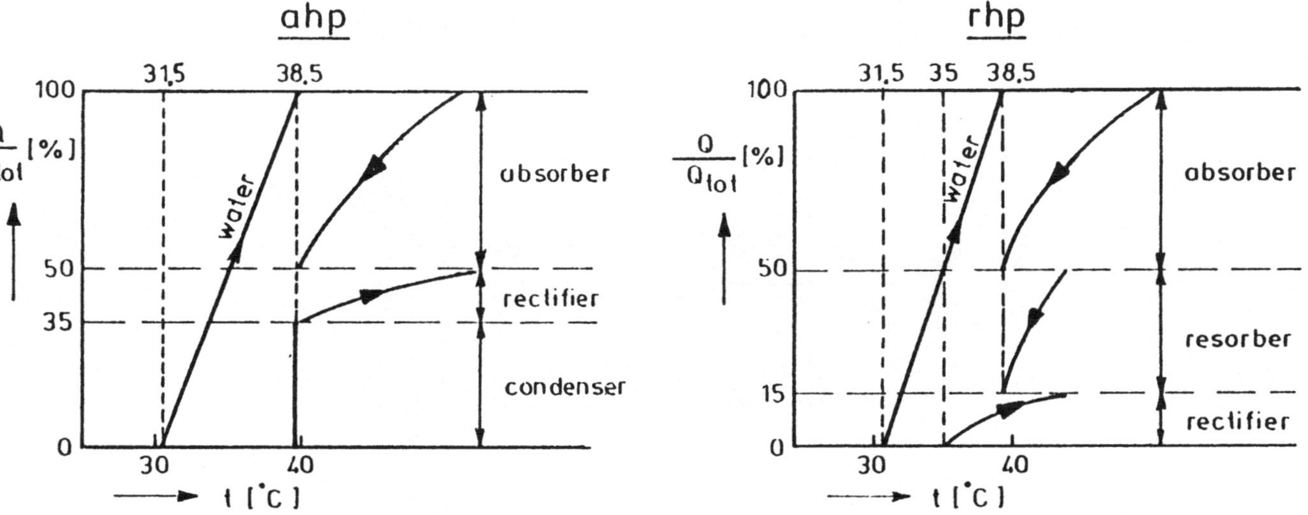

Fig. 11 Extracted heat, liquid temperature - diagrams for ahp 1 and rhp 1 to 4

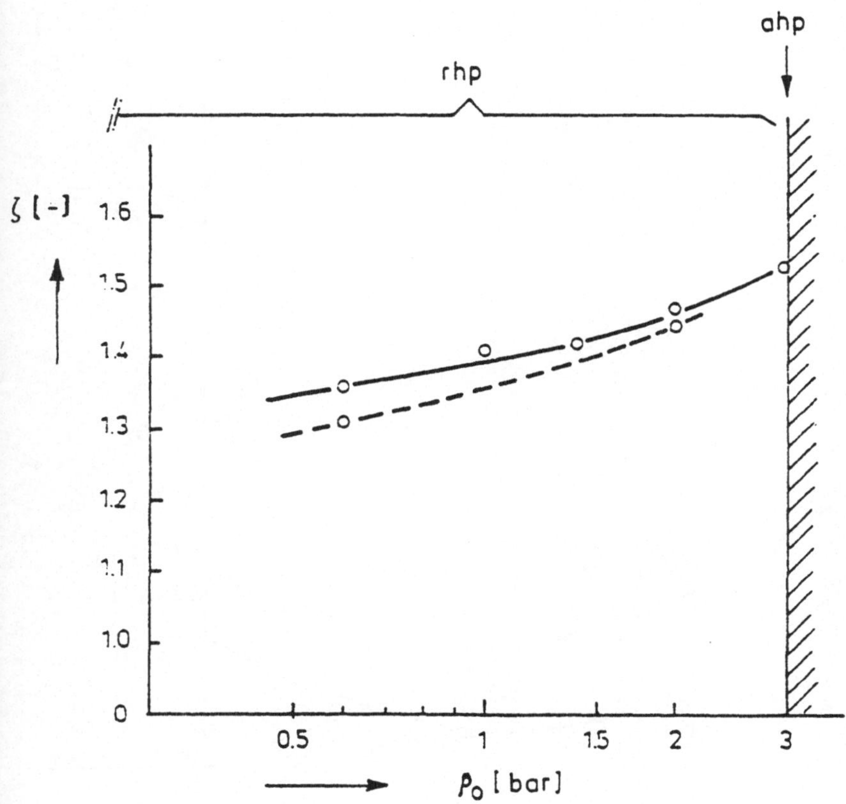

Fig. 12 The heat ratio (ζ) of sorption heat pump examples under fixed external conditions as a function of the desorber/evaporator pressure p_0

——— maximum rectification (table 1)

- - - - - no rectification (table 2)

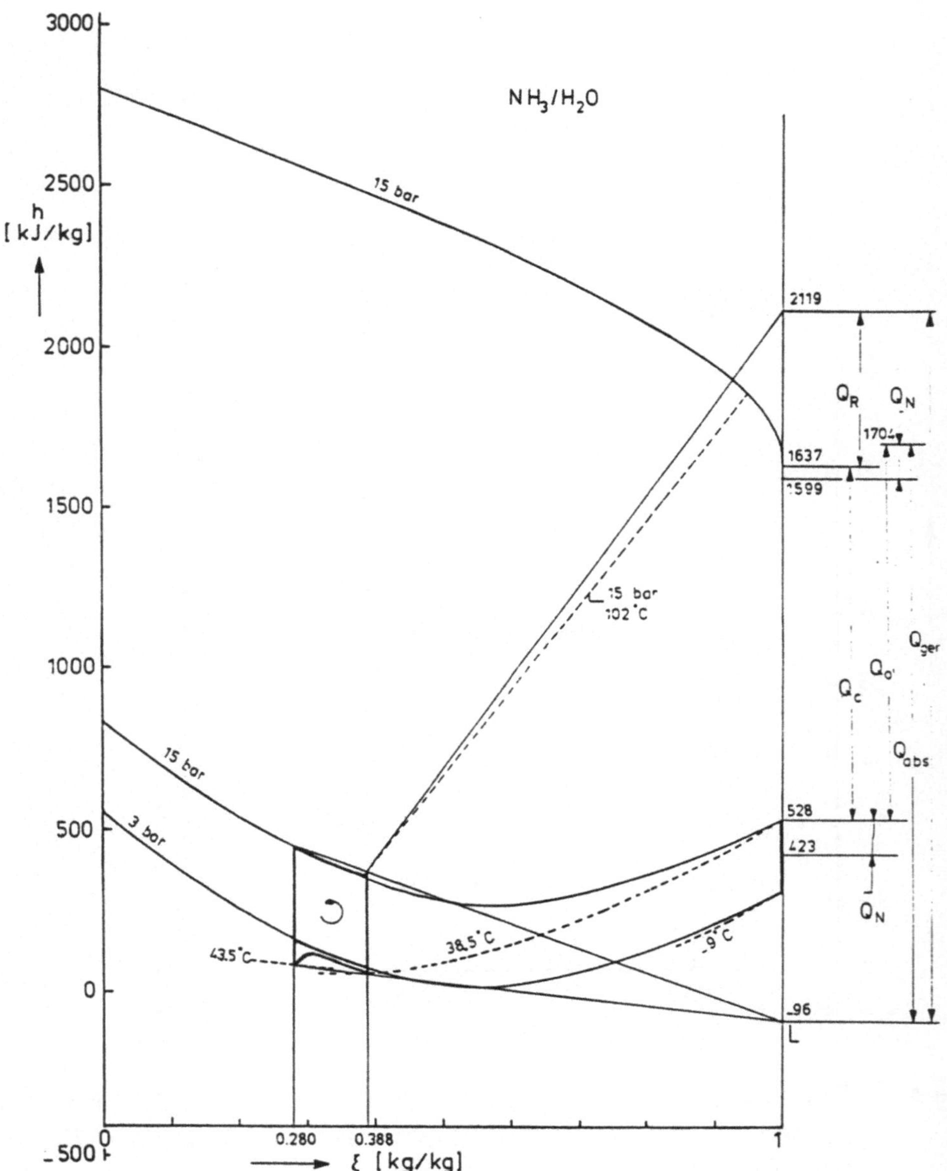

Fig. 13 Plot of ahp 1 in h, ξ - diagram

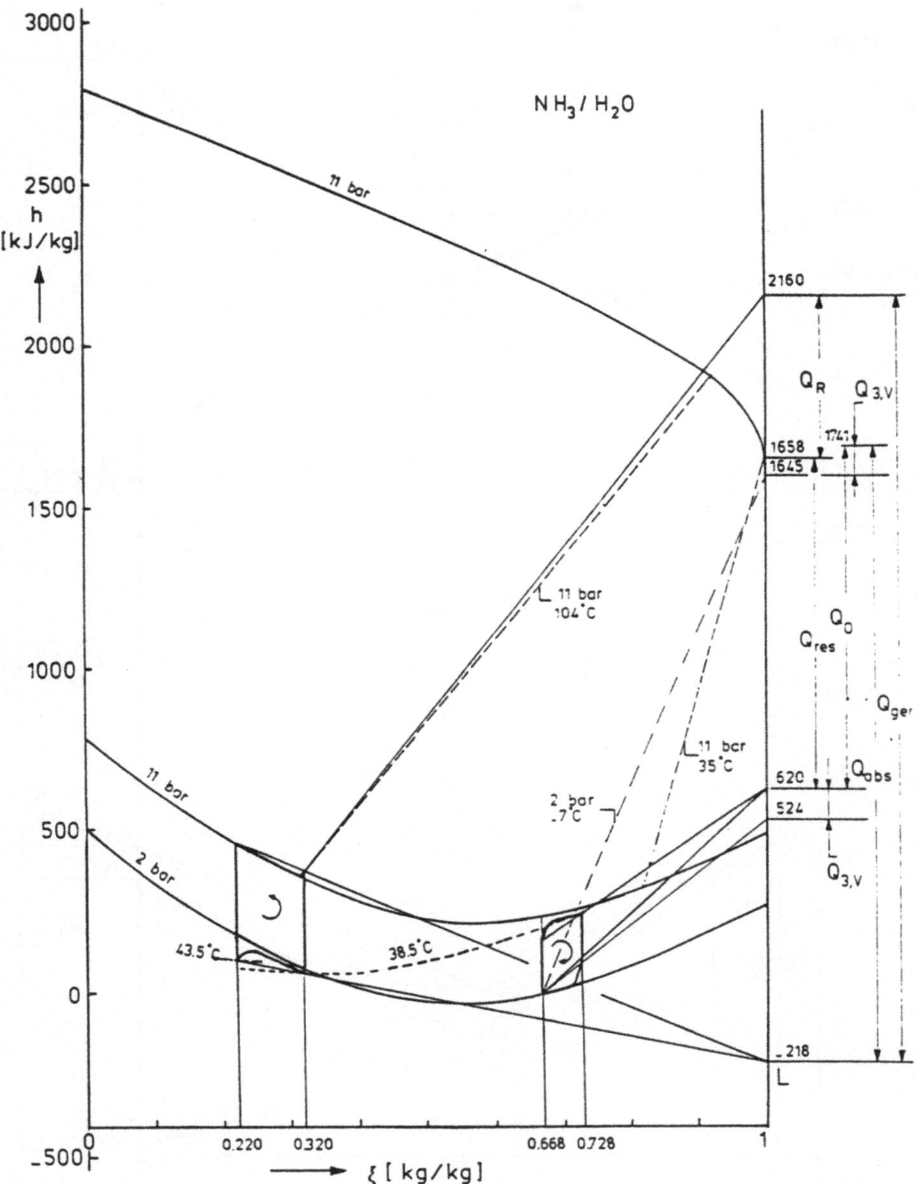

Fig. 14 Plot of rhp 1 in h, ξ - diagram

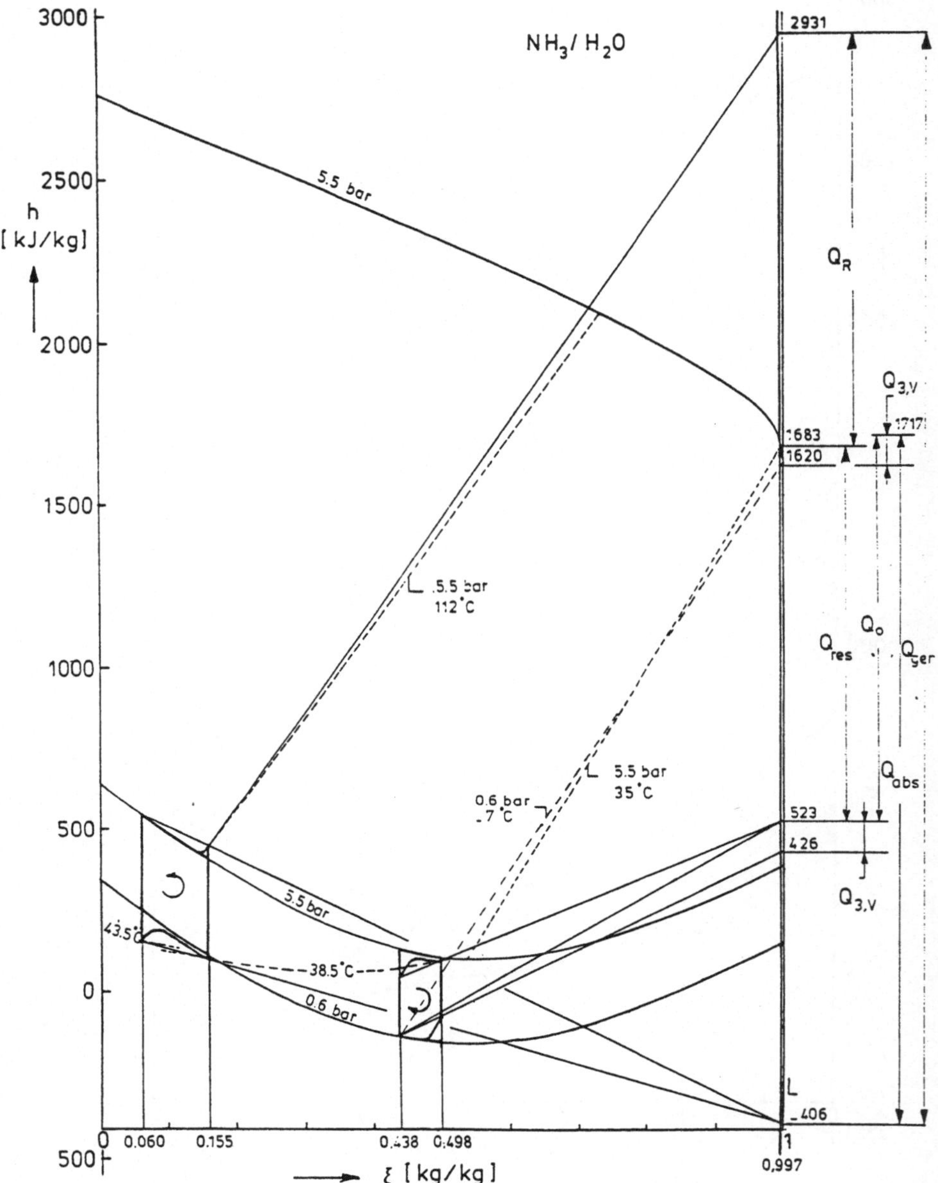

Fig.15 Plot of rhp 4 in h, ξ - diagram

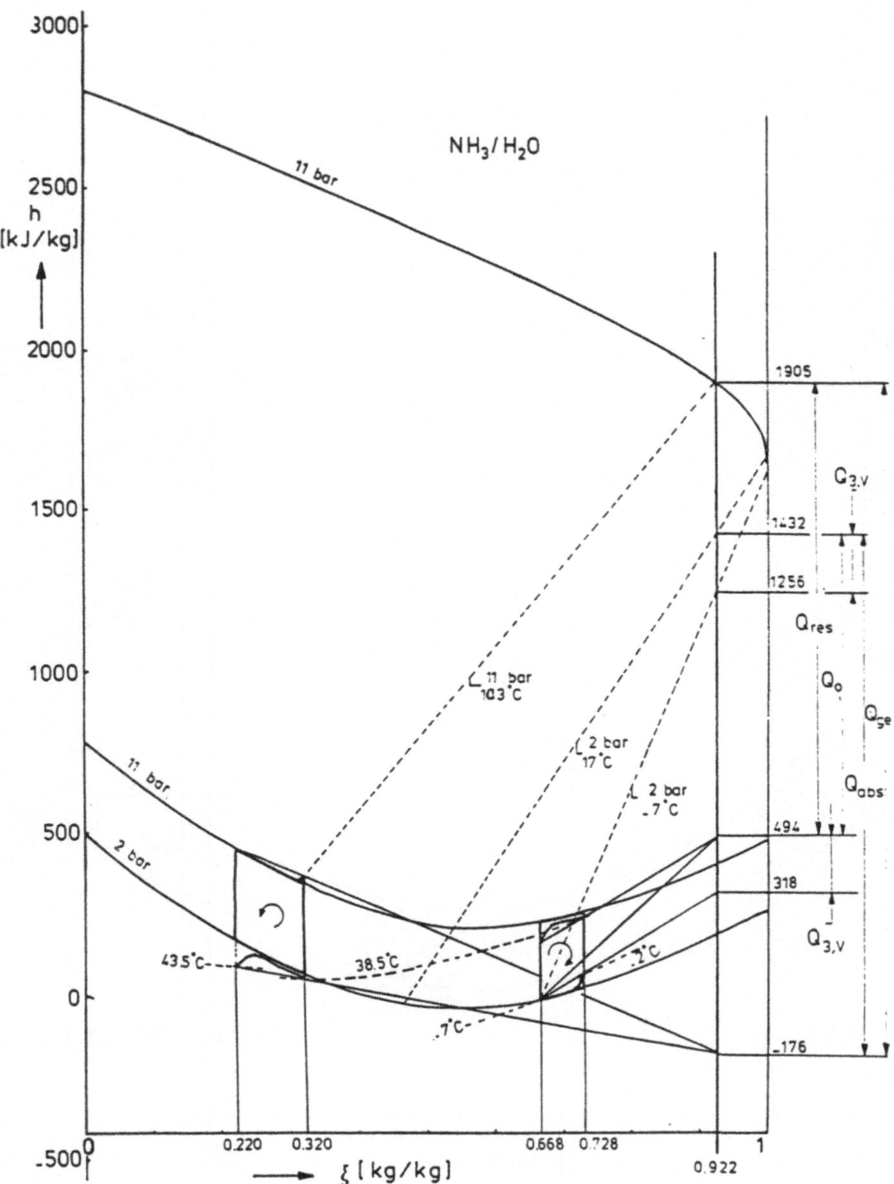

Fig. 16 Plot of rhp 1 in h, ξ-diagram no rectification

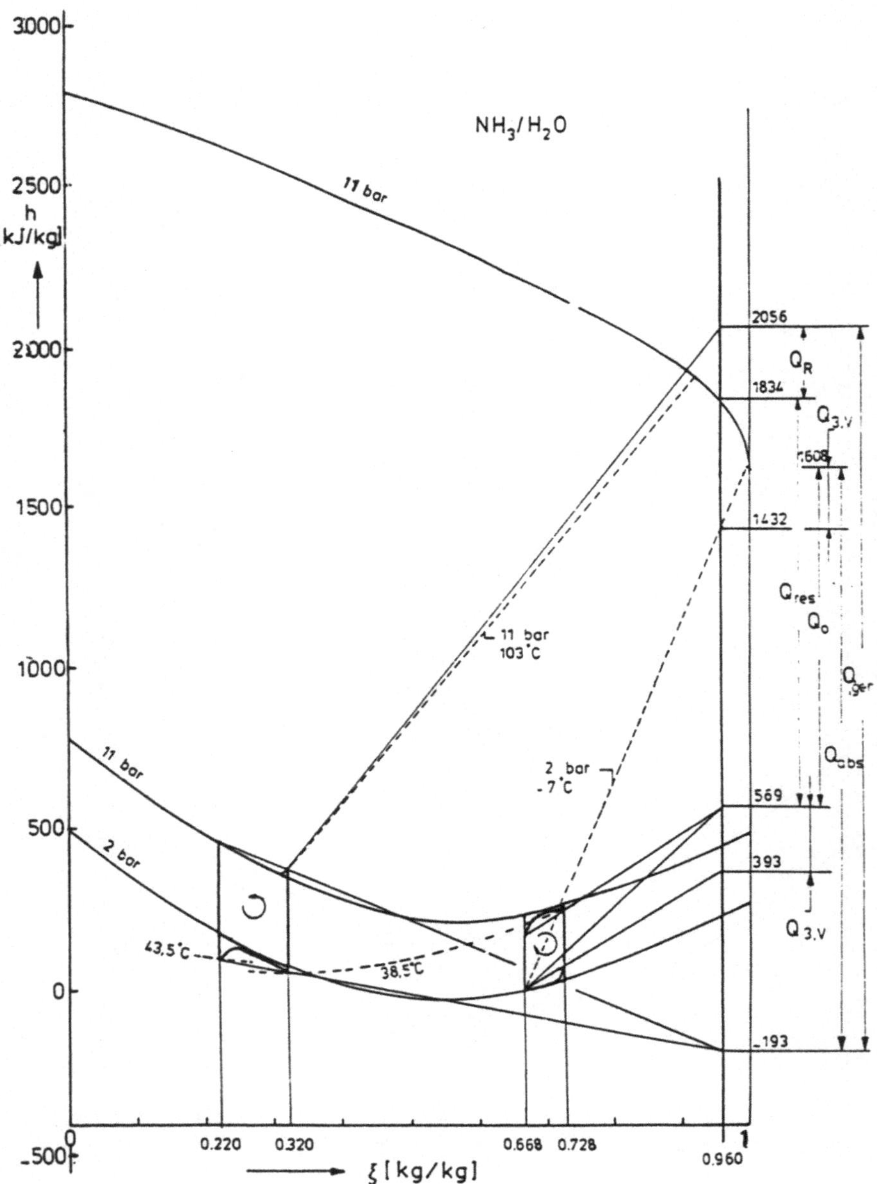

Fig. 17 Plot of rhp 1 in h,ξ-diagram. "medium rectification"

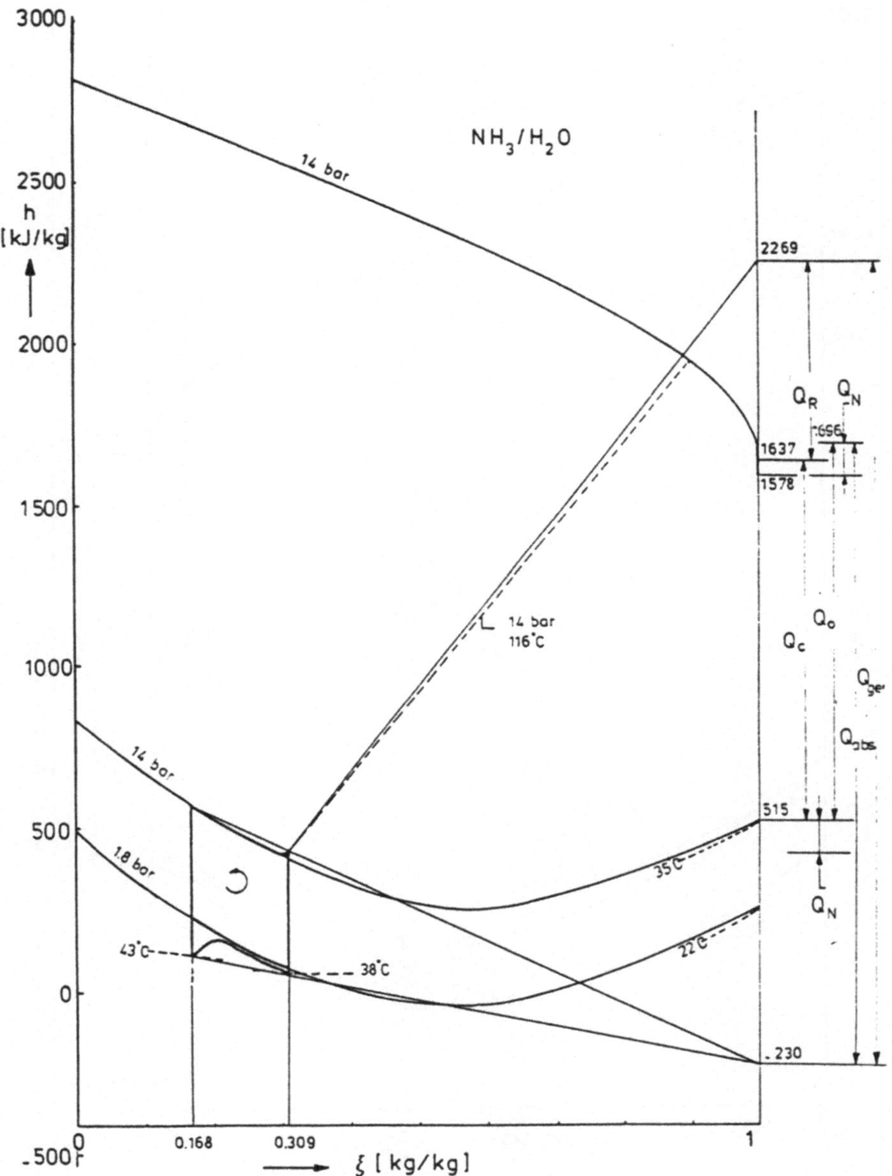

Fig.18 Plot of ahp 2 in h, ξ- diagram

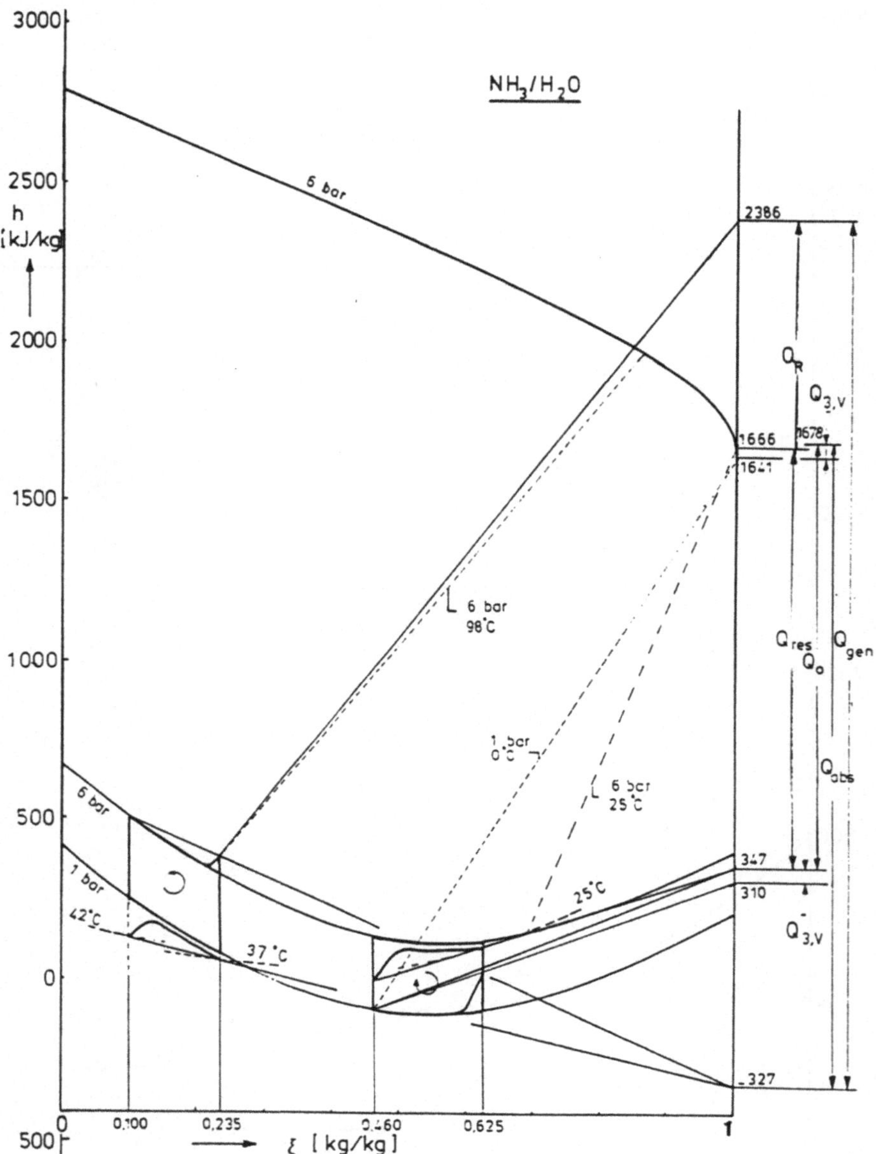

Fig. 19 Plot of rhp 5 in h, ξ - diagram

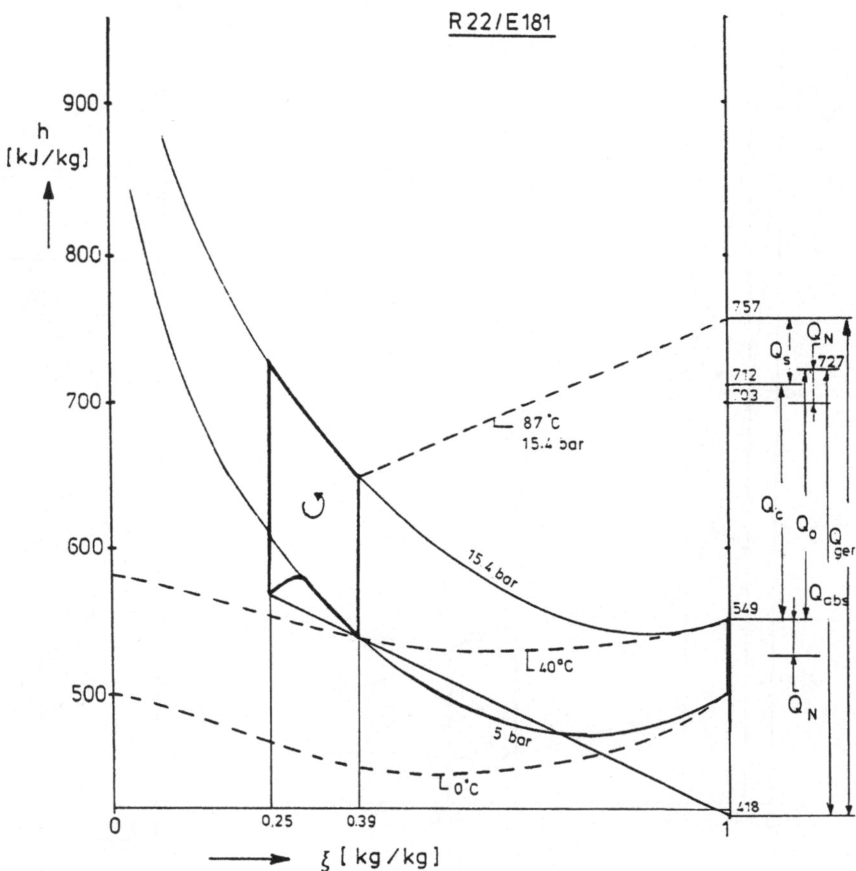

Fig. 20 Plot of ahp 3 in h,ξ-diagram

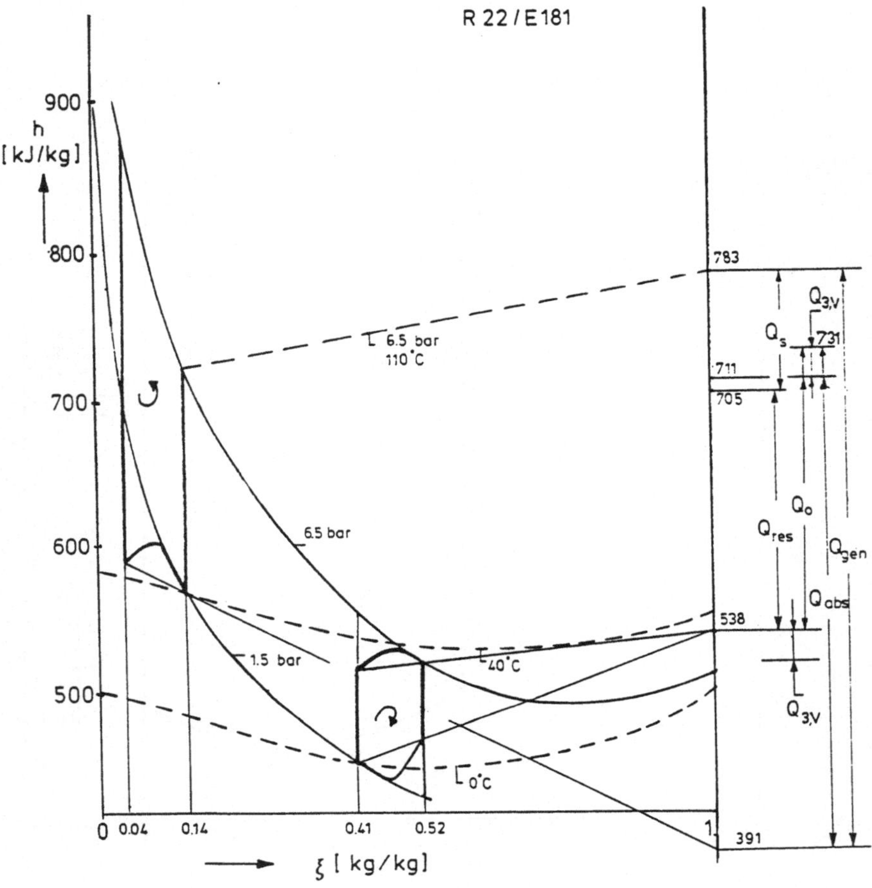

Fig. 21 Plot of rhp 6 in h,ξ- diagram

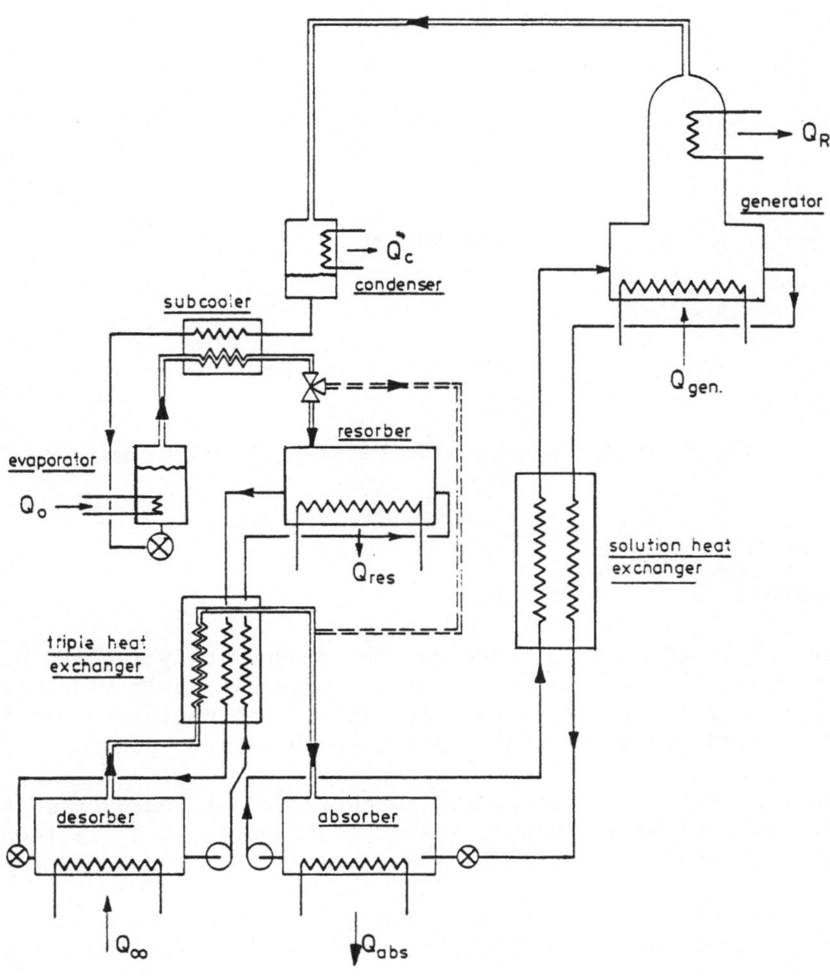

Fig. 22 A two-stage sorption heat pump

ABSORPTION HEAT PUMP DEVELOPMENTS

G. Cohen, A. Rojey

Institut Français du Pétrole, Rueil-Malmaison, France

SUMMARY

The implementation of both new thermodynamic cycles and new suitable fluids makes it possible to considerably widen the capacity to recover and upgrade low level heat contained particularly in industrial thermal wastes, or the surrounding environment.

This will result in improved conservation of energy resources in the leading sectors of energy consumption, i.e. residential, tertiary and industrial.

We present here a brief overview of recent advances in the field of absorption and we describe these possibilities in the form of :

- direct absorption cycles : absorption heat pumps in the field of home and tertiary heating;

- reverse absorption cycles : heat transformers for which the technical advances made and the increase in energy prices should enable them to spread progressively throughout the industrial sector.

1. INTRODUCTION

The amount of energy used either for residential or for industrial heating represents a major portion of the total consumption of primary energy. The use of absorption heat pumps or absorption heat transformers can lead to an appreciable reduction of the consumption of fossil fuels for most of these heating processes.

As compared to electric compression heat pumps, absorption heat pumps offer great advantages. They are more favorable from the primary energy consumption standpoint and do not lead to a high electricity demand at peak-load conditions for the electric network.

Such systems are not yet widespread due to technological difficulties and because the economics have not yet been proven favorable enough. This situation can be greatly improved by developing new and more suitable technology.

We present here a brief overview of recent advances in the field of absorption heat pumps and heat transformers for home heating and industrial applications.

2. PRINCIPLES OF TRITHERMAL CYCLES

Figure 1 gives the flowsheet of a trithermal cycle. For the sake of simplicity we will assume that three sources exchange heat at a given temperature level. This condition can be entirely fulfilled only if there is a change of phase of the external fluid. However when the temperature variation of the sources is slight, each thermal level T_1, T_2, T_3 represents the mean temperature of the external fluid during the exchange.

In Figure 1, T_1 is used to designate the temperature of a source supplying the system with an amount of heat Q_1, and T_2 is the temperature of use of the amount of heat Q_2. Whereas the system does not exchange any work with the exterior, it can operate reversibly only by exchanging heat with a third source at temperature T_3. It is assumed that $T_3 < T_1$ and $T_3 < T_2$ and that the heat exchanges are taken to be positive when heat is received by the system, and negative when heat is given off by the system.

When the system exchanges heat reversibly with three sources, we have :

$$Q_1 \left(1 - \frac{T_3}{T_1}\right) + Q_2 \left(1 - \frac{T_3}{T_2}\right) = 0 \qquad (1)$$

Equation (1) can be used to distinguish types of operation of trithermal cycles.

Case 1 : $T_1 > T_2$

The system transforms an amount of heat Q_1 available at a relatively high temperature level. According to the level of the third source, T_3, it can perform either a heating or a refrigeration function.

a) Heat production

The system transforms Q_1 into a larger amount of heat Q_2 which is supplied at a lower temperature T_2, with the amount of heat $Q_3 = /Q_2/ - Q_1$ being "pumped" from the heat source at temperature T_3.

The performance coefficient, defined as the ratio Q_2/Q_1, will improve when the amount of heat Q_3 is recovered at the highest possible temperature.

Figure 2 gives the theoretical performance coefficient derived from equation (1) versus the difference $T_2 - T_3$ for some values of the difference $T_1 - T_2$. This graph was compiled for T_3 = 303 K, but the Q_2/Q_1 ratio depends only on the difference $T_2 - T_3$ and shows little variation with temperature T_3.

This performance coefficient must be distinguished from the one for an electrically driven system because it is not related to the same level of primary energy transformation. Indeed the performance coefficient is directly related to the energy expended.

b) Refrigeration

The system can also be used for refrigeration, in which case the temperature T_3 is lower than the ambient temperature. If the system works only for refrigeration, the intermediate source at T_2 is supplied by the available cooling fluid, air or water.

Case 2 : $T_2 > T_1$

The system transforms an amount of heat Q_1 available at a low level T_1 into a smaller amount of heat Q_2 but which can be used at a higher temperature, with the amount of heat $Q_3 = Q_1 - /Q_2/$ being released to the cold source. This type of operation enables waste heat to be recovered and upgraded to the level desired by the user.

As opposed to the case 1, the performance coefficient Q_2/Q_1 increases when the level of situation in T_3 decreases. If the amount of heat exchanged with the source at T_3 is not used, it is necessary to work at the lowest available temperature T_3 which is defined by the cooling fluid, air or water.

Figure 3 gives the theoretical performance coefficient Q_2/Q_1 derived from equation (1) versus the difference $T_1 - T_3$ for different values of $T_2 - T_1$.

This case is of particular interest from the standpoint of energy conservation. Indeed, heat Q_2 is thus produced solely from low-level heat coming from the recovery of thermal discharges.

Mechanically-driven trithermal cycles

The above two types of operation of trithermal cycles can be achieved either by compression or absorption. Let us briefly demonstrate what such mechanically driven systems look like.

Figure 4 shows the type 1 system. Heat Q_1 available at a relatively high temperature T_1 is transformed into work W and into heat at the temperature level T_2. This work is used to drive a compression heat pump to produce an amount of heat Q_2 greater than Q_1 and at temperature T_2, with the heat being pumped from the low level T_3. According to the level of temperature T_3, we will produce heating or refrigeration. Industrially, work can be produced either by the expansion of steam under pressure in a counterpressure turbine or by an internal-combustion system (gas turbine or diesel engine) equipped with a recovery boiler.

A type 2 system is shown in Figure 5. The heat available at low thermal level is transformed into work in a cycle operating between temperature T_1 and temperature T_0. This work is used to drive a compression heat pump which produces heat at high level T_2 with the heat pumped at the same level as the one which drives the cycle producing the work.

3. ABSORPTION HEAT PUMPS

3.1. General Scheme

Trithermal absorption systems combine drive cycles and a heat pump without the help of rotating machinery producing or using mechanical power. Such a device is derived directly from absorption refrigeration machinery. These pumps use the chemical affinity between two fluids to induce heat and mass transfer. The solute (higher volatility) and the solvent (low volatility or solid in the case of salts) must be miscible in widely varying proportions and have great deviation from ideality. The most frequently used solute-solvent pairs are ammonia-water and water-lithium bromide.

The schematic representation of an absorption heat pump is shown in Figure 6. The working fluid is generated in the form of vapor by extracting heat at temperature level T_1. It is condensed while producing heat at temperature T_2 before being expanded and being vaporized while extracting heat at a low temperature level (T_3). It is then absorbed while producing heat at temperature T_2.

The absorber and the generator are capable of performing the same transformation as a compressor associated with a combined production cycle for both work and heat. This is achieved by compressing the working fluid while receiving heat at temperature level T_1 and giving back heat at temperature level T_2. In the case of the ammonia-water pair, a Q_2/Q_1 ratio of between 1.3 and 1.5 can be obtained at low temperature operation. The operating diagram in Figure 6 is poorly suited for the production of relatively high-temperature heat, and in the 70-100 °C range the Q_2/Q_1 performance coefficient decreases.

The water-lithium bromide pair theoretically achieves higher performances in such cases but raises problems of materials, thus increasing the investments required.

The flowsheet in Figure 6 can be used only for the first type of system described above. Indeed since the working fluid is much more highly concentrated in the condensor than in the generator, the generator temperature is necessarly higher than that of the condensor.

3.2. New Solute-Solvent Pairs for Absorption Heat Pumps

Many solute-solvent pairs can be used in absorption systems, either for heating or refrigeration purposes. The most frequently used pairs have some disadvantages :

- for the water-lithium bromide mixture investments are increased because of the risk of corrosion and the need for cooling by water;
- for the ammonia-water mixture precautions must be taken when used for individual home heating because of the toxic nature of the ammonia, resulting in increased costs.

To define new pairs, the criteria will be examined for choosing the solute, solvent and mixture.

Criteria common to solutes and solvents

- Low degree of toxicity characterized by the detection threshold and the maximum supportable concentration without the solute causing any trouble in an individual upon contact with the fluid eight hours per day.

- Thermal and chemical stability under the operating conditions of the system. Indeed, it is very important for the fluids to integrally maintain their physico-chemical properties during thermal cycling and especially upon contact with the materials of the sytem, in order not to upset the operating of the machinery and decrease its performance.

- The substances must not be corrosive with regard to inexpensive materials used for building heat pumps.

- Low viscosity and high thermal conductivity.

- Industrial availability and low cost.

Properties of the solute

- Low boiling temperature, generally between - 50 °C and + 100 °C, at 1 atm.

- The highest possible molar vaporization heat so as to reduce the solution flow rate. The corollary for this property must be a sufficient distance for the critical temperature under the operating conditions described for the systems shown in Figure 6.

The criteria can be used to single out several groups of constituents having most of these properties at the same time :

- water and alcohols with 1 to 4 carbon atoms;
- ammonia and methyl- or ethylamines;
- freons 11, 12, 21, 22, 31, 113, 114, 124a, 133a, 134;
- compounds derived from SiH_4 such as $SiHClF_2$, $SiHF_2-CH_3$.

Properties of the solvents

- High atmospheric boiling temperature, above 150 °C, characterizing low volatility.

- Sufficiently low crystallization point for the specific needs of use, except if the solubility of the solute contributes to considerably lowering the crystallization point by its pressence in solution.

A great many mineral and organic compounds simultaneously fulfill these criteria. Mention can be made of halides and alkaline-methyl thiocyanates for mineral salts, glycols or cellosolves and their derivatives.

References [1] and [2] give lists of solutes and solvents which fulfill these criteria.

Properties of solute-solvent mixtures

In addition to the high miscibility of both constituents and the considerable deviations from the idealness of the solution obtained, it is desirable for the following properties to be maintained :

- High dissolution rate of the solute in the solvent so as to reduce the size of absorbers.
- High dissolution heat per unit volume of the mixture so as to reduce the mechanical energy consumption by the solution pump.
- A large difference between boiling point of the solute and solvent so as to have the best possible separation between the two constituents inside the generator, because any entrainment of the solvent inside the condensor is detrimental to the efficiency of the heat pump. Nonetheless, the entrainment can be reduced by adding a rectification column on top of the generator, but this adds to the investment costs of the machinery.
- For cold-source temperatures and fixed hot-source temperatures, the choice of the mixture must be made so that the pressure difference between the generator and the absorber is as small as possible.

Considering the number and diversity of these criteria, it is obvious that the "ideal" solute-solvent pair does not exist but that the choice of a working mixture will be the result of a compromise between these different criteria for a given application.

3.3. Current Research on New Pairs.

New pairs have recently been proposed, and prototypes have been built for the heating or air conditioning of individual homes.

R22 - dimethylethertetraethyleneglycol (E181) mixture

A project, within the framework of the research and development program on energy for the European Communities, has been carried out jointly by the Battelle Institute in Frankfurt for the technico-economic aspect and by Stiebel Eltron Gmbh for the design and experimentation aspect of the prototype (References [3] and [4].

The goal is to create a home-heating installation having the following properties :

- the cold source is outside air or water
- the hot source is water
- the generator is heated by the combustion of gas or fuel oil
- the power to be provided is 12 thermal kW for an outside temperature of - 2 °C and to raise the water in the heating circuit to 55 °C
- the machinery will include a special circuit for using the generator as a conventional furnace when the outside temperature is lower than - 12 °C.

The calculated performance of this installation results in a performance coefficient of 1.26 for one year of heating under the climatic conditions found in the Hamburg region. Savings of primary energy should be about 40 percent compared with a standard type of gas furnace.

R133a - ethyltetrahydrofurfuralether mixture

This mixture is used in a mixed heating and air-conditioning system for individual homes. It was developed by the Allied Chemical Company [5] . This system can both heat the house by hot air and provide sanitary hot water. The cold source is outside air. Plans are being drawn up to begin marketing it in 1983. The performances announced are given in Table I.

A few other mixtures have been proposed, but we do not yet have any results enabling us to evaluate their precise significance in different applications. Among others, one may mention the replacement of water by methanol with lithium bromide, the methanol-calcium chloride pair, sodium thiocianate and ammonia, and R22 or methanol in association with dimethylformamide as solvent.

4. ABSORPTION RESORPTION CYCLES

4.1. General

The inclusion of a resorption step in the absorption cycle considerably widens its operational possibilities. The flowsheet for a bithermal resorption cycle is given in Figure 7 (Reference [6]). It should be compared with the absorption cycle yet is allows for a different function.

In the absorption cycle, the working fluid is compressed in an absorption and desorption step while receiving high-level heat and restoring this heat at a lower thermal level (motor cycle). In the resorption cycle, on the other hand, the working fluid is expanded while rejecting heat during the resorption steps at a higher temperature T_2 than temperature T_1 at which the heat is taken during the desorption step. The term resorption, which designates an absorption, is used to avoid any confusion with the process that occurs during the absorption cycle.

Figure 7 represents a design with a compression step provided by mechanical power. Reference [7] compares the performances of this resorption cyle with compression cycles. Table II gives the comparative results for compression heat pumps operating with freons or ammonia and the mechanically-driven resorption heat pump operating with a water-ammonia mixture. This comparison is made on the following basis :

- cold source = water at 40 °C cooled to 10 °C
- hot source = water at 40 °C heated to 80 °C
- compressor having identical properties for all the tests, a swept volume of 650 m^3/hr and electrically driven.

Under these conditions, the best result for freons is obtained for freon 12 both for the power delivered and for the performance coefficient. The use of ammonia requires two compression stages and leads to a very high pressure in the condenser (about 42 bar), which requires special compressors and heavy equipment. The bithermal compression cycle, using the ammonia-water pair, provides the most favorable technical solution for both the performance coefficient and the specific heating capacity.

In estimating investment costs, it should be noted that they have been determined on the basis of the same compressor, whereas the specific capacities obtained are quite different. An economic comparison for equal power will appreciably reduce the investment ratios.

The combination of absorption and resorption cycles leads to the absorption-resorption cycle as shown in Figure 8, which provides great operational flexibility :

- It makes it possible to operate at temperatures that are similar to or even higher than the critical temperature of the solute with pressure levels which can be limited by recirculating the solvent.

- It also makes it possible to operate both as a heat pump and as a thermal energy transformer. When operating as a heat pump, the solute concentration in the resorption loop is higher than that in the absorption loop. When operating as a heat transformer, the opposite is true.

Figure 8 illustrates operation as a heat pump. High-level heat is supplied to the generator, and low-level heat is sent to the desorber. These heats are transformed into medium-level heat which is transmitted to the hot source in resorber and absorber. Intermediate exchangers E1 and E2 improve internal recovery and thus increase the performance coefficient. To ensure the material balance, since the solute concentration is higher in the resorption loop and solvent entrainment is greater at the generator outlet than at the desorber outlet, part of the solution obtained at the desorber must be recycled through the generator.

4.2. Multi-stage Cycles

The basic principle of absorption-resorption cycles can be applied to multi-stage cycles which provides a way of increasing the coefficient of performance or heat ratio. Figure 9 illustrates an absorption system operating with n stages. This system will be described referring to the ammonia-water solute-solvent pair, but other solute-solvent pairs could be used.

The generator operates at pressure P_1 and the absorber at pressure P_n. The first stage works by condensing and evaporating concentrated ammonia. The next stages are resorption stages operating at decreasing pressures and increasing water concentration in the resorption loop. It should be noted that, if all the evaporating stages operate at the same temperature, the entrainment of water will increase slightly when the pressure decreases and that therefore it may be necessary to compensate for this water entrainment by reflux lines which are not represented in Figure 9 for the sake of simplicity.

It can easily be foreseen that for n stages the heat ratio will be multiplied by about the same factor n. In practice, operating with the ammonia-water solute-solvent pair will limit the generator temperature to about 150 °C in order to avoid excessive entrainment of water at the generator and send a high reflux ratio in the rectification zone. Therefore the number of stages will generally be limited to 2 unless temperature T_3 is very close to temperature T_2 or/and heat exchanger surface areas are very large.

Results which were obtained for a refrigerating machine (Reference [8]) can be applied to the case of a water/air or air/heat pumps. The heat ratio figures to be expected in the case of a heat pump used for residential warm air heating purposes are listed in Table III.
Results are obtained for a heating medium inlet temperature of 20 °C. If this inlet temperature becomes equal to 30 °C, the heat ratio drops to 0.81 instead of 1.04 in the first case and to 0.66 instead of 0.93 in the second case. It appears therefore that the heat ratio decreases rapidly when the heating medium inlet temperature increases. Therefore the heat ratio which can be expected when a hydronic system is used for heating is much lower than in the case of warm air heating, especially if water temperatures above 50 °C are needed.

4.3. Recent Developments

The Institut Français du Pétrole is developing a heat pump operating with a patented process (Reference [9]) on a principle derived from the one that has just been described. The goal is space heating (housing projects, businesses, etc...) having medium power between 100 and 300 thermal kW and having the following main features :

- cold source = outside air or water-table water
- hot source = water
- fuel = natural gas or light oil fraction
- the ammonia-water mixture is used
- the installation is designed without requiring any back-up energy when the outside temperature is low. For the prototype now being investigated, a heating power of 100 kW has been set for an outside temperature of - 7 °C.

The experimental laboratory unit has proved the feasibility of the operation and has been used to test the computing methods which have been developed to simulate the process. The estimate of the annual performance shows an annual saving of fuel of 50 percent taking into account consumption by auxiliaries, pumps and ventilators. These values are based upon the mean value of the degree days in the Paris area. The prototype being tested with a French heatpump manufacturing company will serve to refine the fuel saving measurements and to develop a control system.

Applications to higer-powered units have been carried out on the same basic principle for drying in the agro-food sector. After an initial heat recovery by direct heat exchange, the drying loop supplies 'used' air coming out with a humidity of 100 percent and used as the cold source for the absorption pump. New outside air, no matter what the outdoor climatic conditions may be, arrives at the condensor at between 21 and 25 °C. Two operational cases are proposed :

a) Heating the entire air flow to 60 °C, with the remainder being supplied by a make-up heater. The saving in fuel, compared with the operation without a heat pump is 25 percent for a power of 5.25 MW.

b) Heating the entire air flow to 75 °C while being able to provide a "peak" for finishing drying at 85 °C with three quarters of the flow. For a supply power of 5.25 thermal MW, 3.7 thermal MW must be fed to the furnace, and with the assumption of unvaried furnace efficiency this makes for a saving of 30 percent in the consumption of primary energy.

5. REVERSE CYCLES - HEAT TRANSFORMERS

These machines, which operate in the "opposite direction" from the heat pumps described in the preceding section, make use of :

- a medium-level heat source made up of waste heat taken from the effluents from different industrial sectors;
- a cold source to which is transmitted the heat which has not been transformed into work or high-level heat and which is generally water or air.

In some favorable cases, the heat imparted to the cold source may also be used and upgraded. The medium-level heat is either transformed into higher-level heat or converted into mechanical power.

5.1. Reverse Absorption Cycles

Figure 10 shows a flowsheet for such machines which are here described assuming ammonia-water to be the working pair (see also Reference [10]). The generator receives an expanded ammonia solution from the absorber. The medium-level heat generates a highly solute-rich vapor which condenses in contact with the cold source at T_0. The condensed solute is extracted by a pump and passed through the evaporator where it receives the medium-level heat T_1, which returns it to the vapor state but at a higher pressure. This vapor is then sent to the absorber and mixed with the impoverished liquid coming from the generator so as to produce heat at the highest thermal level T_2.

Improvements in this basic system have been proposed (Reference [11]). They consist in including internal heat exchanges to increase heat recovery and the performance coefficient and to put two stages in series so as to attain even higher temperatures.

Table IV, taken from reference [11], gives performance-coefficient values calculated under the following operating conditions: source of heat to be upgraded = 40 °C, and temperature difference at the exchangers = 3 °C.

Krupp, in association with the University of Stuttgart in West Germany, has undertaken research for the purpose of building a prototype making use of the principle of reverse absorption cycles.

5.2. Absorption-Resorption Systems

As has already been mentioned in the previous section, the inclusion of a resorption cycle considerably increases the possibilities of absorption systems. Figure 11 shows the absorption-resorption cycle of a heat transformer. Generator and desorber receive heat $Q_l = Q_{lA} + Q_{lB}$ at the average heat level. Such a system can be used at relatively high temperatures (above 100 °) with a water-ammonia mixture. In this case, the vapor coming from the desorber, and thus it is richer in ammonia. Therefore, between the generator and desorber, a reflux flow must be circulated, which is taken from the solution coming from the generator. Nonetheless, considerable thermodynamic irreversibilities remain at the level of the heat exchangers which reduce the performance.

5.3. Thermosorb Process

A research and development program has been undertaken at the Institut Français du Pétrole in the field of both residential and industrial heating. The solute-solvent pair that has been chosen is the ammonia-water mixture because, up to now, no other pair has provided the same advantages (excellent thermal stability, good performance, no problems of crystallization or corrosion, industrial availability and low price) in the fields being investigated.

A new process operating with this mixture has been found to raise the heat level of thermal waste in industry (References [12] and [13]). This new process which is derived from absorption-resorption trithermal cycles is covered by several patents.

The flowsheet of the process is represented in Figure 12, and four heat exchanges can be seen to occur during the cycle :

- supplying of heat recovered from exchangers ES and ER;
- cooling provided by condensor EC;
- supplying high-level heat to exchanger EA.

The countercurrent exchange zones S and A serve to reduce the irreversibilities as much as possible. Heat exchanges and suitable chosen internal flux recycling are included in zone T. They enable the efficiency of high-level heat production to be increased. The use of an auxiliary desorption fluid makes it possible to operate satisfactorily when the heat to be upgraded has a low thermal level.

The process has been experimentally verified. Operation and control have been tested on a reduced-power experimental unit under continuous operating conditions. The process is capable of producing heat up to about 200 to 220 °C. The thermal capacity aimed at is between 1000 and 10.000 therms per hour. However, these levels are not the limit. For common industrial applications, the thermal efficiency is between 35 and 55 percent. Maintenance costs are reduced as the process has no turbine or compressors. Investment payout times of between 2 and 3 years can be obtained for relatively large heat capacities.

There are a great many industrial applications and we will describe two of them in detail. These are taken from the industrial sector of refining and petrochemicals.

1) The first case, shown in Figure 13a, has to do with the production of styrene by a catalytic reaction between benzene and ethylene. This exothermal reaction is maintained at constant temperature by the continuous extraction of reaction heat by way of water vaporization. Steam and light hydrocarbons are recovered at the reactor outlet. This thermal waste (which would otherwise be cooled in an air condensor) which is between 85 and 80 °C provides a power equal to Q_1 = 21.2 MW which is used to supply power to the Thermosorb process. Part of this heat (Q_2 = 8.1 thermal MW) has a high heating level of up to 110 °C. This heat is sent to a distillation-column reboiler for the styrene which would otherwise have to be heated by low-pressure steam. In this way the thermal efficiency is 38.6 %. The remainder of the non-upgraded heat Q_0 is extracted by water. This case illustrates what the minimum temperature level is at which heat can be economically upgraded.

2) The second case corresponds to the recovery of the heat contained in vapor from light crude-oil fractions that are gathered in the overhead area of a refinery topping column. In this investigation, the available heat (Q_1) between 140 and 105 °C is equal to 17,5 MW.

   A first possibility of heat recovery by the Thermosorb process is shown in Figure 13b and results in the production, with an efficiency of 40.5 %, of an amount of heat Q_2 having 7.1 thermal MW for producing low-pressure vapor. The remainder of this heat is discharged into the outside environment by an air-cooled exchanger followed by water cooling.

Yet additional upgrading results in better recovery of Q_1 as shown in Figure 13c. Part of the discharged heat has been used to preheat the combustion air of the crude-oil vaporization furnace at the entrance to the topping column. This upgrading of 2.9 thermal MW improves the overall heat efficiency of the system by raising it to 57 %.

6. CONCLUSION

Large improvements can be expected in the field of absorption systems. These improvements mainly concern :

- the maximum temperature at which heat is delivered, which will become higher;
- the heat ratio which will be increased.

These improvements can be achieved by using the well-known water-ammonia solute-solvent pair but with a proper design of the process. The investment level and the optimum operating range in the case of residential heating applications still have to be determined for mass production. They represent a crucial issue before such system will become widespread. In the future, new solute-solvent pairs should become available. They are of special value for small individual heat pumps, but at the present time the solute-solvent pairs already known still have considerable limitations.

REFERENCES.

1. Macriss, R.A., Cole, J.T., Kono, M.T. and Zawacki, T.S. *Analysis of advances conceptual designs for single-family size absorption chiller*, Project 9101, Final Report - Institute of Gas Technology (Chicago, 1979).

2. Stephan, K., Seher, D. *Klima u Kälte - Ingénieur* 1, 6, 865-876 (1980).

3. Janssen, H.K. and Oelert, G. "Development of a primary energy driven absorption heat pump for domestic heating" *New ways to save energy - Commission of the European Communities* pp. 198-220, D. Reidel Publishing Company (1979).

4. Kohnke, H.J. "Bau und Versuchsbetriebe von Absorptionswärmepumpen Prototypen" *New Ways to save Energy* Commission of the European Communities, D. Reidel, 209-220 (1979).

5. Allied Chemical - personnal communication.

6. Altenkirch, E. *Kältetechnik* Vol. 3, N° 8, pp. 201-205, N° 9, pp. 229-234, N° 10, pp. 255-259, Springer Verlag (Berlin, 1951).

7. Novotny *International Journal of Refrigeration* (May 1979).

8. Richter, K.H. *Journal of Refrigeration* pp. 105-110 (Sept.-Oct. 1962).

9. Rojey, A. and Cohen, G. "Brevet d'Invention EN 80/03460" France.

10. Franzen, P. "Wärmetransformator erzeugt Nutzwärme aus Abwärme Erdol und Kohle" *Erdgas Petrochemie* 32, 11, (November 1979).

11. Stephan, K. "Absorption Heating Transformer Cycles" NATO ASI : Heat Pump Fundamentals (Espinho, 1980).

12. Cohen, G., Salvat, J., Rojey, A., *Entropie* N° 84, p. 31-37 (1978).

13. Cohen, G., Salvat, J., Rojey, A. *Proceedings of the 14th Intersociety Energy Conversion Engineering Conference* Vol 2, pp. 1720-1724, (Boston, 1979).

TABLE I

Expected Performance of the Allied Chemical Heat Pump [5]

	Heating		Refrigeration
Ambient temperature (°C)	8.3	-8.3	35
COP	1.25	1.11	0.5
Capacity (kW)	26.3	23.4	10.5
Water temperature (°C)	56	53	10
Electricity consumption (kW)	1.0	1.1	1.0

TABLE II

Comparison of compression heat pumps and the bithermal resorption cycle [7]

Coolant	Type of heat pump	Compression ratio	Specific capacity W/m^3	Performance coefficient	Investment ratio
R21	Compression	8	114	1.95	1.0
R11	Compression	8.2	102	2.99	1.0
R12	Compression	5.1	568	3.23	1.0
NH_3	Compression	6.7	821	3.28	1.5
NH_3 + H_2O	Compression + resorption	4.0	816	4.29	1.8

TABLE III

Heat Source	Heating medium	Heat ratio
Ground water (Outlet temperature = 5 °C)	Internal air	1.04
External air (Average outside temperature = 4 °C)	Internal air	0.93 *

* No defrosting taken into account

TABLE IV

Performance coefficients calculated from reference [11]

	Cold source T	0 °C	5 °C	10 °C	15 °C
Stage 1	High T	65 °C	59 °C	54 °C	~ 40°
	COP	0.446	0.45	0.449	-
Stage 2	High T	117 °C	99 °C	82 °C	-
	COP	0.417	0.428	0.44	-
Total	Overall COP *	0.186	0.192	0.197	-

* Product of the COP_s of stages 1 and 2.

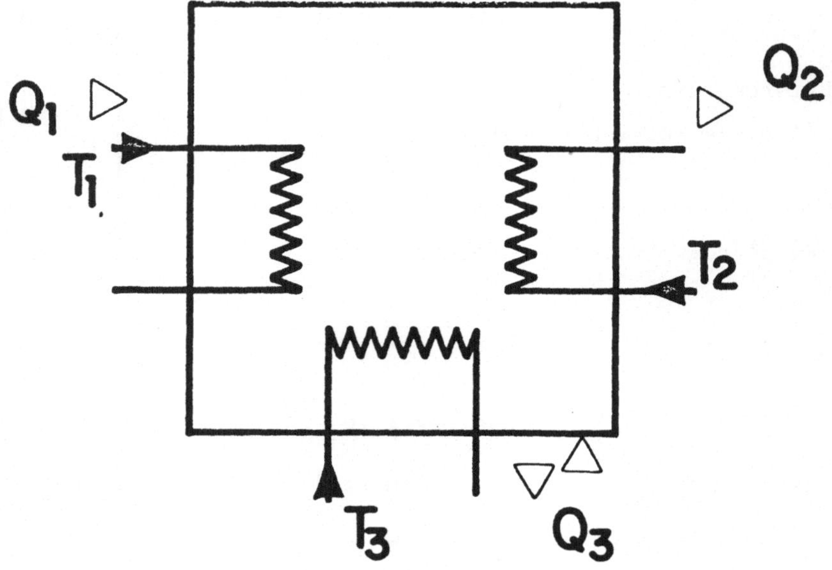

THREE SOURCE CYCLES
General scheme

FIGURE 1

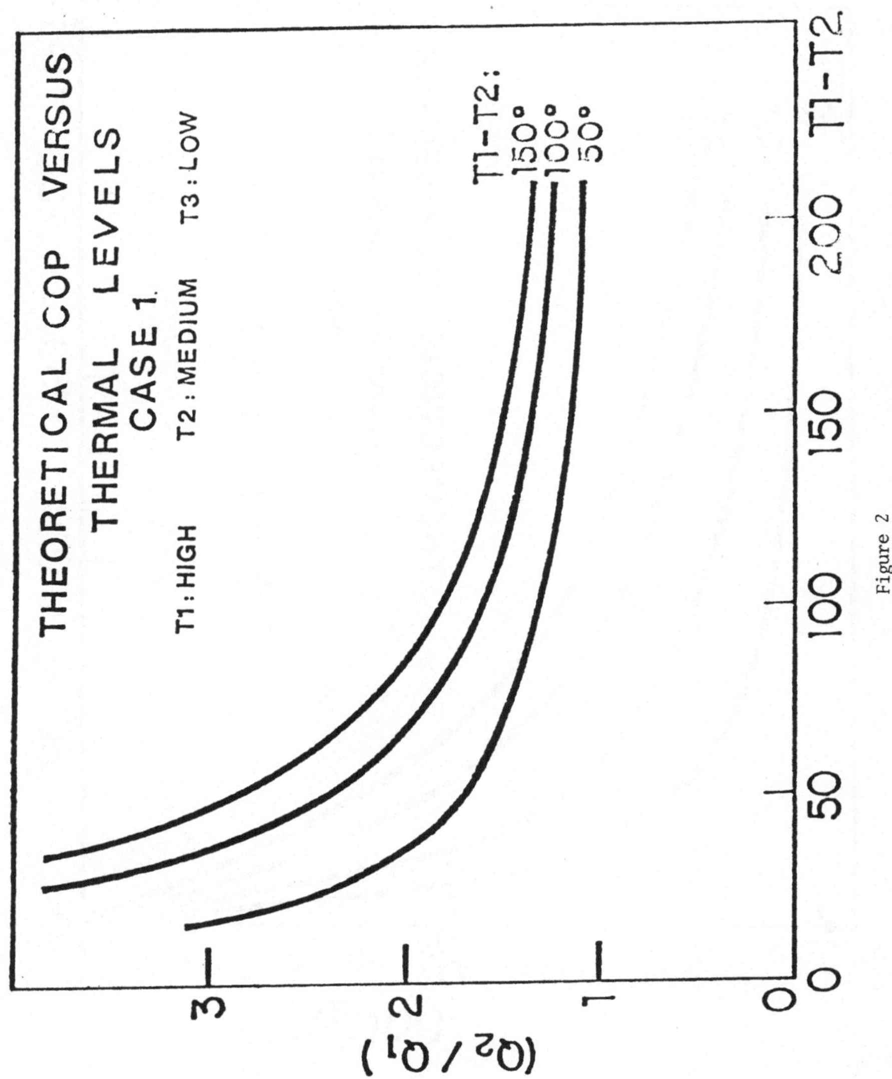
THEORETICAL COP VERSUS
THERMAL LEVELS
CASE 1.
T1 : HIGH
T2 : MEDIUM
T3 : LOW
T1-T2:
150°
100°
50°
(Q2 / Q1)
0
1
2
3
0
50
100
150
200
T1-T2

Figure 2

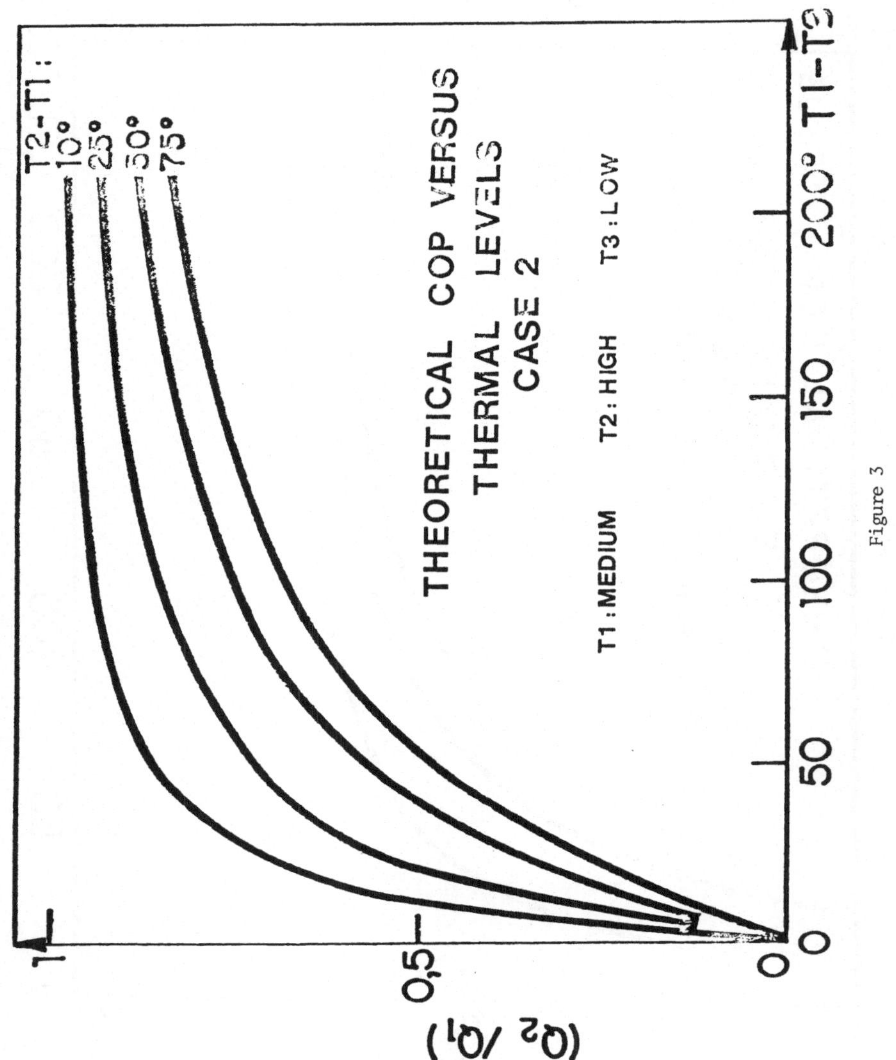

THEORETICAL COP VERSUS
THERMAL LEVELS
CASE 2
T1 : MEDIUM
T2 : HIGH
T3 : LOW
T2-T1 :
10°
25°
50°
75°
(Q2 /Q1)
1
0,5
0
0
50
100
150
200° T1-T3

Figure 3

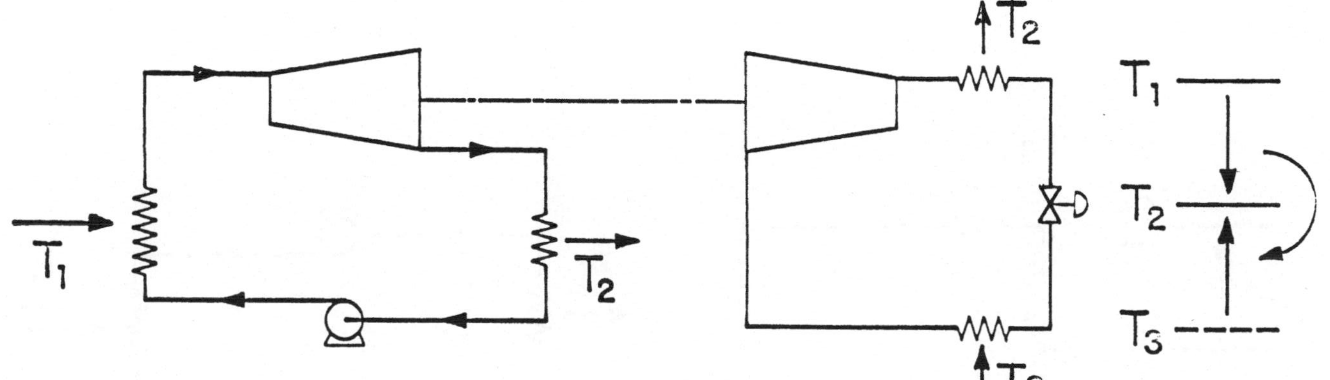

THREE SOURCE CYCLE – Case 1

FIGURE 4

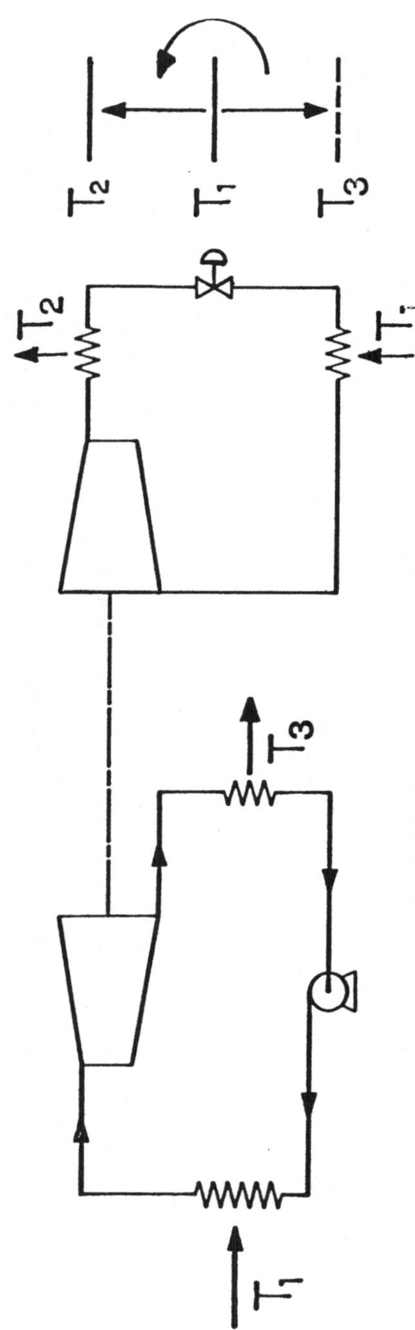

THREE SOURCE CYCLE - Case 2

FIGURE 5

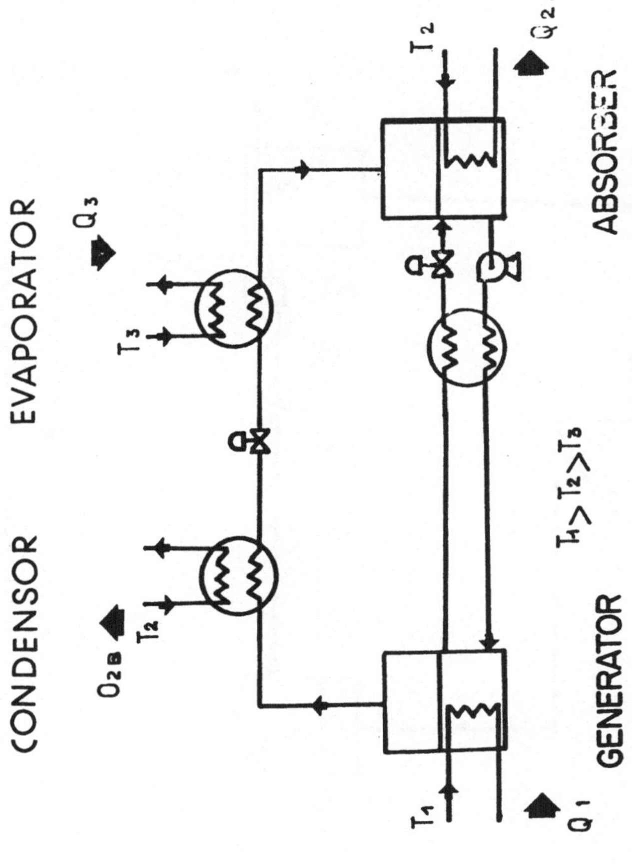

FIGURE 6

Absorption heat pump - simplified scheme

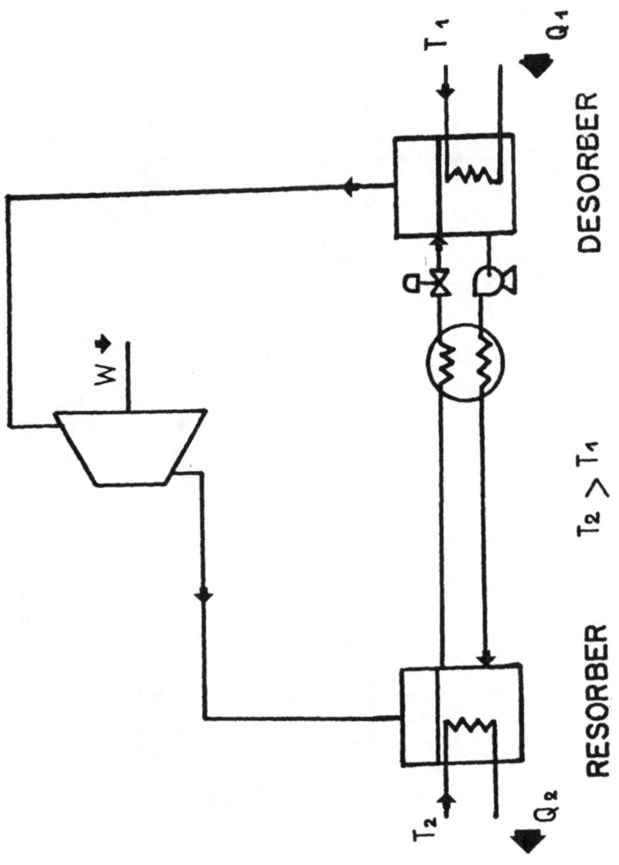

FIGURE 7

Bithermal resorption cycle with compression

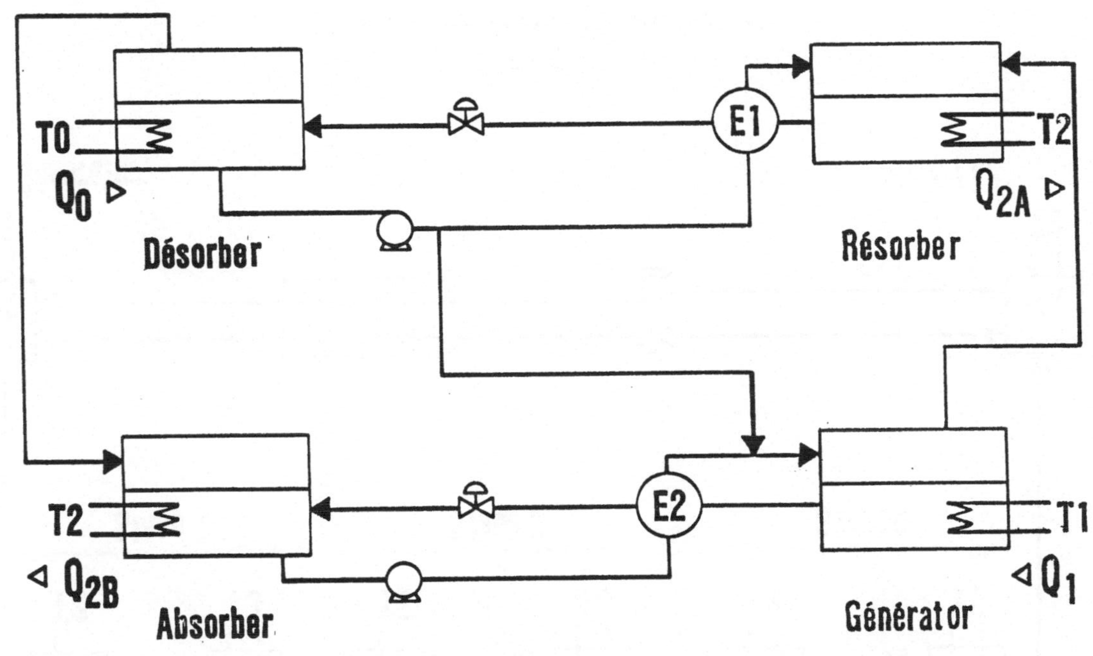

FIGURE 8

Absorption-resorption cycle operating as a heat pump

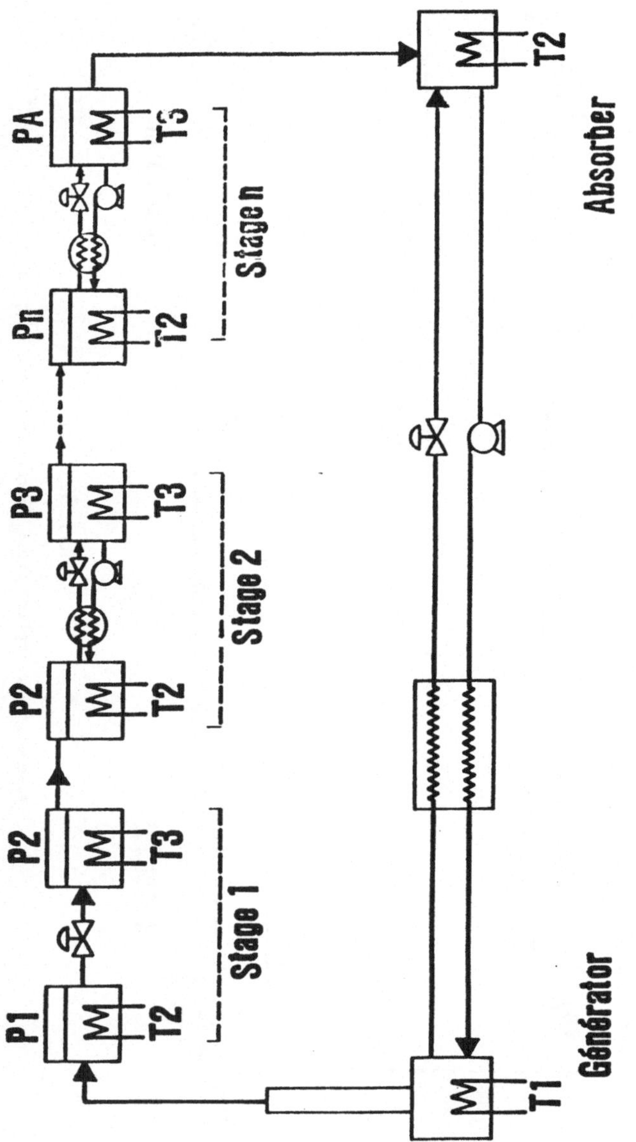

FIGURE 9

Absorption system operating with n stages

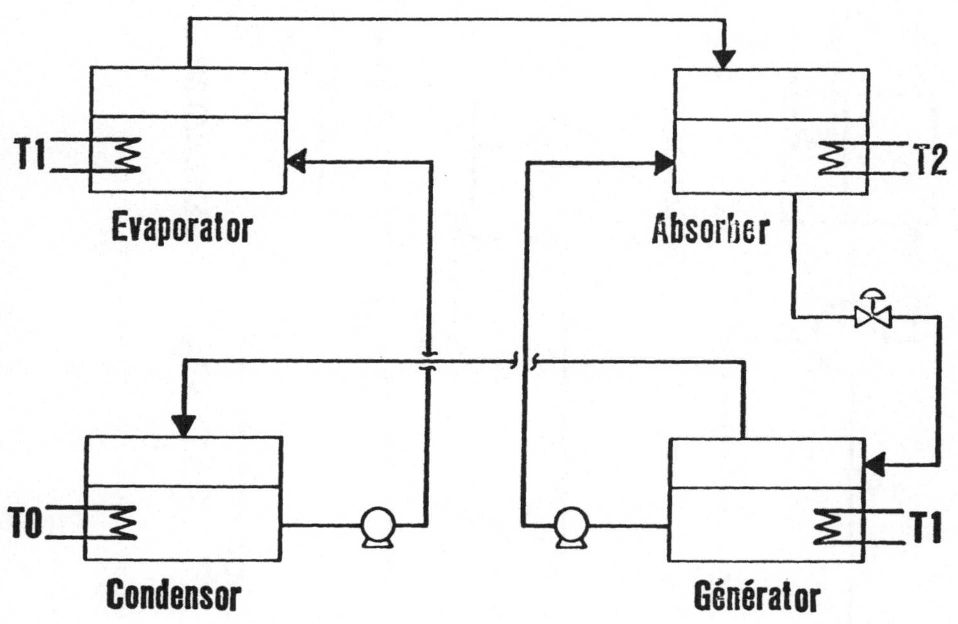

FIGURE 10

Absorption heat transformer (Ref. 10)

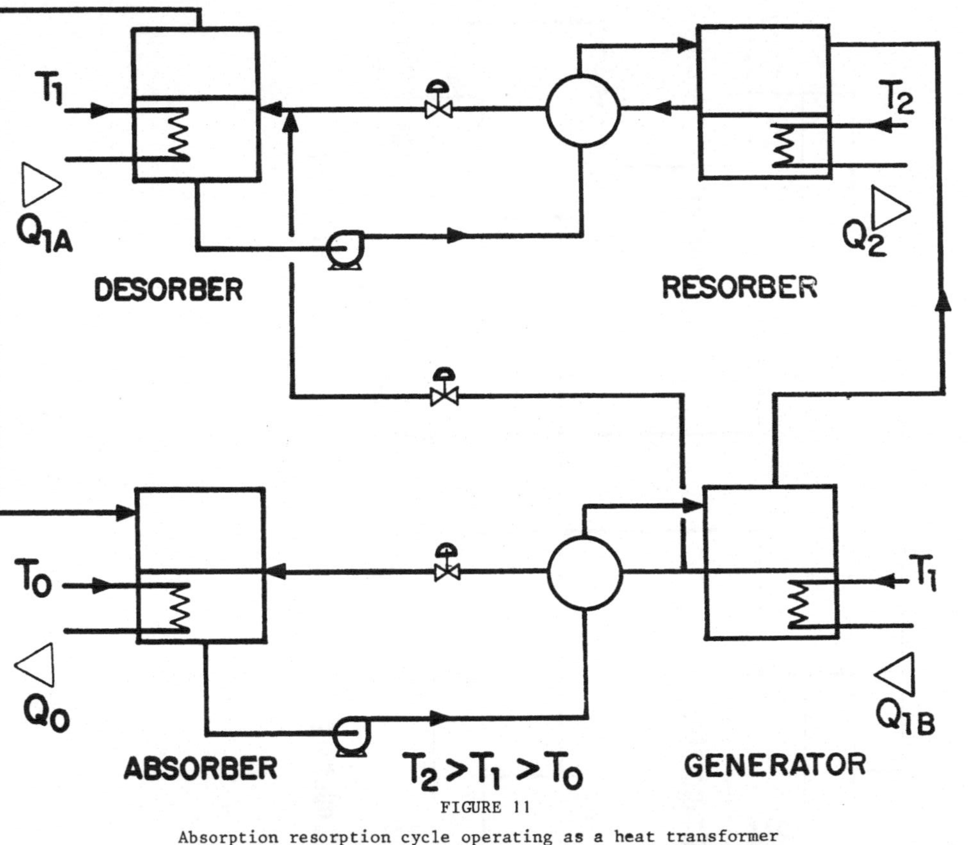

FIGURE 11

Absorption resorption cycle operating as a heat transformer

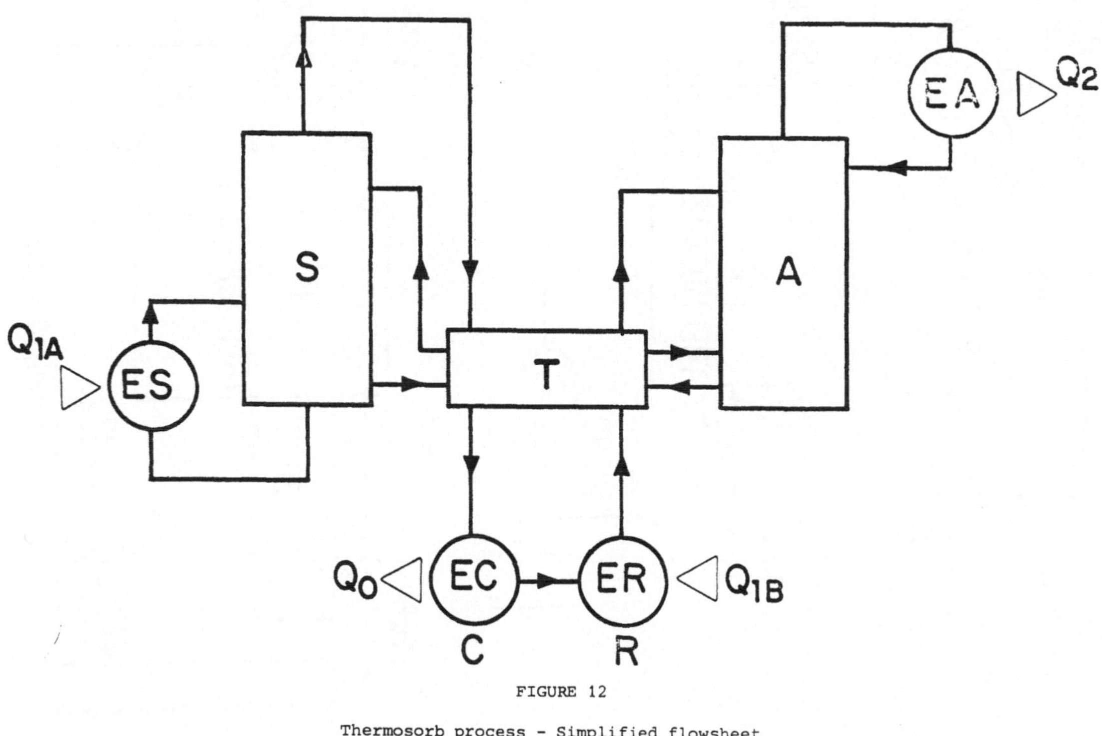

FIGURE 12

Thermosorb process - Simplified flowsheet

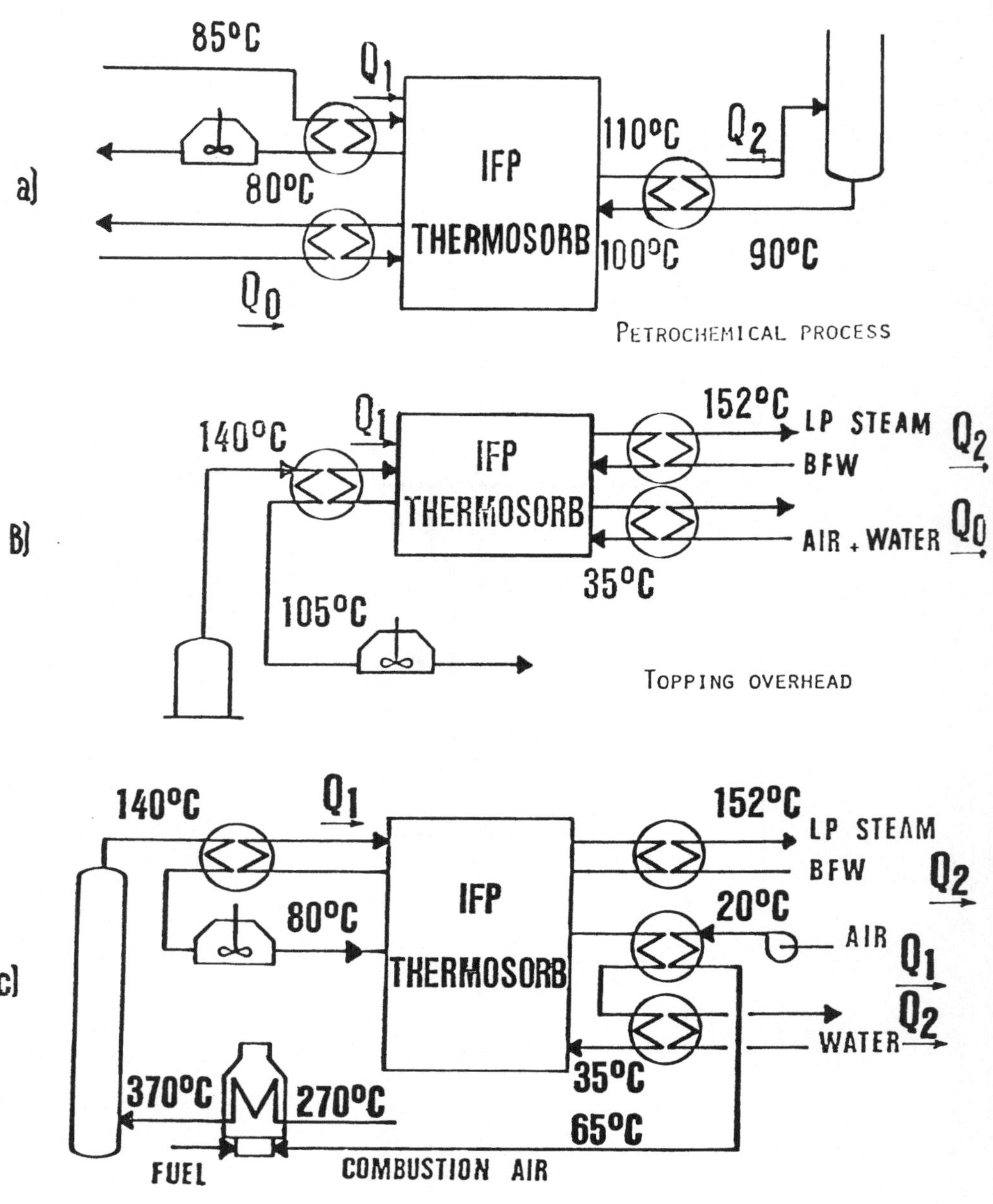

FIGURE 13

Thermosorb process applications